Peacocks and Associated Martingales, with Explicit Constructions

B&SS – Bocconi & Springer Series

Francis Hirsch · Christophe Profeta ·
Bernard Roynette · Marc Yor

Peacocks and Associated Martingales, with Explicit Constructions

Bocconi
University
Press

Springer

Francis Hirsch
Laboratoire d'Analyse et Probabilités
Université d'Évry-Val d'Essonne
francis.hirsch@univ-evry.fr

Bernard Roynette
Institut Élie Cartan
Université Henri Poincaré, Nancy
bernard.roynette@iecn.u-nancy.fr

Christophe Profeta
Institut Élie Cartan
Université Henri Poincaré, Nancy
christophe.profeta@iecn.u-nancy.fr

Marc Yor
Laboratoire de Probabilités et
Modèles Aléatoires
Université Pierre et Marie Curie, Paris
Institut Universitaire de France
deaproba@proba.jussieu.fr

B&SS – Bocconi & Springer Series
ISSN print edition: 2039-1471
ISSN electronic edition: 2039-148X

ISBN 978-88-470-1907-2
ISBN 978-88-470-1908-9 (eBook)
DOI 10.1007/978-88-470-1908-9

Library of Congress Control Number: 2011920258

Springer Milan Dordrecht Heidelberg London New York

9 8 7 6 5 4 3 2 1

Cover-Design: Beatrice ꓤ., Milan
Frontmatter image: Maruyama Ōkyo, Peacock and Peahen, 1781. Hanging scroll, Color on silk, 86,2 cm × 133,2 cm

Typesetting with LaTeX: PTP-Berlin, Protago TeX-Production GmbH, Germany (www.ptp-berlin.eu)

Springer-Verlag Italia srl – Via Decembrio 28 – 20137 Milano
Springer is a part of Springer Science+Business Media (www.springer.com)

Contents

A Few Words of Explanation

Associated Processes

For two processes $(X_t, t \geq 0)$ and $(Y_t, t \geq 0)$, we write:

$$X_t \overset{(1.d)}{=} Y_t$$

if $(X_t, t \geq 0)$ and $(Y_t, t \geq 0)$ have the same one-dimensional marginals and we say that X and Y are associated.

It is a theorem due to Kellerer [106] that a process $(X_t, t \geq 0)$ is associated to a *martingale* if and only if the one-dimensional marginals $(\mu_t, t \geq 0)$ of $(X_t, t \geq 0)$ are increasing in the convex order.

We call such a process a *peacock*: this is a pun coming from the French term: Processus Croissant pour l'Ordre Convexe, which yields the acronym PCOC.

Diverse Appreciations of Peacocks

This monograph is mainly about peacocks. Although this term comes as a pun, it is no more exotic as the term martingale, and we believe the reader will soon have no difficulty with that concept.

Pushing the metaphor a little further, we note that in Asia, a peacock is considered as a symbol of beauty and grace (it is in fact the national bird of India), whereas in Europe, it is considered as a boastful bird. This may be connected with Mathematics, being considered by some (mathematicians!) as one of the most beautiful topics, whereas other scientists (in some applied sciences) may consider that the role of mathematics is sometimes pushed too far.

In Kyoto, during the SAFFF (Stoch. Ana. for and from Finance) conference in August 2009, Pr. Y. Takahashi wrote a Haïku[1] to complement the beautiful painting by Maruyama Ōkyo (1747–1821).

A proud peacock spreads

Its tail pretending to be

A martingale.

孔雀鳥（ピーコック）羽根に粧うマルチンゲール

We could not resist trying for some French version:

Il est un fier paon

Martingale prétendant

Et la roue faisant.

---

[1] A Haïku contains 17 syllables (5+7+5=17), and must make some reference to a season or to Nature. M. Basho (1644–1694) is the most famous Haïku poet. His masterpiece is: “The Narrow Path to the Deep North”.

Introduction

With the help of numerous examples, D. Williams [176] has shown that many questions of probability may be solved once the adequate martingale for a given question has been found.

This ubiquitous role of martingales shows up again, although differently, in the following discussion about modelization. Modelizing the evolution in time of a random phenomenon which is measured and described in terms of a random process $(X_t, t \geq 0)$ is a difficult art, and the authors of this monograph do not claim to be artists of this kind. One of the main difficulties of this art is that, usually, one can only access a limited amount of distributional facts concerning the law of $(X_t, t \geq 0)$.

In this monograph, we only assume knowledge of the set $\mu = (\mu_t, t \geq 0)$ of the one-dimensional marginals of $(X_t, t \geq 0)$. We are concerned here with the time evolution which happens in some kind of "equilibrium", i.e. the μ_t's are the one-dimensional marginals of a martingale. It is a (deep) theorem due to Kellerer [106] that this is *equivalent* to the increase property[2] in the convex order for the μ_t's, i.e.:

$$\forall \psi \text{ convex}, \quad t \longmapsto \int \psi(x)\mu_t(dx) \ (\in]-\infty, +\infty]) \text{ is increasing.}$$

We assume $\int |x|\mu_t(dx) < \infty$ for any t.

Thus, a clear program lies in front of us: given a family $\mu = (\mu_t, t \geq 0)$ of probabilities on $\mathbb{R}$, which are increasing in the convex order, find – as explicitly as possible – a martingale which admits μ as the set of its marginals. As the reader shall see throughout the volume, even for some "simple" μ's, the construction of a related martingale is not that easy. This corroborates the fact that Kellerer's proof is not constructive.

A few words about the title of this monograph: *Peacocks and associated Martingales*, may now be in order. In French, an increasing process for the convex order reads: un Processus Croissant pour l'Ordre Convexe, which yields the acronym PCOC. This being pronounced "peacock", we adopt this name for such processes.

[2] That this condition is necessary is a simple application of Jensen's inequality. It is the sufficiency assertion which is a deep fact.

Hirsch F., Profeta C., Roynette B., Yor M.:
Peacocks and Associated Martingales, with Explicit Constructions

Thus our program now reads: given a peacock, find an associated martingale, that is a martingale with the same one-dimensional marginals as the peacock.

Actually, our motivation comes partly from Mathematical Finance. In the following paragraph, we shall present some possible connections of our work with this field.

Peacocks, Associated Martingales and Mathematical Finance

1) Let $(\Omega, \mathscr{F}, (\mathscr{F}_t, t \geq 0), (X_t^{(0)}, t \geq 0), (X_t, t \geq 0), \mathbb{P})$ be a financial market such that:

- the asset price process $(X_t, t \geq 0)$ is a $\mathbb{R}_+$-valued, $(\mathscr{F}_t, t \geq 0)$-adapted process,
- for every $t \geq 0$, $\mathbb{E}[X_t]$ is finite and does not depend on t,
- for every $t \geq 0$, the numeraire $X_t^{(0)}$ is equal to 1.

For every $K \geq 0$, let $C_X(t,K) := \mathbb{E}[(X_t - K)^+]$ denotes the price of a European call, resp. $\Pi_X(t,K) := \mathbb{E}[(K - X_t)^+]$ the price of a European put, with strike K and maturity t. Since these quantities correspond to risky options, we may ask the quite natural question: under which conditions on X do these quantities increase as functions of t? This seems (arguably!) to be a desirable property, as options prices may (should?) increase with maturity. In our set-up, since the functions $x \longmapsto (x-K)^+$ and $x \longrightarrow (K-x)^+$ are convex, this property is satisfied as soon as $(X_t, t \geq 0)$ is a peacock. In fact (cf. Exercise 1.7), if, for every $t \geq 0$, $\mathbb{E}[X_t] < \infty$ and $\mathbb{E}[X_t]$ does not depend on t, this property is equivalent to the fact that $(X_t, t \geq 0)$ is a peacock (and then, $t \longmapsto \mathbb{E}[\psi(X_t)]$ is increasing for every convex function ψ). Therefore, we assume from now on that $(X_t, t \geq 0)$ is a peacock.

The First Fundamental Theorem of Mathematical Finance states that a market is arbitrage free "essentially" if and only if there exists a probability $\mathbb{Q}$ equivalent to $\mathbb{P}$ on every σ-algebra $\mathscr{F}_t$ such that, under $\mathbb{Q}$, $(X_t, t \geq 0)$ is a martingale (we refer the reader to [31, Theorem 10.5, p.141] for more precisions about the term "essentially"). In this framework, an important problem is then to exhibit, if it exists, such a martingale-probability.

Our approach in this monograph is somewhat different. In many cases, given a peacock $(X_t, t \geq 0)$ – not necessarily $\mathbb{R}_+$-valued –, we are able to exhibit a martingale $(M_t, t \geq 0)$ (which may be defined on another probability space $(\widetilde{\Omega}, \widetilde{\mathscr{F}}, (\widetilde{\mathscr{F}}_t, t \geq 0), \widetilde{\mathbb{P}})$) such that, for every $t \geq 0$ fixed, M_t has the same law as X_t. This implies of course that the prices of the call and put options built on the risky assets $(X_t, t \geq 0)$ and $(M_t, t \geq 0)$ are the same:

$$C_M(t,K) = C_X(t,K) \quad \text{and} \quad \Pi_M(t,K) = \Pi_X(t,K).$$

In other words, when $(X_t, t \geq 0)$ is a peacock with values in $\mathbb{R}_+$, the new market $(\widetilde{\Omega}, \widetilde{\mathscr{F}}, (\widetilde{\mathscr{F}}_t, t \geq 0), (X_t^{(0)} = 1, t \geq 0), (M_t, t \geq 0), \widetilde{\mathbb{P}})$ is an arbitrage free market in which the prices of all call and put options, and more generally of all simple claims, are the same as in the initial market.

2) In this monograph, we restrict ourselves to real-valued peacocks. In fact, many constructions we present in Chapters 2-7 may be extended to processes $(X_t := (X_t^{(1)},\dots,X_t^{(d)}), t \geq 0)$ taking values in $\mathbb{R}^d$ and such that, for every $i = 1,\dots,d$, $(X_t^{(i)}, t \geq 0)$ is a peacock. We then obtain an explicit description of an associated vector-valued martingale $(M_t := (M_t^{(1)},\dots,M_t^{(d)}), t \geq 0)$ such that, for every $t \geq 0$ fixed, the random vectors $(X_t^{(1)},\dots,X_t^{(d)})$ and $(M_t^{(1)},\dots,M_t^{(d)})$ have the same law. For instance, Theorem 2.2 may be generalized in the following way: with the notation of this theorem, let $(h_i, i = 1,\dots,d)$ be d space-time harmonic functions and define the process $(X_t = (X_t^{(1)},\dots,X_t^{(d)}), t \geq 0)$ where $X_t^{(i)} := \frac{1}{t}\int_0^t h_i(s,B_s)ds$. Then, the vector-valued martingale $(M_t := (M_t^{(1)},\dots,M_t^{(d)}), t \geq 0)$, with $M_t^{(i)} := \int_0^1 h_i(tu, W_{u,t})du$, is such that, for every $t \geq 0$ fixed, $M_t \overset{(law)}{=} X_t$.

In financial terms, this means that, to a financial market $(\Omega, \mathscr{F}, (\mathscr{F}_t, t \geq 0), (X_t^{(i)}, t \geq 0, i = 0,\dots,d), \mathbb{P})$ such that for every $t \geq 0$, $X_t^{(0)} = 1$ and for every $i = 1,\dots,d$ $(X_t^{(i)}, t \geq 0)$ is a $\mathbb{R}_+$-valued peacock, we may associate an arbitrage free market $(\widetilde{\Omega}, \widetilde{\mathscr{F}}, (\widetilde{\mathscr{F}}_t, t \geq 0), (X_t^{(0)} = 1, t \geq 0), (M_t^{(i)}, t \geq 0, i = 1,\dots,d), \widetilde{\mathbb{P}})$ where, for every $i = 1,\dots,d$, $(M_t^{(i)}, t \geq 0)$ is a $\mathbb{R}_+$-valued martingale, in which the prices of simple claims are the same. In our example, the martingales are explicitly known and we have, for every $i = 1,\dots,d$, $C_{X^{(i)}}(t,K) = C_{M^{(i)}}(t,K)$, $\Pi_{X^{(i)}}(t,K) = \Pi_{M^{(i)}}(t,K)$, and more generally $\mathbb{E}[\Phi(t, X_t^{(1)},\dots,X_t^{(d)})] = \mathbb{E}[\Phi(t, M_t^{(1)},\dots,M_t^{(d)})]$ for every contract function Φ.

3) Let $(X_t, t \geq 0)$ be a $\mathbb{R}_+$-valued peacock and let $\Pi(K,t) = \mathbb{E}[(K - X_t)^+]$, $(K \geq 0)$. Let us assume that $X_0 = 1$ and that $X_t \xrightarrow[t\to+\infty]{} 0$ in probability, i.e., for every $\varepsilon > 0$, $\mathbb{P}(X_t > \varepsilon) \xrightarrow[t\to+\infty]{} 0$. Under these conditions, $t \longmapsto \frac{1}{K}\Pi(K,t) = \mathbb{E}\left[\left(1 - \frac{X_t}{K}\right)^+\right]$ is an increasing function which converges to 1 when $t \to +\infty$. Therefore, for every $K \geq 0$, there exists a r.v. G_K such that:

$$\frac{1}{K}\Pi(K,t) = \mathbb{P}(G_K \leq t).$$

This leads to a natural question: given a peacock $(X_t, t \geq 0)$, are we able to exhibit a family of r.v.'s $(G_K, K \geq 0)$ such that $\mathbb{E}[(K - X_t)^+] = K\mathbb{P}(G_K \leq t)$? A tougher question: are we able to exhibit a *process* $(G_K, K \geq 0)$ such that the above relation is satisfied?

The construction of a martingale $(M_t, t \geq 0)$ associated to the peacock $(X_t, t \geq 0)$ makes it possible, under an extra hypothesis, to give a positive answer to this question. Indeed, let us assume that there exists an associated martingale $(M_t, t \geq 0)$ which is continuous (or more generally without positive jumps), and let us define, for every

$K \geq 0$:

$$G_K^{(M)} := \sup\{t \geq 0\,;\, M_t = K\}$$
$$(= 0 \text{ if } \{t \geq 0\,;\, M_t = K\} = \emptyset)$$

$G_K^{(M)}$ is the last passage time at level K of the martingale $(M_t, t \geq 0)$. Observe that $M_0 = 1$ (since $X_0 = 1$) and $M_t \xrightarrow[t \to +\infty]{} 0$ a.s. Indeed, $(M_t, t \geq 0)$ being a positive martingale, it converges a.s. towards a r.v. M_∞, and $M_\infty = 0$ a.s. since $(X_t, t \geq 0)$ (hence $(M_t, t \geq 0)$) converges towards 0 in probability as $t \to +\infty$. Then, it is known (cf. [148, p.22]) that, for every K and $t \geq 0$:

$$\mathbb{E}[(K - M_t)^+] = K\mathbb{P}(G_K^{(M)} \leq t),$$

thus, since $\mathbb{E}[(K - X_t)^+] = \mathbb{E}[(K - M_t)^+]$,

$$\mathbb{E}[(K - X_t)^+] = K\mathbb{P}(G_K^{(M)} \leq t).$$

The process $(G_K^{(M)}, K \geq 0)$ is therefore a solution to the question *ii*) raised above. We obtain similarly ([148, p.25]):

$$\mathbb{E}[(X_t - K)^+] = \mathbb{P}^{(M)}(G_K^{(M)} \leq t),$$

where $\mathbb{P}^{(M)}$ is the probability induced by the relation $\mathbb{P}^{(M)}_{|\mathscr{F}_t} = M_t \cdot \mathbb{P}_{|\mathscr{F}_t}$.

4) Let $(B_s, s \geq 0)$ be a Brownian motion started at 0 and let $(N_s := e^{B_s - \frac{s}{2}}, s \geq 0)$ its exponential martingale. Carr, Ewald and Xiao [43] proved that $\left(X_t := \frac{1}{t}\int_0^t N_s ds, t \geq 0\right)$ is a peacock. More generally, let $(N_s, s \geq 0)$ be a martingale belonging to H^1_{loc}. We prove in Theorem 1.4 a generalization of Carr, Ewald and Xiao's result: $\left(X_t := \frac{1}{t}\int_0^t N_s ds, t \geq 0\right)$ is a peacock, and we give numerous examples (cf. Theorem 2.2) for which we are able to associate a martingale $(M_t, t \geq 0)$ to this peacock. Let now:

$$\Pi_N^{(A)}(t,K) := \mathbb{E}\left[\left(K - \frac{1}{t}\int_0^t N_s ds\right)^+\right]$$

be the price of the Asian put option built on N. We then have

$$\Pi_N^{(A)}(t,K) = \Pi_M(t,K) = \mathbb{E}[(K - M_t)^+].$$

In other words, we prove that the computation of the price of the Asian put option built on N boils down to the computation of the European put option built on M. This problem is a priori easier to solve, since the computation of $\Pi_M(t,K)$ only requires the knowledge of the law of the r.v. M_t, while the computation of $\Pi_N^{(A)}(t,K)$ requires a priori the knowledge of the law of the process $(N_s, s \leq t)$.

We now summarize the contents of each of the eight chapters of this monograph.

Chapter 1: Some Examples of Peacocks

C1-1) Our work in this chapter is comparable to that of an entomologist who collects butterflies; however, instead of pinning insects with multicolored wings, we gather here a collection of peacocks, and try to classify them according to their main characteristics.

The first sample of our collection is borrowed from Carr, Ewald and Xiao [43] who show that $\left(\frac{1}{t}\int_0^t e^{B_s-\frac{s}{2}}ds, t\geq 0\right)$ is a peacock, when $(B_s, s\geq 0)$ is a Brownian motion issued from 0. Throughout the volume, we shall refer to this example as the "guiding example".

C1-2) Inspired by this guiding example, we show that, if $(M_s, s\geq 0)$ is a martingale in H^1_{loc} and if $\alpha:\mathbb{R}_+\longrightarrow\mathbb{R}_+$ is continuous and increasing, then $\left(X_t := \frac{1}{\alpha(t)}\int_0^t M_s d\alpha(s), t\geq 0\right)$ and $\left(\widetilde{X}_t := \int_0^t (M_s - M_0)d\alpha(s), t\geq 0\right)$ are two peacocks. The first one we call a $\mathbf{F}_1$-type peacock, and the second one a $\mathbf{F}_2$-type peacock. We note that if $(M_s, s\geq 0)$ models a risky asset, its arithmetic mean $\left(\frac{1}{t}\int_0^t M_s ds, t\geq 0\right)$ is a peacock (cf. Theorem 1.4); we recall that these means may serve to define prices of Asian options $\mathbb{E}\left[\left(\frac{1}{t}\int_0^t M_s ds - K\right)^+\right]$. Thus, the prices of Asian options are increasing functions of t.

Let X denote a centered integrable r.v. Then, the process $(tX, t\geq 0)$ is a $\mathbf{F}_2$-type peacock (simply take $M_t = X$, see Remark 1.10 and Proposition 1.4). Throughout the monograph, this peacock $(tX, t\geq 0)$ plays a recurrent role.

C1-3) We now increase complexity. Still being inspired by the guiding example, we consider $(Y_{t,\lambda}, t\geq 0, \lambda\in\Lambda)$ a process indexed by two parameters. Starting from a bounded positive measure $\sigma(d\lambda)$ on Λ, we define:

$$X_t^{(\sigma)} = \int_\Lambda \frac{e^{Y_{t,\lambda}}}{\mathbb{E}[e^{Y_{t,\lambda}}]}\sigma(d\lambda), \quad t\geq 0. \tag{0.1}$$

Note that the guiding example may be represented as such, with $\sigma(d\lambda)$ the Lebesgue measure on $[0,1]$ and $Y_{t,\lambda} = B_{t\lambda}$, since:

$$\frac{1}{t}\int_0^t \exp\left(B_s - \frac{s}{2}\right)ds = \int_0^1 \exp\left(B_{t\lambda} - \frac{t\lambda}{2}\right)d\lambda, \quad t\geq 0$$

after making the change of variable $s = t\lambda$. We then try to answer the question: under which condition(s) on Y is $(X_t^{(\sigma)}, t\geq 0)$ a peacock? Under some reasonable hypotheses, we can answer this question in the affirmative, in several contexts:

a) firstly, that of additive processes, i.e. processes with independent, but not necessarily stationary, increments: this is Theorem 1.8,

b) secondly, that of Gaussian processes: see Section 1.3,
c) thirdly, that of Markov processes: see Section 1.4.

C1-4) We now come back to formula (0.1) and define, after exchanging the roles of t and λ:

$$\widehat{X}_{\lambda}^{(\mu)} = \int_{\mathbb{R}_+} \frac{e^{Y_{t,\lambda}}}{\mathbb{E}[e^{Y_{t,\lambda}}]} \mu(dt), \quad \lambda \in \mathbb{R}_+, \tag{0.2}$$

where μ is a finite measure on $\mathbb{R}_+$ and $\Lambda = \mathbb{R}_+$. We now consider that, in the definition of the process $(Y_{t,\lambda};\, t \geq 0, \lambda \in \Lambda)$, the t parameter stands for time, whereas λ represents a variance parameter. Borrowing vocabulary from Mathematical Finance, we then say that the processes $(X_t^{(\sigma)}, t \geq 0)$, resp. $(\widehat{X}_{\lambda}^{(\mu)}, \lambda \geq 0)$, are peacocks with respect to maturity, resp. volatility – assuming we have established their peacock property.

C1-5) In every good collection of certain items, one finds somewhat exotic pieces, i.e. samples which do not really belong to any category. Our peacocks collection obeys this rule. Indeed, the reader shall find in Exercises of Chapter 1 several peacocks which do not belong to the preceding categories, i.e.: the image of a peacock by a deterministic function, solutions of SDE's in the Itô-Stratonovich sense, functions of a r.v. depending on a parameter, integrals of processes with scaling property, and so on...

Apart from Chapter 8, which discusses a related, but somewhat different question, the remainder of this monograph is devoted to the following problem: given a peacock in our collection, find as explicitly as possible one (or several!) associated martingales.

Chapter 2: The Sheet Method

C2-1) In the Foreword to his Selected Papers [88], K. Itô writes: "After some time, it became my habit, even for finite dimensional probabilistic phenomena, to look at an infinite dimensional related set-up, the properties of which may illuminate / explain / those of the finite dimensional set-up considered previously". Below, we refer to this as Itô's motto.

This sentence is, intentionally, not exactly that of Itô, but it expresses precisely the contents and spirit of his sentence, which is intended there to explain how he came to think of excursions of, say, Brownian motion away from 0, as a Poisson Point Process. See K. Itô [88], P.A. Meyer [128] and M. E. Vares and M. Yor [168].

C2-2) In this Chapter 2, we shall show how Itô's motto translates, in our search for martingales associated to a $\mathbf{F}_1$-type peacock. Let us come back to the guiding example, and let $(W_{u,t}, u \geq 0, t \geq 0)$ denote the standard Brownian sheet. Then, it is not difficult to see that:

i) for every fixed $t \geq 0$,

$$\frac{1}{t}\int_0^t e^{B_s - \frac{s}{2}} ds = \int_0^1 e^{B_{ut} - \frac{ut}{2}} du \overset{(law)}{=} \int_0^1 e^{W_{u,t} - \frac{ut}{2}} du,$$

ii) $\left(M_t := \int_0^1 e^{W_{u,t} - \frac{ut}{2}} du, t \geq 0\right)$ is a $(\mathscr{G}_t, t \geq 0)$-martingale, with $\mathscr{G}_t := \sigma(W_{u,v}; u \geq 0, v \leq t)$.

The martingale $(M_t, t \geq 0)$ is therefore associated to $(X_t, t \geq 0)$. This result is essentially Theorem 2.2. This follows closely Baker and Yor [12], who introduced the Brownian sheet to construct a martingale associated to the "guiding peacock" of [43]. This is an illustration of Itô's motto presented in **C2-1)** above: we consider Brownian motion $(B_s, s \geq 0)$ as the value at time 1 of the (infinite dimensional) Brownian sheet $(W_{s,t}, s \geq 0)_{t \geq 0}$:

$$(B_s, s \geq 0) \overset{(law)}{=} (W_{s,1}, s \geq 0)$$

and obtain thus a martingale associated to the "guiding peacock".

C2-3) The different sections of this Chapter 2 consist in illustrating further Itô's motto in various and more complex contexts, which we do by:

a) using the notion of space-time harmonic function for the Brownian sheet and constructing two remarkable applications $\Phi^{(\mathbf{M})}$ and $\Phi^{(\mathbf{P})}$ (cf. Theorem 2.7) taking values, respectively in martingales and peacocks,

b) using the notion of Lévy sheet extending a Lévy process: a Lévy sheet extends a Lévy process the same way Brownian sheet extends Brownian motion (cf. Sections 2.4 and 2.5),

c) defining a Gaussian sheet – a notion which seems to be new – and using this sheet to associate martingales to some "Gaussian peacocks" described in Chapter 1 (cf. Section 2.6).

Chapter 3: The Time Reversal Method

C3-1) Let $h : \mathbb{R}_+ \times \mathbb{R} \longrightarrow \mathbb{R}$ be a space-time harmonic function, i.e.: such that $(h(s, B_s)\ s \geq 0)$ is a martingale. We assume furthermore that this martingale is centered, which is equivalent to $h(0,0) = 0$. Thus, from Point **C1-2)** in this Introduction, $\left(X_t := \int_0^t h(s, B_s) ds, t \geq 0\right)$ is a $\mathbf{F}_2$-type peacock. Due to the time-reversal invariance:

$$\text{for any fixed } t \geq 0, \quad (B_s, 0 \leq s \leq t) \overset{(law)}{=} (B_t - B_{t-s}, 0 \leq s \leq t)$$

the process $\left(M_t^{(h)} := \int_0^t h(s, B_t - B_{t-s}) ds, t \geq 0\right)$ admits the same one-dimensional marginals as $(X_t, t \geq 0)$, and, using the identity $M_t^{(h)} = \int_0^t h(t-u, B_t - B_u) du$, it is not difficult to see that $(M_t^{(h)}, t \geq 0)$ is a martingale, which is therefore associated to the peacock $(X_t, t \geq 0)$. See Theorem 3.2.

Thus, in some sense, time reversal plays the same role for $\mathbf{F}_2$-type Brownian peacocks as the sheet method for $\mathbf{F}_1$-type peacocks. On the other hand, we note that, to the contrary of the sheet method, the time reversal method does not necessitate to enlarge the original probability space.

C3-2) In the same spirit as in the preceding Point **C3-1)**, we now consider $(B_u^{(2)}, u \geq 0)$, resp. $(B_u^{(3)}, u \geq 0)$ a 2-dimensional, resp. 3-dimensional Brownian motion issued from 0. For $y \neq 0$, the process $(\log|B_u^{(2)} - y| - \log|y|, u \geq 0)$, resp. $\left(\frac{1}{|B_u^{(3)} - y|} - \frac{1}{|y|}, u \geq 0\right)$ is a strict local martingale, i.e.: it is a local martingale but <u>not</u> a martingale, issued from 0, since the function $x \longmapsto \log|x|$, resp. $x \longmapsto \frac{1}{|x|}$ is harmonic in $\mathbb{R}^2 \setminus \{0\}$, resp. in $\mathbb{R}^3 \setminus \{0\}$. It is then immediate that the processes $\left(X_t^{(2)} := \int_0^t (\log|B_u^{(2)} - y| - \log|y|)du, u \geq 0\right)$ and $\left(X_t^{(3)} := \int_0^t \left(\frac{1}{|B_u^{(3)} - y|} - \frac{1}{|y|}\right) du, u \geq 0\right)$ are not peacocks, since their expectations depend on time. Nonetheless, we show that after subtracting an adequate multiple of an intersection local time, the renormalized processes $(\widetilde{X}_t^{(2)}, t \geq 0)$ and $(\widetilde{X}_t^{(3)}, t \geq 0)$ are indeed peacocks, and we construct for each of them an associated martingale with the help of the time reversal method. See Theorem 3.5 and 3.6.

C3-3) In Section 3.4, we extend the results of the above Point **C3-1)** by replacing Brownian motion by a Lévy process.

C3-4) In the last Section 3.5 of this chapter, we come back to the Brownian set-up. Let $\mathscr{P}_2^{(\mathrm{loc})}$ the space of predictable processes $(h(u; (B_s, s \leq u)), u \geq 0)$ such that, for every $t \geq 0$, $\mathbb{E}\left[\int_0^t h^2(u; (B_s, s \leq u))du\right] < \infty$. We then construct two maps:

$$X^{(\mathrm{P})} : \mathscr{P}_2^{(\mathrm{loc})} \longrightarrow \{\text{Bounded variation peacocks}\}$$
$$X^{(\mathrm{M})} : \mathscr{P}_2^{(\mathrm{loc})} \longrightarrow \{\text{Centered } L^2\text{-martingales}\}$$

such that, for any $t \geq 0$, $X_t^{(\mathrm{P})}(h) \stackrel{(law)}{=} X_t^{(\mathrm{M})}(h)$. These two maps are related to each other by the time reversal operator (cf. Section 3.5). With these two maps, we are, as the saying goes, killing two birds with the same stone: this allows to construct simultaneously a peacock and an associated martingale. This is the purpose of Theorem 3.8 which is then illustrated with several examples. One of them (see Example 3.1) allows to recover the result described in the above Point **C3-1)**.

Chapter 4: The Time Inversion Method

C4-1) Let $(\Lambda_t, t \geq 0)$ denote an integrable Lévy process. A result by Jacod-Protter [90] establishes that $(t\Lambda_{(1/t)}, t > 0)$ is a martingale in its natural filtration. Furthermore, consider $(Z_t, t \geq 0)$ a "nice" real-valued Markov process which is recurrent and admits a jointly continuous family of local times. Let $(\tau_l, l \geq 0)$ denote the right-continuous inverse of its local time at 0. Then, for every locally bounded Borel function $f : \mathbb{R} \longrightarrow \mathbb{R}_+$, it follows from the strong Markov property that $\left(\Lambda_l^{(f)} := \int_0^{\tau_l} f(Z_s)ds, l \geq 0\right)$ is a subordinator. Gathering these two assertions, we obtain that $(t\Lambda^{(f)}_{(1/t)}, t > 0)$ is a martingale. In particular, if $(B_v, v \geq 0)$ is a Brownian motion, and $(\tau_l, l \geq 0)$ the inverse of its local time at 0, then, for every $\alpha \geq 0$, the process:

$$M_t^{(\alpha)} := t \int_0^{\tau_{1/t}} (B_v)^\alpha 1_{[0,1]}(B_v)dv, \quad t \geq 0 \tag{0.3}$$

is a martingale.

C4-2) Let $(Q_s^{(q,0)}, s \geq 0)$ denote a 0-dimensional squared Bessel process issued from $q > 0$; this process is also often called Feller's process. It is a martingale, and from Point **C1-2)**, $\left(X_t^{(\alpha,q)} := \frac{1}{t^{\alpha+1}} \int_0^t Q_s^{(q,0)} s^\alpha ds, t \geq 0\right)$ is a $\mathbf{F}_1$-type peacock. We then show (see Theorem 4.2) that the martingale $\left(qM^{(\alpha)}_{t/q}, t \geq 0\right)$ (with $M^{(\alpha)}$ as defined in (0.3)) is associated with this peacock. Our proof of this assertion hinges upon the second Ray-Knight Theorem for Brownian local times. Thus, some martingales in the family described in Point **C4-1)** above are "naturally" associated with certain $\mathbf{F}_1$-type peacocks.

C4-3) In Section 4.3, we generalize (see Theorem 4.4) the preceding Theorem 4.2 by replacing the squared Bessel process $\left(Q_s^{(q,0)}, s \geq 0\right)$ by a continuous state space branching process (CSBP), which is stable with index $\gamma \in]1,2]$. See Le Gall [114] for a description of this process. Our main tool for the proof of this generalization is an extension, which is also due to Le Gall of the second Ray-Knight Theorem for Brownian local times, which allows to realize the CSPB with index γ as the local times process of the "height process". Theorem 4.2, as described in Point **C4-2)** above, corresponds here to the particular case $\gamma = 2$.

C4-4) The last Section 4.4 of this Chapter 4 exploits the scaling property of some processes (e.g. CSPB with index $\gamma \in]1,2]$, symmetric stable processes,...) and takes up, in a slightly different manner, K. Itô's idea, as described in Point **C2-1)** above: precisely, we consider once again a process $(X_x, x \geq 0)$ as the realization at time 1 of a two parameter process $(Z_{x,t}, t \geq 0, x \geq 0)$, i.e.

$$(X_x, x \geq 0) \overset{(law)}{=} (Z_{x,1}, x \geq 0).$$

Together with the scaling property, this writing allows to obtain Theorem 4.5, which generalizes all previous results of this chapter.

Chapter 5: The Sato Process Method

C5-1) Let $H > 0$ denote a self-similarity index. A process $(U_t, t \geq 0)$ is called a *H-Sato process* if it is additive, i.e. it has independent increments, and is H self-similar, i.e.:

$$\forall c > 0, \quad (U_{ct}, t \geq 0) \overset{(law)}{=} (c^H U_t, t \geq 0).$$

A result due to Sato [155] establishes that a r.v. X is self-decomposable if and only if, for every $H > 0$ (or for some $H > 0$), there exists a H-Sato process $(U_t, t \geq 0)$ such that $U_1 \overset{(law)}{=} X$. This (unique in law) process is called the H-Sato process attached to X. Consider now an integrable, centered r.v. X which is self-decomposable. We know (cf. Point **C1-2)**) that $(tX, t \geq 0)$ is a peacock. Let $(U_t, t \geq 0)$ be the 1-Sato process attached to X. It is then clear (cf. Theorem 5.3) that $(U_t, t \geq 0)$ is a martingale associated to $(tX, t \geq 0)$. This is our first use of Sato processes to associate a martingale to a peacock.

C5-2) Let $(R_N^2(t), t \geq 0)$ be a N-dimensional squared Bessel process issued from 0. Since this process is 1 self-similar and $R_N^2(1) - N$ is a centered r.v., $(V_t^{(N)} := R_N^2(t) - Nt, t \geq 0)$ is a peacock: this is one of our exotic peacocks (cf. Proposition 1.4 and Exercise 1.23). Furthermore, $R_N^2(1)$ is a self-decomposable r.v., so that we can attach a 1-Sato process $(U_t^{(N)}, t \geq 0)$ to it. Then, $(M_t^{(N)} := U_t^{(N)} - Nt, t \geq 0)$ is a martingale associated to $(V_t^{(N)}, t \geq 0)$.

Of course, this process $(U_t^{(N)}, t \geq 0)$, or this martingale $(M_t^{(N)}, t \geq 0)$ is – a priori – somewhat abstract. It would be desirable to give a more concrete description of it. However – this is one of the beauties of mathematics! – the object needed for this construction is already there in the mathematical landscape. It is a perturbed Bessel process $(R_{1,\alpha}(t), t \geq 0)$ (with $\alpha < 1$, cf. Subsection 5.4.1.3), a process which is defined as the unique positive strong solution of the equation

$$R_{1,\alpha}(t) = B_t + \frac{1}{2} L_t(R_{1,\alpha}) + \alpha M_t(R_{1,\alpha}), \quad t \geq 0$$

where $(L_t(R_{1,\alpha}), t \geq 0)$ denotes the local time at 0 of the sub-martingale $(R_{1,\alpha}(t), t \geq 0)$, and $M_t(R_{1,\alpha}) := \sup_{s \leq t} R_{1,\alpha}(s)$. The "concrete" description of $(U_t^{(N)}, t \geq 0)$ is then provided by Proposition 5.2.

C5-3) Now, instead of merely considering the martingale $(R_N^2(t) - Nt, t \geq 0)$, let us introduce its integral, i.e.: the peacock $\left(V_N(t) := \int_0^t (R_N^2(s) - Ns) ds, t \geq 0\right)$ and more generally, the peacock $\left(V_{N,K}(t) := \frac{1}{K^2} \int_0^t (R_N^2(s) - Ns) s^{\frac{2(1-K)}{K}} ds, t \geq 0\right)$, so:

$V_N = V_{N,1}$. We then describe, with the help of Sato processes (cf. Corollaries 5.3 and 5.5) a martingale $(M_{N,K}(t), t \geq 0)$ which is associated to $(V_{N,K}(t), t \geq 0)$. Next, still with the help of a suitably perturbed Bessel process, we give a "concrete" realization of this martingale; cf. Theorems 5.7 and 5.9.

C5-4) The peacock $(V_{N,K}(t), t \geq 0)$ which we described previously "resembles" a $\mathbf{F}_2$-type peacock (cf. Point **C1-2)**) since it can be written in the form $\left(\int_0^t Z_u \nu(du), u \geq 0\right)$ where Z is a centered process. On the other hand, the process $\left(\frac{1}{t}\int_0^t (R_N^2(u) - Nu)du, t \geq 0\right)$ "resembles" rather a $\mathbf{F}_1$-type peacock. But, since: $\frac{1}{t}\int_0^t (R_N^2(u) - Nu)du = \int_0^1 (R_N^2(ts) - Nts)ds$ we are led quite naturally to study the general family of processes $\left(V^{(N,\mu)}(t) := \int_{\mathbb{R}_+^*} (R_N^2(ts) - Nts)\mu(ds), t \geq 0\right)$ where μ is a positive measure on $\mathbb{R}_+^*$ such that $\int_{\mathbb{R}_+^*} s\mu(ds) < \infty$. This study is undertaken in Subsection 5.4.4.1, and relies upon the spectral properties of the operator $K^{(\mu)}$ defined by:

$$K^{(\mu)} f(t) := \int_{\mathbb{R}_+^*} f(s)(s \wedge t)\mu(ds)$$

and upon the Karhunen-Loeve decompositions. We then exhibit two martingales associated to $(V^{(N,\mu)}(t), t \geq 0)$ (cf. Corollary 5.9 and Theorem 5.14): the first one is purely discontinuous, the second is continuous.

C5-5) The last section of this chapter is dedicated to the notion of a Sato sheet. To a self-decomposable r.v. X, we attach a Sato sheet $(\widehat{S}^X_{\lambda,t}, \lambda \geq 0, t \geq 0)$, whose properties are summarized in Theorem 5.15. We then use this sheet, just as we did previously with the Lévy sheet, to construct martingales associated to certain peacocks. For instance, we obtain the following result: let X be a self-decomposable r.v. such that $\mathbb{E}[e^{rX}] < \infty$ for every $r \geq 0$ and $\phi(r) := \log \mathbb{E}[\exp(rX)]$. Consider a Borel, bounded function $h : \mathbb{R}_+ \longrightarrow \mathbb{R}_+$ with compact support and σ a signed measure with finite total mass on $\mathbb{R}_+$. Then (cf. Proposition 5.8), if $(L_s, s \geq 0)$ is a Lévy process which extends X, we obtain that:

$$\exp\left(t\int_0^{+\infty} h(s)dL_s - \int_0^{+\infty} \phi(th(s))ds\right), \quad t \geq 0$$

$$\text{resp. } \int \exp(tL_\lambda - \lambda\phi(t))\sigma(d\lambda), \quad t \geq 0$$

is a peacock which admits the associated martingale:

$$\exp\left(\int_0^{+\infty} h(s)d_s\widehat{S}^X_{s,t} - \int_0^{+\infty} \phi(th(s))ds\right), \quad t \geq 0$$

$$\text{resp. } \int \exp\left(\widehat{S}^X_{\lambda,t} - \lambda\phi(t)\right)\sigma(d\lambda), \quad t \geq 0.$$

Chapter 6: The SDE Method

In this chapter, we associate to some peacocks $(X_t, t \geq 0)$ martingales $(M_t, t \geq 0)$ which solve a SDE of the form $Z_t = a + \int_0^t \sigma(s, Z_s) dB_s$.

C6-1) We start by stating and proving a uniqueness theorem for a Fokker-Planck equation which is due to M. Pierre. This proof may be omitted in a first reading.

C6-2) With the help of M. Pierre's theorem, we prove a general uniqueness result for a class of probabilities on $\mathbb{R}$ satisfying a Fokker-Planck equation, see Theorem 6.2. We then apply this result to associate martingales to some general peacocks in Theorem 6.3. In the two following sections, we particularize this theorem.

C6-3) Let $\varphi : \mathbb{R} \longrightarrow \mathbb{R}$ be an increasing odd function and let $(B_t, t \geq 0)$ be a Brownian motion started from 0. Assume that, for every $t \geq 0$, $\mathbb{E}[|\varphi(B_t)|] < \infty$. We know from Theorem 1.3 (see also Exercise 1.23) that $(\varphi(B_t), t \geq 0)$ is a peacock: this is one of our exotic peacocks. Our aim now is to find a martingale which solves a SDE and admits the same marginals as $(\varphi(B_t), t \geq 0)$. This is the purpose of Theorem 6.4. Under some reasonable hypotheses, there exists a function $\sigma : \mathbb{R}_+ \times \mathbb{R} \longrightarrow \mathbb{R}_+$ such that:

i) the SDE: $Z_t = \int_0^t \sigma(s, Z_s) dB_s$ admits a unique weak solution,
ii) let $(M_t, t \geq 0)$ be this solution; then $(M_t, t \geq 0)$ is associated to the peacock $(\varphi(B_t), t \geq 0)$.

C6-4) Consider a SDE of the form:

$$Y_t = \int_0^t \tau(Y_s) dB_s + \int_0^t b(Y_s) ds.$$

Under simple hypotheses (among which the evenness of τ and the oddness of b), this SDE admits a unique strong solution $(Y_t, t \geq 0)$ which is a peacock (cf. Exercise 1.27). We prove that (cf. Theorem 6.5), once again under suitable hypotheses, there exists a function $\sigma : \mathbb{R}_+ \times \mathbb{R} \longrightarrow \mathbb{R}$ such that the SDE: $Z_t = \int_0^t \sigma(s, Z_s) dB_s$ admits a unique strong solution $(M_t, t \geq 0)$: this is a martingale which is associated to $(Y_t, t \geq 0)$.

C6-5) We state a variant of Theorem 6.2 for probability measures on $\mathbb{R}_+$, and then use this result to associate a martingale to the peacock $\left(\frac{e^{tX}}{\mathbb{E}[e^{tX}]}, t \geq 0\right)$ under a suitable hypothesis on the law of X (log-concavity).

C6-6) Let X be an integrable, centered r.v. We look for a martingale $(M_t, t \geq 0)$ which is associated to the peacock $(\sqrt{t}X, t \geq 0)$ and which also solves a SDE. Let us assume that X admits a density h, and define $a_h : \mathbb{R} \longrightarrow \mathbb{R}_+$ by:

$$a_h(y) = \frac{1}{h(y)} \int_y^{+\infty} z h(z) dz$$

(cf. (6.49) for a more rigorous definition of a_h). Then, if a_h is Lipschitz continuous, the SDE:

$$Z_t := \int_0^t \sqrt{a_h}\left(\frac{Z_s}{\sqrt{s}}\right) dB_s$$

admits a unique weak solution. This solution $(M_t, t \geq 0)$ is a martingale, which is a self-similar process of order $\frac{1}{2}$, i.e.: $\forall c > 0$, $(M_{ct}, t \geq 0) \overset{(law)}{=} (\sqrt{c}M_t, t \geq 0)$, and is associated to the peacock $(\sqrt{t}X, t \geq 0)$; cf. Theorem 6.11.

C6-7) Chapter 6 ends (Section 6.7) with a (solved!) problem which establishes a link between the porous media equation and certain peacocks of the form $(t^\beta X, t \geq 0)$.

Chapter 7: The Skorokhod Embedding (SE) Method

C7-1) Consider $(X_t, t \geq 0)$ an integrable centered peacock. We denote by ν_t the law of the r.v. X_t. Assume that for each t, we know how to realize a Skorokhod embedding of ν_t (either randomized or not, see Subsection 7.1.1 for the definition of a Skorokhod embedding) with the help of a stopping time τ_t such that:

1. $B_{\tau_t} \sim \nu_t$, i.e. B_{τ_t} has law ν_t.
2. For every $t \geq 0$, $(B_{u \wedge \tau_t}, u \geq 0)$ is a uniformly integrable martingale.
3. The family of stopping times $(\tau_t, t \geq 0)$ is a.s. increasing,
   i.e.: for every $s \leq t$, $\tau_s \leq \tau_t$ a.s.

Then, $(M_t := B_{\tau_t}, t \geq 0)$ is a martingale which is associated to $(X_t, t \geq 0)$. In this chapter, we illustrate this ideal program, whose main difficulty is to find an *increasing* family of stopping times $(\tau_t, t \geq 0)$.

C7-2) In Section 7.2, we present a seemingly new SE. This randomized SE makes it possible to associate a martingale to any peacock of the form $(\sqrt{t}X, t \geq 0)$ where X is an integrable centered r.v. Moreover, this martingale $(M_t, t \geq 0)$ enjoys the Brownian scaling property:

$$\forall c > 0, \quad (M_{c^2 t}, t \geq 0) \overset{(law)}{=} (cM_t, t \geq 0). \tag{0.4}$$

C7-3) In Section 7.3, we present an SE method which may be attributed to Hall and Breiman. This method also allows to associate to any peacock of the form $(\sqrt{t}X, t \geq 0)$, with X an integrable centered r.v, a martingale $(M_t, t \geq 0)$ which enjoys the Brownian scaling property (0.4).

C7-4) In Section 7.4, we use the Azéma-Yor SE method. It is a non-randomized SE, which under some extra-hypotheses on the law of X, makes it possible to associate to the peacock $(\sqrt{t}X, t \geq 0)$, with X an integrable centered r.v, a martingale $(M_t, t \geq 0)$ which enjoys the Brownian scaling property (7.1). We present a necessary and sufficient condition for the family $(\tau_t, t \geq 0)$ to be increasing, and give several sufficient conditions.

C7-5) To begin Section 7.5, we use Bass' embedding method in order to associate a martingale to the peacock $(\varphi(B_t), t \geq 0)$, where φ is an odd increasing function such that, for every $t \geq 0$, $\mathbb{E}[|\varphi(B_t)|] < \infty$.

This problem was already tackled in Chapter 6, see **C6-1)**, but solved there only under some extra hypotheses on φ.

This chapter ends with three more sections written in a different spirit:

- at the end of Section 7.5, a relationship between Bougerol's identity in law and Bass' SE method is established and studied,
- Section 7.6 uses the notion of Ocone martingale to associate such martingales to certain peacocks,
- in the last Section 7.7, "Student's peacocks" are studied in the framework of a solved problem.

The Bertoin-Le Jan SE method is also discussed in Exercise 7.14. In particular, we prove that a log-concavity hypothesis allows to obtain an increasing family of stopping times $(\tau_t, t \geq 0)$ such that, for every $t \geq 0$ fixed: $B_{\tau_t} \stackrel{(law)}{=} \sqrt{t}X$.

Chapter 8: Comparing Multidimensional Marginals

This chapter differs notably from the preceding ones: here we do not seek for martingales associated to given peacocks.

C8-1) In order to explain the aim of this chapter, we need to introduce the following notation. Let $(B_u, u \geq 0)$ a Brownian motion issued from 0, and consider again the guiding example:

$$X_t := \int_0^1 \exp\left(B_{ut} - \frac{ut}{2}\right) du, t \geq 0.$$

We further introduce:

$$X_t^{(S)} := \int_0^1 \exp\left(W_{u,t} - \frac{ut}{2}\right) du, \quad t \geq 0$$

where S stands for the Brownian sheet $(W_{s,t}, s \geq 0, t \geq 0)$,

$$X_t^{(R)} := \int_0^1 \exp\left(B_t - B_{(1-u)t} - \frac{ut}{2}\right) du, \quad t \geq 0$$

where R stands for time reversal,

$$X_t^{(Sc)} := \int_0^1 \exp\left(\sqrt{t}B_u - \frac{ut}{2}\right) du, \quad t \geq 0$$

where Sc stands for scaling.

Those four processes admit the same one-dimensional marginals. What can be said of their multidimensional marginals? In particular, can we compare them? For this purpose, we need to use an order relation which bears upon multidimensional distributions. We shall use the upper orthant order which we shall denote by $\underset{u.o}{\geq}$ (see Section 8.1.2 for a precise definition). First, we obtain a general comparison theorem for a large class of functionals of Gaussian processes, see Theorem 8.1. Then, we look at the particular case of this theorem when applied to the processes X, $X^{(S)}$, $X^{(R)}$ and $X^{(Sc)}$. We obtain (cf. Theorem 8.3):

$i)$ $(X_t, t \geq 0) \underset{u.o}{\geq} (X_t^{(S)}, t \geq 0) \underset{u.o}{\geq} (X_t^{(R)}, t \geq 0)$,

$ii)$ $(X_t^{(Sc)}, t \geq 0) \underset{u.o}{\geq} (X_t^{(S)}, t \geq 0) \underset{u.o}{\geq} (X_t^{(R)}, t \geq 0)$.

On the other hand, the processes $(X_t, t \geq 0)$ and $(X_t^{(Sc)}, t \geq 0)$ are not u.o. comparable. In Sections 8.1–8.4, we also obtain:

- comparison results for the lower orthant order, as defined at the beginning of Section 8.3,
- inequalities between the quantities: $\mathbb{E}[(k-X_t)^+]$, $\mathbb{E}\left[(k-X_t^{(S)})^+\right]$, $\mathbb{E}\left[(k-X_t^{(R)})^+\right]$, $\mathbb{E}\left[(k-X_t^{(Sc)})^+\right]$ and other similar relations (cf. Corollary 8.3),
- inequalities between the quantities: $\mathbb{E}\left[|X_t - X_s|^q\right]$, $\mathbb{E}\left[|X_t^{(S)} - X_s^{(S)}|^q\right]$, $\mathbb{E}\left[|X_t^{(R)} - X_s^{(R)}|^q\right]$ and $\mathbb{E}\left[|X_t^{(Sc)} - X_s^{(Sc)}|^q\right]$.

C8-2) The last Section 8.5 of this Chapter 8 bears upon a different question. Let μ and ν be two probability measures on $\mathbb{R}_+$ such that μ dominates ν stochastically, i.e. such that, for all $t \geq 0$

$$F_\mu(t) := \mu([0,t]) \leq F_\nu(t) := \nu([0,t]).$$

We then consider a process $(X_s, s \geq 0)$ and ask under which conditions is it true that

$$A_X^{(\mu)} := \int_0^{+\infty} X_s \mu(ds) \overset{(c)}{\geq} A_X^{(\nu)} := \int_0^{+\infty} X_s \nu(ds)\ ?$$

(where $\overset{(c)}{\geq}$ denotes inequality in the convex order sense). In fact, this question has been motivated by Theorem 1.4. Indeed, let $(M_s, s \geq 0)$ be a martingale in H^1_{loc} and, for $s < t$, define $\mu(du) := \frac{1}{t} 1_{[0,t]}(u)du$ and $\nu(du) := \frac{1}{s} 1_{[0,s]}(u)du$. It is then clear that $\mu \overset{(st)}{\geq} \nu$ and that $A_M^{(\mu)} \overset{(c)}{\geq} A_M^{(\nu)}$ since $\left(\frac{1}{t}\int_0^t M_u du, t \geq 0\right)$ is a peacock from Theorem 1.4. Section 8.5 gives a partial answer to the preceding question. Theorem 8.6 presents a set of sufficient conditions on X which ensure that $A_X^{(\mu)} \overset{(c)}{\geq} A_X^{(\nu)}$. Many examples show that these sufficient conditions are often satisfied.

- As the reader noticed when reading the above summary of the eight chapters of this monograph, we often find ourselves in the following situation: starting from a random object R, we "create" a related random object $\widetilde{R}$ which plays an important role. As these random objects are rather different, we do not want to always use the terminology: $\widetilde{R}$ is associated to R. Thus, we have devoted the next three pages about the various terms we use for these different "associations".
- The volume ends with a list of open problems.

As a temporary conclusion of our investigations so far, we believe this topic of *Peacocks and associated martingales* is quite rich; despite the length of our volume, we are still far from having exhausted the field; in particular, we have only been interested in the existence question, i.e. to find a martingale $(M_t, t \geq 0)$ with a given marginals set $\mu = (\mu_t, t \geq 0)$, but we have not touched the study of the set $\mathscr{M}_\mu$ of all laws of martingales with this set μ of marginals. As D. Baker pointed out to us, there exists a fast developing literature about this, with authors such as: Albin [2], Baker, Donati-Martin and Yor [11], Lowther [120]... For example, G. Lowther proves, under some suitable conditions, uniqueness of the martingale associated to a given peacock. Note that this result is not true in full generality; indeed, we shall give in this monograph numerous examples of peacocks which admit several different associated martingales. We hope to be able to contribute to this other aspect of the topic in the near future.

Acknowledgements

Obviously, we owe a lot of gratitude to Carr-Ewald-Xiao for providing the example of a peacock which has been our guiding example throughout the volume. We are also grateful to D. Baker and D. Madan for joint works and for their comments, as well as to Y. Takahashi for offering us a Haïku about peacocks. Our special thanks go finally to M. Pierre for providing us with the proof of the uniqueness theorem given in Section 6.1.

Terminology about: Association, Attachment, Extension

Associated Processes

- For two r.v.'s U and V, we write $U \overset{(law)}{=} V$ if U and V have the same law.
- For two processes $(X_t, t \geq 0)$ and $(Y_t, t \geq 0)$, we write:

$$(X_t, t \geq 0) \overset{(law)}{=} (Y_t, t \geq 0)$$

  if these processes have the same law, i.e.: the same finite dimensional marginals.
- For two processes $(X_t, t \geq 0)$ and $(Y_t, t \geq 0)$, we write:

$$\boxed{X_t \overset{(1.d)}{=} Y_t}$$

  if $(X_t, t \geq 0)$ and $(Y_t, t \geq 0)$ have the same one-dimensional marginals. We shall say that such two processes are *associated*.
- A process $(X_t, t \geq 0)$ is called a *1-martingale* if there exists a martingale $(M_t, t \geq 0)$ which is associated with X, i.e.: $X_t \overset{(1.d)}{=} M_t$.

Attached Processes

- *Gaussian sheet attached to a family of Gaussian processes (see Section 2.6)*

Let $(G_{\lambda,t};\ \lambda \in \Lambda, t \geq 0)$ be a family of centered Gaussian processes, that is, for each $t \geq 0$, $G_{\bullet,t} := (G_{\lambda,t}, \lambda \in \Lambda)$ is a centered Gaussian process. Let

$$c_{\lambda,\mu}(t) := \mathbb{E}[G_{\lambda,t} G_{\mu,t}], \quad (\lambda, \mu \in \Lambda, t \geq 0).$$

Let us suppose that, for every $n \geq 0$, every $\lambda_1, \dots, \lambda_n \in \Lambda$, the matrix function: $t \in \mathbb{R}_+ \longrightarrow (c_{\lambda_j,\lambda_k}(t))_{1 \leq j,k \leq n} \in \mathbb{S}_n$, the space of symmetric $n \times n$ matrices, is increasing with respect to the order on $\mathbb{S}_n$ defined by the cone $\mathbb{S}_n^+$ of symmetric positive matrices.

Hirsch F., Profeta C., Roynette B., Yor M.:
Peacocks and Associated Martingales, with Explicit Constructions

Then, under some extra regularity hypotheses, there exists a centered Gaussian process $(\Gamma_{\lambda,t};\ \lambda \in \Lambda, t \geq 0)$ such that

$$\mathbb{E}[\Gamma_{\lambda,s}\Gamma_{\mu,t}] = c_{\lambda,\mu}(s \wedge t), \quad (\lambda,\mu \in \Lambda; t,s \geq 0).$$

This process $(\Gamma_{\lambda,t};\ \lambda \in \Lambda, t \geq 0)$ is said to be *attached* to the family of Gaussian processes $(G_{\lambda,t};\ \lambda \in \Lambda, t \geq 0)$. It satisfies, in particular,

i) for every $t \geq 0$ fixed: $(\Gamma_{\lambda,t},\ \lambda \in \Lambda) \overset{(law)}{=} (G_{\lambda,t},\ \lambda \in \Lambda)$,
ii) $(\Gamma_{\bullet,t}, t \geq 0)$ has independent increments.

• *H-Sato process attached to X (see Section 5.2)*
Let X be a self-decomposable r.v. For every $H > 0$, there exists a H-Sato process $(S_t^{H,X}, t \geq 0)$, which is unique in law, and is characterized by:

i) $S_1^{H,X} \overset{(law)}{=} X$,
ii) $(S_t^{H,X}, t \geq 0)$ has independent increments,
iii) $(S_t^{H,X}, t \geq 0)$ is H-self-similar, i.e.:

$$\forall c > 0, \quad (S_{ct}^{H,X}, t \geq 0) \overset{(law)}{=} (c^H S_t^{H,X}, t \geq 0).$$

Such a process $(S_t^{H,X}, t \geq 0)$ is said *attached* to X.
If $H = 1$, $S^{H,X}$ is simply denoted by S^X.

• *Sato sheet attached to a self-decomposable r.v. X (see Section 5.5)*
Let X be a self-decomposable r.v. There exists a process $(\widehat{S}^X_{\lambda,t};\ \lambda, t \geq 0)$, unique in law, characterized by:

i) $(\widehat{S}^X_{\bullet,t}, t \geq 0)$ is a $\mathbb{D}_0$-valued 1-Sato process, i.e.:
 – $(\widehat{S}^X_{\bullet,t}, t \geq 0)$ is a $\mathbb{D}_0$-valued process with independent increments,
 – $(\widehat{S}^X_{\bullet,t}, t \geq 0)$ is a $\mathbb{D}_0$-valued 1-self similar process:

$$\forall c \geq 0, \quad (\widehat{S}^X_{\bullet,ct}, t \geq 0) \overset{(law)}{=} (c\widehat{S}^X_{\bullet,t}, t \geq 0),$$

ii) for every $t \geq 0$ fixed, $(\widehat{S}^X_{\lambda,t}, \lambda \geq 0)$ is a Lévy process,
iii) $\widehat{S}^X_{1,1} \overset{(law)}{=} X$

$(\widehat{S}^X_{\lambda,t};\ \lambda, t \geq 0)$ is called the Sato sheet *attached* to X.

Extending Processes

• *Lévy sheet extending a Lévy process (see Section 2.4)*
Let $(L_t, t \geq 0)$ be a Lévy process started from 0. The Lévy sheet *extending* L, denoted by $(W_{s,t}^{(L)};\ s,t \geq 0)$, is the process, which is unique in law, and is characterized by:

i) $(W^{(L)}_{\bullet,t}, t \geq 0)$ is a $\mathbb{D}_0$-valued Lévy process, where $\mathbb{D}_0$ denotes the set of càdlàg functions $\varepsilon : \mathbb{R}_+ \longrightarrow \mathbb{R}$ such that $\varepsilon(0) = 0$,

ii) for every $t \geq 0$, $(W^{(L)}_{s,t}, s \geq 0) \overset{(law)}{=} (L_{ts}, s \geq 0)$.

• *The Brownian sheet (see Section 2.2)*
In the particular case where $(L_s, s \geq 0)$ is a Brownian motion started at 0, $(W^{(L)}_{s,t}; s,t \geq 0)$ is called the standard Brownian sheet. We denote it by $(W_{s,t}; s,t \geq 0)$.

• *Sato sheet extending a self-decomposable Lévy process (see Section 5.5)*
Let L be a Lévy process such that L_1 is self-decomposable (L is called a self-decomposable Lévy process). We then denote $\widehat{S}^{L_1}$ by $\widehat{S}^{(L)}$. One has:

$$\left(\widehat{S}^{(L)}_{\lambda,1}, \lambda \geq 0\right) \overset{(law)}{=} (L_\lambda, \lambda \geq 0)$$

and $\left(\widehat{S}^{(L)}_{\lambda,t}; \lambda \geq 0, t \geq 0\right)$ is called the Sato sheet *extending* the self-decomposable Lévy process L.

1

Some Examples of Peacocks

Abstract We exhibit several classes of processes which are increasing in the convex order. Among these, an important class consists of the arithmetic means of martingales (which are closely related with Asian options).

1.1 Introduction

In this first section, we present basic facts about processes which increase in the convex order.

1.1.1 Processes Increasing in the Convex Order and 1-Martingales

Definition 1.1. Let X and Y be two real-valued r.v.'s. X is said to be dominated by Y for the convex order if, for every convex function $\psi : \mathbb{R} \to \mathbb{R}$ such that $\mathbb{E}[|\psi(X)|] < \infty$ and $\mathbb{E}[|\psi(Y)|] < \infty$, one has:

$$\mathbb{E}[\psi(X)] \leq \mathbb{E}[\psi(Y)]. \tag{1.1}$$

We denote this order by:

$$X \overset{(c)}{\leq} Y. \tag{1.2}$$

The class $\boldsymbol{C}$

$\boldsymbol{C}$ denotes the class of convex $\mathscr{C}^2$-functions $\psi : \mathbb{R} \longrightarrow \mathbb{R}$ such that ψ'' has a compact support. We note that if $\psi \in \boldsymbol{C}$:

- $|\psi'|$ is a bounded function,
- there exist k_1 and $k_2 \geq 0$ such that:

$$|\psi(x)| \leq k_1 + k_2|x|. \tag{1.3}$$

The class $\boldsymbol{C}_+$

We denote by $\boldsymbol{C}_+$ the class of convex functions $\psi \in \boldsymbol{C}$ such that ψ is positive and increasing.

Hirsch F., Profeta C., Roynette B., Yor M.:
Peacocks and Associated Martingales, with Explicit Constructions

Remark 1.1. Let X and Y be two integrable real-valued r.v.'s. Then, the following are equivalent:

1) $X \overset{(c)}{\leq} Y$.
2) For every $\psi \in \boldsymbol{C}$: $\mathbb{E}[\psi(X)] \leq \mathbb{E}[\psi(Y)]$.
3) $\mathbb{E}[X] = \mathbb{E}[Y]$ and for every $\psi \in \boldsymbol{C}_+$: $\mathbb{E}[\psi(X)] \leq \mathbb{E}[\psi(Y)]$.

Proof. i) We prove that 1) $\Rightarrow$ 3). Assume that $X \overset{(c)}{\leq} Y$. Applying (1.1) first with $\psi(x) = x$, then with $\psi(x) = -x$, we deduce that $\mathbb{E}[X] = \mathbb{E}[Y]$. We also note that if $\psi \in \boldsymbol{C}_+$:

$$\mathbb{E}[|\psi(X)|] \leq k_1 + k_2\mathbb{E}[|X|] < \infty\,, \quad \text{(from (1.3))}$$

and similarly for Y.
ii) If $\psi \in \boldsymbol{C}$, then there exist a and b such that $x \longmapsto a + bx + \psi(x)$ belongs to $\boldsymbol{C}_+$. This shows 3) $\Rightarrow$ 2).
iii) Since any convex function ψ is the envelope from below of the affine functions which are smaller than ψ, one sees that any convex function is an increasing limit of a sequence of functions in $\boldsymbol{C}$. Then, 2) $\Rightarrow$ 1) follows from the monotone convergence theorem. □

Remark 1.2. It is easy to show that, in Remark 1.1, we may replace the class $\boldsymbol{C}$ by the class $\boldsymbol{C}_a$: $\boldsymbol{C}_a$ is the class of convex functions ψ such that ψ is affine by parts, or by the class $\boldsymbol{C}_\infty$, consisting of elements of $\boldsymbol{C}$ which are $\mathscr{C}^\infty$.

Definition 1.2. A process $(X_t, t \geq 0)$ is said to be increasing for the convex order if, for every $s \leq t$, $X_s \overset{(c)}{\leq} X_t$.

Definition 1.3. A process $(X_t, t \geq 0)$ is said to be integrable if, for every $t \geq 0$, $\mathbb{E}[|X_t|] < \infty$. An integrable process which is increasing in the convex order will be called a peacock.

This pun comes from the French: *P*rocessus *C*roissant pour l'*O*rdre *C*onvexe, the acronym of which: "PCOC" may be pronounced as "peacock".

Definition 1.4 (1-martingale). We call a process $(X_t, t \geq 0)$ a 1-martingale if there exists a martingale $(M_t, t \geq 0)$ (defined possibly on another probability space) such that, for every fixed $t \geq 0$:

$$X_t \overset{(law)}{=} M_t. \tag{1.4}$$

Remark 1.3. 1) In Definition 1.4, only equality in law for fixed t is assumed, which is much weaker than equality in law between processes. In the sequel, when two processes $(U_t, t \geq 0)$ and $(V_t, t \geq 0)$ have the same 1-dimensional marginals, we shall write:

$$\boxed{U_t \overset{(1.d)}{=} V_t} \tag{1.5}$$

and say that $(U_t, t \geq 0)$ and $(V_t, t \geq 0)$ are *associated.*
2) The martingale $(M_t, t \geq 0)$ found in (1.4) is neither necessarily unique (in law) nor defined on the same probability space as the process $(X_t, t \geq 0)$. We shall call such a martingale a *martingale associated* to the process $(X_t, t \geq 0)$.
3) In Definitions 1.2, 1.3 and 1.4, only the 1-dimensional marginals of the processes involved are featured. Thus, regularity properties of these processes are not relevant to our discussion.

The main result in this Subsection 1.1.1 is the following:

Theorem 1.1 (H.G. Kellerer [106]). *The two following properties are equivalent:*
1) $(X_t, t \geq 0)$ *is a peacock.*
2) $(X_t, t \geq 0)$ *is a 1-martingale.*

Remark 1.4. The implication: 2) $\Rightarrow$ 1) is easy. Indeed, for $\psi \in \boldsymbol{C}$ and $s \leq t$:

$$\begin{aligned} \mathbb{E}[\psi(X_t)] &= \mathbb{E}[\psi(M_t)] = \mathbb{E}[\mathbb{E}[\psi(M_t)|\mathscr{F}_s]] \\ &\geq \mathbb{E}[\psi(\mathbb{E}[M_t|\mathscr{F}_s])] \quad \text{(from Jensen's inequality)} \\ &= \mathbb{E}[\psi(M_s)] = \mathbb{E}[\psi(X_s)]. \end{aligned}$$

On the contrary, the implication 1) $\Rightarrow$ 2) is difficult to establish. It was obtained through the succession of works of V. Strassen ([163]), J.L. Doob ([56]) and H.G. Kellerer ([106]) to which we refer the reader.

However, let us note that the main result of Kellerer (Theorem 3 p.129 in [106]) is not exactly Theorem 1.1 above. Indeed, Kellerer proves the existence of a sub-martingale $(M_t, t \geq 0)$ such that $X_t \overset{(1.d)}{=} M_t$. But Kellerer does not use exactly the same notion of convex order as the one presented in Definitions 1.1 and 1.2. In particular, for Kellerer, the function $\psi(x) := -x$ is not a "reference" function. But it is not difficult to see that Theorem 3, p.110 of [106] implies Theorem 1.1.

On the other hand, the process $(M_t, t \geq 0)$ such that $X_t \overset{(1.d)}{=} M_t$, constructed by Kellerer, can be chosen to be Markovian.

Kellerer's proof of Theorem 1.1 – the identity between peacocks and 1-martingales – is not constructive. It is an interesting question, given a peacock $(X_t, t \geq 0)$, whose family of one-dimensional marginals is $(\mu_t, t \geq 0)$, to find, as explicitly and concretely as possible, a martingale $(M_t, t \geq 0)$ which admits the same one-dimensional marginals.

Most of this monograph is dedicated to this question.

Exercise 1.1 (Around uniform integrability of peacocks). Let $(X_t, t \geq 0)$ be a peacock and $(M_t, t \geq 0)$ an associated $(\mathscr{F}_t, t \geq 0)$-martingale. Since, for $s \leq t$, $M_s = \mathbb{E}[M_t|\mathscr{F}_s]$, the family $(M_s, s \leq t)$ is uniformly integrable, hence so is the family $(X_s, s \leq t)$.

1) *i*) Prove directly – i.e. without Kellerer's theorem – that $(X_s, s \leq t)$ is a uniformly integrable family.
(*Hint*: use the inequality: $|x|1_{\{|x|\geq a\}} \leq (2|x| - a)^+$ for every $x \in \mathbb{R}$ and $a \geq 0$.)
ii) Deduce that there exists a positive, increasing and convex function G such that $\frac{G(x)}{x} \xrightarrow[x \to +\infty]{} +\infty$ and $\sup_{s \leq t} \mathbb{E}[G(|X_s|)] < \infty$. (*Hint*: use de la Vallée-Poussin's Theorem, see [127, Chapitre 2, Th. 22, p.38].)
2) Another proof of the uniform integrability of $(X_s, s \leq t)$.
i) Prove directly the existence of a positive, increasing and convex function G such that $\frac{G(x)}{x} \xrightarrow[x \to +\infty]{} +\infty$ and $\mathbb{E}[G(|X_t|)] < \infty$.
ii) Deduce that the family $(X_s, s \leq t)$ is uniformly integrable.

Remark 1.5. We say that a r.v. X is smaller than a r.v. Y for the stochastic order (and we denote $X \overset{(st)}{\leq} Y$) if, for every $\lambda \in \mathbb{R}$,

$$\mathbb{P}(X \geq \lambda) \leq \mathbb{P}(Y \geq \lambda).$$

A process $(X_t, t \geq 0)$ is said to be *increasing in the stochastic order* if for every $s \leq t$, $X_s \overset{(st)}{\leq} X_t$. Here is now a theorem due to Kamae and Krengel, in the same spirit as Kellerer's theorem, where the convex order is replaced by the stochastic order:

Theorem 1.2 (Kamae-Krengel [104]). *The two following properties are equivalent:*
1) $(X_t, t \geq 0)$ *is increasing in the stochastic order.*
2) There exists an increasing process $(A_t, t \geq 0)$ *such that*

$$X_t \overset{(1.d)}{=} A_t.$$

A discussion about some links between Kellerer's and Kamae-Krengel's theorems is undertaken in Exercise 1.26.

Exercise 1.2 (A proof of Kamae-Krengel Theorem). Let $(\Omega, \mathscr{F}, m)$ be the probability space $(]0,1[, \mathscr{B}(]0,1[, m)$ where m is the Lebesgue measure. Let also X be a $\mathbb{R}$-valued r.v. and $F_X : \mathbb{R} \longmapsto [0,1]$ its distribution function defined by $F_X(\lambda) := \mathbb{P}(X < \lambda)$. We define, for every $\omega \in]0,1[$, $A_X(\omega) := \inf\{\lambda; F_X(\lambda) > \omega\}$.

1) Prove that $m(A_X < \lambda) = \mathbb{P}(X < \lambda)$ and deduce that $A_X \overset{(law)}{=} X$.
2) Let X and Y be two r.v.'s such that $Y \overset{(st)}{\geq} X$. Show that $A_Y \geq A_X$.
3) Deduce the Kamae-Krengel Theorem.

Exercise 1.3 ($\mathbb{R}_+$-valued peacocks and stochastic order). Let $(X_t, t \geq 0)$ be a $\mathbb{R}_+$-valued peacock.

1) Prove that there exists an increasing process $(A_t, t \geq 0)$ such that

$$\frac{\mathfrak{e}}{X_t} \overset{(1.d)}{=} A_t,$$

where $\mathfrak{e}$ is an exponential r.v. independent of X_t. (*Hint*: use the convex functions $x \longmapsto e^{-\lambda x}$ $(\lambda \geq 0)$ and Kamae-Krengel Theorem.)

2) Prove that, in general, the process $\left(\frac{1}{X_t}, t \geq 0\right)$ is not stochastically increasing. (*Hint*: consider a peacock of the form $\left(\frac{e^{tX}}{\mathbb{E}[e^{tX}]}, t \geq 0\right)$ for a suitable r.v. X, see Corollary 1.2.)

3) Let Z be a $\mathbb{R}_+$-valued r.v. such that $\mathbb{P}(Z > 0) = 1$. Prove that there exists a continuous and strictly increasing function $\psi : \mathbb{R}_+ \longrightarrow \mathbb{R}_+$ such that $\psi(0) = 0$, $\psi(+\infty) = \infty$ and

$$\mathbb{E}[e^{-\lambda Z}] = \exp(-\psi(\lambda)).$$

4) Deduce that, if for every $t \geq 0$, $\mathbb{P}(X_t > 0) = 1$, then one can chose

$$A_t = \psi_t^{-1}(\mathfrak{e})$$

where $\mathfrak{e}$ is an exponential r.v., and $\psi_t^{-1} : \mathbb{R}_+ \longrightarrow \mathbb{R}_+$ is the inverse in the sense of composition of functions of ψ_t, the function attached to the r.v. X_t by question 3).

Remark 1.6. 1) Similarly to the definition of a 1-martingale as given above, the notion of a 2-martingale may be defined in the following way: a process $(X_t, t \geq 0)$ is a 2-martingale if there exists a martingale $(M_t, t \geq 0)$ such that, for every t_1 and t_2 fixed:

$$(X_{t_1}, X_{t_2}) \overset{(law)}{=} (M_{t_1}, M_{t_2}). \tag{1.6}$$

Throughout this monograph, most of the studied peacocks are bounded variation processes (see Theorem 1.4 for example). In this context of bounded variation processes, the notion of a 2-martingale is of little interest. Indeed, we state and prove:

Proposition 1.1. *Let $(X_t, t \leq T)$ be a continuous process with bounded variation such that $\mathbb{E}\left[\left(\int_0^T |dX_s|\right)^2\right] < \infty$. Assume furthermore that $(X_t, t \leq T)$ has the same 2-dimensional marginals as a square integrable, right-continuous martingale $(M_t, t \leq T)$. Then, $(M_t, t \leq T)$ and $(X_t, t \leq T)$ are a.s. constant.*

Proof. Let $(\sigma_n, n \geq 0)$ be a sequence of subdivisions of $[0, T]$ whose meshes tend to 0 as $n \longrightarrow \infty$. Then:

$$\begin{aligned}\mathbb{E}\left[\sum_{\sigma_n}(X_{t_{i+1}} - X_{t_i})^2\right] &= \mathbb{E}\left[\sum_{\sigma_n}(M_{t_{i+1}} - M_{t_i})^2\right] \text{ (since } X_t \overset{(2.d)}{=} M_t). \\ &= \mathbb{E}\left[(M_T - M_0)^2\right]\end{aligned}$$

On the other hand:

$$\sum_{\sigma_n}(X_{t_{i+1}} - X_{t_i})^2 \le \sup_{\sigma_n} |X_{t_{i+1}} - X_{t_i}| \left(\int_0^T |dX_s|\right) \underset{n\to\infty}{\longrightarrow} 0 \text{ a.s.}$$

$$\text{and} \quad \sum_{\sigma_n}(X_{t_{i+1}} - X_{t_i})^2 \le \left(\int_0^T |dX_s|\right)^2.$$

Thus, from the dominated convergence theorem, $\mathbb{E}\left[(M_T - M_0)^2\right] = 0$ and $(M_t, t \le T)$, $(X_t, t \le T)$ are constant a.s. □

2) The previous Point 1) of this Remark 1.6 may be compared with a result by H. Föllmer, C.T. Wu and M. Yor (see [67], and in particular Proposition 7.4), which we now describe: these authors construct, for every fixed integer $k \ge 1$, a process $(X_t^{(k)}, 0 \le t \le 1)$ such that:

a) for every $0 \le t_1 < ... < t_k \le 1$ fixed:

$$(X_{t_1}^{(k)}, ..., X_{t_k}^{(k)}) \overset{(law)}{=} (B_{t_1}, ..., B_{t_k}) \tag{1.7}$$

where $(B_t, t \ge 0)$ is a standard Brownian motion,

b) the law $P^{(k)}$ of $(X_t^{(k)}, t \le 1)$ is equivalent to the Wiener measure W on $\mathscr{C}([0,1] \to \mathbb{R})$,

c) $(X_t^{(k)}, t \le 1)$ is not a martingale and, in particular, $P^{(k)} \ne W$.

Of course, from point *b*) for $k \ge 1$, $(X_t^{(k)}, t \le 1)$ has unbounded variation.
Note that, as soon as $k \ge 4$:

- $X_{t_2}^{(k)} - X_{t_1}^{(k)}$ is a centered Gaussian r.v. with variance $|t_2 - t_1|$,
- for $t_1 \le t_2 \le t_3 \le t_4$, $(X_{t_2}^{(k)} - X_{t_1}^{(k)})$ is independent from $(X_{t_4}^{(k)} - X_{t_3}^{(k)})$.

3) The implication "1-martingale $\Longrightarrow$ peacock" in Theorem 1.1 shows that a martingale is a peacock. Let us now consider an integrable process $(X_t, t \ge 0)$ such that, for every $s \le t$:

$$\mathbb{E}[X_t | X_s] = X_s \tag{1.8}$$

(for ease of reference, we shall call such a process a fair 2-process). Then $(X_t, t \ge 0)$ is a peacock. Indeed, for $\psi \in \boldsymbol{C}$:

$$\begin{aligned}
\mathbb{E}[\psi(X_t)] &= \mathbb{E}[\mathbb{E}[\psi(X_t)|X_s]] \\
&\ge \mathbb{E}[\psi(\mathbb{E}[X_t|X_s])] \quad \text{(from Jensen's inequality)} \\
&= \mathbb{E}[\psi(X_s)].
\end{aligned}$$

Of course, if $(X_t, t \ge 0)$ is a martingale, the relation (1.8) is satisfied, but the converse is not always true (see Exercise 1.4).

Exercise 1.4 (Fair 2-processes and martingales).

1) Give an example of an integrable process $(X_t, t \ge 0)$ which satisfies (1.8) and which is not a martingale.

(*Hint*: the process $(X_t^{(k)}, t \geq 0)$, for $k \geq 2$ defined in Point 2) of Remark 1.6 satisfies (1.8); more generally, it satisfies, for every $0 \leq t_1 \leq t_2 ... \leq t_k \leq 1$: $\mathbb{E}[X_{t_k} | X_{t_1}, ... X_{t_{k-1}}] = X_{t_{k-1}}$. Again, for ease of reference, we shall call such a process a fair k-process.)

2) If $(X_t, t \geq 0)$ is a centered Gaussian process, prove that the following properties are equivalent:
   i) for every $s < t$, $\mathbb{E}[X_t | X_s] = X_s$, i.e. X is a fair 2-process,
   ii) $(X_t, t \geq 0)$ is a martingale, i.e., for $s < t$, $\mathbb{E}[X_t | \mathscr{F}_s] = X_s$ with $\mathscr{F}_s = \sigma(X_u, u \leq s)$

Exercise 1.5 (A fair k-process for every k is a martingale). Using the terminology introduced above, prove that $(X_t, t \geq 0)$ is a fair k-process for every $k \in \mathbb{N}$ if and only if it is a martingale.

Exercise 1.6 (Decreasing processes in the convex order and inverse martingales). An integrable process $(M_v, v \geq 0)$ is an *inverse martingale* if, for every $s \leq t$:

$$\mathbb{E}[M_s | \mathscr{F}_t^+] = M_t$$

where $\mathscr{F}_t^+ \supseteq \sigma(M_v, v \geq t)$. Let $(X_t, t \geq 0)$ be an integrable process decreasing in the convex order. Show that there exists an inverse martingale $(M_t, t \geq 0)$ such that $X_t \overset{(1.d)}{=} M_t$. (*Hint*: apply Kellerer's Theorem to the peacock $(X_{1/t}, t > 0)$, and note that if $(M_t, t \geq 0)$ is a martingale in its natural filtration, then $(\widehat{M}_u := M_{1/u}, u \geq 0)$ is an inverse martingale.)

Exercise 1.7 (Peacocks and a sub-class of convex functions). Let $(X_t, t \geq 0)$ be an integrable process such that $\mathbb{E}[X_t]$ does not depend on t. Prove that $(X_t, t \geq 0)$ is a peacock if and only if either of the following properties is satisfied:

i) for every real K, $\mathbb{E}[(K - X_t)^+]$ is an increasing function of t,
ii) for every real K, $\mathbb{E}[(X_t - K)^+]$ is an increasing function of t,
iii) for every real K, $\mathbb{E}[|X_t - K|]$ is an increasing function of t.

Exercise 1.8 (Some conditions for $X \overset{(c)}{\geq} Y$). Let X and Y be two integrable r.v.'s whose laws are respectively denoted μ and ν. We define, for every $K \in \mathbb{R}$:

$$\begin{aligned} \Pi^{(\mu)}(K) &= \mathbb{E}[(K - X)^+], & C^{(\mu)}(K) &= \mathbb{E}[(X - K)^+], \\ \overline{\mu}(K) &:= \mu([K, +\infty[), & \underline{\mu}(K) &:= \mu(]-\infty, K]) \\ \overline{\overline{\mu}}(K) &:= \int_K^{+\infty} \overline{\mu}(x)dx & \underline{\underline{\mu}}(K) &:= \int_{-\infty}^K \underline{\mu}(x)dx, \end{aligned}$$

and similar quantities for ν.

1) *i*) Prove that $C^{(\mu)}(K) = \overline{\overline{\mu}}(K)$ and $\Pi^{(\mu)}(K) = \underline{\underline{\mu}}(K)$.
ii) Deduce, using Exercise 1.7, that, if $\mathbb{E}[X] = \mathbb{E}[Y]$, then:

$$X \overset{(c)}{\geq} Y \iff \forall K \in \mathbb{R},\ \overline{\overline{\mu}}(K) \geq \overline{\overline{\nu}}(K) \iff \forall K \in \mathbb{R},\ \underline{\underline{\mu}}(K) \geq \underline{\underline{\nu}}(K). \qquad (1.9)$$

2) We now assume that $\mathbb{E}[X] = \mathbb{E}[Y]$ and

$$\begin{array}{ll} a)\ \forall K > 0, & \overline{\mu}(K) \geq \overline{\nu}(K) \\ b)\ \forall K < 0, & \underline{\mu}(K) \geq \underline{\nu}(K) \end{array} \qquad (1.10)$$

Prove that $X \overset{(c)}{\geq} Y$.
(*Hint*: use relation (1.9). You may separate the cases $K \geq 0$ and $K \leq 0$ and observe that $\Pi^{(\mu)}(K) - C^{(\mu)}(K) = K - \mathbb{E}[X]$.)

3) Let Z_1 and Z_2 be two positive r.v.'s. Z_1 is said to be greater than Z_2 for the stochastic order (notation: $Z_1 \overset{(st)}{\geq} Z_2$) if, for every $a \geq 0$, $\mathbb{P}(Z_1 \geq a) \geq \mathbb{P}(Z_2 \geq a)$.

Prove that Condition (1.10) is equivalent to $\{X^+ \overset{(st)}{\geq} Y^+$ and $X^- \overset{(st)}{\geq} Y^-\}$. Thus, if $\mathbb{E}[X] = \mathbb{E}[Y]$ and $\{X^+ \overset{(st)}{\geq} Y^+, X^- \overset{(st)}{\geq} Y^-\}$, then $X \overset{(c)}{\geq} Y$.

4) *Application*: let Z be an integrable and centered r.v. Prove that $(tZ, t \geq 0)$ is a peacock. (*Hint*: apply Question 3) with, for $s < t$, $X = tZ$ and $Y = sZ$, see also Proposition 1.4 and Remark 1.10.)

Exercise 1.9 (A prototypic strict local martingale).
1) Give an example of a local martingale which is not a peacock. (*Hint*: let $(R_t, t \geq 0)$ be a 3-dimensional Bessel process started at 1; then $\left(\frac{1}{R_t}, t \geq 0\right)$ is a local martingale such that, for every $t \geq 0$, $\mathbb{E}\left[\frac{1}{R_t}\right] < \infty$ and $\mathbb{E}\left[\frac{1}{R_t}\right]$ is not constant).

2) Prove that a $\mathbb{R}_+$-valued local martingale is a *strict* local martingale (i.e.: it is not a martingale) if and only if $(\mathbb{E}[X_t], t \geq 0)$ is not constant.

See also [150, Exercise 2.13, Chapter. V, p.194 and Exercise 1.16, Chapter XI, p. 449], Pal-Protter [137] or Elworthy-Li-Yor [64, 65] for similar computations.

Exercise 1.10 (Dominated limit of peacocks: examples and counterexamples).
1) Let $(X_t^{(n)}, t \geq 0, n \geq 0)$ be a sequence of processes such that:

i) for every $t \geq 0$, $\mathbb{E}\left[\sup_{n \geq 0} |X_t^{(n)}|\right] < \infty$,
ii) for every n, $(X_t^{(n)}, t \geq 0)$ is a peacock,
iii) for every $t \geq 0$, $X_t^{(n)} \underset{n \to \infty}{\longrightarrow} X_t$ a.s.

Prove that $(X_t, t \geq 0)$ is a peacock.

2) Let $(R_t, t \geq 0)$ be a 3-dimensional Bessel process started at 1 and let $T_{1/n} := \inf\{t \geq 0; R_t = 1/n\}$.

a) Prove that, for every $n \geq 1$, $\left(Y_t^{(n)} := \frac{1}{R_{t \wedge T_{1/n}}}\right)$ is a peacock. (*Hint*: for every $n \geq 1$, $Y_t^{(n)}$ is a bounded local martingale, hence a martingale.)

b) Deduce from Exercise 1.9 and from Point 1) of the present exercise that $\mathbb{E}\left[\sup_{n\geq 1} \frac{1}{R_{t\wedge T_{1/n}}}\right] = \infty$. (Observe that $Y_t^{(n)} \underset{n\to\infty}{\longrightarrow} \frac{1}{R_t}$ a.s.)

3) We now provide a direct proof that:

$$\mathbb{E}\left[\sup_{n\geq 1} \frac{1}{R_{t\wedge T_{1/n}}}\right] = \infty. \tag{1.11}$$

Define: $I_t := \inf_{s\leq t} R_s$.

a) Prove that

$$\operatorname{int}\left(\frac{1}{I_t}\right) \leq \sup_{n\geq 1} \frac{1}{R_{t\wedge T_{1/n}}} = \frac{1}{\inf_{n\geq 1} R_{t\wedge T_{1/n}}} \leq \frac{1}{I_t} \tag{1.12}$$

where $\operatorname{int}(x)$ denotes the integer part of $x \in \mathbb{R}_+$.

b) Deduce from (1.12) that $\mathbb{E}\left[\sup_{n\geq 1} \frac{1}{R_{t\wedge T_{1/n}}}\right] = \infty$ holds if and only if $\mathbb{E}\left[\frac{1}{I_t}\right] = \infty$.

c) Prove that $\mathbb{E}\left[\frac{1}{I_t}\right] = \infty$. (*Hint*: let $(B_t, t \geq 0)$ be a Brownian motion started at 1. From the absolute continuity formula between BES(3) and Brownian motion killed when reaching 0 (or Doob's h-transform, see [150, Chapter XI, Exercise (1.22), p.450]):

$$\mathbb{E}\left[\frac{1}{I_t}\right] = \mathbb{E}\left[\frac{1}{I_t(B)} B_{t\wedge T_0}\right]$$

(with $I_t(B) := \inf_{s\leq t} B_s$, $T_0 := \inf\{s \geq 0, B_s = 0\}$) and from:

$$\mathbb{E}\left[\frac{1}{I_t(B)} B_{t\wedge T_0}\right] = 1 + \int_1^\infty \mathbb{E}\left[1_{\{I_t(B)<\frac{1}{a}\}} B_{t\wedge T_0}\right] da$$

deduce that:

$$\begin{aligned}
\mathbb{E}\left[\frac{1}{I_t}\right] &= 1 + \int_1^\infty \mathbb{E}\left[1_{\{T_{1/a}(B)<t\}} B_{t\wedge T_0}\right] da \\
&= 1 + \int_1^\infty \frac{da}{a} \mathbb{P}(T_{1/a}(B) < t).
\end{aligned}$$

(*Hint*: use the identity $\{I_t(B) < \frac{1}{a}\} = \{T_{1/a}(B) < t\}$ and $\mathbb{E}\left[1_{\{T_{1/a}(B)<t\}} B_{t\wedge T_0}\right] = \frac{1}{a}\mathbb{P}\left(T_{1/a}(B) < t\right)$, from the martingale property). Conclude, using $\mathbb{P}(T_{1/a}(B) < t) \underset{a\to\infty}{\searrow} \mathbb{P}(T_0(B) < t) > 0$).

d) Prove that, for any increasing function $\varphi : [1,+\infty[\longrightarrow \mathbb{R}_+$,

$$\mathbb{E}\left[\varphi\left(\frac{1}{I_t}\right)\right] = \varphi(1) + \int_1^{+\infty} \frac{dx}{x}\varphi'(x)\mathbb{P}\left(T_{1/x}(B) < t\right).$$

Deduce that:

$$\mathbb{E}\left[\varphi\left(\sup_{n\geq 1}\frac{1}{R_{t\wedge T_{1/n}}}\right)\right] < \infty$$

if and only if:

$$\int^{+\infty} \frac{dx}{x}\varphi'(x) < \infty.$$

In particular, prove that, for every $r < 1$, $\mathbb{E}\left[\left(\sup_{n\geq 1}\frac{1}{R_{t\wedge T_{1/n}}}\right)^r\right] < \infty$.

Exercise 1.11 (Increasing limit of peacocks). Let $(X_t^{(n)}, t \geq 0)$ be a sequence of processes such that:

i) for every n, $(X_t^{(n)}, t \geq 0)$ is an integrable peacock,
ii) $(X_t^{(n)}, t \geq 0)$ is an increasing sequence of processes with

$$\text{for every } t \geq 0, \qquad \lim_{n\to\infty} X_t^{(n)} := X_t \text{ a.s.},$$

iii) $(X_t, t \geq 0)$ is an integrable process.

Prove that $(X_t, t \geq 0)$ is a peacock. (*Hint*: it suffices to prove that, for $\psi \in \boldsymbol{C}_+$, $\mathbb{E}[\psi(X_s)] \leq \mathbb{E}[\psi(X_t)]$ $(s \leq t)$.)

Exercise 1.12 (Sum, product and subordination of peacocks). Let $(X_t, t \geq 0)$ and $(Y_t, t \geq 0)$ be two independent peacocks.

1) Prove that $(X_t + Y_t, t \geq 0)$ and $(X_t Y_t, t \geq 0)$ are two peacocks.
2) Let $(\tau_t, t \geq 0)$ be a $\mathbb{R}_+$-valued increasing process, independent of $(X_u, u \geq 0)$. Prove that, if $(X_{\tau_t}, t \geq 0)$ is integrable, then $(X_{\tau_t}, t \geq 0)$ is a peacock. See Exercise 1.26 for a more general statement.

This exercise may be solved by two different methods: either by using Kellerer's theorem, or by direct proofs.

Exercise 1.13 (Sums of independent r.v's and convex order). Let $(X_i)_{i\in\mathbb{N}}$ (resp. $(Y_i)_{i\in\mathbb{N}}$) be a sequence of integrable independent r.v.'s such that, for every $i \in \mathbb{N}$, $X_i \overset{(c)}{\geq} Y_i$.
1) Prove that for every $k \in \mathbb{N}^*$,

$$X_1 + \ldots + X_k \overset{(c)}{\geq} Y_1 + \ldots + Y_k.$$

2) Let N be a $\mathbb{N}$-valued r.v. independent from $(X_i)_{i\in\mathbb{N}}$ and $(Y_i)_{i\in\mathbb{N}}$. Prove that:

$$\sum_{i=1}^{N} X_i \overset{(c)}{\geq} \sum_{i=1}^{N} Y_i.$$

3) *Application (Example of infinitely divisible r.v.'s ordered in the convex order)*: let $(N_t, t \geq 0)$ be a Poisson process with intensity λ. Let $(X_i)_{i\in\mathbb{N}}$ (resp. $(Y_i)_{i\in\mathbb{N}}$) be a sequence of integrable i.i.d r.v.'s with law μ (resp. ν) independent from $(N_t, t \geq 0)$, and define

$$L_t^{(\mu)} := \sum_{i=1}^{N_t} X_i \qquad \left(\text{resp. } L_t^{(\nu)} := \sum_{i=1}^{N_t} Y_i\right).$$

$(L_t^{(\mu)}, t \geq 0)$ and $(L_t^{(\nu)}, t \geq 0)$ are two compound Poisson processes with respective Lévy measure $\lambda\mu(dx)$ and $\lambda\nu(dx)$. We assume that $\mu \overset{(c)}{\geq} \nu$. Deduce that, for every $t \geq 0$, $L_t^{(\mu)} \overset{(c)}{\geq} L_t^{(\nu)}$. (*Hint*: use the property that, conditionally on $\{N_t = k\}$, $L_t^{(\mu)}$ is distributed as $X_1 + \ldots + X_k$).

Exercise 1.14 (Convex order and Brownian local time). Let G and H be two integrable r.v.'s and let $(B_s, s \geq 0)$ be a Brownian motion independent from G and H.
1) Show that the following are equivalent:

i) $GB_1 \overset{(c)}{\leq} HB_1$,
ii) for every $y \in \mathbb{R}$, $\mathbb{E}\left[L^y_{G^2}\right] \leq \mathbb{E}\left[L^y_{H^2}\right]$, where $(L^y_t, t \geq 0, y \in \mathbb{R})$ denotes the bicontinuous family of local times of B.

(*Hint*: for $\psi \in \mathbf{C}$, $\mathbb{E}[\psi(HB_1)] = \mathbb{E}[\psi(B_{H^2})] = \psi(0) + \frac{1}{2}\mathbb{E}\left[\int_0^{H^2} \psi''(B_s)ds\right] = \psi(0) + \frac{1}{2}\int_{\mathbb{R}} \mathbb{E}\left[L^y_{H^2}\right] \psi''(y)dy$, from the occupation times density formula.)
2) Prove that, for every $y \in \mathbb{R}$, $h \to \mathbb{E}\left[L^y_{h^2}\right]$ is convex.
(*Hint*: use (cf. [143]) $L^y_{h^2} \overset{(law)}{=} \left(L^0_{h^2} - |y|\right)^+ \overset{(law)}{=} (|h|L^0_1 - |y|)^+$). Then, deduce that:

$$G \overset{(c)}{\leq} H \quad \Longrightarrow \quad \forall\, y \in \mathbb{R},\ \mathbb{E}\left[L^y_{G^2}\right] \leq \mathbb{E}\left[L^y_{H^2}\right].)$$

3) We now assume that $G \overset{(c)}{\leq} H$. Let X be an integrable r.v., independent from G and H. Prove that $GX \overset{(c)}{\leq} HX$.

Exercise 1.15 (Peacock, martingale and increasing rearrangement). Let $(\Omega, \mathcal{F}, \mathbb{P})$ be the probability space $([0,1], \mathcal{B}([0,1]), \mathbb{P})$ where $\mathbb{P}$ is the Lebesgue measure. For any r.v. $Z : [0,1] \longrightarrow \mathbb{R}$, Z^* denotes its increasing rearrangement, characterized by: Z^* is increasing, right-continuous and such that, for every a real, $\mathbb{P}(Z \geq a) = \mathbb{P}(Z^* \geq a)$.

1) For $0 \leq t \leq 1$, let $\mathscr{F}_t$ be the smallest sub-σ-field of $\mathscr{F}$ containing the Borel subsets of $[0,t]$ and the negligible sets of $[0,1]$. Let $Y \in L^1(\Omega, \mathscr{F}, \mathbb{P})$. We set $(Y_t, 0 \leq t \leq 1)$ the right-continuous version of $(\mathbb{E}[Y|\mathscr{F}_t], 0 \leq t \leq 1)$ (see [150, Chapter II, Exercise (3.15)] for some properties of this martingale).

i) Prove that, if $a > \frac{1}{1-t}\int_t^1 Y(x)dx$, then:

$$\mathbb{P}(Y_t \geq a) = \mathbb{P}([0,t] \cap \{Y \geq a\}),$$

and, if $a \leq \frac{1}{1-t}\int_t^1 Y(x)dx$, then:

$$\mathbb{P}(Y_t \geq a) = \mathbb{P}([0,t] \cap \{Y \geq a\}) + 1 - t.$$

ii) Let $X, Y \in L^1(\Omega, \mathscr{F}, \mathbb{P})$ and suppose that, for every $t \in [0,1]$, $X_t \overset{(law)}{=} Y_t$. Prove that $Y = X$ $\mathbb{P}$-a.s.
(*Hint*: denote for $n \geq 0$ by (H_n) the property:

$$(H_n) \qquad X(x) = Y(x) \quad \mathbb{P}-\text{a.s. on } [0, 1-2^{-n}].$$

Suppose that (H_n) holds. Deduce from 1) *i*) that, if for some $t \geq 1 - 2^{-n}$,

$$\int_t^1 X(x)dx \neq \int_t^1 Y(x)dx,$$

then:

$$(1-t) \leq t - (1-2^{-n}) \quad \text{and hence} \quad t \geq 1 - 2^{-n-1}.$$

Conclude that, for every $t \in [0, 1-2^{-n-1}]$,

$$\int_t^1 X(x)dx = \int_t^1 Y(x)dx$$

and hence, (H_{n+1}) holds).

iii) Let $X \in L^1(\Omega, \mathscr{F}, \mathbb{P})$.

a) Prove that $((X_t)^*, 0 \leq t \leq 1)$ is a peacock associated to the martingale $(X_t, 0 \leq t \leq 1)$. Note that, in general, $((X_t)^*, 0 \leq t \leq 1)$ is not a $(\mathscr{F}_t)$-adapted process.

b) Prove that $(X_t)^* \underset{t \to 1}{\longrightarrow} X^*$ in L^1.
(*Hint*: $X_t \underset{t \to 1}{\longrightarrow} X$ a.s. and in L^1 and, if U and V are two r.v.'s defined on $(\Omega, \mathscr{F}, \mathbb{P})$, for every $p \in [1, +\infty]$, $\mathbb{E}[|U^* - V^*|^p] \leq \mathbb{E}[|U - V|^p]$, see [108, p.8-13].)

c) Prove that $((X_t)^*, 0 \leq t \leq 1)$ is a $(\mathscr{F}_t, 0 \leq t \leq 1)$-martingale if and only if X is a.s. equal to an increasing function.
(*Hint*: show that, if $((X_t)^*, 0 \leq t \leq 1)$ is a $(\mathscr{F}_t)$-martingale, then $(X_t)^* = (X^*)_t$, and in particular, $X_t \overset{(law)}{=} (X^*)_t$. Deduce from 1) *ii*) that $X = X^*$ a.s.)

2) Let, for every integer n, $\mathscr{G}_n$ be the σ-algebra generated by the intervals $[0, \frac{1}{2^n}[$, $[\frac{1}{2^n}, \frac{2}{2^n}[$, ..., $[\frac{2^n-1}{2^n}, 1]$. Let $Y \in L^1(\Omega, \mathscr{F}, \mathbb{P})$.

i) Prove that $(Y_n := \mathbb{E}[Y|\mathscr{G}_n], n \geq 0)$ is a martingale, and that $((Y_n)^*, n \geq 0)$ is an associated peacock.

ii) Prove that $Y_n \xrightarrow[n\to+\infty]{} Y$ a.s. and in L^1, and that $(Y_n)^* \xrightarrow[n\to+\infty]{} Y^*$ in L^1.

iii) Prove that if Y is a.s. increasing or decreasing, then $((Y_n)^*, n \geq 0)$ is a $(\mathscr{G}_n, n \geq 0)$-martingale.

iv) Prove that, in general, $((Y_n)^*, n \geq 0)$ is a $(\mathscr{G}_n, n \geq 0)$ adapted process, but is not a $(\mathscr{G}_n, n \geq 0)$-martingale.

Exercise 1.16 (Anticipating martingales: peacocks or not peacocks?). A process $(M_t, t \geq 0)$ is called an *anticipating martingale* with respect to a filtration $(\mathscr{F}_t, t \geq 0)$ if, for every $t \geq 0$, $\mathbb{E}[|M_t|] < \infty$ and if, for every $s < t$:

$$\mathbb{E}[M_t|\mathscr{F}_s] = \mathbb{E}[M_s|\mathscr{F}_s].$$

Note that in this definition, we do not assume that $(M_t, t \geq 0)$ is $(\mathscr{F}_t, t \geq 0)$ adapted. Observe nevertheless that $(\mathbb{E}[M_t|\mathscr{F}_t], t \geq 0)$ is a $(\mathscr{F}_t, t \geq 0)$-martingale.

1) Let $(M_t, t \geq 0)$ be a $(\mathscr{F}_t, t \geq 0)$ anticipating martingale. Prove that, for every $\psi \in \mathbf{C}$ and $s \leq t$:

$$\mathbb{E}[\psi(\mathbb{E}[M_s|\mathscr{F}_s])] = \mathbb{E}[\psi(\mathbb{E}[M_t|\mathscr{F}_s])] \leq \mathbb{E}[\psi(M_s)] \wedge \mathbb{E}[\psi(M_t)].$$

(*Hint*: use Jensen's inequality.)

2) Let $(Y_t, 0 \leq t \leq 1)$ be a $(\mathscr{Y}_t, 0 \leq t \leq 1)$-martingale (not a.s. constant).

i) We set $(M_t := Y_t - Y_1, 0 \leq t \leq 1)$. Prove that $(M_t, 0 \leq t \leq 1)$ is a $(\mathscr{Y}_t, 0 \leq t \leq 1)$ anticipating martingale, but is not a peacock.

ii) Let $(\mathscr{X}_t, 0 \leq t \leq 1)$ be a sub-filtration of $(\mathscr{Y}_t, 0 \leq t \leq 1)$. Prove that $(Y_t, 0 \leq t \leq 1)$ is a $(\mathscr{X}_t, 0 \leq t \leq 1)$-anticipating martingale and a peacock.

In the following, we shall exhibit others anticipating martingales which are also peacocks.

3) Let $(\Omega, (\mathscr{F}_t, t \geq 0), \mathscr{F} := \bigvee_{t\geq 0} \mathscr{F}_t, \mathbb{P})$ be a probability space such that $(\mathscr{F}_t, t \geq 0)$ is a Brownian filtration. Let $\Lambda : \Omega \longrightarrow \mathbb{R}_+$ be the end of a predictable set Σ, namely $\Lambda := \sup\{t \geq 0; (t, \omega) \in \Sigma\}$. We assume furthermore that Λ avoids the $(\mathscr{F}_t, t \geq 0)$-stopping times, i.e. for every $(\mathscr{F}_t, t \geq 0)$-stopping time T, $\mathbb{P}(\Lambda = T) = 0$. Let $(Z_t, t \geq 0)$ be the càdlàg version of the Azéma supermartingale $(\mathbb{P}(\Lambda > t|\mathscr{F}_t), t \geq 0)$ (see [125, p.12-14]), and let

$$Z_t = M_t - A_t$$

its Doob-Meyer decomposition (in the filtration $(\mathscr{F}_t, t \geq 0)$). Let $(\mathscr{G}_t, t \geq 0)$ be the smallest filtration which makes Λ a stopping time and contains $(\mathscr{F}_t, t \geq 0)$.

i) Let us define $(V_t := 1_{\{\Lambda > t\}} + A_t, t \geq 0)$. Prove that $(V_t, t \geq 0)$ is a $(\mathscr{F}_t, t \geq 0)$ anticipating martingale. In particular, the expectation of V_t does not depend on t.

ii) Prove that $(V_t, t \geq 0)$ is a $(\mathscr{G}_t, t \geq 0)$ martingale, hence a peacock.

(*Hint*: use the two following properties (see [95], [99] or [125, p.14]):

- Let $(J_u, u \geq 0)$ be a $(\mathscr{G}_u, u \geq 0)$ predictable process; then there exist two positive $(\mathscr{F}_u, u \geq 0)$ predictable processes $(J_u^+, u \geq 0)$ and $(J_u^-, u \geq 0)$ such that:

$$\forall u \geq 0, \quad J_u = J_u^- 1_{[0,\Lambda]}(u) + J_u^+ 1_{]\Lambda,+\infty[}(u).$$

- The support of the measure dA_t is included in $[0, \Lambda]$.)

We now particularize the framework of this Question. Let $(N_t, t \geq 0)$ be a continuous positive martingale such that $N_0 = 1$ and $\lim_{t\to+\infty} N_t = 0$ a.s. We set $\overline{N}_t := \sup_{s \leq t} N_s$ and $\Lambda := \sup\{t \geq 0; \overline{N}_t = N_t\}$. From [125, p.32], $A_t = \log(\overline{N}_t)$. We shall now prove directly that $(V_t, t \geq 0)$ is a peacock (without using the fact that $(V_t, t \geq 0)$ is a $(\mathscr{G}_t, t \geq 0)$-martingale).

iii) Let $\psi \in \boldsymbol{C}$. Prove that:

$$\mathbb{E}[\psi(V_t)] = \mathbb{E}\left[\frac{N_t}{\overline{N}_t}\psi(1+\log(\overline{N}_t)) + \left(1 - \frac{N_t}{\overline{N}_t}\right)\psi(\log(\overline{N}_t))\right].$$

(Hint: $\{\Lambda \geq t\} = \{\sup_{s \geq t} N_s \geq \overline{N}_t\}$ and observe that, conditionally on $\mathscr{F}_t$, $\sup_{s\geq t} N_s \overset{(law)}{=} \frac{N_t}{U}$ where U is a uniform r.v. on $[0,1]$, independent from $\mathscr{F}_t$, see [125, p.3].)

iv) Prove that:

$$\begin{aligned}\mathbb{E}[\psi(V_t)] &= \psi(1)\\ &\quad + \mathbb{E}\left[\int_0^t \frac{1}{\overline{N}_u}\left(\psi'(1+\log(\overline{N}_u)) - \psi(1+\log(\overline{N}_u)) + \psi(\log(\overline{N}_u))\right) d\overline{N}_u\right].\end{aligned}$$

(*Hint*: use Itô formula and the fact that $d\overline{N}_u$ only charges the set $\{u \geq 0; N_u = \overline{N}_u\}$.)

v) Deduce that $(V_t, t \geq 0)$ is a peacock. (*Hint*: use the convexity of ψ.)

We refer the reader to [125] for numerous examples in the set-up of this exercise.

1.1.2 How to Prove the Peacock Property

We now give some sufficient conditions on $(X_t, t \geq 0)$ which imply that $(X_t, t \geq 0)$ is a peacock.

Proposition 1.2. *Let $(X_t, t \geq 0)$ be an integrable process. Then, the following are equivalent:*

1) $(X_t, t \geq 0)$ is a peacock.

2) For every $\psi \in \boldsymbol{C}$, for every $s < t$, there exists a r.v. $T_{s,t}^{\psi}$ such that:

$$\mathbb{E}\left[T_{s,t}^{\psi}\right] = 0 \tag{1.13}$$

$$\psi(X_t) - \psi(X_s) - T_{s,t}^{\psi} \geq 0 \qquad a.s. \tag{1.14}$$

Definition 1.5. The r.v. $T_{s,t}^{\psi}$ is called a trading cash flow.

In general, if it exists, $T_{s,t}^{\psi}$ is not unique.

(We thank D. Madan for introducing us to the notion of trading cash flows.)

Proof. $\Rightarrow$ 1) is a consequence of Remark 1.1. We now prove 1) $\Rightarrow$ 2). We first observe that an integrable r.v. Z_1 satisfies $\mathbb{E}[Z_1] \geq 0$ if and only if there exists an integrable r.v. Z_2 such that $Z_1 - Z_2 \geq 0$ and $\mathbb{E}[Z_2] = 0$. Then we apply this remark to $Z_1 = \psi(X_t) - \psi(X_s)$ and we define $T_{s,t}^{\psi} = Z_2$. □

Remark 1.7. Despite the obvious character of Proposition 1.2, this proposition may be thought of as a useful "guideline" to show that a given process $(X_t, t \geq 0)$ is indeed a peacock. Namely, we are led to "decompose" $\psi(X_t) - \psi(X_s)$ as the sum of a trading cash flow $T_{s,t}^{\psi}$ and a positive r.v. We shall illustrate this method in the next section.

Exercise 1.17 (Centered minorant of a r.v.). Let X be a r.v. such that $\mathbb{E}[X] \geq 0$. "Describe" all Y's such that $X - Y \geq 0$ a.s. and $\mathbb{E}[Y] = 0$.

(Solution: $Y = X - \frac{Z}{\mathbb{E}[Z]}\mathbb{E}[X]$ for Z a positive r.v., $Z \neq 0$.)

Proposition 1.3. *Let $(X_t, t \geq 0)$ be a real valued process satisfying the following hypotheses:*
i) the process X is a.s. continuous on $[0, +\infty[$ and differentiable on $]0, +\infty[$, its derivative being denoted by $\dfrac{\partial X_t}{\partial t}$,
ii) for every $a > 0$:

$$\mathbb{E}\left[\sup_{t\in[0,a]} |X_t|\right] < \infty \tag{1.15}$$

and, for every $0 < a < b$:

$$\mathbb{E}\left[\sup_{t\in[a,b]} \left|\frac{\partial X_t}{\partial t}\right|\right] < \infty. \tag{1.16}$$

Then, the process $(X_t, t \geq 0)$ is a peacock if and only if the two following properties hold:

$$a) \quad \mathbb{E}[X_t] \text{ does not depend on } t \geq 0, \tag{1.17}$$

$$b) \quad \text{for every real } c \text{ and } t > 0\text{: } \mathbb{E}\left[1_{\{X_t \geq c\}} \frac{\partial X_t}{\partial t}\right] \geq 0. \tag{1.18}$$

Proof. Suppose first that X satisfies i), ii), a) and b). Let $\psi \in \boldsymbol{C}$. To prove that $(X_t, t \geq 0)$ is a peacock, it suffices to show that:

$$f : t \to \mathbb{E}[\psi(X_t)] \tag{1.19}$$

is increasing. Since $\psi \in \boldsymbol{C}$, there exists k_1, k_2 such that:

$$|\psi(x)| \leq k_1 + k_2|x|.$$

Therefore, by assumption *ii)* f is continuous on $[0,+\infty[$ and differentiable on $]0,+\infty[$ and:

$$f'(t) = \mathbb{E}\left[\psi'(X_t)\frac{\partial X_t}{\partial t}\right] \tag{1.20}$$

$$= \mathbb{E}\left[(\psi'(X_t)+k)\frac{\partial X_t}{\partial t}\right] \tag{1.21}$$

for any constant k, since $\mathbb{E}\left[\frac{\partial X_t}{\partial t}\right] = 0$, from a). But, since $\psi \in \boldsymbol{C}$, $\psi' + k$ is continuous, positive, (for k large enough), bounded and increasing and hence may be approximated pointwise by finite sums of functions of the type $d1_{[c,+\infty[}$, with $d \geq 0$ and $c \in \mathbb{R}$. Thus, $f'(t) \geq 0$ and $(X_t, t \geq 0)$ is a peacock.

Suppose now that $(X_t, t \geq 0)$ *satisfies i), ii) and is a peacock.* Then, property a) is obviously satisfied. We set, for $c \in \mathbb{R}$ and $\eta > 0$:

$$h_{c,\eta}(x) = \frac{1}{\eta}\int_x^{x+\eta}(y-c)^+ dy.$$

Since $h_{c,\eta}$ is a convex $\mathscr{C}^1$-function:

$$\mathbb{E}\left[h'_{c,\eta}(X_t)\frac{\partial X_t}{\partial t}\right] \geq 0.$$

Letting then η tend to 0, we obtain property b). □

The following lemma will often be useful to prove Property (1.18).

Lemma 1.1. *Let U be a real-valued integrable r.v. Then, the following properties are equivalent:*
1) For every real c, $\mathbb{E}[1_{\{U\geq c\}}U] \geq 0$.
2) For bounded and increasing function $h : \mathbb{R} \to \mathbb{R}_+$*:*

$$\mathbb{E}[h(U)U] \geq 0.$$

3) $\mathbb{E}[U] \geq 0$.

Proof. By approximation, the equivalence between 1) and 2) is clear. From the dominated convergence theorem, we have:

$$\lim_{c \to -\infty} \mathbb{E}[1_{\{U\geq c\}}U] = \mathbb{E}[U].$$

Hence 1) $\Rightarrow$ 3). We now prove that 3) $\Rightarrow$ 1). Assume $\mathbb{E}[U] \geq 0$. If $c \geq 0$, $\mathbb{E}[1_{\{U\geq c\}}U] \geq 0$. If $c < 0$, $\mathbb{E}[1_{\{U\geq c\}}U] = \mathbb{E}[U] + \left(-\mathbb{E}[1_{\{U<c\}}U]\right)$ and the RHS is non negative as the sum of two elements of $\mathbb{R}_+$. □

Exercise 1.18. (Necessary and sufficient conditions for $\mathbb{E}[U]=0$). Let U be a real integrable r.v. Prove the equivalence of the three following properties:
1) $\mathbb{E}[U]=0$.
2) For every c real:

$$\mathbb{E}[1_{\{U\leq c\}}U]\leq 0\leq \mathbb{E}[1_{\{U\geq c\}}U].$$

3) For every bounded increasing function $h:\mathbb{R}\longrightarrow\mathbb{R}$, $\mathbb{E}[h(U)U]\geq 0$.

Exercise 1.19 (Some one-parameter transforms of r.v.'s are peacocks).
1) Let X be a r.v. and $F:\mathbb{R}_+\times\mathbb{R}\to\mathbb{R}$ such that:

i) $(X_t:=F(t,X),t\geq 0)$ satisfies properties *i)* and *ii)* and *a)* of Proposition 1.3,
ii) there exists $C:]0,+\infty[\times\mathbb{R}\longrightarrow\mathbb{R}$ such that

$$\forall t>0,\quad y\in\mathbb{R}\longmapsto C(t,y) \text{ is strictly increasing}$$

and such that F solves the following equation:

$$\forall t>0,\ \forall x\in\mathbb{R},\quad \frac{\partial F}{\partial t}(t,x)=C(t,F(t,x)).$$

Prove that $(X_t,t\geq 0)$ is a peacock. (*Hint*: use Proposition 1.3 and Lemma 1.1.)

2) Let X be a symmetric r.v. (i.e. which satisfies $X\overset{(law)}{=}-X$), $\varphi:\mathbb{R}\longrightarrow\mathbb{R}$ Lipschitz continuous, odd and strictly increasing and $\theta:\mathbb{R}\longrightarrow\mathbb{R}$ odd and such that $\mathbb{E}[|\theta(X)|]<\infty$. Prove that $(X_t:=F(t,X),t\geq 0)$ is a peacock, where F is solution of:

$$\begin{cases}\dfrac{\partial F}{\partial t}(t,x) &= C(t,F(t,x))\\ F(0,x) &= \theta(x)\end{cases}$$

with $C(t,y):=\varphi(ty)$. (*Hint*: prove that $F(t,x)=-F(t,-x)$ and apply Point 1).)

3) Let X be a symmetric r.v. and $\varphi:\mathbb{R}\longrightarrow\mathbb{R}$ $\mathscr{C}^2$, odd, with $\varphi'>0$ and $\operatorname{sgn}\varphi''(x)=\operatorname{sgn}(x)$. We assume furthermore that, for every $t\geq 0$ $\mathbb{E}[|X|\varphi'(tX)]<\infty$. Prove that $(\varphi(tX),t\geq 0)$ is a peacock.
(*Hint*: prove that $y\longrightarrow y\varphi'(y)$ is strictly increasing and apply Point 1).) See Question 1) of Exercise 1.23 for another proof of this result.

The aim of the following Section 1.2 is to give significative examples of peacocks. There are two kinds of peacocks:

- peacocks whose parameter is time – in financial terms maturity – called peacocks in maturity,
- peacocks depending of a dilation parameter – in financial terms volatility – called peacocks in volatility.

1.2 Examples of Peacocks

1.2.1 Composition of Some Peacocks by a Function

Definition 1.6. Let us denote:

$$\mathbf{I} := \{\varphi : \mathbb{R} \to \mathbb{R}, \varphi \text{ is increasing}\}$$
$$\mathbf{J} := \{\varphi : \mathbb{R} \to \mathbb{R}, \varphi \text{ is increasing, convex on } \mathbb{R}_+ \text{ and concave on } \mathbb{R}_-\}$$

Clearly, $\mathbf{J} \subset \mathbf{I}$.

Theorem 1.3. 1) *Let $\varphi \in \mathbf{I}$ and X a r.v. such that, for every $t \geq 0$, $\varphi(tX)$ is integrable and $\mathbb{E}[\varphi(tX)]$ does not depend on t, i.e. $\mathbb{E}[\varphi(tX)] = \varphi(0)$. Then $(\varphi(tX), t \geq 0)$ is a peacock.*
2) *Let $\varphi \in \mathbf{J}$ and $(X_t, t \geq 0)$ be a peacock such that, for every $t \geq 0$, $\varphi(X_t)$ is integrable and $\mathbb{E}[\varphi(X_t)]$ does not depend on t, i.e. $\mathbb{E}[\varphi(X_t)] = \mathbb{E}[\varphi(X_0)]$. Then $(\varphi(X_t), t \geq 0)$ is a peacock.*

Proof. We first prove Point 1). Upon replacing φ by $\varphi - \varphi(0)$, we may, without loss of generality, assume that $\varphi(0) = 0$. We must prove that, for $\psi \in \boldsymbol{C}$, $\mathbb{E}[\psi(\varphi(tX))]$ is an increasing function of t. But, since $\mathbb{E}[\varphi(tX)]$ does not depend on t, upon replacing $\psi(x)$ by $\psi(x) - \psi(0) - \psi'(0)x$, we may suppose that $\psi(0) = 0$ and that ψ is positive. Then, for $0 \leq s \leq t$, we have by convexity, since $\varphi(sX) \in [0, \varphi(tX)]$:

$$\psi(\varphi(sX)) \leq \lambda\psi(0) + (1-\lambda)\psi(\varphi(tX)) \leq \psi(\varphi(tX)) \quad (\lambda \in [0,1]).$$

We obtain Point 1) by taking the expectations of both sides.

We now prove Point 2). We may once again assume, without loss of generality, that $\varphi(0) = 0$. Since φ is increasing, we have $x\varphi(x) \geq 0$ for every $x \in \mathbb{R}$. Let $\psi \in \boldsymbol{C}$. We define:

$$\theta(x) := \psi(x) - \psi(0) - \psi'(0)x \quad x \in \mathbb{R}.$$

Observe that θ is positive, increasing on $[0, +\infty[$, and decreasing on $]-\infty, 0]$. Now, since $\mathbb{E}[\varphi(X_t)]$ does not depend on t, proving that $t \longmapsto \mathbb{E}[\psi(\varphi(X_t))]$ is increasing is equivalent to proving that $\mathbb{E}[\theta(\varphi(X_t))]$ is increasing. To this end, $(X_t, t \geq 0)$ being a peacock, it is sufficient to prove that $\theta \circ \varphi$ is convex.

- If $0 \leq x \leq y$, since φ is convex on $[0, +\infty[$:

$$0 \leq \varphi(\lambda x + (1-\lambda)y) \leq \lambda\varphi(x) + (1-\lambda)\varphi(y) \quad (\lambda \in [0,1])$$

  and, θ being increasing and convex on $[0, +\infty[$:

$$\begin{aligned}\theta \circ \varphi(\lambda x + (1-\lambda)y) &\leq \theta(\lambda\varphi(x) + (1-\lambda)\varphi(y)) \\ &\leq \lambda\theta \circ \varphi(x) + (1-\lambda)\theta \circ \varphi(y).\end{aligned}$$

- If $x \le 0 \le y$ and $\lambda x + (1-\lambda)y \ge 0$, then:

$$0 \le \varphi(\lambda x + (1-\lambda)y) \le \varphi((1-\lambda)y) \le (1-\lambda)\varphi(y)$$

and

$$\begin{aligned}\theta \circ \varphi(\lambda x + (1-\lambda)y) &\le \theta\left((1-\lambda)\varphi(y)\right) \\ &\le (1-\lambda)\theta\left(\varphi(y)\right) \quad \le (1-\lambda)\theta \circ \varphi(y) + \lambda \theta \circ \varphi(x).\end{aligned}$$

The other cases can be dealt with in the same way. □

A direct consequence of Point 1) in Theorem 1.3 with $\varphi(x) = x$ is the following:

Proposition 1.4. *Let X be an integrable and centered r.v. Then:*

$$(X_t := tX, t \ge 0) \textit{ is a peacock.}$$

Corollary 1.1. *Let $X_1, X_2, ... X_n$ be n independent, integrable and centered r.v.'s. Let $\alpha_1, \alpha_2, ... \alpha_n$ be n positive and increasing functions (from $\mathbb{R}_+$ to $\mathbb{R}_+$). Then:*

$$\left(\sum_{i=1}^{n} \alpha_i(t) X_i, t \ge 0\right) \textit{ is a peacock.}$$

Proof. The result follows by iteration from Proposition 1.4 and from Point 1) in Exercise 1.12. □

Exercise 1.20. Give another proof of Proposition 1.4, based on Proposition 1.3 and Lemma 1.1.

Exercise 1.21 (If $\mathbb{E}[X|Y] = 0$, then $(Y + tX, t \ge 0)$ is a peacock). Let X and Y be two integrable r.v.'s such that $\mathbb{E}[X|Y] = 0$. Prove that $(Y + tX, t \ge 0)$ is a peacock.
(*Hint*: let ψ be a convex function and $0 \le s < t$. Prove that:

$$\psi(Y + sX) \le \left(1 - \frac{s}{t}\right) \psi(Y) + \frac{s}{t} \psi(Y + tX)$$

and

$$\psi(Y) \le \mathbb{E}[\psi(Y + tX)|Y];$$

then take the expectations.

Alternative proof: let $\psi \in \boldsymbol{C}$. Then $\varphi : t \in \mathbb{R} \longmapsto \mathbb{E}[\psi(Y + tX)]$ is a convex function and $\varphi'(0) = \mathbb{E}[X \psi'(Y)] = 0$.)

Exercise 1.22 (Weak self-similarity and peacocks). Let $(X_t, t \ge 0)$ be a process such that:

$$X_t \overset{(1.d)}{=} t^r X_1$$

for some $r \ge 0$. Let $\varphi \in \mathbf{I}$ such that the process $(\varphi(X_t), t \ge 0)$ is integrable and $\mathbb{E}[\varphi(X_t)]$ does not depend on t. Prove that $(\varphi(X_t), t \ge 0)$ is a peacock.

Exercise 1.23 (Peacocks and symmetry). 1) Let $\varphi \in \mathbf{I}$ such that φ is an odd function. Let X be a symmetric r.v. (i.e. $-X \overset{(law)}{=} X$) such that, for every $t \geq 0$, $\mathbb{E}[|\varphi(tX)|] < \infty$. Prove that $(\varphi(tX), t \geq 0)$ is a peacock.

2) Let $\varphi \in \mathbf{I}$ such that φ is an odd function and let $r \geq 0$. Consider a process $(X_t, t \geq 0)$ such that:

i) $X_t \overset{(1.d)}{=} t^r X_1$,
ii) X_1 is symmetric,
iii) for every $t \geq 0$, $\mathbb{E}[|\varphi(X_t)|] < \infty$.

Prove that $(\varphi(X_t), t \geq 0)$ is a peacock.

3) Let $\varphi \in \mathbf{J}$ such that φ is an odd function and let $(X_t, t \geq 0)$ be a peacock such that, for every $t \geq 0$, X_t is symmetric and $\mathbb{E}[|\varphi(X_t)|] < \infty$. Prove that $(\varphi(X_t), t \geq 0)$ is a peacock.

Exercise 1.24 (Gaussian peacocks). We consider a process $(G_t, t \geq 0)$ such that, for every $t \geq 0$, G_t is a centered Gaussian r.v. We set

$$k(t) = \mathbb{E}[G_t^2], \quad t \geq 0.$$

1) Prove that $(G_t, t \geq 0)$ is a peacock if and only if k is increasing.
(*Hint*: $G_t \overset{(law)}{=} \sqrt{k(t)}G$, with G a reduced normal r.v.)

2) Let $(B_t, t \geq 0)$ be a standard Brownian motion starting from 0. Prove that:

$$G_t \overset{(1.d)}{=} B_{k(t)}$$

and, if k is increasing, then $(B_{k(t)}, t \geq 0)$ is a martingale which is associated to the peacock $(G_t, t \geq 0)$.

3) Apply the previous results to $(G_t, t \geq 0)$ an Ornstein-Uhlenbeck process, solution of

$$X_t = B_t - c\int_0^t X_s ds$$

with $c \in \mathbb{R}$.
(*Hint*: prove first that $(X_t, t \geq 0) \overset{(law)}{=} (e^{-ct}\int_0^t e^{cs} dB_s, t \geq 0)$.)

Exercise 1.25 (Product of a peacock by a stochastically increasing process). Let $(X_t, t \geq 0)$ be an integrable process taking values in $\mathbb{R}_+$, which is stochastically increasing, i.e. such that for every $0 \leq s < t$ and every $a \geq 0$:

$$\mathbb{P}(X_t > a) \geq \mathbb{P}(X_s > a).$$

1) Prove that, for every increasing function $f : \mathbb{R}_+ \longrightarrow \mathbb{R}_+$, the function $t \longmapsto \mathbb{E}[f(X_t)]$ is increasing.

2) Let Y be an integrable and centered r.v. independent from $(X_t, t \geq 0)$. Prove that $(Z_t := Y \cdot X_t, t \geq 0)$ is a peacock.
(*Hint*: let $\psi \in \boldsymbol{C}$ and define $\theta(x) = \psi(x) - \psi(0) - x\psi'(0)$ (see proof of Point 1) of Theorem 1.3); prove that $t \longmapsto \mathbb{E}[\theta(Z_t)]$ is increasing.)

3) More generally, let $(Y_t, t \geq 0)$ be a centered peacock independent from $(X_t, t \geq 0)$. Prove that $(U_t := Y_t \cdot X_t, t \geq 0)$ is a peacock.

4) Let $(B_t, t \geq 0)$ be a standard Brownian motion started from 0, and independent from $(X_t, t \geq 0)$. Assume furthermore that $(X_t, t \geq 0)$ is an increasing process and set, for $\alpha \geq 0$,

$$M_t^{(\alpha)} = B_{t^\alpha X_t^2}, \quad t \geq 0.$$

Prove that $(M_t^{(\alpha)}, t \geq 0)$ is a martingale and:

$$M_t^{(\alpha)} \overset{(1.d)}{=} B_{t^\alpha} X_t.$$

Exercise 1.26 (Using Kamae-Krengel's Theorem to prove that some processes are peacocks). Let $(X_u, u \geq 0)$ and $(Y_t, t \geq 0)$ be two independent processes such that:

a) $(X_u, u \geq 0)$ is a peacock,
b) $(Y_t, t \geq 0)$ is a stochastically increasing $\mathbb{R}_+$-valued process,
c) for every $t \geq 0$, $\mathbb{E}[|X_{Y_t}|] < +\infty$.

Prove, by two different methods, that $(X_{Y_t}, t \geq 0)$ is a peacock.
(*Hint*. First method: let $\psi \in \boldsymbol{C}$. Write $\mathbb{E}[\psi(X_{Y_t})] = \int_0^{+\infty} \mathbb{E}[\psi(X_u)]\mu_t(du)$ where μ_t denotes the law of Y_t and deduce from Question 1) of Exercise 1.25 that $(X_{Y_t}, t \geq 0)$ is a peacock. Second method: let $(M_u, u \geq 0)$ (resp. $(A_t, t \geq 0)$) be a martingale (resp. an increasing process) associated to $(X_u, u \geq 0)$ (resp. $(Y_t, t \geq 0)$) according to Theorems 1.1 and 1.2. Observe then that $X_{Y_t} \overset{(1.d)}{=} M_{A_t}$, with M and A independent.)

Exercise 1.27 (Solutions of some SDE's are peacocks). Let $\sigma : \mathbb{R}_+ \times \mathbb{R} \to \mathbb{R}$ and $b : \mathbb{R}_+ \times \mathbb{R} \to \mathbb{R}$ denote two Borel functions such that $\sigma_s(x) := \sigma(s,x)$ and $b_s(x) := b(s,x)$ are Lipschitz continuous with respect to x, locally uniformly with respect to s. Let (X_t) be the solution of:

$$X_t = \int_0^t \sigma(s, X_s)dB_s + \int_0^t b(s, X_s)ds \tag{1.22}$$

where $(B_s, s \geq 0)$ denotes a standard Brownian motion started at 0. Furthermore, we assume:

- for every $s \geq 0$, σ_s is an even (or odd) function, (1.23)
- for every $s \geq 0$, b_s is an odd function such that: (1.24)

$$\operatorname{sgn}(b_s(x)) = \operatorname{sgn}(x).$$

1) Prove that $(X_t, t \geq 0)$, the solution of (1.22), is a peacock. (*Hint*: from Itô's formula, for $\psi \in \mathbf{C}$:

$$\frac{\partial}{\partial t}\mathbb{E}[\psi(X_t)] \geq \mathbb{E}[\psi'(X_t)b_t(X_t)] = \mathbb{E}\left[b_t(X_t)\left[\psi'(0) + \int_0^{X_t}\psi''(u)du\right]\right]$$

and prove that:

$$\mathbb{E}[b_t(X_t)] = 0 \text{ and } \mathbb{E}\left[b_t(X_t)\int_0^{X_t}\psi''(u)du\right] \geq 0).$$

2) Let $\varphi \in \mathbf{J}$ be an odd function such that the process $(\varphi(X_t), t \geq 0)$ is integrable. Prove that $(\varphi(X_t), t \geq 0)$ is a peacock.

3) Let $(X_t, t \geq 0)$ be the Ornstein-Uhlenbeck process, defined by:

$$X_t = B_t - c\int_0^t X_s ds.$$

a) use Point 1) to prove that, if $c \leq 0$, $(X_t, t \geq 0)$ is a peacock,

b) prove that, for every c real, $(X_t, t \geq 0)$ is a peacock, (*Hint*: see Exercise 1.24 above.)

c) deduce from b) that hypothesis (1.24) is not necessary to obtain that $(X_t, t \geq 0)$ is a peacock,

d) find a martingale M_t such that $X_t \overset{(1.d)}{=} M_t$.

4) Let $\varphi \in \mathbf{I}$ be an odd function such that $(\varphi(B_t), t \geq 0)$ is integrable. We know from Point 2) in Exercise 1.23 with $r = \frac{1}{2}$ that $(\varphi(B_t), t \geq 0)$ is a peacock. Suppose now moreover that φ is strictly increasing and of $\mathscr{C}^2$ class.

Prove that $X_t = \varphi(B_t)$ satisfies (1.22) with $\sigma(s,x) = \varphi' \circ \varphi^{-1}(x)$ and $b(s,x) = \frac{1}{2}\varphi'' \circ \varphi^{-1}(x)$. Show that Condition (1.23) is satisfied and Condition (1.24) is satisfied if and only if $\varphi \in \mathbf{J}$.

Exercise 1.28 (Peacocks and Markov processes). Let $((X_t, t \geq 0), (\mathbb{P}_x, x \in \mathbb{R}))$ be a Fellerian $\mathbb{R}$-valued Markov process. We assume that, for every $x \in \mathbb{R}$ and $t \geq 0$, $\mathbb{E}[|X_t|] < +\infty$. More precisely, we assume that there exists two continuous functions $x \longmapsto b(x)$ and $x \longmapsto \sigma(x)$ and a family of positive and finite measures $(N(x,du), x \in \mathbb{R})$ satisfying $N(x,\{x\}) = 0$ and $\int_{\mathbb{R}}|u|N(x,du) < +\infty$ for every $x \in \mathbb{R}$, such that the infinitesimal generator L of X is given by:

$$Lf(x) = \frac{1}{2}\sigma^2(x)f''(x) + b(x)f'(x) + \int_{\mathbb{R}}\left(f(x+u) - f(x) - \frac{u}{1+u^2}f'(x)\right)\frac{1+u^2}{u^2}N(x,du)$$

for all $\mathscr{C}^2$-functions f with sublinear growth at infinity.

1) Prove that the following properties are equivalent:
   a) $\forall x \in \mathbb{R}$, $\forall t \geq 0$, $\mathbb{E}_x[X_t] = x$,
   b) for every $x \in \mathbb{R}$, $(X_t, t \geq 0)$ is a $\mathbb{P}_x$-martingale,
   c) for every $x \in \mathbb{R}$, $b(x) + \int_{\mathbb{R}} uN(x, du) = 0$,
   d) for every $x \in \mathbb{R}$, $(X_t, t \geq 0)$ is a peacock under $\mathbb{P}_x$.

2) Let $x_0 \in \mathbb{R}$ and assume that, for every $t \geq 0$:

$$\mathbb{E}_{x_0}[X_t] = x_0.$$

i) Prove that, for every $t \geq 0$, $\mathbb{E}_{x_0}\left[b(X_t) + \int_{\mathbb{R}} uN(X_t, du)\right] = 0$. In particular, $b(x_0) + \int_{\mathbb{R}} uN(x_0, du) = 0$.

ii) Assume moreover that, for every $x \in \mathbb{R}$,

$$\operatorname{sgn}\left(b(x) + \int_{\mathbb{R}} uN(x, du)\right) = \operatorname{sgn}(x - x_0).$$

Prove that, under $\mathbb{P}_{x_0}$, $(X_t, t \geq 0)$ is a peacock.
(*Hint*: let $\psi \in \boldsymbol{C}$,

$$\frac{d}{dt}\mathbb{E}_{x_0}[\psi(X_t)] \geq \mathbb{E}_{x_0}\left[\psi'(X_t)\left(b(X_t) + \int_{\mathbb{R}} uN(X_t, du)\right)\right]$$

and use the same argument as in Exercise 1.27.)

3) Observe that if $(X_t, t \geq 0)$ is an integrable $\mathbb{R}$-valued Lévy process starting from 0, we recover the easy-to-prove equivalence:
   a) $(X_t, t \geq 0)$ is a martingale,
   b) $(X_t, t \geq 0)$ is a peacock,
   c) $\mathbb{E}[X_t]$ does not depend on t.

Exercise 1.29 (Conditions for a semimartingale to be a peacock). Let $(X_t, t \geq 0)$ be a continuous semimartingale whose Doob-Meyer's decomposition is given by $X_t = M_t + V_t$, with

$$\langle M \rangle_t = \int_0^t \sigma_s^2 ds \quad \text{and} \quad V_t := \int_0^t b_s ds.$$

We assume that:

i) for every $t \geq 0$:

$$\mathbb{E}\left[\left(\int_0^t \sigma_s^2 ds\right)^{1/2} + \int_0^t |b_s| ds\right] < \infty,$$

ii) the mappings $s \longmapsto b_s$ and $s \longmapsto \sigma_s^2$ are continuous in L^1.

1) Prove that $(X_t, t \geq 0)$ is a peacock if and only if, for every $\psi \in \boldsymbol{C}$ and every $s \geq 0$:

$$\mathbb{E}\left[\psi'(X_s) b_s\right] + \frac{1}{2}\mathbb{E}\left[\psi''(X_s)\sigma_s^2\right] \geq 0.$$

2) For every $t \geq 0$, we denote by $p(t,dx)$ the law of X_t and we define:

$$\widehat{b}(t,X_t) := \mathbb{E}[b_t|X_t] \quad \text{and} \quad \widehat{\sigma}^2(t,X_t) := \mathbb{E}[\sigma_t^2|X_t].$$

Prove that $(X_t, t \geq 0)$ is a peacock if and only if: for every $t \geq 0$, $\mathbb{E}[b_t] = 0$ and the measure

$$\left(\int_y^{+\infty} \widehat{b}(t,x)p(t,dx)\right) dy + \frac{1}{2}\widehat{\sigma}^2(t,y)p(t,dy)$$

is positive. Deduce that, if for every $t \geq 0$, $\mathbb{E}[b_t] = 0$ and $p(t,dx)$-a.s., $\widehat{b}(t,x)$ has the same sign as x, then $(X_t, t \geq 0)$ is a peacock.

3) Let, with the hypotheses and notation of Exercise 1.27, $(X_t, t \geq 0)$ be the unique strong solution of the SDE:

$$X_t = \int_0^t \sigma(s,X_s)dB_s + \int_0^t b(s,X_s)ds.$$

Prove, applying the result of the previous question, that $(X_t, t \geq 0)$ is a peacock.

4) Prove that, if $(B_t, t \geq 0)$ is a Brownian motion started from 0, then $(\sinh(B_t), t \geq 0)$ is a peacock.
(*Hint*: $\sinh(B_t) := Y_t = \int_0^t \sqrt{1+Y_s^2}dB_s + \frac{1}{2}\int_0^t Y_s ds$.) See Subsection 7.5.4 for a detailed study of this peacock.

Exercise 1.30 (Solutions of some SDE's in the sense of Stratonovitch are peacocks). Let σ be a strictly positive, even $\mathscr{C}^1$-function such that σ and σ' are Lipschitz continuous. Let $(X_t, t \geq 0)$ be the unique solution of:

$$X_t = \int_0^t \sigma(X_s) \circ dB_s \quad \left(= \int_0^t \sigma(X_s)dB_s + \frac{1}{2}\int_0^t \sigma\sigma'(X_s)ds\right)$$

where $\circ dB_s$ indicates Stratonovitch integration with respect to Brownian motion $(B_s, s \geq 0)$.

Prove that $(X_t, t \geq 0)$ is a peacock.
(*Hint*: let $h : \mathbb{R} \to \mathbb{R}$ be defined by $h(x) := \int_0^x \frac{dy}{\sigma(y)}$; prove that $(h(X_t), t \geq 0) = (B_t, t \geq 0)$ and that $X_t = h^{-1}(B_t)$, with h^{-1} ($\in \mathbf{I}$) an odd function. Then use Question 2) in Exercise 1.23). The change of variable $h(x) := \int_0^x \frac{dy}{\sigma(y)}$ seems to have a long history going back at least to Doeblin [53] and Kolmogorov [111].)

Exercise 1.31 (Peacock after a change of probability). Let $(\Omega = \mathscr{C}([0,+\infty[,\mathbb{R}),$ $(X_t, \mathscr{F}_t)_{t\geq 0}, \mathbb{P})$ be the canonical realization of the Brownian motion started from 0, and let $\mathscr{F}_\infty := \bigvee_{t\geq 0} \mathscr{F}_t$.

1) Let $\theta : \mathbb{R} \longrightarrow \mathbb{R}_+$ be a function of $\mathscr{C}^2$ class which is strictly positive, such that θ' is sub-exponential at $\pm\infty$ and $\theta(0) = 1$. We assume furthermore that $\dfrac{\theta'}{\theta}$ is Lipschitz continuous and $-\inf\left\{\dfrac{\theta''}{\theta}, 0\right\}$ is bounded.

i) Prove that $\left(M_t^{(\theta)} := \theta(X_t)\exp\left(-\frac{1}{2}\int_0^t \frac{\theta''}{\theta}(X_s)ds\right), t \geq 0\right)$ is a positive $(\mathscr{F}_t, t \geq 0)$-martingale such that $M_0^{(\theta)} = 1$.

ii) Let $\mathbb{Q}^{(\theta)}$ be the probability on $(\Omega, \mathscr{F}_\infty)$ characterized by

$$\mathbb{Q}^{(\theta)}_{|\mathscr{F}_t} = M_t^{(\theta)} \centerdot \mathbb{P}_{|\mathscr{F}_t}.$$

Prove that there exists a $\mathbb{Q}^{(\theta)}$-Brownian motion $(B_t, t \geq 0)$ such that, under $\mathbb{Q}^{(\theta)}$, we have:

$$X_t = B_t + \int_0^t \frac{\theta'}{\theta}(X_s)ds. \tag{1.25}$$

(*Hint*: apply Girsanov's Theorem.)

2) Let $\varphi : \mathbb{R} \longrightarrow \mathbb{R}$ be a function of $\mathscr{C}^2$ class, which is increasing, odd, such φ' is sub-exponential at $\pm\infty$, and $-\inf\left\{\dfrac{\varphi''}{\varphi}, 0\right\}$ is bounded.

i) Prove that $\left(M_t^{(\varphi)} := \varphi(X_t)\exp\left(-\frac{1}{2}\int_0^t \frac{\varphi''}{\varphi}(X_s)ds\right), t \geq 0\right)$ is a $((\mathscr{F}_t, t \geq 0), \mathbb{P})$-martingale but is not, in general, a $((\mathscr{F}_t, t \geq 0), \mathbb{Q}^{(\theta)})$-martingale.

We now assume furthermore that θ is even and that $\operatorname{sgn} \theta'(x) = \operatorname{sgn} x$.

ii) Prove that, for every $t \geq 0$, $\mathbb{E}_{\mathbb{Q}^{(\theta)}}[M_t^{(\varphi)}] = 0$.
(*Hint*: from Itô's formula

$$\begin{aligned}\mathbb{E}_{\mathbb{Q}^{(\theta)}}[M_t^{(\varphi)}] &= \mathbb{E}_{\mathbb{P}}[M_t^{(\theta)} M_t^{(\varphi)}] \\ &= \mathbb{E}_{\mathbb{P}}\left[\int_0^t ds\,\theta'(X_s)\varphi'(X_s)\exp\left(-\frac{1}{2}\int_0^s\left(\frac{\theta''}{\theta} + \frac{\varphi''}{\varphi}\right)(X_u)du\right)\right]\end{aligned}$$

and use the evenness (resp. oddness) of θ (resp. φ).)

iii) Prove that, under $\mathbb{Q}^{(\theta)}$, $(M_t^{(\varphi)}, t \geq 0)$ is a peacock. (*Hint*: let $\psi \in \boldsymbol{C}$, justify that, upon replacing ψ by $x \longmapsto \psi(x) - \psi(0) - x\psi'(0)$, we may assume that ψ is positive and $\operatorname{sgn} \psi'(x) = \operatorname{sgn} x$; then apply Itô's formula.)

iv) Let $(Z_t, t \geq 0)$ be the solution of (1.25). Prove that

$$\left(\varphi(Z_t)\exp\left(-\frac{1}{2}\int_0^t \frac{\varphi''}{\varphi}(Z_s)ds\right), t \geq 0\right)$$

is a peacock.

Exercise 1.32 (Peacock built with a Markovian semi-group). Let $(\Omega, \mathscr{F}, \mathbb{P})$ be a probability space and denote by $(T_u, u \geq 0)$ a Markovian semi-group on $L^1(\mathbb{P})$ such that:

$$\forall X \in L^1(\mathbb{P}), \forall u \geq 0, \quad \mathbb{E}[T_u X] = \mathbb{E}[X].$$

1) Prove that, for every $X \in L^1(\mathbb{P})$, $(T_{-\log t}X, 0 < t \leq 1)$ is a peacock.
(*Hint*: prove first that, if $X \in L^1(\mathbb{P})$ and ψ is a convex function, then $\psi(T_u X) \leq T_u(\psi(X))$.)

2) Let $X \in L^1(\mathbb{P})$. We define $(P_t X := e^{-t}X + (1-e^{-t})\mathbb{E}[X], t \geq 0)$. Prove that $(P_t, t \geq 0)$ is a Markovian semi-group, and recover the result of Proposition 1.4: $(t(X - \mathbb{E}[X]), t \geq 0)$ is a peacock.

1.2.2 Peacocks with Respect to Maturity

Here we denote by Asian process any process $(X_t, t \geq 0)$ of the form:

$$\left(X_t = \frac{1}{t}\int_0^t M_s ds, t \geq 0\right) \tag{1.26}$$

where $(M_s, s \geq 0)$ is a right-continuous martingale. In the following, the martingales we shall consider will always be assumed to be right-continuous. We shall not mention this hypothesis anymore.

Definition 1.7. Let $(M_t, t \geq 0)$ be a martingale with respect to the filtration $(\mathscr{F}_t, t \geq 0)$. This martingale belongs to H^1_{loc} if, for every $t \geq 0$:

$$\mathbb{E}\left[\sup_{s \leq t} |M_s|\right] < \infty. \tag{1.27}$$

If $M \in H^1_{\text{loc}}$, then, for every predictable and bounded process $(\phi_u, u \geq 0)$,

$$\left(\int_0^t \phi_u dM_u, t \geq 0\right) \tag{1.28}$$

is a martingale. Indeed, in this situation, the local martingale $\left(\int_0^t \phi_u dM_u, t \geq 0\right)$ also belongs, from Burkholder-Davis-Gundy inequalities, to H^1_{loc}, hence it is a true martingale.

Theorem 1.4 (Asian processes are peacocks). *Let $(M_t, t \geq 0)$ be a $(\mathscr{F}_t, t \geq 0)$-martingale in H^1_{loc} and $\alpha : \mathbb{R}_+ \to \mathbb{R}_+$ an increasing and continuous function such that $\alpha(0) = 0$. Then:*

$$1) \quad \left(X_t := \frac{1}{\alpha(t)}\int_0^t M_s d\alpha(s), t \geq 0\right) \text{ is a peacock} \tag{1.29}$$

$$2) \quad \left(\widetilde{X}_t := \int_0^t (M_s - M_0)d\alpha(s), t \geq 0\right) \textit{ is a peacock} \tag{1.30}$$

(in (1.29), by convention, $X_0 := M_0$).

Definition 1.8. We call a peacock defined by (1.29) a $\mathbf{F}_1$-type peacock and a peacock defined by (1.30) a $\mathbf{F}_2$-type peacock.

Remark 1.8. 1) *a*) Let $\sigma := \inf\{s \geq 0; \alpha(s) > 0\}$. In case $\sigma > 0$, one should define $X_t = M_t$ for $t \leq \sigma$.
b) Concerning either Point 1) or Point 2) of Theorem 1.4, there is no loss of generality in assuming $\alpha(t) = t$ since, assuming the result holds in this case, it also holds for a general α because $(M_{\alpha^{-1}(s)}, s \geq 0)$ is still a martingale (α^{-1} being the right-continuous inverse of $\alpha : \alpha^{-1}(u) := \inf\{s \geq 0, \alpha(s) > u\}$).

2) *a*) If $\alpha(t) = t$, then, for every real K, since $x \to (x - K)^+$ is convex, Point 1) of Theorem 1.4 yields that $\mathbb{E}\left[\left(\frac{1}{t}\int_0^t M_s ds - K\right)^+\right]$ is an increasing function of t. In other terms, the price of an Asian option is an increasing function of t.
b) On the other hand, from Theorem 1.1, there exists a martingale $(\widetilde{M}_t, t \geq 0)$ such that

$$\frac{1}{t}\int_0^t M_s ds \overset{(1.d)}{=} \widetilde{M}_t. \tag{1.31}$$

Hence, for every K real:

$$\mathbb{E}\left[\left(\frac{1}{t}\int_0^t M_s ds - K\right)^+\right] = \mathbb{E}\left[(\widetilde{M}_t - K)^+\right]. \tag{1.32}$$

The price of an Asian option is the price of an European option, after a change of the martingale of reference. Let us assume that $M_0 = 1$ and, $(M_s, s \geq 0)$ is positive and $\lim\limits_{s\to\infty} M_s = 0$ a.s. As a consequence, we have $\widetilde{M}_0 = 1$, $(\widetilde{M}_t, t \geq 0)$ is positive and $\lim\limits_{t\to\infty} \widetilde{M}_t = 0$ a.s. Now, we assume furthermore that $(\widetilde{M}_t, t \geq 0)$ is continuous. Then, from [122] or [148]:

$$\begin{aligned} \mathbb{E}\left[\left(K - \frac{1}{t}\int_0^t M_s ds\right)^+\right] &= \mathbb{E}\left[(K - \widetilde{M}_t)^+\right] \\ &= K\mathbb{P}\left(\mathscr{G}_K^{(\widetilde{M})} \leq t\right) \end{aligned} \tag{1.33}$$

for $K \geq 0$ and with $\mathscr{G}_K^{(\widetilde{M})} := \sup\{u \geq 0; \widetilde{M}_u = K\}$ the last passage time at level K of the martingale $(\widetilde{M}_u, u \geq 0)$. Thus, we obtain with this last formula another view of the increasing property of the LHS of (1.33).

3) Let $(B_t, t \geq 0)$ denote a Brownian motion starting from 0, $\nu \in \mathbb{R}$ and define $\left(\mathscr{E}_t^{(\nu)} := \exp\left(\nu B_t - \frac{\nu^2}{2}t\right), t \geq 0\right)$ to be the exponential Brownian martingale with parameter ν. A result due to Carr, Ewald and Xiao (see [43]) establishes that

$$\left(X_t^{(\nu)} := \frac{1}{t}\int_0^t \mathscr{E}_s^{(\nu)} ds, t \geq 0\right) \quad \text{is a peacock.} \tag{1.34}$$

Thus, Point 1) of Theorem 1.4 is an extension of Carr, Ewald and Xiao's result.

From Theorem 1.1, $(X_t, t \geq 0)$ defined by (1.29) and $(\widetilde{X}_t, t \geq 0)$ defined by (1.30) are 1-martingales. However:

- $(\widetilde{X}_t, t \geq 0)$ has bounded variation on any interval $[0, T]$,
- $(X_t, t \geq 0)$ has bounded variation on every interval $[\varepsilon, T]$, with $\varepsilon > 0$ (as soon as $\sigma := \inf\{s \geq 0; \alpha(s) > 0\} = 0$).

We now give two proofs of Theorem 1.4.

First proof of Point 1) of Theorem 1.4. An integration by parts yields:

$$\begin{aligned} X_t &= \frac{1}{\alpha(t)}\int_0^t M_s d\alpha(s) = \frac{1}{\alpha(t)}\left[M_t\alpha(t) - \int_0^t \alpha(s) dM_s\right] \\ &= M_t - \frac{M_t^{(\alpha)}}{\alpha(t)} \end{aligned} \tag{1.35}$$

with

$$M_t^{(\alpha)} := \int_0^t \alpha(s) dM_s.$$

Hence:

$$dX_t = dM_t + \frac{M_t^{(\alpha)}}{\alpha^2(t)} d\alpha(t) - \frac{dM_t^{(\alpha)}}{\alpha(t)} = M_t^{(\alpha)} \frac{d\alpha(t)}{\alpha^2(t)}. \tag{1.36}$$

On the other hand, for $0 < s \leq t$:

$$\begin{aligned} \mathbb{E}[X_t | \mathscr{F}_s] &= M_s - \frac{1}{\alpha(t)} M_s^{(\alpha)} \\ &= M_s - \frac{1}{\alpha(s)} M_s^{(\alpha)} + \left(\frac{1}{\alpha(s)} - \frac{1}{\alpha(t)}\right) M_s^{(\alpha)} \\ &= X_s + \left(\frac{1}{\alpha(s)} - \frac{1}{\alpha(t)}\right) M_s^{(\alpha)}. \end{aligned} \tag{1.37}$$

Now, let $\psi \in \mathbf{C}$. Then, for $0 < s \leq t$:

$$\begin{aligned} \mathbb{E}[\psi(X_t)] &= \mathbb{E}[\mathbb{E}[\psi(X_t) | \mathscr{F}_s]] \\ &\geq \mathbb{E}[\psi(\mathbb{E}[X_t | \mathscr{F}_s])] \quad \text{(from Jensen's inequality)} \\ &= \mathbb{E}\left[\psi\left(X_s + \left(\frac{1}{\alpha(s)} - \frac{1}{\alpha(t)}\right) M_s^{(\alpha)}\right)\right] \quad \text{(from (1.37))} \end{aligned}$$

$$\geq \mathbb{E}\left[\psi(X_s)+\left(\frac{1}{\alpha(s)}-\frac{1}{\alpha(t)}\right)\psi'(X_s)M_s^{(\alpha)}\right] \quad \text{(by convexity)}$$
$$= \mathbb{E}[\psi(X_s)]+\left(\frac{1}{\alpha(s)}-\frac{1}{\alpha(t)}\right)\mathbb{E}[\psi'(X_s)M_s^{(\alpha)}]. \tag{1.38}$$

Hence, Point 1) of Theorem 1.4 will be proven if we show that $\mathbb{E}[\psi'(X_s)M_s^{(\alpha)}] \geq 0$. It follows from Itô's formula, since $(X_s, s > \varepsilon)$ has bounded variation, that, for $0 < \varepsilon < s$:

$$\begin{aligned}\mathbb{E}[\psi'(X_s)M_s^{(\alpha)}] &= \mathbb{E}[\psi'(X_\varepsilon)M_\varepsilon^{(\alpha)}]+\mathbb{E}\left[\int_\varepsilon^s M_u^{(\alpha)}\psi''(X_u)dX_u\right]\\ &\quad +\mathbb{E}\left[\int_\varepsilon^s \psi'(X_u)dM_u^{(\alpha)}\right]\\ &= \mathbb{E}[\psi'(X_\varepsilon)M_\varepsilon^{(\alpha)}]+\mathbb{E}\left[\int_\varepsilon^s M_u^{(\alpha)}\psi''(X_u)dX_u\right].\end{aligned}$$

Since $M_0^{(\alpha)} = 0$ and ψ' is bounded (see definition of the class $\boldsymbol{C}$) $\mathbb{E}[\psi'(X_\varepsilon)M_\varepsilon^{(\alpha)}] \underset{\varepsilon\to 0}{\longrightarrow} 0$ and it suffices to show that $\mathbb{E}[M_u^{(\alpha)}\psi''(X_u)dX_u] \geq 0$. However:

$$M_u^{(\alpha)}\psi''(X_u)dX_u = (M_u^{(\alpha)})^2\psi''(X_u)\frac{d\alpha(u)}{\alpha^2(u)} \geq 0 \tag{1.39}$$

from (1.36).

Actually, we have just proven the peacock property on $]0,+\infty[$, but the result holds on $[0,+\infty[$ because of the continuity at 0 of the process $(X_t, t \geq 0)$. □

Second proof of Point 1) of Theorem 1.4 (Existence of trading cash flow). For $\psi \in \boldsymbol{C}$ and $0 < s < t$, we have, since $(X_u, s \leq u \leq t)$ has bounded variation:

$$\begin{aligned}\psi(X_t)-\psi(X_s) &= \int_s^t \psi'(X_u)dX_u\\ &= \int_s^t \psi'(X_u)M_u^{(\alpha)}\frac{d\alpha(u)}{\alpha^2(u)} \quad \text{(from (1.36))}.\end{aligned}$$

But, from Itô's formula and (1.36):

$$\psi'(X_u)M_u^{(\alpha)} = \int_0^u \psi''(X_v)(M_v^{(\alpha)})^2\frac{d\alpha(v)}{\alpha^2(v)}+\int_0^u \psi'(X_v)dM_v^{(\alpha)}.$$

Gathering the two previous formulae we obtain:

$$\psi(X_t)-\psi(X_s) = \int_s^t \frac{d\alpha(u)}{\alpha^2(u)}\left[\int_0^u \psi''(X_v)(M_v^{(\alpha)})^2\frac{d\alpha(v)}{\alpha^2(v)}+\int_0^u \psi'(X_v)dM_v^{(\alpha)}\right].$$

Since:

$$\int_s^t \frac{d\alpha(u)}{\alpha^2(u)}\int_0^u \psi''(X_v)(M_v^{(\alpha)})^2\frac{d\alpha(v)}{\alpha^2(v)} \geq 0$$

and

$$\mathbb{E}\left[\int_s^t \frac{d\alpha(u)}{\alpha^2(u)}\left(\int_0^u \psi'(X_v)dM_v^{(\alpha)}\right)\right] = 0,$$

$T_{s,t}^{\psi} := \int_s^t \frac{d\alpha(u)}{\alpha^2(u)} \int_0^u \psi'(X_v)dM_v^{(\alpha)}$ is a trading cash flow, and Point 1) of Theorem 1.4 is a consequence of Proposition 1.2. □

First proof of Point 2) of Theorem 1.4. Without loss of generality, we suppose that $M_0 = 0$.

By definition, $\widetilde{X}_t := \int_0^t M_s d\alpha(s)$, and we get, for $s \le t$:

$$\begin{aligned}
\mathbb{E}[\widetilde{X}_t|\mathscr{F}_s] &= \mathbb{E}\left[\int_0^t M_u d\alpha(u)|\mathscr{F}_s\right] \\
&= \int_0^s \mathbb{E}[M_u|\mathscr{F}_s]d\alpha(u) + \int_s^t \mathbb{E}[M_u|\mathscr{F}_s]d\alpha(u) \\
&= \int_0^s M_u d\alpha(u) + (\alpha(t)-\alpha(s))M_s \\
&= \widetilde{X}_s + (\alpha(t)-\alpha(s))M_s. \qquad (1.40)
\end{aligned}$$

Thus, for $\psi \in \boldsymbol{C}$:

$$\begin{aligned}
\mathbb{E}[\psi(\widetilde{X}_t)] &= \mathbb{E}[\mathbb{E}[\psi(\widetilde{X}_t)|\mathscr{F}_s]] \\
&\ge \mathbb{E}[\psi(\mathbb{E}[\widetilde{X}_t|\mathscr{F}_s])] \quad \text{(from Jensen's inequality)} \\
&= \mathbb{E}[\psi(\widetilde{X}_s + (\alpha(t)-\alpha(s))M_s)] \quad \text{(from (1.40))} \\
&\ge \mathbb{E}[\psi(\widetilde{X}_s)] + (\alpha(t)-\alpha(s))\mathbb{E}[\psi'(\widetilde{X}_s)M_s] \quad \text{(by convexity).}
\end{aligned}$$

Thus, Point 2) of Theorem 1.4 will be established once we show:

$$\mathbb{E}[\psi'(\widetilde{X}_s)M_s] \ge 0.$$

But, from Itô's formula and since $(\widetilde{X}_s, s \ge 0)$ has bounded variation:

$$\begin{aligned}
\mathbb{E}[\psi'(\widetilde{X}_s)M_s] &= \mathbb{E}[\psi'(0)M_0] + \mathbb{E}\left[\int_0^s M_u \psi''(\widetilde{X}_u)d\widetilde{X}_u\right] \\
&= \mathbb{E}\left[\int_0^s \psi''(\widetilde{X}_u)M_u^2 d\alpha(u)\right] \ge 0
\end{aligned}$$

since $d\widetilde{X}_u = M_u d\alpha(u)$ and $M_0 = 0$. □

Second proof of Point 2) of Theorem 1.4 (Existence of trading cash flow). For $\psi \in \boldsymbol{C}$ and $s < t$, we have, from Itô's formula:

$$\psi(\widetilde{X}_t) - \psi(\widetilde{X}_s) = \int_s^t \psi'(\widetilde{X}_u)(M_u - M_0)d\alpha(u)$$

and
$$\psi'(\widetilde{X}_u)(M_u - M_0) = \int_0^u \psi'(\widetilde{X}_v)dM_v + \int_0^u (M_v - M_0)^2 \psi''(\widetilde{X}_v)d\alpha(v).$$

Hence:

$$\psi(\widetilde{X}_t) - \psi(\widetilde{X}_s) - \int_s^t d\alpha(u) \int_0^u \psi'(\widetilde{X}_v)dM_v$$
$$= \int_s^t d\alpha(u) \int_0^u \psi''(\widetilde{X}_v)(M_v - M_0)^2 d\alpha(v)$$

and
$$T_{s,t}^{\psi} := \int_s^t d\alpha(u) \int_0^u \psi'(\widetilde{X}_v)dM_v$$

is a trading cash flow since:

$$\mathbb{E}\left[\int_s^t d\alpha(u) \int_0^u \psi'(\widetilde{X}_v)dM_v\right] = 0$$

and
$$\int_s^t d\alpha(u) \int_0^u \psi''(\widetilde{X}_v)(M_v - M_0)^2 d\alpha(v) \geq 0.$$

□

Remark 1.9. 1) *a*) In the first proof of Point 1) of Theorem 1.4, we were careful to state that $(X_t, t > 0)$ has bounded variation. Indeed, $(X_t, t \geq 0)$ may not have bounded variation in $t = 0$. However, as we showed, this does not prevent us to develop our argument, for $t > 0$.

b) An example for which the integral $\displaystyle\int_0^t M_s^{(\alpha)} \frac{d\alpha(s)}{\alpha^2(s)}$ is not absolutely convergent at 0 may be obtained from
$$M_s := \int_0^s f(u)dB_u \tag{1.41}$$

with $s \leq 1$ and $f \in L^2([0,1])$, with $\alpha(t) = t$. Indeed, one has:

$$X_t = X_0 + \int_0^t \frac{ds}{s^2} \int_0^s u dM_u = X_0 + \int_0^t \frac{ds}{s^2} \int_0^s u f(u)dB_u \quad (t \leq 1).$$

The quantity:
$$\int_0^1 \frac{ds}{s^2} \left| \int_0^s u f(u)dB_u \right|$$

is finite a.s. if and only if, from Jeulin's lemma (see [96])

$$\int_0^1 \frac{ds}{s^2} \left(\int_0^s u^2 f^2(u)du \right)^{1/2} < \infty. \tag{1.42}$$

By Hölder's inequality, this condition is satisfied if $f \in L^q([0,1])$ for some $q > 2$. On the contrary, it is easy to prove that, for $f^2(u) = \dfrac{1}{u(1+\log 1/u)^\beta}$, with $\beta \in]1,2[$, $f^2 \in L^1([0,1])$ and $\displaystyle\int_0^1 \frac{ds}{s^2}\left(\int_0^s u^2 f^2(u)du\right)^{1/2} = +\infty$.

2) Our Theorem 1.4 shows a little better than the increasing property of $(X_t, t \geq 0)$ and $(\widetilde{X}_t, t \geq 0)$ for the convex order since we have established in fact:

$$\begin{aligned}
\mathbb{E}[\psi(X_t)] - \mathbb{E}[\psi(X_s)] &\geq \mathbb{E}\left[\int_s^t \frac{d\alpha(u)}{\alpha^2(u)} \int_0^u \psi''(X_v)(M_v^{(\alpha)})^2 \frac{d\alpha(v)}{\alpha^2(v)}\right] \\
&\geq \left(\frac{1}{\alpha(s)} - \frac{1}{\alpha(t)}\right)\mathbb{E}\left[\int_0^s \psi''(X_v)(M_v^{(\alpha)})^2 \frac{d\alpha(v)}{\alpha^2(v)}\right] \qquad (1.43)
\end{aligned}$$

and:

$$\begin{aligned}
\mathbb{E}[\psi(\widetilde{X}_t)] - \mathbb{E}[\psi(\widetilde{X}_s)] &\geq \mathbb{E}\left[\int_s^t d\alpha(u) \int_0^u \psi''(\widetilde{X}_v)(M_v - M_0)^2 d\alpha(v)\right] \\
&\geq (\alpha(t) - \alpha(s))\mathbb{E}\left[\int_0^s \psi''(\widetilde{X}_v)(M_v - M_0)^2 d\alpha(v)\right]. \qquad (1.44)
\end{aligned}$$

3) In fact, Point 1) of Theorem 1.4 is a consequence of Point 2) of this theorem since, from (1.36)

$$X_t = X_s + \int_s^t M_u^{(\alpha)} \frac{d\alpha(u)}{\alpha^2(u)}$$

a formula which should be taken with some care for $s = 0$ – see Point 1) of this remark.

4) From Point 1) of Theorem 1.5, there exists a martingale $(\widetilde{M}_s, s \geq 0)$ such that:

$$\frac{1}{t}\int_0^t M_s ds \overset{(1.d)}{=} \widetilde{M}_t.$$

Then, we have: for every $t \geq 0$,

$$M_t \overset{(c)}{\geq} \widetilde{M}_t. \qquad (1.45)$$

Proof of (1.45). For $\psi \in \boldsymbol{C}$, we have:

$$\begin{aligned}
\mathbb{E}[\psi(\widetilde{M}_t)] &= \mathbb{E}\left[\psi\left(\frac{1}{t}\int_0^t M_s ds\right)\right] \\
&\leq \mathbb{E}\left[\frac{1}{t}\int_0^t \psi(M_s)ds\right] \quad \text{(by convexity)} \\
&= \frac{1}{t}\int_0^t \mathbb{E}[\psi(M_s)]ds \quad \text{(by Fubini)} \\
&\leq \frac{1}{t}\int_0^t \mathbb{E}[\psi(M_t)]ds
\end{aligned}$$

since $(M_t, t \geq 0)$ is a martingale, hence a peacock.

Thus:

$$\mathbb{E}\left[\psi(\widetilde{M}_t)\right] \leq \mathbb{E}\left[\psi(M_t)\right]. \qquad \square$$

Remark 1.10. Let X be an integrable and centered r.v. Let us define:

$$M_s = X \quad \text{for } s \geq 1 \qquad \text{and} \qquad M_s = 0 \quad \text{for } 0 \leq s < 1.$$

Then, $(M_s, s \geq 0)$ is a (right-continuous) martingale in its natural filtration ($\mathscr{M}_s = \sigma(X)$ for $s \geq 1$, and $\mathscr{M}_s$ is trivial for $s < 1$) and, from Point 2) of Theorem 1.4:

$$\left(X_t := \int_0^t M_s ds = \int_1^{1\wedge t} X ds = (t-1)^+ X\,,\ t \geq 0\right) \tag{1.46}$$

is a peacock (since $(hX, h \geq 0)$ is a peacock).

This is a new proof of Proposition 1.4.

We shall call a peacock of the form $(tX, t \geq 0)$ (or $(\alpha(t)X, t \geq 0)$ with α an increasing function) a $\mathbf{F}_2'$-type peacock. Of course, from (1.46):

$$\mathbf{F}_2' \subset \mathbf{F}_2. \tag{1.47}$$

More generally, there is the following result which is a variant of Point 2) of Theorem 1.4: under the preceding hypotheses of Theorem 1.4,

$$\left(\int_0^t (M_s - \mathbb{E}[M_0])d\alpha(s),\ t \geq 0\right) \text{ is a peacock.}$$

Indeed, applying Point 2) of Theorem 1.4 to the martingale:

$$M_t^\varepsilon := M_t 1_{[\varepsilon,+\infty[}(t) + \mathbb{E}[M_0] 1_{[0,\varepsilon[}(t)$$

we have, for any $\varepsilon > 0$,

$$\left(1_{[\varepsilon,+\infty[}(t) \int_\varepsilon^t (M_s - \mathbb{E}[M_0])d\alpha(s),\ t \geq 0\right) \text{ is a peacock,}$$

which entails the result, letting ε tend to 0.

We now deduce from Remark 1.10 (or from Proposition 1.4) that "some integral processes with scale property" are peacocks.

Definition 1.9. A process $(Z_t, t \geq 0)$ satisfies the scaling property of order β if: for every $c > 0$,

$$(Z_{cu}, u \geq 0) \overset{(law)}{=} (c^\beta Z_u, u \geq 0). \tag{1.48}$$

Theorem 1.5 (Scaling property and peacocks). *Let $(Z_u, u \geq 0)$ be a centered process satisfying the scaling property of order β. Then, for every α and γ such that:*

$$\beta + \alpha + 1 \geq \gamma \quad \textit{and} \quad \beta + \alpha + 1 > 0 \tag{1.49}$$

$\left(X_t^{\alpha,\gamma} := \dfrac{1}{t^\gamma}\displaystyle\int_0^t Z_s s^\alpha ds\,,\ t \geq 0\right)$ *is a peacock.*

Proof. We have:

$$\begin{aligned} X_t^{\alpha,\gamma} &:= \frac{1}{t^\gamma}\int_0^t Z_s s^\alpha ds = \frac{1}{t^\gamma}\int_0^1 Z_{tu} t^{\alpha+1} u^\alpha du \\ &\quad \text{(after the change of variable } s = tu) \\ &\overset{(1.d)}{=} \frac{1}{t^\gamma}\int_0^1 t^{\beta+\alpha+1} Z_u u^\alpha du \quad \text{(from the scaling property)} \\ &= t^{\beta+\alpha+1-\gamma}\int_0^1 Z_u u^\alpha du. \end{aligned} \tag{1.50}$$

Since by hypothesis Z is a centered process, we get $\mathbb{E}\left[\int_0^1 Z_u u^\alpha du\right] = \int_0^1 u^\alpha \mathbb{E}[Z_u] du = 0$. Now, Theorem 1.5 is a consequence of (1.46). ($\int_0^1 Z_u u^\alpha du$ is, by scaling, integrable as soon as $\alpha + \beta + 1 > 0$). □

Example 1.1. Let $(Q_t^{(\delta)}, t \geq 0)$ be a squared Bessel process of dimension $\delta > 0$ started at 0 (see [150], Chap XI). Then $(M_t^{(\delta)} := Q_t^{(\delta)} - \delta t, t \geq 0)$ is a centered martingale which satisfies the scaling property of order 1. Thus, for every $\alpha > -2$:

$$\left(X_t^{(\delta)} := \int_0^t M_s^{(\delta)} s^\alpha ds, t \geq 0\right)$$

is a $\mathbf{F}_2'$-type peacock, hence a $\mathbf{F}_2$-type peacock.

We now deduce from Theorem 1.4 a discrete version.

Theorem 1.6. *Let $(M_n, n \geq 1)$ be a martingale indexed by $\mathbb{N}^*$ and $(a_k, k \geq 1)$ a sequence of positive reals; we note $\alpha(n) = \sum_{k=1}^n a_k$.*

1) $\left(X_n := \frac{1}{\alpha(n)} \sum_{k=1}^n a_k M_k,\ n \geq 1\right)$ is a random sequence which increases in the convex order.

2) Assume furthermore that $\mathbb{E}[M_1] = 0$. Then $\left(\widetilde{X}_n := \sum_{k=1}^n a_k M_k,\ n \geq 1\right)$ is a random sequence which increases in the convex order.

Theorem 1.6 is a consequence of Theorem 1.4, by using the martingale $\left(M_t := \sum_{k=0}^{\infty} 1_{[k,k+1[}(t) M_k,\ t \geq 0\right)$.

Exercise 1.33 (Galton-Watson peacocks). Let $(X_k, k \geq 0)$ be a Galton-Watson chain such that $X_0 = 1$ and $\mathbb{E}[X_1] = m$. Let $S_n := \sum_{k=0}^n X_k$ the number of offsprings from the

ancestor until the nth-generation. Prove that for $m \neq 1$, $\left(\frac{1-m}{1-m^{n+1}} S_n, n \geq 0\right)$ is a peacock, while, for $m = 1$, $\left(\frac{1}{n+1} S_n, n \geq 0\right)$ is a peacock.
(*Hint*: $\left(\frac{X_n}{m^n}, n \geq 0\right)$ is a martingale.)

Exercise 1.34 (On some Wiener integrals of a Lévy process which are peacocks).
1) Let $(L_s, s \geq 0)$ denote a right-continuous process with independent increments (not necessarily stationary) such that $L_0 = 0$ and, for every $s \geq 0$:

$$\mathbb{E}[L_s^2] < \infty,\ \mathbb{E}[L_s] = 0.$$

Let $a : \mathbb{R}_+ \times \mathbb{R}_+ \to \mathbb{R}_+$ be a Borel function such that:

$$\begin{aligned} &i) \quad \int_0^\infty a^2(t,s) d\mathbb{E}[L_s^2] < \infty \\ &ii) \quad \text{for every } s \geq 0,\ t \to a(t,s) \text{ is increasing.} \end{aligned} \tag{1.51}$$

Let us define:

$$\left(X_t := \int_0^\infty a(t,s) dL_s, t \geq 0 \right).$$

a) Prove that $(X_t, t \geq 0)$ is well defined in L^2.
b) Prove that $(X_t, t \geq 0)$ is a peacock
(*Hint*: approximate X_t by $\left(\sum_{k=0}^\infty \left(n \int_{\frac{k}{n}}^{\frac{k+1}{n}} a(t,s) ds \right) \left(L_{\frac{k+1}{n}} - L_{\frac{k}{n}} \right) \right)$ and use Corollary 1.1.)

2) We now assume that $(L_s, s \geq 0)$ is a Lévy process. Deduce, from the classical result (see J. Jacod and P. Protter [90], and Chapter 4 of this monograph):

$$\forall\, 0 < s \leq t, \quad \mathbb{E}\left[\frac{L_s}{s} \middle| \mathscr{G}_t^+ \right] = \frac{L_t}{t} \quad \text{where } \mathscr{G}_t^+ := \sigma(L_u, u \geq t)$$

that $\left(\frac{L_t}{t}, t \geq 0\right)$ is *decreasing* for the convex order. Note that $\left(\frac{L_t}{t}, t \geq 0\right)$ may be written:

$$\frac{L_t}{t} = \int_0^\infty a(t,s) dL_s, \text{ with } a(t,s) = \frac{1}{t} 1_{\{s \leq t\}}.$$

Of course, this function a does not satisfy (1.51).

Exercise 1.35 (Integrals of Gaussian processes: examples and counterexamples).
1) Let $(G_s, s \geq 0)$ be a centered Gaussian process. Prove that $(G_s, s \geq 0)$ is a peacock if and only if $k(s) := \mathbb{E}[G_s^2]$ is an increasing function.
(*Hint*: see Point 1) of Exercise 1.24.)

2) (Non-transitivity of Theorem 1.4)
a) Find a process $(\Pi_s, s \geq 0)$ such that:

- $(\Pi_s, s \geq 0)$ is a peacock,
- $\left(\frac{1}{t}\int_0^t \Pi_s ds, t \geq 0\right)$ is not a peacock.

(*Hint*: let G be a reduced normal r.v. and $\Pi_s := \alpha(s)G$ with $\alpha(s) = 1$ for $s \leq 1$ and $\alpha(s) = -2$ for $s > 1$.)

b) Find a centered process $(\Pi_s, s \geq 0)$ such that:

- $(\Pi_s, s \geq 0)$ is a peacock,
- $\left(\int_0^t \Pi_s ds, t \geq 0\right)$ is not a peacock.

3) Find a process $(\Pi_s, s \geq 0)$ such that:

- $(\Pi_s, s \geq 0)$ is not a peacock,
- $\int_0^t \Pi_s ds, t \geq 0$ is a peacock.

(*Hint*: we may choose $\Pi_s = \beta(s)G$, with G a reduced normal r.v. and β a $\mathbb{R}_+$ valued decreasing function.)

4) Let X and Y be two independent reduced normal r.v.'s. Let us define:

$$\begin{aligned} G_s &:= X 1_{[0,1]}(s) - 2X 1_{]1,+\infty[}(s) \qquad (s \geq 0) \\ \widetilde{G}_s &:= X 1_{[0,1]}(s) - 2Y 1_{]1,+\infty[}(s) \qquad (s \geq 0). \end{aligned}$$

Prove that:

$$\begin{aligned} &a) \quad G_s \overset{(1.d)}{=} \widetilde{G}_s \\ &b) \quad \left(\int_0^t G_s ds, t \geq 0\right) \quad \text{is not a peacock} \\ &c) \quad \left(\int_0^t \widetilde{G}_s ds, t \geq 0\right) \quad \text{is a peacock.} \end{aligned}$$

5) Let $(G_s, s \geq 0)$ be a centered gaussian process such that $k(s) := \mathbb{E}[G_s^2]$ is an increasing function such that $k(s_0) > 0$ for some $s_0 \geq 0$. Hence, from question 1), $(G_s, s \geq 0)$ is a peacock. Prove that there exist two Gaussian (centered) processes $(G_s^{(1)}, s \geq 0)$ and $(G_s^{(2)}, s \geq 0)$ such that:

$$\begin{aligned} &a) \quad G_s^{(1)} \overset{(1.d)}{=} G_s^{(2)} \overset{(1.d)}{=} G_s \\ &b) \quad \left(\int_0^t G_s^{(1)} ds, t \geq 0\right) \quad \text{is a peacock} \\ &c) \quad \left(\int_0^t G_s^{(2)} ds, t \geq 0\right) \quad \text{is not a peacock.} \end{aligned}$$

(*Hint*: we may choose $G_s^{(1)} = \sqrt{k(s)}G$ with G a reduced normal r.v. and $G_s^{(2)} = \sqrt{k(s)}G 1_{[0,s_1]}(s) - \sqrt{k(s)}G 1_{]s_1,\infty]}(s)$ for $s_1 > s_0$.)

6) Let $(X_t^{(1)}, t \geq 0)$ and $(X_t^{(2)}, t \geq 0)$ be two peacocks defined on the same probability space. Give an example of peacocks $X^{(1)}$ and $X^{(2)}$ such that $(X_t^{(1)} + X_t^{(2)}, t \geq 0)$ is not a peacock. (*Hint*: let $(B_t, t \geq 0)$ be a Brownian motion started at 0, and $\alpha : \mathbb{R}_+ \to \mathbb{R}_+$ be an increasing function such that $\alpha(t) = 0$ if $t \in [0,1]$ and $\alpha(t) = 1$ if $t \geq 2$. Define $X_t^{(1)} = -B_t$, $X_t^{(2)} = \alpha(t) B_t$. Note that $(X_t^{(1)}, t \geq 0)$ and $(X_t^{(2)}, t \geq 0)$ are two peacocks and that: $\mathbb{E}\left[(X_t^{(1)} + X_t^{(2)})^2\right] = t$ if $t \leq 1$ and $\mathbb{E}\left[(X_t^{(1)} + X_t^{(2)})^2\right] = 0$ if $t \geq 2$.)

Exercise 1.36 (Some examples of peacocks as integrals of self-similar processes). Let $(\theta_u, u \geq 0)$ be a continuous, integrable and centered process which satisfies the scaling property of order h $(h > -1)$, i.e. for every $c > 0$, $(\theta_{cu}, u \geq 0) \overset{(law)}{=} (c^h \theta_u, u \geq 0)$. Let us define:

$$\left(X_t := \int_0^t \theta_u du, t \geq 0 \right).$$

1) Use Theorem 1.5 to prove that $(X_t, t \geq 0)$ is a peacock.
2) Show that

$$\mathbb{E}[\theta_t | X_t] = \frac{h+1}{t} X_t. \tag{1.52}$$

(*Hint*: let $\varphi : \mathbb{R} \to \mathbb{R}$ be a regular function and $\phi(x) := \int_0^x \varphi(y) dy$; prove that $\mathbb{E}\left[\varphi\left(\int_0^t \theta_u du\right) \theta_t\right] = \frac{\partial}{\partial t} \mathbb{E}\left[\phi\left(\int_0^t \theta_u du\right)\right] = \mathbb{E}\left[\varphi\left(\int_0^t \theta_u du\right) \frac{h+1}{t} X_t\right]$, by the scaling property.)

3) Use Proposition 1.3, (1.52) and Lemma 1.1 to obtain another proof of the peacock property of $(X_t, t \geq 0)$.

4) *Example 1* $(\theta_u := Q_u^{(\delta)} - \delta u, u \geq 0)$ where $(Q_u^{(\delta)}, u \geq 0)$ is the square of a Bessel process of dimension $\delta > 0$ started at 0. $(Q_u^{(\delta)}, u \geq 0)$ satisfies the scaling property of order 1 and is a martingale. Use Theorem 1.4 for a third proof that $\left(X_t := \int_0^t (Q_u^{(\delta)} - \delta u) du, t \geq 0 \right)$ is a peacock.

5) *Example 2* $\left(\theta_u := 1_{\{B_u > 0\}} - \frac{1}{2}, u \geq 0\right)$ where $(B_u, u \geq 0)$ is a standard Brownian motion started at 0. $(\theta_u, u \geq 0)$ satisfies the scaling property of order 0. Thus $\left(X_t := \int_0^t 1_{\{B_u > 0\}} du - \frac{t}{2}, t \geq 0\right)$ is a peacock.

a) Prove that $X_t \overset{(1.d)}{=} t\left(g_1 - \frac{1}{2}\right)$ (compare with Proposition 1.4) where g_1 is arcsine distributed, i.e.:

$$P(g_1 \in du) = \frac{1}{\pi \sqrt{u(1-u)}} du \qquad (0 \leq u \leq 1).$$

b) Let $G_t := \sup\{s \leq t; B_s = 0\}$. We recall that

$$\left(M_1(t) = \mathrm{sgn}(B_t) \sqrt{t - G_t}, t \geq 0 \right)$$

and

$$\left(M_2(t) := t - G_t - \frac{t}{2} = (M_1(t))^2 - \langle M_1, M_1 \rangle_t,\ t \geq 0\right)$$

are two martingales (M_1 is Azéma's martingale, see [9]). Find a martingale M such that $X_t \overset{(1.d)}{=} M_t$. (*Hint*: $G_t \overset{(law)}{=} t g_1$).

6) *Example 3* $\left(\theta_u^{(\alpha)} := 1_{\{B_u^{(\alpha)} > 0\}} - \frac{1}{2}, u \geq 0\right)$ where $(B_t^{(\alpha)}, t \geq 0)$ is a symmetric Bessel process of dimension $\delta = 2(1-\alpha)$, with $0 < \delta < 2$, or $0 < \alpha < 1$, started at 0 (see Barlow, Pitman, Yor [14] for a precise definition of this process).

Prove that $(\theta_u^{(\alpha)}, u \geq 0)$ satisfies the scaling property of order 0. Then $\left(X_t^{(\alpha)} := \int_0^t \left(1_{\{B_u^{(\alpha)} > 0\}} - \frac{1}{2}\right) du, t \geq 0\right)$ is a peacock. From [14], we have:

$$X_t^{(\alpha)} \overset{(1.d)}{=} t\left(\frac{T_\alpha}{T_\alpha + T'_\alpha} - \frac{1}{2}\right)$$

where T_α and T'_α are two independent copies of a positive (unilateral) stable r.v. of index α.

Exercise 1.37 (Convex order of martingales and local times). Let $(M_t, t \geq 0)$ and $(\widetilde{M}_t, t \geq 0)$ be two continuous martingales such that $M_0 \overset{(law)}{=} \widetilde{M}_0$.
Prove that

$$M_t \overset{(c)}{\leq} \widetilde{M}_t \iff \forall a \in \mathbb{R},\ \mathbb{E}[L_t^a] \leq \mathbb{E}[\widetilde{L}_t^a]$$

where $(L_t^a;\ a \in \mathbb{R}, t \geq 0)$ (resp. $(\widetilde{L}_t^a;\ a \in \mathbb{R}, t \geq 0)$) denotes the bicontinuous family of local times of M (resp. $\widetilde{M}$).

Exercise 1.38 ($\mathbf{F}_1$ and $\mathbf{F}_2$-type peacocks built on the Poisson and gamma processes). Let us consider (see e.g. Schoutens [156]) the following generating function:

$$(1+\lambda)^u \exp(-\lambda a) = \sum_{n=0}^{\infty} \frac{\lambda^n}{n!} C_n(u,a) \qquad u, a, \lambda \geq 0$$

where $(C_n(u,a), n \geq 0)$ denotes the sequence of Charlier polynomials (in both variables u and a).

1) Let $(N_t, t \geq 0)$ be a standard Poisson process (with parameter 1).
a) Show that, for every $\lambda \geq 0$:

$$\left((1+\lambda)^{N_t} \exp(-\lambda t), t \geq 0\right)$$

is a martingale.
b) Deduce that, for every $n \geq 1$:

$$\left(\frac{1}{t}\int_0^t C_n(N_s, s) ds, t \geq 0\right), \quad \text{resp.} \quad \left(\int_0^t C_n(N_s, s) ds, t \geq 0\right),$$

is a $\mathbf{F}_1$-type peacock, resp. a $\mathbf{F}_2$-type peacock.

2) Let $(\gamma_t, t \geq 0)$ be the standard gamma subordinator, i.e. the subordinator such that:

$$\mathbb{E}\left[e^{-\lambda \gamma_t}\right] = \exp\left(-t \int_0^{\infty} (1 - e^{-\lambda x}) \frac{dx}{x} e^{-x}\right) \quad (\lambda \geq 0).$$

Prove that γ_t is a gamma r.v. with parameter t.

a) Show that, for every $\lambda \geq 0$:

$$((1+\lambda)^t \exp(-\lambda \gamma_t), t \geq 0)$$

is a martingale.

b) Deduce that, for every $n \geq 1$:

$$\left(\frac{1}{t}\int_0^t C_n(s, \gamma_s) ds, t \geq 0\right), \quad \text{resp.} \left(\int_0^t C_n(s, \gamma_s) ds, t \geq 0\right),$$

is a $\mathbf{F}_1$-type peacock, resp. a $\mathbf{F}_2$-type peacock.

1.2.3 Peacocks with Respect to Volatility

Theorem 1.4 shows that if $(M_t, t \geq 0)$ is a martingale (resp. $(\widetilde{M}_t, t \geq 0)$ is a martingale such that $\widetilde{M}_0 = 0$), then:

$$\left(X_t := \frac{1}{t}\int_0^t M_s ds\,,\ t \geq 0\right) \quad \text{is a peacock} \tag{1.53}$$

$$\text{resp.} \left(\widetilde{X}_t := \int_0^t \widetilde{M}_s ds\,,\ t \geq 0\right) \quad \text{is a peacock.} \tag{1.54}$$

One of the arguments used – and required – in order to prove (1.53) (resp. (1.54)) is that $\mathbb{E}[M_s]$ does not depend on s (resp. $\mathbb{E}[\widetilde{M}_s] = 0$).

We now look for replacing in (1.53) (resp. (1.54)) the martingale $(M_s, s \geq 0)$ by a process $(\Pi_s, s \geq 0)$ (resp. the martingale $(\widetilde{M}_s, s \geq 0)$ by a process $(\widetilde{\Pi}_s, s \geq 0)$) such that:

$$\left(X_t^{\Pi} := \frac{1}{t}\int_0^t \Pi_s ds\,,\ t \geq 0\right) \quad \text{is still a peacock}$$

$$\text{resp.} \left(\widetilde{X}_t^{\widetilde{\Pi}} := \int_0^t \widetilde{\Pi}_s ds\,,\ t \geq 0\right) \quad \text{is still a peacock.}$$

Inspired by the situation of the Brownian exponential martingale:

$$\begin{aligned} M_s &= \exp\left(\nu B_s - \frac{\nu^2}{2}s\right) \overset{(1.d)}{=} \exp\left(\nu\sqrt{s}G - \frac{\nu^2}{2}s\right) \\ &\overset{(1.d)}{=} \frac{\exp(\nu\sqrt{s}G)}{\mathbb{E}[e^{\nu\sqrt{s}G}]} \end{aligned}$$

(where G is a reduced gaussian r.v.) and:

$$\widetilde{M}_s = \exp\left(\nu B_s - \frac{\nu^2}{2}s\right) - 1 \overset{(1.d)}{=} \frac{\exp(\nu\sqrt{s}G)}{\mathbb{E}\left[e^{\nu\sqrt{s}G}\right]} - 1,$$

let us choose:

$$\Pi_s = \frac{\exp sX}{\mathbb{E}\,[\exp sX]} \quad \text{and} \quad \widetilde{\Pi}_s = \frac{\exp sX}{\mathbb{E}[\exp sX]} - 1$$

where X is an r.v. admitting positive exponential moments. Thus, one has:

$$X_t^{\Pi} := \frac{1}{t}\int_0^t \frac{\exp sX}{\mathbb{E}[\exp sX]}ds = \int_0^1 \frac{\exp tuX}{\mathbb{E}[\exp tuX]}du \tag{1.55}$$

(after the change of variable $s = tu$) and

$$\widetilde{X}_t^{\widetilde{\Pi}} := \int_0^t \left(\frac{\exp sX}{\mathbb{E}[\exp sX]} - 1\right) ds. \tag{1.56}$$

Note that, in the above formulae, the parameters s and t appear as dilation (or volatility) parameters rather than time parameters. On the other hand, it seems natural to replace in (1.55) (resp. (1.56)) the integral with respect to du (resp. ds) by an integral with respect to a positive measure of finite total mass. Hence, there is the following theorem, which yields examples of peacocks that we call “peacocks with respect to volatility”.

Theorem 1.7. *Let X be a random variable such that, for every $s \geq 0$,*

$$\mathbb{E}[e^{sX}] < \infty. \tag{1.57}$$

Let σ be a positive and finite measure on $\mathbb{R}_+$. Then:

1) $$\left(X_t := \int_0^\infty \frac{\exp tuX}{\mathbb{E}[\exp tuX]}\sigma(du), t \geq 0\right) \text{ is a peacock.} \tag{1.58}$$

2) $$\left(\widetilde{X}_t := \int_0^t \left(\frac{\exp uX}{\mathbb{E}[\exp uX]} - 1\right)\sigma(du), t \geq 0\right) \text{ is a peacock} \tag{1.59}$$

(we denote here, and in the sequel, $\int_0^t \ldots \sigma(du)$ for $\int 1_{[0,t]}(u) \ldots \sigma(du)$).

Proof. *We first prove Point 1).* *i)* Let $0 < a < b$. The following elementary inequalities hold:

$$\text{For every } t \in [0,b], \quad \exp(tX) \leq 1 + \exp(bX) \tag{1.60}$$

$$\text{For every } t \in [a,b], \quad |X|\exp(tX) \leq \frac{1}{ea} + \exp((b+1)X). \tag{1.61}$$

As a consequence, the function: $t \to \mathbb{E}\left[e^{tX}\right]$ is continuous on $[0,+\infty[$ and derivable on $]0,+\infty[$.

ii) Now suppose that the measure σ has a compact support, which is contained in $]0,+\infty[$. Then, by the above step *i*), the process $(X_t, t \geq 0)$ satisfies properties *i*) and *ii*) of Proposition 1.3. We now prove that $(X_t, t \geq 0)$ also satisfies properties *a*) and *b*) of this Proposition 1.3. Property *a*) is clear since:

$$\begin{aligned} \mathbb{E}[X_t] &= \mathbb{E}\left[\int_0^\infty \frac{\exp tuX}{\mathbb{E}\left[\exp tuX\right]}\sigma(du)\right] \\ &= \int_0^\infty \mathbb{E}\left[\frac{\exp tuX}{\mathbb{E}\left[\exp tuX\right]}\right]\sigma(du) \quad \text{(by Fubini)} \\ &= \int_0^\infty \sigma(du) \quad (< \infty \text{ by hypothesis}). \end{aligned}$$

We now prove property *b*), i.e.: for every c real,

$$\mathbb{E}\left[1_{\{X_t \geq c\}} \frac{\partial X_t}{\partial t}\right] \geq 0. \tag{1.62}$$

For $r \geq 0$, we denote by $\mathbb{E}^{(r)}$ the expectation with respect to the probability measure:

$$\mathbb{P}^{(r)} := \frac{\exp rX}{\mathbb{E}\left[\exp rX\right]} \cdot \mathbb{P}. \tag{1.63}$$

We define the function $h_X : \mathbb{R}_+ \to \mathbb{R}$ by

$$h_X(r) := \log\left(\mathbb{E}[\exp(rX)]\right) \tag{1.64}$$

and we have:

$$\begin{aligned} 0 &= \frac{\partial}{\partial r}\mathbb{E}\left[\frac{\exp rX}{\mathbb{E}[\exp rX]}\right] = \frac{\partial}{\partial r}\mathbb{E}[\exp(rX - h_X(r))] \\ &= \mathbb{E}[(X - h'_X(r))\exp(rX - h_X(r))] \\ &= \mathbb{E}^{(r)}[X - h'_X(r)]. \end{aligned} \tag{1.65}$$

Hence, from (1.58), (1.63) and (1.64):

$$\begin{aligned} \mathbb{E}\left[1_{\{X_t \geq c\}} \frac{\partial X_t}{\partial t}\right] &= \mathbb{E}\left[1_{\{X_t \geq c\}} \int_0^\infty u(X - h'_X(tu))e^{utX - h_X(tu)}\sigma(du)\right] \\ &= \int_0^\infty u\mathbb{E}^{(tu)}\left[1_{\{X_t \geq c\}}(X - h'_X(ut))\right]\sigma(du) \\ &\quad \text{(by Fubini).} \end{aligned} \tag{1.66}$$

We set:

$$\Gamma_t(x) := \int_0^\infty \exp(tux - h_X(tu))\sigma(du). \tag{1.67}$$

The function Γ_t is clearly increasing. Hence, for t and u fixed, for every c real, there exists $\widetilde{c} \equiv \widetilde{c}(t,u)$ such that:

$$\{X_t \geq c\} = \{X - h'_X(ut) \geq \widetilde{c}\}. \tag{1.68}$$

Since, from (1.65), $\mathbb{E}^{(tu)}[X - h'_X(ut)] = 0$, Lemma 1.1 implies:

$$\mathbb{E}^{(tu)}[1_{\{X_t \geq c\}}(X - h'_X(ut))] = \mathbb{E}^{(tu)}[1_{\{X - h'_X(ut) \geq \widetilde{c}\}}(X - h'_X(ut))] \geq 0. \tag{1.69}$$

From (1.66) and (1.69), we obtain:

$$\mathbb{E}\left[1_{\{X_t \geq c\}} \frac{\partial X_t}{\partial t}\right] \geq 0\,,$$

i.e. hypothesis $b)$ of Proposition 1.3.

$iii)$ We now drop the hypothesis that supp σ is compact. Let, for every $n \in \mathbb{N}^*$:

$$\begin{aligned} \sigma_n(du) &= 1_{[1/n,n]}(u)\sigma(du) + \sigma(\{0\})\delta_{1/n} \\ \text{and} \quad & \left(X_t^{(n)} := \int_0^\infty \frac{\exp(tuX)}{\mathbb{E}[\exp tuX]}\sigma_n(du), t \geq 0\right). \end{aligned}$$

From the previous Point $ii)$ $(X_t^{(n)}, t \geq 0)$ is a peacock and we conclude by applying Point $1)$ of Exercise 1.10.

We now prove Point 2) of Theorem 1.7. By density, we may assume that σ is absolutely continuous with respect to the Lebesgue measure and admits a continuous Radon-Nikodym density:

$$\sigma(dt) = \sigma(t)dt. \tag{1.70}$$

Since:

$$\frac{\partial \widetilde{X}_t}{\partial t} = \{\exp(tX - h_X(t)) - 1\}\sigma(t) \tag{1.71}$$

with, as previously, $h_X(t) := \log \mathbb{E}[\exp(tX)]$, it suffices, from Proposition 1.3, to prove: for every c real

$$\mathbb{E}\left[1_{\{\widetilde{X}_t \geq c\}}(e^{tX - h_X(t)} - 1)\right] \geq 0. \tag{1.72}$$

But $x \longrightarrow \int_0^t (\exp(ux - h_X(u)) - 1)\,\sigma(du)$ is an increasing function. Thus, for every c real and t fixed, there exists $\widetilde{c}(t)$ and $\widetilde{\widetilde{c}}(t)$ such that:

$$\{\widetilde{X}_t \geq c\} = \{X \geq \widetilde{c}(t)\} = \{e^{tX - h_X(t)} - 1 \geq \widetilde{\widetilde{c}}(t)\}. \tag{1.73}$$

Hence:

$$\mathbb{E}\left[1_{\{\widetilde{X}_t \geq c\}}(e^{tX - h_X(t)} - 1)\right] = \mathbb{E}\left[1_{\{e^{tX - h_X(t)} - 1 \geq \widetilde{\widetilde{c}}(t)\}}(e^{tX - h_X(t)} - 1)\right] \geq 0$$

from Lemma 1.1. □

Corollary 1.2. *Let X be a r.v. such that for every $t \geq 0$, $\mathbb{E}\left[e^{tX}\right] < \infty$. Then:*

$$\left(\frac{e^{tX}}{\mathbb{E}\left[e^{tX}\right]}, t \geq 0\right) \tag{1.74}$$

is a peacock.

This Corollary is obtained by taking $\sigma(du) = \delta_1(du)$, the Dirac measure at 1, in (1.58).

Remark 1.11. i) From Point 1) of Theorem 1.7, if σ is a positive and finite measure on $\mathbb{R}_+$, $\left(X_t := \int_0^\infty \frac{e^{tuX}}{\mathbb{E}\left[e^{tuX}\right]}\,\sigma(du), t \geq 0\right)$ is a peacock. The following example shows that this may not be true if σ is a *signed* measure. Indeed, let X be a symmetric Bernoulli random variable:

$$P(X = 1) = P(X = -1) = 1/2 \tag{1.75}$$

and $\sigma(du) = \delta_1(du) - a\delta_2(du)$ with $a > 1/2$.

Then:

$$X_t = \frac{e^{tX}}{\cosh t} - a\frac{e^{2tX}}{\cosh 2t}$$

and $t \longrightarrow \mathbb{E}[X_t^2]$ is not increasing. Indeed:

$$\frac{\partial}{\partial t}\mathbb{E}[X_t^2] = 2\sinh(t)\left[4a^2\frac{\cosh t}{(\cosh 2t)^3} - 4a\frac{\cosh t}{(\cosh 2t)^2} + \frac{1}{(\cosh 2t)^3}\right].$$

For t such that $\cosh 2t = 2a$, we obtain:

$$\frac{\partial}{\partial t}\mathbb{E}\left[X_t^2\right] = -\frac{(2a-1)^{3/2}}{2a(2a+1)^{3/2}} < 0.$$

ii) On the contrary, we shall prove, in Chapter 5 Theorem 5.5, that if X is *self-decomposable* and σ is a finite *signed* measure, then

$$\left(X_t = \int_0^\infty \frac{e^{tuX}}{\mathbb{E}\left[e^{tuX}\right]}\sigma(du), t \geq 0\right) \quad \text{is still a peacock.}$$

Then, a natural question is: in (1.74), may the process $(uX, u \geq 0)$ be replaced by a "general process" $(X_u, u \geq 0)$? Here is an answer to this question, which gives another family of peacocks "with respect to volatility". In Chapter 5 Theorem 5.5, another motivation, related to the notion of self-decomposability, leading to Theorem 1.8 may be found.

Theorem 1.8. *Let $(X_u, u \geq 0)$ be a real valued, right continuous process with independent increments[1] (not necessarily stationary). We assume that $X_0 = 0$ and for every $t \geq 0$, for every $u \geq 0$,*

$$\mathbb{E}[e^{tX_u}] < \infty. \tag{1.76}$$

[1] In the literature, such a process is often called an additive process.

We set:

$$h_{X_u}(t) = \log(\mathbb{E}[\exp(tX_u)]) \tag{1.77}$$

Let σ be a signed *and finite measure on $\mathbb{R}_+$. Then:*

$$\left(Y_t := \int_0^\infty \exp(tX_u - h_{X_u}(t))\sigma(du), t \geq 0\right)$$

is a peacock.

Proof. 1) We begin the proof with some preliminaries.
a) The hypotheses on X ensure that, for every $t \geq 0$, the process $(\exp(tX_u - h_{X_u}(t)), u \geq 0)$ is a martingale with respect to a filtration satisfying the usual hypotheses. Consequently, this process admits a right-continuous version. This easily implies that the function

$$u \longrightarrow h_{X_u}(t)$$

defined on $\mathbb{R}_+$ is right-continuous, and therefore, for every $t \geq 0$, the process $(\exp(tX_u - h_{X_u}(t)), u \geq 0)$ is a right-continuous martingale.
b) The above martingale is clearly in L^p for every $p \in]1, +\infty[$. Therefore, Doob's L^p inequality entails: for every $t \geq 0$, for every $a > 0$,

$$\mathbb{E}\left[\sup_{u \in [0,a]} \exp(tX_u - h_{X_u}(t))\right] < \infty.$$

c) By the independence of the increments of X_u, if $0 \leq u \leq v$ and $t \geq 0$, then:

$$\mathbb{E}[\exp t(X_v - X_u)] = \exp(h_{X_v}(t) - h_{X_u}(t)).$$

By Corollary 1.2:

$$\left(\exp\left(t(X_v - X_u) - (h_{X_v}(t) - h_{X_u}(t))\right), t \geq 0\right) \tag{1.78}$$

is a peacock.

2) We shall now prove by induction, for $k \geq 1$, the following property (P_k): for every $0 \leq u_1 \leq u_2 ... \leq u_k$ and $d_1, d_2, ...d_k \in \mathbb{R}$, the process $\left(\sum_{j=1}^{k} d_j \exp\left(tX_{u_j} - h_{X_{u_j}}(t)\right), t \geq 0\right)$ is a peacock.

Property (P_1) is a direct consequence of Corollary 1.2. Suppose now that (P_k) is satisfied. Then:

$$\begin{aligned}
&\sum_{j=1}^{k+1} d_j \exp(tX_{u_j} - h_{X_{u_j}}(t)) \\
&\qquad = \sum_{j=1}^{k} d_j \exp(tX_{u_j} - h_{X_{u_j}}(t)) + d_{k+1} \exp\left(tX_{u_k} - h_{X_{u_k}}(t)\right) \\
&\qquad\qquad \times \left(\exp\left(t(X_{u_{k+1}} - X_{u_k}) - (h_{X_{u_{k+1}}}(t) - h_{X_{u_k}}(t))\right)\right).
\end{aligned} \tag{1.79}$$

We define, for $t,s \geq 0$:

$$Y_{t,s} := \sum_{j=1}^{k} d_j \exp(tX_{u_j} - h_{X_{u_j}}(t)) + d_{k+1} \exp(tX_{u_k} - h_{X_{u_k}}(t)) \times \left(\exp\left(s(X_{u_{k+1}} - X_{u_k}) - (h_{X_{u_{k+1}}}(s) - h_{X_{u_k}}(s))\right)\right). \tag{1.80}$$

Now let $\psi \in \boldsymbol{C}$. Using the independence of the increments of $(X_u, u \geq 0)$, we obtain that $\mathbb{E}[\psi(Y_{t,s})]$ is separately increasing with respect:

- to t (from (P_k)),
- to s (from the above property (1.78)).

Consequently, the function $t \longmapsto \mathbb{E}[\psi(Y_{t,t})]$ is increasing, which is the property (P_{k+1}).

3) Approximating weakly the measure σ by a sequence of linear combinations of Dirac measures, we deduce from Step 2, using $a)$ and $b)$ of Step 1, that, if the support of σ is compact then:

$$\left(\int_0^\infty \exp(tX_u - h_{X_u}(t))\sigma(du), t \geq 0\right)$$

is a peacock. Then, the general result follows by approximating σ by measures with compact support, as in Step *iii*) of the proof of Theorem 1.7. □

Another approach to Point 2) of the previous proof. With the notations of the previous proof, we shall prove by induction for $k \geq 1$ the following property (P'_k): for every $0 \leq u_1 \leq ... \leq u_k$ and $d_1, d_2, ... d_k \in \mathbb{R}$ the process:

$$\left(\exp(tX_{u_1} - h_{X_{u_1}}(t)), \exp(tX_{u_2} - h_{X_{u_2}}(t)), ..., \exp(tX_{u_k} - h_{X_{u_k}}(t)), t \geq 0\right)$$

is a $\mathbb{R}^k$-valued 1-martingale.

Of course, as a consequence of Jensen's inequality and since a linear combination of martingales is a martingale, (P'_k) implies (P_k).

By Kellerer's Theorem 1.1, (P'_1) follows from Corollary 1.2. Let us denote:

$$M_t^{(u)} := \exp(tX_u - h_{X_u}(t)).$$

Assume (P'_k) is satisfied. We have:

$$(M_t^{(u_1)}, ... M_t^{(u_k)}, M_t^{(u_{k+1})}) = (M_t^{(u_1)}, ... M_t^{(u_k)}, M_t^{(u_k)} M_t^{(u_k,u_{k+1})}) \tag{1.81}$$

with

$$M_t^{(u_k,u_{k+1})} := \exp\left(t(X_{u_{k+1}} - X_{u_k}) - (h_{X_{u_{k+1}}}(t) - h_{X_{u_k}}(t))\right).$$

Using again Kellerer's theorem, we deduce from (1.78) that, $(M_t^{(u_k,u_{k+1})}, t \geq 0)$ is a 1-martingale. Hence, $(M_t^{(u_k)} M_t^{(u_k,u_{k+1})}, t \geq 0)$ is still a 1-martingale, thanks to the independence of increments of $(X_u, u \geq 0)$ and the fact that the product of two independent martingales – in L^p for every $p \geq 1$ – is still a martingale. We have obtained (P'_{k+1}). □

In the next Section 1.3, we shall prove, in a Gaussian setting, an analogue of Theorem 1.8, without making independence hypotheses for the increments.

Exercise 1.39 (Some double integrals of exponentials of additive processes are peacocks). Let $(X_u, u \geq 0)$ be a real-valued, right-continuous process with independent increments (not necessarily stationary). We assume that $X_0 = 0$ and for every $t \geq 0$, $\mathbb{E}\left[e^{tX_u}\right] < \infty$. We set: $h_{X_u}(t) := \log(\mathbb{E}[\exp(tX_u)])$. Let μ be a signed finite measure on $\mathbb{R}_+ \times \mathbb{R}_+$. Show that:

$$\left(\int_0^\infty \int_0^\infty \exp(tvX_u - h_{X_u}(tv))\mu(dv, du),\ t \geq 0\right)$$

is a peacock. (*Hint*: it suffices to prove that, for every k,

$$\left(\sum_{j=1}^{k} \int_0^\infty \sigma_j(dv) \exp(tvX_{u_j} - h_{X_{u_j}}(tv)),\ t \geq 0\right)$$

is a peacock, with $u_1 < \ldots < u_k$ and $\sigma_j, j = 1, \ldots, k$ signed finite measures on $\mathbb{R}_+$; then operate as in the proof of Theorem 1.8, Point 2).)

Exercise 1.40 (A martingale associated to $\left(\frac{e^{a\varepsilon}}{\mathbb{E}[e^{a\varepsilon}]}, a \geq 0\right)$). Let ε be a symmetric Bernoulli r.v. (i.e. $\mathbb{P}(\varepsilon = 1) = \mathbb{P}(\varepsilon = -1) = \frac{1}{2}$) and consider the peacock $\left(\dfrac{e^{a\varepsilon}}{\mathbb{E}\left[e^{a\varepsilon}\right]}, a \geq 0\right)$.
1) Prove that:

$$\frac{e^{a\varepsilon}}{\mathbb{E}\left[e^{a\varepsilon}\right]} = 1 + \varepsilon \tanh a.$$

2) Let $(B_t, t \geq 0)$ be a Brownian motion started from 1 and $T_a := \inf\{u \geq 0,\ |B_u - 1| = \tanh a\}$. Prove that $(B_{T_a}, a \geq 0)$ is a martingale associated to this peacock (the above solution was communicated to us by J. Bertoin).

1.3 Peacocks in a Gaussian Setting

In this section, we shall particularize our framework to the Gaussian set-up, and exhibit some peacocks involving Gaussian processes (cf. Subsections 1.3.5 to 1.3.7). To this end, we shall need some general results about Gaussian processes, in particular, a weak form of Itô's formula, see Subsections 1.3.1 to 1.3.4.

1.3.1 A Weak Itô Formula

In this subsection, we consider a family of $\mathbb{R}^n$-valued centered Gaussian variables:

$$(G_t, t \in [a, b])$$

where $[a,b]$ denotes a compact interval of $\mathbb{R}$. We denote, for $t \in [a,b]$, by $(G_t^{(1)},\cdots,G_t^{(n)})$ the components of the vector G_t, and by

$$C(t) = \left(c_{j,k}(t)\right)_{1\leq j,k\leq n}$$

the matrix of G_t.

We assume that, for all $1 \leq j,k \leq n$,

$$t \in [a,b] \longrightarrow c_{j,k}(t) \in \mathbb{R}$$

is a continuous function with finite variation.

Note that $C(t) \in \mathbb{S}_n$, the space of $n \times n$ symmetric matrices with real entries. We also define $\mathbb{S}_n^+$ to be the convex cone in $\mathbb{S}_n$ consisting of positive matrices. Thus, a matrix $M = (m_{j,k})_{1\leq j,k\leq n}$ belongs to $\mathbb{S}_n^+$ if M belongs to $\mathbb{S}_n$ and

$$\forall a_1,\cdots,a_n \in \mathbb{R} \qquad \sum_{1\leq j,k\leq n} a_j a_k m_{j,k} \geq 0\,.$$

In the sequel, $\mathbb{S}_n$ is assumed to be equipped with the following partial order, induced by the convex cone $\mathbb{S}_n^+$:

$$\forall M,N \in \mathbb{S}_n \qquad M \leq N \Longleftrightarrow (N-M) \in \mathbb{S}_n^+\,.$$

The main result of this subsection is the following weak form of Itô formula.

Theorem 1.9 (Weak Itô formula). *Let*

$$F : (t,x) \in [a,b] \times \mathbb{R}^n \longrightarrow F(t,x) \in \mathbb{R}$$

be a $\mathscr{C}^{1,2}$-function whose derivatives of order 2 with respect to x: F''_{x_j,x_k}, $1 \leq j,k \leq n$, and whose derivative of order 1 with respect to t: F'_t, grow sub-exponentially at infinity with respect to x, uniformly with respect to $t \in [a,b]$. Then, for every s,t with $a \leq s \leq t \leq b$,

$$\begin{aligned}\mathbb{E}[F(t,G_t)] &= \mathbb{E}[F(s,G_s)] + \int_s^t \mathbb{E}[F'_t(u,G_u)]du \\ &\quad + \frac{1}{2}\sum_{j,k}\int_s^t \mathbb{E}[F''_{x_j,x_k}(u,G_u)]dc_{j,k}(u)\,.\end{aligned} \tag{1.82}$$

Proof. The proof is based on the following lemma:

Lemma 1.2. *Denote by $\mu_t(\mathrm{d}x)$ the law of the Gaussian variable G_t. Then, for every s,t with $a \leq s \leq t \leq b$, there is the following identity, in the sense of Schwartz distributions:*

$$\mu_t = \mu_s + \frac{1}{2}\sum_{j,k}\int_s^t (\mu_u)''_{x_j,x_k} dc_{j,k}(u)\,. \tag{1.83}$$

□

Proof. Let $\gamma = (\gamma_1, \cdots, \gamma_n) \in \mathbb{R}^n$, and denote by $< x, y >$ the scalar product of $x, y \in \mathbb{R}^n$. Set, for $t \in [a,b]$,

$$\varphi_\gamma(t) = \int_{\mathbb{R}^n} \mathrm{e}^{\mathrm{i} <\gamma,x>} \, \mu_t(\mathrm{d}x) \, .$$

We have, for $t \in [a,b]$,

$$\varphi_\gamma(t) = \mathbb{E}[\exp(\mathrm{i} < \gamma, G_t >)] = \exp\left(-\frac{1}{2} \sum_{j,k} \gamma_j \, \gamma_k \, c_{j,k}(t) \right) .$$

Then, for $a \leq s \leq t \leq b$,

$$\varphi_\gamma(t) = \varphi_\gamma(s) - \frac{1}{2} \sum_{j,k} \int_s^t \varphi_\gamma(u) \, \gamma_j \, \gamma_k d c_{j,k}(u) \, .$$

Now,

$$\varphi_\gamma(u) \, \gamma_j \, \gamma_k = \int_{\mathbb{R}^n} \mathrm{e}^{\mathrm{i} <\gamma,x>} \, \gamma_j \, \gamma_k \, \mu_u(\mathrm{d}x) = - \int_{\mathbb{R}^n} \frac{\partial^2}{\partial x_j \partial x_k} \mathrm{e}^{\mathrm{i} <\gamma,x>} \, \mu_u(\mathrm{d}x) \, .$$

Finally, we obtain that the Fourier transforms, in the sense of distributions, of both sides in (1.83) are equal, hence the desired result follows thanks to the injectivity of the Fourier transform. □

Proof of Theorem 1.9. We still denote the law of G_t by $\mu_t(\mathrm{d}x)$. We first suppose that

$$F(t,x) = g(t) \, h(x)$$

with $h \in \mathscr{C}^2(\mathbb{R}^n)$ and $g \in \mathscr{C}^1([a,b])$. We also assume that h has compact support. Then for $t \in [a,b]$,

$$\mathbb{E}[F(t, G_t)] = g(t) \int h(x) \, \mu_t(\mathrm{d}x) \, .$$

Lemma 1.2 ensures that, for $a \leq s \leq t \leq b$,

$$\begin{aligned} \mathbb{E}[F(t,G_t)] \; &= \; \mathbb{E}[F(s,G_s)] + \int_s^t g'(u) \left(\int h(x) \, \mu_u(\mathrm{d}x) \right) \mathrm{d}u \qquad (1.84) \\ &+ \frac{1}{2} \sum_{j,k} \int_s^t g(u) \left(\int h''_{x_j,x_k}(x) \, \mu_u(\mathrm{d}x) \right) \mathrm{d}c_{j,k}(u) \, . \end{aligned}$$

Clearly, equality (1.84) yields formula (1.82) in this case.

Finally, if F satisfies the hypotheses of Theorem 1.9, the result follows easily by approximating F by linear combinations of functions of the previous type. □

Remark 1.12.

• In formula (1.82), only the law of G_t for *each* $t \in [a,b]$, and consequently only the matrices $C(t)$, are involved. This explains why this formula is well adapted to the study of peacocks.

• Suppose that

$$t \in [a,b] \longrightarrow C(t) \in \mathbb{S}_n$$

is an absolutely continuous function on $[a,b]$ and that the derivative $C'(t)$ is, for almost every $t \in [a,b]$, a positive symmetric matrix. Then, there exists a measurable function

$$t \in [a,b] \longrightarrow D(t) \in \mathbb{S}_n^+$$

such that, for almost every $t \in [a,b]$,

$$[D(t)]^2 = C'(t)\,.$$

In particular,

$$t \in [a,b] \longrightarrow D(t) \in \mathbb{S}_n$$

is a square integrable function. We set

$$M_t = G_a + \int_a^t D(s)dB_s$$

where B denotes a standard $\mathbb{R}^n$-valued Brownian motion starting from 0, independent of G_a. Then $(M_t, t \in [a,b])$ is a continuous Gaussian martingale, and, for any $t \in [a,b]$, the covariance matrix of M_t is $C(t)$. Consequently, to prove (1.82) in this case, we may replace $(G_t, t \in [a,b])$ by $(M_t, t \in [a,b])$, but then (1.82) is a direct consequence of the classical Itô formula.

• Theorem 1.9 states a weak Itô formula. There exist many "strong" Itô formulae for Gaussian processes in the literature, which take the form:

$$F(t,G_t) = F(s,G_s) + \int_s^t F_t'(u,G_u)du + \frac{1}{2}\int_s^t F_{x^2}''(u,G_u)dc(u) + \int_s^t F_x'(u,G_u)dG_u.$$

In this formula, it is necessary to give an appropriate definition of the last integral. We refer the reader to Biagini-Hu-Oksendal-Zhang [27], Mishura [132], Bender [20], Pipiras-Taqqu [141], Carmona-Coutin-Montseny [42], Russo-Vallois [154], Alòs-Mazet-Nualart [4].

1.3.2 Examples

In this subsection, we present some examples of application of our Itô-like formula in the scalar case $n = 1$. In particular, when $(G_t, t \in [a,b])$ has the same one-marginals as a semimartingale, we compare our formula (1.82) with the one obtained by application of the classical Itô formula.

1.3.2.1 Time Changed Brownian Motion

We consider continuous functions u and v from an interval $[a,b]$ into $\mathbb{R}$ and we suppose that u has a finite variation and v is increasing and nonnegative. Let $(B_t, t \geq 0)$ be a standard linear Brownian motion starting from 0 and consider the process:

$$G_t = u(t)\,B_{v(t)} \quad , \quad t \in [a,b] \,.$$

Recall this is the general form of a Gaussian Markovian process (see, for instance, [133]).

We have $c(t) = \mathbb{E}[G_t^2] = u^2(t)\,v(t)$. Let F be a $\mathscr{C}^2$-function on $\mathbb{R}$ with compact support. Our formula (1.82) yields:

$$\mathrm{d}_t\mathbb{E}[F(G_t)] = \frac{1}{2}\,\mathbb{E}[F''(G_t)]\,\big(2u(t)v(t)\,\mathrm{d}u(t) + u^2(t)\,\mathrm{d}v(t)\big)\,,$$

whereas the application of the classical Itô formula gives:

$$\mathrm{d}_t\mathbb{E}[F(G_t)] = \mathbb{E}[F'(G_t)\,B_{v(t)}]\,\mathrm{d}u(t) + \frac{1}{2}\,\mathbb{E}[F''(G_t)]\,u^2(t)\,\mathrm{d}v(t)\,.$$

Consequently, we obtain:

$$\mathbb{E}[F''(u(t)\,B_{v(t)})]\,u(t)\,v(t) = \mathbb{E}[F'(u(t)\,B_{v(t)})\,B_{v(t)}]\,.$$

The above equality may be written down as:

$$\mathbb{E}[F''(\alpha B_1)]\,\alpha = \mathbb{E}[F'(\alpha B_1)\,B_1] \tag{1.85}$$

with $\alpha = u(t)\,\sqrt{v(t)}$. Obviously, α may be taken equal to 1, and (1.85) is a well-known characterization of the law of B_1. In fact, formula (1.85) with g instead of F', is the starting point of Stein's method (see [162]); see also [165, Appendix A.6: Gaussian r.v.] and [166, Appendix A4: Gaussian integration by parts].

The following examples 1.3.2.2 and 1.3.2.3 may be considered as particular cases of the previous example 1.3.2.1.

1.3.2.2 Ornstein-Uhlenbeck Process

Besides Brownian motion, the most famous example of a Gaussian process is the Ornstein-Uhlenbeck process. For $\lambda \in \mathbb{R}$, let $(U_t^\lambda, t \geq 0)$ be the scalar Ornstein-Uhlenbeck process with parameter λ, starting from 0. Thus, U^λ is solution to the SDE:

$$U_t^\lambda = B_t + \lambda \int_0^t U_s^\lambda\,ds\,. \tag{1.86}$$

This SDE solves explicitly, as:

$$\forall t \geq 0, \qquad U_t^\lambda = \mathrm{e}^{\lambda t}\int_0^t \mathrm{e}^{-\lambda s}\,dB_s\,.$$

Consequently, there exists a Brownian motion (β_u) such that:

$$\forall t \geq 0 \qquad U_t^{\lambda} = \mathrm{e}^{\lambda t}\, \beta_{v(t)}$$

with

$$v(t) = \int_0^t \mathrm{e}^{-2\lambda s} ds\,.$$

We are therefore brought back to Example 1.3.2.1.

On the other hand, denoting by $\mu^{(\lambda)}(t,x)$ the density of $U_t^{(\lambda)}$, the Fokker-Planck equation corresponding to the SDE (1.86) yields:

$$\frac{\partial \mu^{(\lambda)}}{\partial t}(t,x) = -\lambda\, \frac{\partial}{\partial x}(x\,\mu^{(\lambda)}(t,x)) + \frac{1}{2}\,\frac{\partial^2 \mu^{(\lambda)}}{\partial x^2}(t,x)\,,$$

whereas our Lemma 1.2 leads to:

$$\frac{\partial \mu^{(\lambda)}}{\partial t}(t,x) = \frac{1}{2}\,\mathrm{e}^{2\lambda t}\,\frac{\partial^2 \mu^{(\lambda)}}{\partial x^2}(t,x)\,.$$

Consequently, we obtain:

$$\frac{\partial}{\partial x}(x\,\mu^{(\lambda)}(t,x)) = \frac{1-\mathrm{e}^{2\lambda t}}{2\lambda}\,\frac{\partial^2 \mu^{(\lambda)}}{\partial x^2}(t,x)\,,$$

which is easy to verify directly.

1.3.2.3 Brownian Bridge

Let $(b_t, 0 \leq t \leq 1)$ be the standard Brownian bridge satisfying $b_0 = b_1 = 0$. It may be obtained as solution to the SDE:

$$X_t = B_t - \int_0^t \frac{X_s}{1-s} ds \tag{1.87}$$

and one has: $\mathbb{E}[b_t^2] = t\,(1-t)$. Let F be a $\mathscr{C}^2$-function on $\mathbb{R}$ with compact support. Our formula (1.82) yields:

$$\frac{\mathrm{d}}{\mathrm{d}t}\mathbb{E}[F(b_t)] = \frac{1-2t}{2}\,\mathbb{E}[F''(b_t)]\,,$$

whereas the application of the classical Itô formula gives:

$$\frac{\mathrm{d}}{\mathrm{d}t}\mathbb{E}[F(b_t)] = -\frac{1}{1-t}\,\mathbb{E}[F'(b_t)\,b_t] + \frac{1}{2}\,\mathbb{E}[F''(b_t)]\,.$$

Consequently, we obtain:

$$t\,(1-t)\,\mathbb{E}[F''(b_t)] = \mathbb{E}[F'(b_t)\,b_t]\,,$$

which is equivalent to (1.85) with $\alpha = \sqrt{t(1-t)}$.

On the other hand, there exists a Brownian motion (β_u) such that:

$$\forall t \in [0,1) \qquad b_t = (1-t)\,\beta_{t/(1-t)}\,.$$

Thus, here again, we are brought back to Example 1.3.2.1.

Besides, denoting by $\mu(t,x)$ the density of b_t for $0 < t < 1$, the Fokker-Planck equation corresponding to the SDE (1.87) yields:

$$\frac{\partial \mu}{\partial t}(t,x) = \frac{1}{1-t}\frac{\partial}{\partial x}(x\,\mu(t,x)) + \frac{1}{2}\frac{\partial^2 \mu}{\partial x^2}(t,x)\,,$$

whereas our Lemma 1.2 leads to:

$$\frac{\partial \mu}{\partial t}(t,x) = \frac{1-2t}{2}\frac{\partial^2 \mu}{\partial x^2}(t,x)\,.$$

Consequently, we obtain:

$$\frac{\partial}{\partial x}(x\,\mu(t,x)) = -t\,(1-t)\,\frac{\partial^2 \mu}{\partial x^2}(t,x)\,,$$

which is easy to verify directly. We note that, more generally than (1.87), we might consider the ε-generalized Brownian bridges on the time interval $[0,T]$, which solve:

$$X_t = B_t - \varepsilon \int_0^t \frac{X_s}{T-s}ds \quad , \quad t < T.$$

These processes are also particular cases of Example 1.3.2.1 (see Exercise 1.42 below). They have been considered in [124].

1.3.2.4 Fractional Brownian Motion

Let $(B_t^H, t \geq 0)$ be the fractional Brownian motion with Hurst index $H \in (0,1)$. It is a continuous centered Gaussian process such that, for any $s,t \geq 0$, $\mathbb{E}\left[(B_t^H - B_s^H)^2\right] = |t-s|^{2H}$ and $B_0^H = 0$. In particular $\mathbb{E}\left[(B_t^H)^2\right] = t^{2H}$. See for instance [28] for further information. Let F be a $\mathscr{C}^2$-function on $\mathbb{R}$ with compact support. Our formula (1.82) yields, for $0 \leq s \leq t$:

$$\mathbb{E}\left[F(B_t^H)\right] = \mathbb{E}[F(B_s^H)] + H\int_s^t \mathbb{E}[F''(B_u^H)]\,u^{2H-1}du\,. \tag{1.88}$$

Although the fractional Brownian motion (for $H \neq \frac{1}{2}$) is not Markovian, and hence is not a particular case of example 1.3.2.1, there exists a Brownian motion (β_u) such that,

$$B_t^H \overset{(1.d)}{=} \beta_{t^{2H}}\,.$$

Then formula (1.88) also follows from the classical Itô formula applied to the martingale $(\beta_{t^{2H}})$.

Exercise 1.41 (Gaussian integration by parts formula). 1) Prove formula (1.85), i.e.: for any real-valued, centered Gaussian r.v. g and for any $F \in \mathscr{C}^1(\mathbb{R})$ with compact support:

$$\mathbb{E}[g^2]\mathbb{E}[F'(g)] = \mathbb{E}[gF(g)].$$

2) Let $(g_1, \dots, g_n, g)$ be an $\mathbb{R}^{n+1}$-valued, centered Gaussian vector, and let $F \in \mathscr{C}^1(\mathbb{R}^n)$ with compact support. Prove that:

$$\sum_{j=1}^{n} \mathbb{E}[gg_j]\mathbb{E}\left[F'_{x_j}(g_1, \dots, g_n)\right] = \mathbb{E}\left[gF(g_1, \dots, g_n)\right].$$

(*Hint*: suppose $g \neq 0$ and set $\widetilde{g}_j = g_j - \dfrac{\mathbb{E}[gg_j]}{\mathbb{E}[g^2]} g$. Then $F(g_1, \dots, g_n) = F(\dots, \widetilde{g}_j + \dfrac{\mathbb{E}[gg_j]}{\mathbb{E}[g^2]} g, \dots)$. Apply Point 1) to this function of g, using the fact that, by construction, g is independent of the vector: $(\widetilde{g}_1, \dots, \widetilde{g}_n)$.)

Exercise 1.42 (ε-generalized Brownian bridges). Let, for $\varepsilon > 0$, $(X_t, t \in [0, T[)$ be the ε-generalized Brownian bridge on $[0, T]$ which solves:

$$X_t = B_t - \varepsilon \int_0^t \frac{X_s}{T - s} ds, \quad t < T.$$

Prove that there exist two continuous functions u and v from $[0, T[$ into $\mathbb{R}$, u with finite variation and v increasing and non-negative, and a Brownian motion β such that:

$$X_t = u(t)\beta_{v(t)}, \quad t < T.$$

(*Solution:* if $\varepsilon \neq \frac{1}{2}$, $u(t) = (T - t)^{\varepsilon}$ and $v(t) = (1 - 2\varepsilon)^{-1}(T^{1-2\varepsilon} - (T - t)^{1-2\varepsilon})$. If $\varepsilon = \frac{1}{2}$, $u(t) = \sqrt{T - t}$ and $v(t) = -\log\left(1 - \frac{t}{T}\right)$.)

1.3.3 An Application of the Weak Itô Formula

As an application of Theorem 1.9, we now present a simple proof of the Gordon-Slepian Lemma, the statement of which we recall below.

Proposition 1.5 (Gordon-Slepian Lemma, see [118], p. 74). *Let $X = (X_1, \cdots, X_n)$ and $Y = (Y_1, \cdots, Y_n)$ be two centered Gaussian vectors in $\mathbb{R}^n$, and let A and B be two subsets of $\{1, \cdots, n\} \times \{1, \cdots, n\}$. We assume:*

$$\begin{aligned} \mathbb{E}[X_j X_k] &\leq \mathbb{E}[Y_j Y_k] \quad \textit{if } (j,k) \in A \\ \mathbb{E}[X_j X_k] &\geq \mathbb{E}[Y_j Y_k] \quad \textit{if } (j,k) \in B \\ \mathbb{E}[X_j X_k] &= \mathbb{E}[Y_j Y_k] \quad \textit{if } (j,k) \notin A \cup B. \end{aligned}$$

Let $F : \mathbb{R}^n \longrightarrow \mathbb{R}$ be a $\mathscr{C}^2$-function whose derivatives of order 2 grow sub-exponentially at infinity. We assume:

$$F''_{x_j,x_k} \geq 0 \quad \textit{if} \quad (j,k) \in A \qquad \textit{and} \qquad F''_{x_j,x_k} \leq 0 \quad \textit{if} \quad (j,k) \in B\,.$$

Then

$$\mathbb{E}[F(X)] \leq \mathbb{E}[F(Y)]\,.$$

Proof. As in Kahane's proof ([102]), we set, for $t \in [0,1]$,

$$G_t = \sqrt{t}\,Y + \sqrt{1-t}\,X$$

where X and Y are assumed to be independent (a special case of the *smart path method* used, again and again, in [166]). Then, by Theorem 1.9,

$$\mathbb{E}[F(G_1)] = \mathbb{E}[F(G_0)] + \frac{1}{2}\sum_{j,k}\int_0^1 \mathbb{E}[F''_{x_j,x_k}(G_u)]\,(\mathbb{E}[Y_jY_k] - \mathbb{E}[X_jX_k])du\,.$$

Now, by hypothesis, for every (j,k),

$$(\mathbb{E}[Y_jY_k] - \mathbb{E}[X_jX_k])\,F''_{x_j,x_k} \geq 0\,.$$

□

Exercise 1.43 (An alternative proof of the weak Itô formula). We keep the notation and hypotheses of Subsection 1.3.1, and we define, for $a \leq s \leq t \leq b$ and $0 \leq v \leq 1$, $\widehat{G}_{s,t}(v)$ by the equality in law:

$$\widehat{G}_{s,t}(v) \overset{(law)}{=} \sqrt{v}\,G_t + \sqrt{1-v}\,G_s$$

where G_s and G_t are assumed to be independent.

Let F be a $\mathscr{C}^{1,2}$-function on $[a,b] \times \mathbb{R}^n$ with compact support.

1) Prove, differentiating the function of v:

$$v \in [0,1] \longmapsto \mathbb{E}\left[F(vt + (1-v)s,\, \widehat{G}_{s,t}(v))\right]$$

and then using the Gaussian integration by parts formula (Exercise 1.41), that:

$$\mathbb{E}\,[F(t,G_t)] = \mathbb{E}\,[F(s,G_s)] + (t-s)\int_0^1 \mathbb{E}\left[F'_t(vt+(1-v)s,\, \widehat{G}_{s,t}(v))\right]dv$$
$$+\frac{1}{2}\sum_{j,k}(c_{j,k}(t) - c_{j,k(s)})\int_0^1 \mathbb{E}\left[F''_{x_j,x_k}(vt+(1-v)s,\, \widehat{G}_{s,t}(v))\right]dv.$$

2) Consider a subdivision $w = (w_1,\dots,w_r)$ of the interval $[s,t]$ $(w_1 = s, w_r = t)$. Writing down successively the above formula for $(s,t) = (w_l, w_{l+1})$, $1 \leq l \leq r-1$, and then telescoping, prove that:

$$\mathbb{E}\,[F(t,G_t)] = \mathbb{E}\,[F(s,G_s)] + \int_s^t H^w(u)du + \frac{1}{2}\sum_{j,k}\int_s^t K^w_{j,k}(u)dc_{j,k}(u)$$

where H^w and $K^w_{i,j}$ are functions which are constant on each interval $[w_l, w_{l+1}[$.

3) Conclude that formula (1.82) holds.
(*Hint*: consider $(w^N, N \geq 1)$ a sequence of subdivisions whose meshes tend to 0 as N tends to $+\infty$, and use the fact that, if $u \in [w^N_{l_N}, w^N_{l_N+1}[$, then, for any $v \in [0,1]$, $\widehat{G}_{w^N_{l_N}, w^N_{l_N+1}}(v)$ converges in law to G_u as N tends to $+\infty$.)

1.3.4 A Variant of the Weak Itô Formula

In this subsection, we keep the framework, hypotheses and notation of Subsection 1.3.1. Let $h(t,x) : \mathbb{R}_+ \times \mathbb{R} \longrightarrow \mathbb{R}$ be a $\mathscr{C}^{1,2}$-function satisfying, on $\mathbb{R}_+ \times \mathbb{R}$ the equation:

$$\frac{\partial h}{\partial t} + \frac{1}{2}\frac{\partial^2 h}{\partial x^2} = 0 . \tag{1.89}$$

Note that under an extra integration assumption, h is a space-time harmonic function. We shall define this notion precisely in Chapter 2.

Theorem 1.10. *Let $h^{(1)}, \cdots, h^{(n)}$ be n functions satisfying (1.89), and define the function $H = (H_1, \cdots, H_n)$, from $[a,b] \times \mathbb{R}^n$ into $\mathbb{R}^n$, by*

$$H_j(t,x) = h^{(j)}(c_{j,j}(t), x_j) .$$

Let $\Phi : \mathbb{R}^n \longrightarrow \mathbb{R}$ be a $\mathscr{C}^2$-function such that the derivatives of order 2 with respect to x of the function F defined by

$$F(t,x) = \Phi(H(t,x)) ,$$

grow sub-exponentially at infinity with respect to x, uniformly with respect to $t \in [a,b]$. Then, for every s,t with $a \leq s \leq t \leq b$,

$$\mathbb{E}[\Phi(H(t,G_t))] = \mathbb{E}[\Phi(H(s,G_s))] +$$

$$\sum_{j,k} \int_s^t \mathbb{E}\left[\Phi''_{x_j,x_k}(H(u,G_u)) \frac{\partial h^{(j)}}{\partial x}(c_{j,j}(u), G_u^{(j)}) \frac{\partial h^{(k)}}{\partial x}(c_{k,k}(u), G_u^{(k)})\right] \frac{dc_{j,k}(u)}{2} .$$

Proof. If the covariance matrix C is a $\mathscr{C}^1$-function, the function F is of class $\mathscr{C}^{1,2}$ and Theorem 1.10 follows by a direct application of Theorem 1.9, after simplifications which are consequences of (1.89). Actually, the general case may be treated by a slight adaptation of the proof of Theorem 1.9. □

We now state two easy corollaries.

Corollary 1.3. *Let $h^{(1)}, \cdots, h^{(n)}$ be n functions satisfying (1.89) and $a_1, \cdots, a_n \in \mathbb{R}$. Define the function k, from $[a,b] \times \mathbb{R}^n$ into $\mathbb{R}$, by*

$$k(t,x) = \sum_j a_j h^{(j)}(c_{j,j}(t), x_j) .$$

Let $\varphi : \mathbb{R} \longrightarrow \mathbb{R}$ *be a* $\mathscr{C}^2$*-function such that the derivatives of order* 2 *with respect to* x *of the function* F *defined by*

$$F(t,x) = \varphi(k(t,x)) \ ,$$

grow sub-exponentially at infinity with respect to x*, uniformly with respect to* $t \in [a,b]$*. Then, for every* s,t *with* $a \leq s \leq t \leq b$*,*

$$\mathbb{E}[\varphi(k(t,G_t))] = \mathbb{E}[\varphi(k(s,G_s))] +$$

$$\sum_{j,k} a_j a_k \int_s^t \mathbb{E}\left[\varphi''(k(u,G_u)) \frac{\partial h^{(j)}}{\partial x}(c_{j,j}(u), G_u^{(j)}) \frac{\partial h^{(k)}}{\partial x}(c_{k,k}(u), G_u^{(k)})\right] \frac{dc_{j,k}(u)}{2}.$$

Corollary 1.4. *Let* $a_1, \cdots, a_n \in \mathbb{R}$ *and let* $\varphi : \mathbb{R} \longrightarrow \mathbb{R}$ *be a* $\mathscr{C}^2$*-function whose second derivative grows at most polynomially at infinity. We set, for* $1 \leq j \leq n$*,*

$$Y_u^j = \exp\left(G_u^{(j)} - \frac{c_{j,j}(u)}{2}\right)$$

and

$$K_u = \sum_j a_j Y_u^{(j)} \ .$$

Then, for every s,t *with* $a \leq s \leq t \leq b$*,*

$$\mathbb{E}[\varphi(K_t)] = \mathbb{E}[\varphi(K_s)] + \frac{1}{2} \sum_{j,k} a_j a_k \int_s^t \mathbb{E}\left[\varphi''(K_u) Y_u^{(j)} Y_u^{(k)}\right] dc_{j,k}(u) \ .$$

We shall now apply Theorem 1.9 to exhibit new families of peacocks. But before doing so, we introduce some notation which will be in force throughout the end of this Section 1.3.

1.3.5 Notation

- We denote by Λ a measurable space.
- We consider, for each $t \geq 0$, a real valued measurable centered Gaussian process

$$G_{\bullet,t} = (G_{\lambda,t}, \lambda \in \Lambda) \ .$$

- For $\lambda, \mu \in \Lambda$ and $t \geq 0$, we set:

$$c_{\lambda,\mu}(t) = \mathbb{E}[G_{\lambda,t} G_{\mu,t}] \ .$$

- For any signed finite measure σ on Λ, we set, for $t \geq 0$,

$$A_t^{(\sigma)} = \int_\Lambda \exp\left(G_{\lambda,t} - \frac{1}{2} c_{\lambda,\lambda}(t)\right) \sigma(\mathrm{d}\lambda) \ .$$

We now introduce various conditions which will appear in the sequel:

(M_1) Λ is a metric σ-compact space equipped with its Borel σ-field.

(C_1) For all $t \geq 0$, the function:

$$(\lambda, \mu) \in \Lambda \times \Lambda \longrightarrow c_{\lambda,\mu}(t) \in \mathbb{R}$$

is continuous.

(I_1) For every $\lambda, \mu \in \Lambda$, the function

$$t \in \mathbb{R}_+ \longrightarrow c_{\lambda,\mu}(t) \in \mathbb{R}$$

is increasing.

(I_2) For every $n \geq 1$, for every $\lambda_1, \cdots, \lambda_n \in \Lambda$, the matrix function

$$t \in \mathbb{R}_+ \longrightarrow (c_{\lambda_j,\lambda_k}(t))_{1 \leq j,k \leq n} \in \mathbb{S}_n$$

is increasing with respect to the partial order on $\mathbb{S}_n$ induced by the convex cone $\mathbb{S}_n^+$.

Note that conditions (I_1) and (I_2) are not comparable.

1.3.6 Integrals of Log-normal Processes

The next theorem provides sufficient conditions for the process $(A_t^{(\sigma)}, t \geq 0)$ to be a peacock.

Theorem 1.11. *We assume (M_1) and (C_1). If either (I_1) is satisfied and σ is a positive finite measure, or (I_2) is satisfied and σ is a signed finite measure, then $(A_t^{(\sigma)}, t \geq 0)$ is a peacock.*

Proof. The proof of Theorem 1.11 proceeds in three steps.

- We first assume that

$$\sigma = \sum_{j=1}^{n} a_j \, \delta_{\lambda_j}$$

where δ_λ denotes the Dirac measure at λ and $a_1, \cdots, a_n \in \mathbb{R}$.

We have

$$A_t^{(\sigma)} = \sum_{j=1}^{n} a_j \exp\left(G_{\lambda_j,t} - \frac{1}{2} c_{\lambda_j,\lambda_j}(t) \right) .$$

Since for any t, $\mathbb{E}\left[|A_t^{(\sigma)}| \right] \leq \int_\Lambda |\sigma(\mathrm{d}\lambda)| < \infty$, to prove that $(A_t^{(\sigma)}, t \geq 0)$ is a peacock, it suffices to prove (Remark 1.1) that, for any $\psi \in \mathbf{C}$, the function $t \geq 0 \longrightarrow \mathbb{E}\left[\psi(A_t^{(\sigma)}) \right]$ is increasing.

We fix $0 \leq s \leq t$. We set, for $1 \leq j \leq n$ and $u \in [0,1]$,

$$G_u^{(j)} = \sqrt{u}\, G_{\lambda_j,t} + \sqrt{1-u}\, G_{\lambda_j,s}$$

where the Gaussian vectors $(G_{\lambda_1,t},\cdots,G_{\lambda_n,t})$ and $(G_{\lambda_1,s},\cdots,G_{\lambda_n,s})$ are supposed to be independent. This is yet another instance of the smart path method, see [166]. Then, by Corollary 1.4, we have:

$$\mathbb{E}[\psi(K_1)] = \mathbb{E}[\psi(K_0)] +$$
$$\frac{1}{2}\sum_{j,k} a_j a_k \int_0^1 \mathbb{E}\left[\psi''(K_u)Y_u^{(j)}Y_u^{(k)}\right](c_{\lambda_j,\lambda_k}(t) - c_{\lambda_j,\lambda_k}(s))du$$

where, for $1 \leq j \leq n$,

$$Y_u^{(j)} = \exp\left(G_u^{(j)} - \frac{1}{2}\mathbb{E}\left[G_u^{(j)}G_u^{(j)}\right]\right)$$

and

$$K_u = \sum_{j=1}^{n} a_j Y_u^{(j)} .$$

Since

$$K_1 = A_t^{(\sigma)}, \quad K_0 = A_s^{(\sigma)}, \quad \psi'' \geq 0 \quad \text{and} \quad Y_u^{(j)} \geq 0 ,$$

if either (I_1) is satisfied and σ is a positive measure, or (I_2) is satisfied, then, for $0 \leq s \leq t$,

$$\mathbb{E}[\psi(A_t^{(\sigma)})] \geq \mathbb{E}\left[\psi(A_s^{(\sigma)})\right] ,$$

which proves the result.

• By hypothesis (M_1), there exists a sequence $(\Lambda_n)_{n\geq 0}$ of compact subsets of Λ with $\bigcup_{n\geq 0}\Lambda_n = \Lambda$. We now assume that the support of σ is contained in some compact set Λ_{n_0}. Then, there exists a sequence $(\sigma_n, n \geq 0)$, weakly converging to σ, such that, for each n, σ_n is as in step 1 a linear combination of Dirac measures supported by Λ_{n_0}. Besides, we may suppose

$$\forall n, \qquad \int |\sigma_n(d\lambda)| \leq \int |\sigma(d\lambda)| . \tag{1.90}$$

Moreover, if σ is a positive measure, all measures σ_n may be assumed to be positive. Let $\psi \in \boldsymbol{C}$. By step 1, if either (I_1) is satisfied and σ is a positive measure, or (I_2) is satisfied, then for any $n \geq 0$ and $0 \leq s \leq t$,

$$\mathbb{E}\left[\psi(A_t^{(\sigma_n)})\right] \geq \mathbb{E}\left[\psi(A_s^{(\sigma_n)})\right] . \tag{1.91}$$

On the other hand,

$$\mathbb{E}[(A_t^{(\sigma)} - A_t^{(\sigma_n)})^2] = \int\int_{\Lambda_{n_0}^2} e^{c_{\lambda,\mu}(t)} d(\sigma - \sigma_n)(\lambda) d(\sigma - \sigma_n)(\mu) .$$

Consequently, using (1.90) and (C_1), we obtain the convergence, in L^2, of the sequence $(A_t^{(\sigma_n)}, n \geq 0)$ to $A_t^{(\sigma)}$.

Since ψ is affine outside of a compact interval, then ψ is a Lipschitz continuous function, and the sequence $(\psi(A_t^{(\sigma_n)}), n \geq 0)$ converges in L^2 to $\psi(A_t^{(\sigma)})$. We may then pass to the limit in (1.91) and obtain, for $0 \leq s \leq t$,

$$\mathbb{E}[\psi(A_t^{(\sigma)})] \geq \mathbb{E}[\psi(A_s^{(\sigma)})] .$$

Then the desired result follows from Remark 1.1.

• In the general case, we set, for any $n \geq 0$

$$\sigma_n = 1_{\Lambda_n} \sigma .$$

We have, for $t \geq 0$,

$$\lim_{n \to \infty} A_t^{(\sigma_n)} = A_t^{(\sigma)} \text{ a.s.} \quad \text{and} \quad |A_t^{(\sigma_n)}| \leq A_t^{(|\sigma|)} ,$$

which allows to apply step 2 and to pass to the limit. □

Remark 1.13. Suppose that, for all $\lambda, \mu \in \Lambda$, the function $c_{\lambda,\mu}$ is absolutely continuous on $\mathbb{R}_+$. Then, Condition (I_2) may be written as:

For every $n \geq 1$, for every $\lambda_1, \cdots, \lambda_n \in \Lambda$, the matrix

$$(c'_{\lambda_j,\lambda_k}(t))_{1 \leq j,k \leq n}$$

is a positive symmetric matrix for a.e. $t \geq 0$.

1.3.7 Examples

1.3.7.1 Processes $(t\, G_\lambda)$

We assume (M_1) and we consider a real valued measurable centered Gaussian process $(G_\lambda, \lambda \in \Lambda)$. For $\lambda, \mu \in \Lambda$, we set:

$$c(\lambda, \mu) = \mathbb{E}[G_\lambda\, G_\mu] .$$

We assume the following hypothesis:

$(\widetilde{C})$ The function:

$$(\lambda, \mu) \in \Lambda \times \Lambda \longrightarrow c(\lambda, \mu) \in \mathbb{R}$$

is continuous.

We set, for $\lambda \in \Lambda$ and $t \geq 0$,

$$G_{\lambda,t}^{(1)} = t\, G_\lambda .$$

Then, $G^{(1)}$ satisfies (C_1) and (I_2).

1.3.7.2 Processes $(G_{\lambda t})$

Here, we consider the particular case $\Lambda = \mathbb{R}_+$, and a measurable centered Gaussian process $(G_\lambda, \lambda \geq 0)$ satisfying the previous condition $(\widetilde{C})$. We set, for $\lambda \geq 0$ and $t \geq 0$,

$$G^{(2)}_{\lambda,t} = G_{\lambda t} \,.$$

Furthermore, we assume the following hypothesis:

$(\widetilde{I})$ For $\lambda\,, \mu \geq 0$, the function

$$t \geq 0 \longrightarrow c(t\,\lambda, t\,\mu)$$

is increasing.

Then the process $G^{(2)}$ satisfies (C_1) and (I_1). In particular, Theorem 1.11 implies that the process

$$\left(\frac{1}{t}\int_0^t \exp\left(G_\lambda - \frac{1}{2}c(\lambda,\lambda)\right) d\lambda\;,\; t \geq 0\right)$$

is a peacock.

An example of a process $(G_\lambda, \lambda \geq 0)$ satisfying the above properties $(\widetilde{C})$ and $(\widetilde{I})$, is the fractional Brownian motion B^H with Hurst index $H \in (0,1)$. Indeed, then:

$$c(\lambda,\mu) = \frac{1}{2}\left(|\lambda|^{2H} + |\mu|^{2H} - |\lambda - \mu|^{2H}\right) \geq 0$$

and

$$c(t\,\lambda, t\,\mu) = t^{2H}\, c(\lambda,\mu) \,.$$

Actually, for each $t \geq 0$, there is the equality in law:

$$B^H_{t\bullet} \overset{(law)}{=} t^H B^H_{\bullet} \,.$$

Therefore, we may as well apply the results in the previous paragraph 1.3.7.1. Consequently, for any *signed* finite measure σ on $\mathbb{R}_+$,

$$\int_{\mathbb{R}_+} \exp\left(B^H_{t\lambda} - \frac{(t\,\lambda)^{2H}}{2}\right) d\sigma(\lambda) \quad , \quad t \geq 0$$

is a peacock.

We now introduce another example. Let

$$a : (\lambda, s) \in \mathbb{R}_+ \times \mathbb{R}_+ \longrightarrow a(\lambda, s) \in \mathbb{R}_+$$

be a nonnegative measurable function such that:

i) For every $\lambda \geq 0$, $a(\lambda, \bullet) \in L^2(\mathbb{R}_+)$.

ii) The function

$$(\lambda,\mu) \in \mathbb{R}_+ \times \mathbb{R}_+ \longrightarrow \int_0^\infty a(\lambda,s)\,a(\mu,s)ds$$

is continuous.

iii) For any $s \geq 0$, the function

$$\lambda \in \mathbb{R}_+ \longrightarrow a(\lambda,s)$$

is increasing.

Setting

$$G_\lambda = \int_0^\infty a(\lambda,s)dB_s \quad , \quad \lambda \geq 0$$

where (B_s) is a standard Brownian motion, we see that properties $(\widetilde{C})$ and $(\widetilde{I})$ are satisfied.

Exercise 1.44 (Integrals of exponentials of Brownian bridges are not peacocks). Let $\varepsilon \in \mathbb{R}$ and set:

$$G_\lambda = B_{\lambda \wedge 1} - \varepsilon\,(\lambda \wedge 1)\,B_1$$

where B denotes standard Brownian motion.
1) Prove that $(G_\lambda, \lambda \geq 0)$ satisfies $(\widetilde{I})$ if and only if

$$|1-\varepsilon| \geq \frac{1}{\sqrt{2}}\,.$$

2) Assume $\varepsilon = 1$. Then:

$$(G_t = B_t - tB_1, 0 \leq t \leq 1) \overset{(law)}{=} (b_t, 0 \leq t \leq 1)$$

where $(b_t, 0 \leq t \leq 1)$ denotes the standard Brownian bridge. Let $a \in]\frac{1}{2}, 1]$. Consider, for $t \in [0,1]$,

$$\gamma_t^{(1)} = \int_0^1 \exp\left(b_{ut} - \frac{ut(1-ut)}{2}\right) du \quad \text{and} \quad \gamma_t^{(2)} = \exp\left(b_{at} - \frac{at(1-at)}{2}\right).$$

Prove that neither $(\gamma_t^{(1)}, 0 \leq t \leq 1)$ nor $(\gamma_t^{(2)}, 0 \leq t \leq 1)$ is a peacock.
(*Hint*: show that the left derivative at $t = 1$ of $\mathbb{E}\left[(\gamma_t^{(j)})^2\right]$ is strictly negative for $j = 1, 2$.)

1.3.7.3 Brownian Sheet

Let, for $\lambda, t \geq 0$,

$$G_{\lambda,t} = W_{\lambda,t}$$

where W denotes standard Brownian sheet (see Chapter 2, Subsection 2.2.2). We have:

$$c_{\lambda,\mu}(t) = t\,(\lambda \wedge \mu)\,.$$

Then G satisfies hypotheses (C_1), (I_1) and (I_2). In fact, for any $t \geq 0$,

$$G_{\bullet,t} \overset{(law)}{=} \sqrt{t}B_{\bullet}$$

where B denotes standard Brownian motion. We may then consider this example as a particular case of Example 1.3.7.1 (replacing t by $\sqrt{t}$). On the other hand, for any $\lambda \geq 0$,

$$\left(\exp\left(W_{\lambda,t} - \frac{t\lambda}{2}\right), t \geq 0\right)$$

is a $(\mathscr{W}_t)$-martingale, with

$$\mathscr{W}_t = \sigma\{W_{\lambda,s} \,;\, \lambda \geq 0, 0 \leq s \leq t\}\,.$$

Therefore, Theorem 1.11 is obvious in this case since $(A_t^{(\sigma)})$ is a $(\mathscr{W}_t)$-martingale.

1.3.7.4 Stochastic Integrals

We assume (M_1). Let

$$h : (\lambda, s) \in \Lambda \times \mathbb{R}_+ \longrightarrow h(\lambda, s) \in \mathbb{R}$$

be a measurable function such that:

i) for every $\lambda \in \Lambda$, $h(\lambda, \bullet) \in L^2_{loc}(\mathbb{R}_+)$,
ii) for any $t \geq 0$, the function

$$(\lambda, \mu) \in \Lambda \times \Lambda \longrightarrow \int_0^t h(\lambda, s) h(\mu, s) ds$$

is continuous.

We note that, for

$$G_{\lambda,t} = \int_0^t h(\lambda, s) dB_s \quad ; \quad \lambda \in \Lambda\,, t \geq 0$$

where (B_s) is a standard Brownian motion:

$$c_{\lambda,\mu}(t) = \int_0^t h(\lambda, s) h(\mu, s) ds\,.$$

Therefore, G satisfies hypotheses (C_1) and (I_2).

Exercise 1.45 (A Theorem of Kahane). Suppose (M_1) and let

$$X = (X_\lambda, \lambda \in \Lambda) \quad , \quad Y = (Y_\lambda, \lambda \in \Lambda)$$

be two real valued measurable centered Gaussian processes. Set, for $\lambda, \mu \in \Lambda$,

$$c_X(\lambda, \mu) = \mathbb{E}[X_\lambda X_\mu] \quad , \quad c_Y(\lambda, \mu) = \mathbb{E}[Y_\lambda Y_\mu].$$

Assume that:

i) c_X and c_Y are continuous functions on $\Lambda \times \Lambda$,
ii) $\forall \lambda, \mu \in \Lambda$, $c_X(\lambda,\mu) \leq c_Y(\lambda,\mu)$.

Prove that, for any positive and finite measure σ on Λ and for any convex function ψ on $\mathbb{R}$:

$$\mathbb{E}\left[\psi\left\{\int_\Lambda \exp\left(X_\lambda - \frac{c_X(\lambda,\lambda)}{2}\right)\sigma(d\lambda)\right\}\right] \leq \mathbb{E}\left[\psi\left\{\int_\Lambda \exp\left(Y_\lambda - \frac{c_Y(\lambda,\lambda)}{2}\right)\sigma(d\lambda)\right\}\right].$$

(*Hint*: set

$$G_{\lambda,t} = \sqrt{t}Y_\lambda + \sqrt{1-t}X_\lambda, \quad 0 \leq t \leq 1$$

where the processes X and Y are assumed to be independent. Verify that Property (I_1) is satisfied on $[0,1]$ and then apply Theorem 1.11.) We refer the reader to [102] for the original proof of this result.

1.4 Peacocks in a Markovian Set-up

1.4.1 Conditionally Monotone Peacocks

In this section, we introduce and study the notion of conditional monotonicity, which already appears in [157, Chapter 4.B, p.114-126].

Definition 1.10 (Conditional monotonicity). A process $(X_\lambda, \lambda \geq 0)$ is said to be conditionally monotone if, for every $n \in \mathbb{N}^*$, every $i \in \{1,\dots,n\}$, every $0 \leq \lambda_1 < \cdots < \lambda_n$ and every bounded Borel function $\phi : \mathbb{R}^n \to \mathbb{R}$ which increases (resp. decreases) with respect to each of its arguments, we have:

$$\mathbb{E}[\phi(X_{\lambda_1}, X_{\lambda_2}, \dots, X_{\lambda_n})|X_{\lambda_i}] = \phi_i(X_{\lambda_i}), \tag{1.92}$$

where $\phi_i : \mathbb{R} \to \mathbb{R}$ is a bounded increasing (resp. decreasing) function.

Remark 1.14. 1) If there is an interval I of $\mathbb{R}$ such that, for every $\lambda \geq 0$, $X_\lambda \in I$, we may assume in Definition 1.10 that ϕ is merely defined on I^n, and ϕ_i is defined on I.
2) Note that $(X_\lambda, \lambda \geq 0)$ is conditionally monotone if and only if $(-X_\lambda, \lambda \geq 0)$ is conditionally monotone.
3) Let $\theta : \mathbb{R} \longrightarrow \mathbb{R}$ be a strictly monotone and continuous function. It is not difficult to see that if the process $(X_\lambda, \lambda \geq 0)$ is conditionally monotone, then so is $(\theta(X_\lambda), \lambda \geq 0)$.

To prove that a process is conditionally monotone, we can restrict ourselves to bounded Borel functions ϕ increasing with respect to each of their arguments. Indeed, replacing ϕ by $-\phi$, the result then holds also for bounded Borel functions decreasing with respect to each of their arguments.

Definition 1.11. We denote by $\mathscr{E}_n$ the set of bounded Borel functions $\phi : \mathbb{R}^n \longrightarrow \mathbb{R}$ which are increasing with respect to each of their arguments.

Theorem 1.12. *Let $(X_\lambda, \lambda \geq 0)$ a real-valued process which is right-continuous, conditionally monotone and which satisfies the following integrability conditions: For every compact $K \subset \mathbb{R}_+$ and every $t \geq 0$:*

$$\Theta_{K,t} := \sup_{\lambda \in K} \exp(tX_\lambda) = \exp\left(t \sup_{\lambda \in K} X_\lambda\right) \text{ is integrable,} \tag{INT1}$$

and

$$k_{K,t} := \inf_{\lambda \in K} \mathbb{E}\left[\exp(tX_\lambda)\right] > 0. \tag{INT2}$$

We set $h_\lambda(t) = \log \mathbb{E}[\exp(tX_\lambda)]$. Then, for every finite positive measure μ on $\mathbb{R}_+$:

$$\left(A_t^{(\mu)} := \int_0^\infty e^{tX_\lambda - h_\lambda(t)} \mu(d\lambda)\,,\, t \geq 0\right)$$

is a peacock.

Proof. *i*) By (INT1), for every $\lambda \geq 0$ and every $t \geq 0$, $\mathbb{E}\left[\exp(tX_\lambda)\right] < \infty$. Hence, we see as in the proof of Theorem 1.7, that h_λ is continuous on $\mathbb{R}_+$, differentiable on $]0, +\infty[$, and

$$h'_\lambda(t)e^{h_\lambda(t)} = \mathbb{E}\left[X_\lambda e^{tX_\lambda}\right]. \tag{1.93}$$

Since $\mathbb{E}\left[e^{tX_\lambda - h_\lambda(t)}\right] = 1$, we obtain from (1.93):

$$\mathbb{E}\left[(X_\lambda - h'_\lambda(t))e^{tX_\lambda - h_\lambda(t)}\right] = 0. \tag{1.94}$$

Moreover, we also deduce from (INT1) that, for every $t \geq 0$, the function $\lambda \geq 0 \longmapsto h_\lambda(t)$ is right-continuous.

ii) We first consider the case

$$\mu = \sum_{i=1}^n a_i \delta_{\lambda_i}$$

where $n \in \mathbb{N}^*$, $a_1 \geq 0, \dots, a_n \geq 0$, $\lambda_1 \geq 0, \dots, \lambda_n \geq 0$ and, to simplify, we write $A_t^{(\mu)}$ for A_t. Let $\psi \in \boldsymbol{C}$. For $t > 0$, we have:

$$\frac{\partial}{\partial t}\mathbb{E}[\psi(A_t)] = \mathbb{E}\left[\psi'(A_t)\sum_{i=1}^n a_i \left(X_{\lambda_i} - h'_{\lambda_i}(t)\right)\exp\left(tX_{\lambda_i} - h_{\lambda_i}(t)\right)\right].$$

Setting for $i \in \{1, \dots, n\}$,

$$\Delta_i = \mathbb{E}\left[\psi'(A_t)\left(X_{\lambda_i} - h'_{\lambda_i}(t)\right)\exp\left(tX_{\lambda_i} - h_{\lambda_i}(t)\right)\right]$$

we shall show that $\Delta_i \geq 0$ for every $i \in \{1,\dots,n\}$. Let us notice that the function

$$(x_1,\dots,x_n) \longmapsto \psi'\left(\sum_{j=1}^{n} a_j \exp\left(tx_j - h_{\lambda_j}(t)\right)\right)$$

increases with respect to each of its arguments. Hence, from the conditional monotonicity property of $(X_\lambda, \lambda \geq 0)$:

$$\begin{aligned}\Delta_i &= \mathbb{E}\left[\mathbb{E}\left[\psi'(A_t)(X_{\lambda_i} - h'_{\lambda_i}(t))e^{tX_{\lambda_i} - h_{\lambda_i}(t)} | X_{\lambda_i}\right]\right] \\ &= \mathbb{E}\left[(X_{\lambda_i} - h'_{\lambda_i}(t))e^{tX_{\lambda_i} - h_{\lambda_i}(t)} \phi_i(X_{\lambda_i})\right]\end{aligned}$$

where ϕ_i is a bounded increasing function. Besides, we have,

$$(X_{\lambda_i} - h'_{\lambda_i}(t))\left(\phi_i(X_{\lambda_i}) - \phi_i\left(h'_{\lambda_i}(t)\right)\right) \geq 0.$$

Therefore,

$$\begin{aligned}\Delta_i &\geq \phi_i\left(h'_{\lambda_i}(t)\right)\mathbb{E}\left[(X_{\lambda_i} - h'_{\lambda_i}(t))e^{tX_{\lambda_i} - h_{\lambda_i}(t)}\right] \\ &= 0 \quad \text{from (1.94).}\end{aligned}$$

iii) We now assume that μ has a compact support contained in a compact interval K. Since the function $\lambda \longmapsto \exp(tX_\lambda - h_\lambda(t))$ is right-continuous and bounded from above by $k_{K,t}^{-1}\Theta_{K,t}$ which is finite a.s., there exists a sequence $(\mu_n, n \geq 0)$ of measures of the type considered in *ii*), with supp $\mu_n \subset K$, $\int \mu_n(d\lambda) = \int \mu(d\lambda)$ and for every $t \geq 0$, $\lim_{n \to +\infty} A_t^{(\mu_n)} = A_t^{(\mu)}$ a.s. Moreover:

$$|A_t^{(\mu_n)}| \leq \frac{\theta_{K,t}}{k_{K,t}} \int \mu(d\lambda).$$

We then deduce from Exercice 1.10 that $\left(A_t^{(\mu)}, t \geq 0\right)$ is a peacock.

iv) In the general case, we set $\mu_n(d\lambda) = 1_{[0,n]}(\lambda)\mu(d\lambda)$ and we apply Exercise 1.11 to conclude. □

Remark 1.15. Let $\theta : \mathbb{R} \longrightarrow \mathbb{R}$ be a strictly monotone and continuous function, and μ denote a finite positive measure. From Remark 1.14, under the assumption that $(\theta(X_\lambda), \lambda \geq 0)$ still satisfies conditions (INT1) and (INT2), we obtain, denoting $h_{\lambda,\theta}(t) = \log \mathbb{E}[\exp(t\theta(X_\lambda))]$, that the process

$$\left(A_t^{(\theta,\mu)} := \int_0^\infty e^{t\theta(X_\lambda) - h_{\lambda,\theta}(t)} \mu(d\lambda)\,,\, t \geq 0\right)$$

is a peacock. Note that θ only needs to be continuous and strictly monotone on an interval containing the image of X_λ for every $\lambda \geq 0$.

Of course, Theorem 1.12 may only have some practical interest if we are able to exhibit enough examples of processes which enjoy the conditional monotonicity (1.92) property. Below, we shall see that there exists a large class of diffusions which enjoy this property. But to start with, let us first give a few examples which consist of processes with independent increments and Lévy processes in particular.

1.4.2 Conditionally Monotone Processes with Independent Increments

We first give an assertion which is equivalent to (1.92) when dealing with additive processes, i.e. processes with independent (not necessarily time-homogeneous) increments.

Proposition 1.6. *Let $(X_\lambda, \lambda \geq 0)$ be a process with independent increments. Then, the conditional monotonicity hypothesis (1.92) is equivalent to the following:*

For every $n \in \mathbb{N}^*$, every $0 \leq \lambda_1 < \cdots < \lambda_n$ and every function $\phi : \mathbb{R}^n \longrightarrow \mathbb{R}$ in $\mathscr{E}_n$, we have:

$$\mathbb{E}\left[\phi\left(X_{\lambda_1}, \ldots, X_{\lambda_n}\right) |X_{\lambda_n}\right] = \phi_n(X_{\lambda_n}) \tag{$\widetilde{\text{CM}}$}$$

where ϕ_n is an increasing bounded function.

Proof. The proof is straightforward. Indeed, let $\phi \in \mathscr{E}_n$. For $i \in \{1, \ldots, n\}$, the hypothesis of independent increments implies:

$$\begin{aligned}
&\mathbb{E}\left[\phi\left(X_{\lambda_1}, \ldots, X_{\lambda_n}\right) |X_{\lambda_i}\right] \\
&\quad = \mathbb{E}\left[\mathbb{E}\left[\phi\left(X_{\lambda_1}, \ldots, X_{\lambda_n}\right) |\mathscr{F}_{\lambda_i}\right] |X_{\lambda_i}\right] \\
&\quad = \mathbb{E}\left[\mathbb{E}\left[\phi\left(X_{\lambda_1}, \ldots, X_{\lambda_i}, X_{\lambda_{i+1}} - X_{\lambda_i} + X_{\lambda_i}, \ldots, X_{\lambda_n} - X_{\lambda_i} + X_{\lambda_i}\right) |\mathscr{F}_{\lambda_i}\right] |X_{\lambda_i}\right] \\
&\quad = \mathbb{E}\left[\widetilde{\phi}\left(X_{\lambda_1}, \ldots, X_{\lambda_i}\right) |X_{\lambda_i}\right]
\end{aligned}$$

where

$$\widetilde{\phi}(x_1, \ldots, x_i) = \mathbb{E}\left[\phi\left(x_1, \ldots, x_i, X_{\lambda_{i+1}} - X_{\lambda_i} + x_i, \ldots, X_{\lambda_n} - X_{\lambda_i} + x_i\right)\right]$$

belongs to $\mathscr{E}_i$. □

1.4.2.1 The Gamma Subordinator is Conditionally Monotone

The Gamma subordinator $(\gamma_\lambda, \lambda \geq 0)$ is characterized by:

$$\mathbb{E}\left[e^{-t\gamma_\lambda}\right] = \frac{1}{(1+t)^\lambda} = \exp\left(-\lambda \int_0^\infty (1 - e^{-tx}) \frac{e^{-x}}{x} dx\right).$$

In particular, γ_λ is a gamma random variable with parameter λ. From ($\widetilde{\text{CM}}$), we wish to show that for every $n \in \mathbb{N}^*$, every $0 \leq \lambda_1 < \cdots < \lambda_n$ and every function $\phi : \mathbb{R}^n \longrightarrow \mathbb{R}$ in $\mathscr{E}_n$:

$$\mathbb{E}[\phi(\gamma_{\lambda_1}, \ldots, \gamma_{\lambda_n}) | \gamma_{\lambda_n}] = \phi_n(\gamma_{\lambda_n}), \tag{1.95}$$

where ϕ_n is an increasing function. The explicit knowledge of the law of γ_λ and the fact that $(\gamma_\lambda, \lambda \geq 0)$ has time-homogeneous independent increments imply the well-known result that, given $\{\gamma_{\lambda_n} = x\}$, the vector $(\gamma_{\lambda_1}, \gamma_{\lambda_2} - \gamma_{\lambda_1}, \dots, \gamma_{\lambda_n} - \gamma_{\lambda_{n-1}})$ follows the Dirichlet law with parameters $(\lambda_1, \lambda_2 - \lambda_1, \dots, \lambda_n - \lambda_{n-1})$ on $[0,x]$. In other words, the density f_n^x of $(\gamma_{\lambda_1}, \gamma_{\lambda_2}, \dots, \gamma_{\lambda_{n-1}})$ conditionally on $\{\gamma_{\lambda_n} = x\}$ equals:

$$f_n^x(x_1,\dots,x_{n-1}) = \frac{C}{x^{\lambda_n - 1}} x_1^{\lambda_1 - 1}(x_2 - x_1)^{\lambda_2 - \lambda_1 - 1}\dots$$
$$(x_{n-1} - x_{n-2})^{\lambda_{n-1} - \lambda_{n-2} - 1}(x - x_{n-1})^{\lambda_n - \lambda_{n-1} - 1} 1_{\mathbb{S}^{n,x}},$$

where $C := C(\lambda_1, \dots, \lambda_n)$ is a positive constant and

$$\mathbb{S}^{n,x} = \{(x_1,\dots,x_{n-1}) \in \mathbb{R}^{n-1} : 0 \leq x_1 \leq \dots \leq x_{n-1} \leq x\}.$$

Hence,

$$\mathbb{E}[\phi(\gamma_{\lambda_1},\dots,\gamma_{\lambda_n}) | \gamma_{\lambda_n} = x]$$
$$= \int_{\mathbb{S}^{n,x}} \phi(x_1,\dots,x_{n-1},x) f_n^x(x_1,\dots,x_{n-1}) dx_1 \dots dx_{n-1}$$
$$= C \int_{\mathbb{S}^{n,1}} \phi(xy_1,\dots,xy_{n-1},x) y_1^{\lambda_1 - 1}(y_2 - y_1)^{\lambda_2 - \lambda_1 - 1}\dots$$
$$(y_{n-1} - y_{n-2})^{\lambda_{n-1} - \lambda_{n-2} - 1}(1 - y_{n-1})^{\lambda_n - \lambda_{n-1} - 1} dy_1 \dots dy_{n-1}$$

after the change of variables: $x_i = xy_i$, $i = 1,\dots,n-1$. It is then clear that since ϕ increases with respect to each of its arguments, this last expression is an increasing function with respect to x.

Corollary 1.5. *Let $(\gamma_\lambda, \lambda \geq 0)$ be the gamma subordinator. Then, for every finite positive measure μ on $\mathbb{R}_+$, and for every $p > 0$, the process:*

$$\left(A_t^{(\mu,p)} := \int_0^\infty e^{-t(\gamma_\lambda)^p - h_{\lambda,p}(t)} \mu(d\lambda) \, , \, t \geq 0 \right) \tag{1.96}$$

is a peacock. Here, the function $h_{\lambda,p}$ is defined as:

$$h_{\lambda,p}(t) = \log \mathbb{E}\left[\exp(-t(\gamma_\lambda)^p)\right].$$

Proof. By Remark 1.15 with $\theta(x) = -x^p$ for $x \geq 0$, the process $(X_\lambda := -\gamma_\lambda^p, \lambda \geq 0)$ is conditionally monotone. Since it is a negative process, (INT1) is obviously satisfied. Moreover, since $(\gamma_\lambda, \lambda \geq 0)$ is an increasing process, (INT2) is easily verified. Finally, Theorem 1.12 holds. □

Remark 1.16. One can compare Corollary 1.5 with Theorem 1.8 applied with $(X_u, u \geq 0) = (-\gamma_\lambda^p, \lambda \geq 0)$.

- for $p = 1$, Theorem 1.8 holds if σ is a signed measure, while in Corollary 1.5, σ is positive,
- on the contrary, for $p > 0$, $p \neq 1$, Theorem 1.8 does not apply.

1.4.2.2 The Simple Random Walk is Conditionally Monotone

Let $(\varepsilon_i, i \in \mathbb{N}^*)$ be a sequence of independent and identically distributed r. v.'s such that, for every $i \in \mathbb{N}^*$:

$$\mathbb{P}(\varepsilon_i = 1) = p, \quad \mathbb{P}(\varepsilon_i = -1) = q \qquad \text{with } p,q > 0 \text{ and } p+q = 1.$$

Let $(S_n, n \in \mathbb{N})$ be the random walk defined by: $S_0 = 0$ and

$$S_n = \sum_{i=1}^{n} \varepsilon_i, \qquad \text{for every } n \in \mathbb{N}^*.$$

We shall prove that $(S_n, n \in \mathbb{N})$ is conditionally monotone; i.e: for every $r \in [\![2,+\infty[\![$, every $0 \leq n_1 < n_2 < \cdots < n_r < +\infty$ and every function $\phi : \mathbb{R}^{r-1} \longrightarrow \mathbb{R}$ in $\mathscr{E}_{r-1}$,

$$k \in I_{n_r} \longmapsto \mathbb{E}[\phi(S_{n_1}, S_{n_2}, \dots, S_{n_{r-1}})|S_{n_r} = k] \text{ is an increasing function on } I_{n_r} \quad (1.97)$$

where $I_x \subset [\![-x,x]\!]$ denotes the set of all the values the r.v. S_x can take. It is not difficult to see that (1.97) holds if and only if: for every $N \in [\![2,+\infty[\![$ and every function $\phi : \mathbb{R}^{N-1} \longrightarrow \mathbb{R}$ in $\mathscr{E}_{N-1}$:

$$k \in I_N \longmapsto \mathbb{E}[\phi(S_1, \dots, S_{N-1})|S_N = k] \quad \text{is an increasing function on } I_N. \quad (1.98)$$

We shall distinguish two cases:

1) If N and k are even, we set $N = 2n$ ($n \in [\![1,+\infty[\![$) and $k = 2x$ ($x \in [\![-n,n]\!]$). For every $n \in [\![1,+\infty[\![$ and every $x \in [\![-n,n]\!]$, let us denote by $\mathscr{J}_{2n}^{2x}$, the set of polygonal lines $\omega := (\omega_i, i \in [\![0,2n]\!])$ such that $\omega_0 = 0$, $\omega_{p+1} = \omega_p \pm 1$, ($p \in [\![0,2n-1]\!]$) and $\omega_{2n} = 2x$. Observe that any $\omega \in \mathscr{J}_{2n}^{2x}$ has $n+x$ positive slopes and $n-x$ negative ones. This implies that:

$$|\mathscr{J}_{2n}^{2x}| = C_{2n}^{n+x},$$

where $|\cdot|$ denotes cardinality. It is well known that, conditionally on $\{S_{2n} = 2x\}$, the law of the random vector $(S_0, S_1, \dots, S_{2n})$ is the uniform law on $\mathscr{J}_{2n}^{2x}$.

Let $n \in [\![1,+\infty[\![$ and $x \in [\![-n,n]\!]$ be fixed and consider, for every $i \in [\![1,n+x+1]\!]$ the map:

$$\Pi_i : \mathscr{J}_{2n}^{2x+2} \longrightarrow \mathscr{J}_{2n}^{2x}$$

defined by: for every $\omega \in \mathscr{J}_{2n}^{2x+2}$, $\Pi_i(\omega)$ has the same negative slopes and the same positive slopes as ω except the ith positive slope which is replaced by a negative one. For every $\omega \in \mathscr{J}_{2n}^{2x+2}$ and every function $\phi : \mathbb{R}^{2n} \longrightarrow \mathbb{R}$ in $\mathscr{E}_{2n}$,

$$\phi(\omega) \geq \phi(\Pi_i(\omega)).$$

Summing both side of this relation, we obtain:

$$(n+x+1) \sum_{\omega \in \mathscr{J}_{2n}^{2x+2}} \phi(\omega) \geq \sum_{\omega \in \mathscr{J}_{2n}^{2x+2}} \sum_{i=1}^{n+x+1} \phi(\Pi_i(\omega))$$

$$= \sum_{\omega \in \mathscr{I}_{2n}^{2x}} \sum_{i=1}^{n+x+1} |\Pi_i^{-1}(\omega)| \phi(\omega)$$
$$= (n-x) \sum_{\omega \in \mathscr{I}_{2n}^{2x}} \phi(\omega).$$

Thus, we have proved the following:

Lemma 1.3. *For every $n \in \mathbb{N}^*$ and every $\phi : \mathbb{R}^{2n} \longrightarrow \mathbb{R}$ in $\mathscr{E}_{2n}$,*

$$\frac{1}{|\mathscr{I}_{2n}^{2x+2}|} \sum_{\omega \in \mathscr{I}_{2n}^{2x+2}} \phi(\omega) \geq \frac{1}{|\mathscr{I}_{2n}^{2x}|} \sum_{\omega \in \mathscr{I}_{2n}^{2x}} \phi(\omega), \tag{1.99}$$

which means that $(S_{2n}, n \in \mathbb{N})$ is conditionally monotone.

2) It is not difficult to establish a similar result when k and N are odd.

Corollary 1.6. *For every odd and positive integer p, and for every positive finite measure $\sum_{n \in \mathbb{N}} a_n \delta_n$ on $\mathbb{N}$:*

$$\left(\sum_{n=0}^{+\infty} a_n e^{-t(S_n)^p - h_{n,p}(t)}, t \geq 0 \right) \quad \textit{is a peacock.}$$

Here, the function $h_{n,p}$ is defined by: $h_{n,p}(t) = \log \mathbb{E}\left[\exp\left(-t(S_n)^p\right)\right]$.

1.4.2.3 The Processes with Independent Log-concave Increments are Conditionally Monotone

We first introduce the notions of PF_2 and log-concave random variables (see [49]).

Definition 1.12 ($\mathbb{R}$-valued PF_2 r.v.'s). An $\mathbb{R}$-valued random variable X is said to be PF_2 if:

1) the law of X admits a probability density f,
2) for every $x_2 \geq x_1$, $y_2 \geq y_1$,

$$\det \begin{pmatrix} f(x_1 - y_1) & f(x_1 - y_2) \\ f(x_2 - y_1) & f(x_2 - y_2) \end{pmatrix} \geq 0.$$

Definition 1.13 ($\mathbb{Z}$-valued PF_2 r.v.'s). A $\mathbb{Z}$-valued random variable X is said to be PF_2 if, setting $f(x) = \mathbb{P}(X = x)$ $(x \in \mathbb{Z})$, one has: for every $x_2 \geq x_1$, $y_2 \geq y_1$,

$$\det \begin{pmatrix} f(x_1 - y_1) & f(x_1 - y_2) \\ f(x_2 - y_1) & f(x_2 - y_2) \end{pmatrix} \geq 0.$$

Definition 1.14 ($\mathbb{R}$-valued log-concave r.v.'s). An $\mathbb{R}$-valued random variable X is said to be log-concave if:

1) the law of X admits a probability density f,
2) f is a log-concave function on $\mathbb{R}$, i.e. for every $x, y \in \mathbb{R}$ and every $\theta \in]0,1[$,

$$f(\theta x + (1-\theta)y) \geq (f(x))^{\theta} (f(y))^{1-\theta}.$$

Obviously f is a log-concave function on $\mathbb{R}$ if and only if the set $S_f := \{f > 0\}$ is an interval and $\log f$ is concave on S_f.

Definition 1.15 ($\mathbb{Z}$-valued log-concave r.v.'s). A $\mathbb{Z}$-valued random variable X is said to be log-concave if, with $f(x) = \mathbb{P}(X = x)$, $(x \in \mathbb{Z})$, f is a log-concave function on $\mathbb{Z}$, i.e. for every $n \in \mathbb{Z}$

$$f^2(n) \geq f(n-1)f(n+1).$$

Here again, f is a log-concave function on $\mathbb{Z}$ if and only if the set $S_f := \{f > 0\}$ is an interval of $\mathbb{Z}$, and $\log f$ is a (discrete) concave function on S_f.

We recall the following equivalence:

Theorem 1.13 (see [5] or [49]). *An $\mathbb{R}$-valued (or $\mathbb{Z}$-valued) random variable is PF_2 if and only if it is* log-*concave.*

Example 1.2. Many common density functions on $\mathbb{R}$ (or $\mathbb{Z}$) are PF_2. Indeed, the normal density, the uniform density, the exponential density, the negative binomial density, the Poisson density and the geometric density are PF_2. We refer to [5] for more examples. Note that:

a) a gamma random variable of parameter $a > 0$ (with density $f_a(x) = \dfrac{1}{\Gamma(a)} e^{-x} x^{a-1}, x > 0$) is not PF_2 if $a < 1$,
b) a Bernoulli random variable X such that $\mathbb{P}(X = 1) = p = 1 - \mathbb{P}(X = -1)$ is not PF_2.

The following result is due to Efron [62] (see also [159]).

Theorem 1.14. *Let $n \in [\![1, +\infty[\![$, $X_1, X_2, \ldots, X_n$ be independent $\mathbb{R}$-valued (or $\mathbb{Z}$-valued) PF_2 random variables, $S_n = \sum_{i=1}^{n} X_i$, and $\phi : \mathbb{R}^n \to \mathbb{R}$ belonging to $\mathscr{E}_n$. Then,*

$$\mathbb{E}[\phi(X_1, X_2, \ldots, X_n) | S_n = x] \quad \textit{is increasing in } x.$$

Thanks to Theorem 1.14, we obtain the following result:

Theorem 1.15. *Let $(Z_\lambda, \lambda \in \mathbb{R}_+$ or $\lambda \in \mathbb{N})$ be a $\mathbb{R}$-valued (or $\mathbb{Z}$-valued) process satisfying (INT1) and (INT2), with independent (not necessarily time-homogeneous) PF_2*

increments. Then, $(Z_\lambda, \lambda \geq 0)$ is conditionally monotone, and for every positive measure μ on $\mathbb{R}_+$ (or $\mathbb{N}$) with finite total mass,

$$\left(\int_0^{+\infty} e^{tZ_\lambda - h_\lambda(t)}\mu(d\lambda), t \geq 0\right) \quad \textit{is a peacock,}$$

where the function h_λ is defined by: $h_\lambda(t) = \log \mathbb{E}\left[e^{tZ_\lambda}\right]$.

Proof. It suffices to show that $(Z_\lambda, \lambda \geq 0)$ satisfies $(\widetilde{CM})$. Let $n \in [\![1, +\infty[\![$ and $\phi : \mathbb{R}^n \to \mathbb{R}$ belonging to $\mathscr{E}_n$. For every $0 \leq \lambda_1 < \lambda_2 < \cdots < \lambda_n$ and $k \in \mathbb{R}$ (or $\mathbb{Z}$),

$$\mathbb{E}[\phi(Z_{\lambda_1}, Z_{\lambda_2}, \ldots, Z_{\lambda_n})|Z_{\lambda_n} = k] = \mathbb{E}\left[\widehat{\phi}(Z_{\lambda_1}, Z_{\lambda_2} - Z_{\lambda_1}, \ldots, Z_{\lambda_n} - Z_{\lambda_{n-1}})|Z_{\lambda_n} = k\right],$$

where the function $\widehat{\phi}$ is given by:

$$\widehat{\phi}(x_1, x_2, \ldots, x_n) = \phi(x_1, x_1 + x_2, \ldots, x_1 + x_2 + \cdots + x_n).$$

It is obvious that $\widehat{\phi}$ belongs to $\mathscr{E}_n$. Thus, applying Theorem 1.14 with: $X_1 = Z_{\lambda_1}$ and $X_{i+1} = Z_{\lambda_{i+1}} - Z_{\lambda_i}$ $i = 1, \ldots n-1$, one obtains the desired result. □

Remark 1.17. 1) Theorem 1.15 applies neither in the case of the Gamma subordinator, nor in the case of the random walk whose increments are Bernoulli with values in $\{-1, 1\}$. Nevertheless, its conclusion remains true in these cases, see Subsections 1.4.2.1 and 1.4.2.2.

2) We deduce from Theorem 1.15 that the Poisson process and the random walk with geometric increments are conditionally monotone. We shall give a direct proof below, i.e. without using Theorem 1.14.

1.4.2.4 The Poisson Process is Conditionally Monotone

Let $(N_\lambda, \lambda \geq 0)$ be a Poisson process with parameter 1 and let $(T_n, n \geq 1)$ denote its successive jump times. Then

$$N_\lambda = \#\{i \geq 1 : T_i \leq \lambda\}.$$

In order to prove that $(N_\lambda, \lambda \geq 0)$ is conditionally monotone, we shall show that for every $0 \leq \lambda_1 < \cdots < \lambda_n$ and every function $\phi : \mathbb{R}^n \longrightarrow \mathbb{R}$ in $\mathscr{E}_n$, we have:

$$\mathbb{E}[\phi(N_{\lambda_1}, \ldots, N_{\lambda_n})|N_{\lambda_n}] = \phi_n(N_{\lambda_n}), \tag{1.100}$$

where $\phi_n : \mathbb{R} \longrightarrow \mathbb{R}$ increases. But, conditionally on $\{N_{\lambda_n} = k\}$, the random vector $(T_1, \ldots, T_k)$ is distributed as $(U_1, \ldots, U_k)$, $U_1, \ldots, U_k$ being the increasing rearrangement of k independent random variables, uniformly distributed on $[0, \lambda_n]$. We go from k to $k+1$ by adding one more point. Thus, with obvious notation, it is clear that: for all $\lambda \in [0, \lambda_n]$, $N_\lambda^{(k+1)} \geq N_\lambda^{(k)}$. Then, the conditional monotonicity property follows immediately.

Corollary 1.7. *Let $(N_\lambda, \lambda \geq 0)$ be a Poisson process and let μ be a finite positive measure on $\mathbb{R}_+$. Then, for every $p > 0$, the process:*

$$\left(A_t^{(\mu,p)} := \int_0^\infty e^{-t(N_\lambda)^p - h_{\lambda,p}(t)} \mu(d\lambda), t \geq 0\right) \tag{1.101}$$

is a peacock with:

$$h_{\lambda,p}(t) = \log \mathbb{E}\left[\exp\left(-t(N_\lambda)^p\right)\right].$$

1.4.2.5 The Random Walk with Geometric Increments is Conditionally Monotone

Let $(\varepsilon_i, i \in [\![1, +\infty[\![)$ be a sequence of independent geometric variables with the same parameter p; i.e, such that:

$$\mathbb{P}(\varepsilon_i = k) = p^k(1-p) \qquad (k \geq 0, 0 < p < 1).$$

We consider the random walk $(S_n, n \in \mathbb{N})$ defined by:

$$S_0 = 0 \quad \text{and} \quad S_n = \sum_{i=1}^n \varepsilon_i, \text{ for every } n \in \mathbb{N}^*.$$

For $n \in \mathbb{N}^*$, S_n is distributed as a negative binomial random variable with parameters n and p; more precisely:

$$\mathbb{P}(S_n = k) = C_{n+k-1}^k p^k (1-p)^n, \text{ for every } k \in \mathbb{N}.$$

As in Subsection 1.4.2.2, we only need to prove that: for every $N \in \mathbb{N}^*$ and every function $\phi : \mathbb{R}^N \longrightarrow \mathbb{R}$ in $\mathscr{E}_N$:

$$k \longmapsto \mathbb{E}[\phi(S_1, \dots, S_N) | S_{N+1} = k] \quad \text{is an increasing function on } \mathbb{N}. \tag{1.102}$$

Let $\mathbf{J}_N^k$ denote the set:

$$\mathbf{J}_N^k := \{(x_1, \dots, x_N) \in \mathbb{N}^N : 0 \leq x_1 \leq \dots \leq x_N \leq k\}. \tag{1.103}$$

For every $k \geq 0$ and $N \geq 1$, it is well known that $|\mathbf{J}_N^k| = C_{N+k}^k$. Now, we have:

$$\begin{aligned}
&\mathbb{E}\left[\phi\left(S_1, \dots, S_N\right) | S_{N+1} = k\right] \\
&= \sum_{(l_1,\dots,l_N) \in \mathbf{J}_N^k} \phi\left(l_1, \dots, l_N\right) \frac{\mathbb{P}(S_1 = l_1, \dots, S_N = l_N, S_{N+1} = k)}{\mathbb{P}(S_{N+1} = k)} \\
&= \sum_{(l_1,\dots,l_N) \in \mathbf{J}_N^k} \phi\left(l_1, \dots, l_N\right) \frac{\mathbb{P}(S_1 = l_1, S_2 - S_1 = l_2 - l_1, \dots, S_{N+1} - S_N = k - l_N)}{\mathbb{P}(S_{N+1} = k)} \\
&= \sum_{(l_1,\dots,l_N) \in \mathbf{J}_N^k} \phi\left(l_1, \dots, l_N\right) \frac{\mathbb{P}(S_1 = l_1)\mathbb{P}(S_2 - S_1 = l_2 - l_1) \dots \mathbb{P}(S_{N+1} - S_N = k - l_N)}{\mathbb{P}(S_{N+1} = k)}
\end{aligned}$$

$$= \sum_{(l_1,\dots,l_N)\in\mathbf{J}_N^k} \phi(l_1,\dots,l_N) \frac{p^{l_1}(1-p)p^{l_2-l_1}(1-p)\dots p^{k-l_N}(1-p)}{C_{N+k}^k p^k(1-p)^{N+1}}$$
$$= \frac{1}{C_{N+k}^k} \sum_{(l_1,\dots,l_N)\in\mathbf{J}_N^k} \phi(l_1,\dots,l_N)$$
$$= \frac{1}{|\mathbf{J}_N^k|} \sum_{(l_1,\dots,l_N)\in\mathbf{J}_N^k} \phi(l_1,\dots,l_N).$$

Therefore, the law of the random vector $(S_1,\dots,S_N)$ conditionally on $\{S_{N+1}=k\}$ is the uniform law on the set $\mathbf{J}_N^k$. Hence, we will obtain (1.102) if we prove that: for every $k\in\mathbb{N}$, every $N\in\mathbb{N}^*$ and every function $\phi:\mathbb{R}^N\longrightarrow\mathbb{R}_+$ in $\mathscr{E}_N$:

$$\frac{1}{|\mathbf{J}_N^k|}\sum_{x\in\mathbf{J}_N^k}\phi(x)\le\frac{1}{|\mathbf{J}_N^{k+1}|}\sum_{x\in\mathbf{J}_N^{k+1}}\phi(x). \tag{1.104}$$

Let us notice that:

$$\mathbf{J}_N^0=\{(\underbrace{0,\dots,0}_{N\text{ times}})\},\qquad \text{for every } N\in[\![1,+\infty[\![$$

and

$$\mathbf{J}_1^k=\{(0),(1),\dots,(k)\},\qquad \text{for every } k\in[\![0,+\infty[\![.$$

For $k\in[\![0,+\infty[\![$ and $N\in[\![1,+\infty[\![$, we define:

$$\mathbf{D}_N^{k+1}:=\mathbf{J}_N^{k+1}\setminus\mathbf{J}_N^k=\{(x_1,\dots,x_N)\in\mathbf{J}_N^{k+1}: x_N=k+1\}. \tag{1.105}$$

and set $\mathbf{D}_N^0=\emptyset$. By Pascal's formula,

$$|\mathbf{D}_N^{k+1}|=C_{k+1+N}^{k+1}-C_{k+N}^k=C_{N+k}^{k+1}=|\mathbf{J}_{N-1}^{k+1}|,\quad (\text{with } N\in[\![2,+\infty[\![).$$

On one hand, we consider, for $N\in[\![2,+\infty[\![$, the map $\Gamma:\mathbf{J}_{N-1}^{k+1}\longrightarrow\mathbf{D}_N^{k+1}$ defined by:

$$\Gamma[(x_1,\dots,x_{N-1})]=(x_1,\dots,x_{N-1},k+1). \tag{1.106}$$

The map Γ is bijective, and for every non empty pair of subsets G and H of $\mathbf{J}_{N-1}^{k+1}$, there is the equivalence:

$$\begin{cases}\forall f:\mathbb{R}^{N-1}\longrightarrow\mathbb{R}\in\mathscr{E}_{N-1},\\[2ex] \dfrac{1}{|G|}\sum_{x\in G}f(x)\le\dfrac{1}{|H|}\sum_{x\in H}f(x)\end{cases}\iff\begin{cases}\forall\phi:\mathbb{R}^N\longrightarrow\mathbb{R}\in\mathscr{E}_N,\\[2ex] \dfrac{1}{|\Gamma(G)|}\sum_{z\in\Gamma(G)}\phi(z)\le\dfrac{1}{|\Gamma(H)|}\sum_{z\in\Gamma(H)}\phi(z).\end{cases}$$

On the other hand, for $N\in[\![2,+\infty[\![$, let $\Lambda:\mathbf{D}_N^k\longrightarrow\mathbf{D}_N^{k+1}$ be the injection given by:

$$\Lambda[(x_1,\dots,x_{N-1},k)]=(x_1,\dots,x_{N-1},k+1). \tag{1.107}$$

For every $z \in \mathbf{D}_N^k$ and function $\phi : \mathbb{R}^N \longrightarrow \mathbb{R}$ in $\mathscr{E}_N$,

$$\phi(z) \leq \phi(\Lambda(z)).$$

Therefore, for every non empty subset K of $\mathbf{D}_N^k$,

$$\frac{1}{|K|} \sum_{z \in K} \phi(z) \leq \frac{1}{|\Lambda(K)|} \sum_{u \in \Lambda(K)} \phi(u). \tag{1.108}$$

since $|K| = |\Lambda(K)|$. Furthermore, one notices that:

$$\Gamma^{-1}[\Lambda(\mathbf{D}_N^k)] = \mathbf{J}_{N-1}^k \quad \text{and} \quad \Gamma^{-1}(\mathbf{D}_N^{k+1}) = \mathbf{J}_{N-1}^{k+1}$$

where Γ^{-1} denotes the inverse map of Γ. Hence, the following is easily obtained:

Lemma 1.4. *Let $k \in [\![1, +\infty[\![$ and $N \in [\![2, +\infty[\![$. Assume that for every function $f : \mathbb{R}^{N-1} \longrightarrow \mathbb{R}$ in $\mathscr{E}_{N-1}$:*

$$\frac{1}{|\mathbf{J}_{N-1}^k|} \sum_{x \in \mathbf{J}_{N-1}^k} f(x) \leq \frac{1}{|\mathbf{J}_{N-1}^{k+1}|} \sum_{x \in \mathbf{J}_{N-1}^{k+1}} f(x). \tag{1.109}$$

Then, for every function $\phi : \mathbb{R}^N \to \mathbb{R}$ in $\mathscr{E}_N$,

$$\frac{1}{|\mathbf{D}_N^k|} \sum_{y \in \mathbf{D}_N^k} \phi(y) \leq \frac{1}{|\mathbf{D}_N^{k+1}|} \sum_{y \in \mathbf{D}_N^{k+1}} \phi(y). \tag{1.110}$$

Now, we are able to prove (1.104) by induction on $N \in [\![1, +\infty[\![$ and $k \in [\![0, +\infty[\![$.

Proposition 1.7. *Let $k \in [\![0, +\infty[\![$, $N \in [\![1, +\infty[\![$ and let $\phi : \mathbb{R}^N \to \mathbb{R}$ be any function in $\mathscr{E}_N$. Then,*

$$\frac{1}{|\mathbf{J}_N^k|} \sum_{z \in \mathbf{J}_N^k} \phi(z) \leq \frac{1}{|\mathbf{J}_N^{k+1}|} \sum_{z \in \mathbf{J}_N^{k+1}} \phi(z); \tag{1.111}$$

in other words, $(S_n, n \in \mathbb{N})$ is conditionally monotone.

Proof. 1) It is obvious that (1.111) holds for $(k,N) \in [\![0, +\infty[\![\times \{1\}$, and for $(k,N) \in \{0\} \times [\![1, +\infty[\![$.
2) Let $(k,N) \in [\![1, +\infty[\![\times [\![2, +\infty[\![$. We assume that:

$$\forall\, (l,m) \in \mathscr{D} := [\![0, k-1]\!] \times [\![1, +\infty[\![\cup \{k\} \times [\![1, N-1]\!]$$

and any function $f : \mathbb{R}^m \to \mathbb{R}$ in $\mathscr{E}_m$:

$$\frac{1}{|\mathbf{J}_m^l|} \sum_{x \in \mathbf{J}_m^l} f(x) \leq \frac{1}{|\mathbf{J}_m^{l+1}|} \sum_{x \in \mathbf{J}_m^{l+1}} f(x). \tag{IH}$$

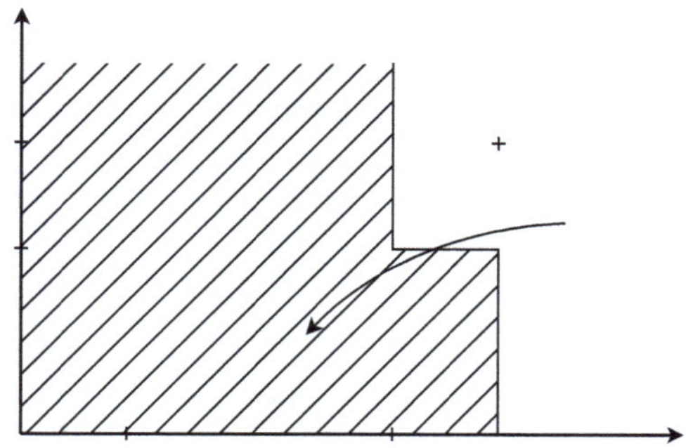

Fig. 1.1. $\mathscr{D} := [\![0,k-1]\!] \times [\![1,+\infty[\![\cup \{k\} \times [\![1,N-1]\!]$

By taking $(l,m) = (k,N-1)$ in (IH), lemma (1.4) yields:

$$\frac{1}{|\mathbf{D}_N^k|}\sum_{y\in\mathbf{D}_N^k}\phi(y) \le \frac{1}{|\mathbf{D}_N^{k+1}|}\sum_{y\in\mathbf{D}_N^{k+1}}\phi(y). \tag{1.112}$$

On the other hand, from the definition of $\mathbf{D}_N^{k+1}$, (1.111) is equivalent to:

$$\frac{1}{|\mathbf{J}_N^k|}\sum_{y\in\mathbf{J}_N^k}\phi(y) \le \frac{1}{|\mathbf{D}_N^{k+1}|}\sum_{y\in\mathbf{D}_N^{k+1}}\phi(y). \tag{1.113}$$

Using (IH) with $(l,m) = (k-1,N)$, we have:

$$\frac{1}{|\mathbf{J}_N^k|}\sum_{y\in\mathbf{J}_N^k}\phi(y) \le \frac{1}{|\mathbf{D}_N^k|}\sum_{y\in\mathbf{D}_N^k}\phi(y). \tag{1.114}$$

The comparison of (1.112) with (1.114) yields (1.113) which is equivalent to (1.111). □

Corollary 1.8. *For every positive finite measure $\sum_{n\in\mathbb{N}} a_n\delta_n$ on $\mathbb{N}$ and every $p>0$:*

$$\left(\sum_{n=0}^{+\infty} a_n e^{-t(S_n)^p - h_{n,p}(t)}, t\ge 0\right) \quad \text{is a peacock,}$$

where the function $h_{n,p}$ is defined by: $h_{n,p}(t) = \log\mathbb{E}\left[\exp\left(-t(S_n)^p\right)\right]$.

Remark 1.18. The result in this example may be compared with that of Subsection 1.4.2.1: we replace the gamma r.v's by geometric ones.

1.4.3 "Well-Reversible" Diffusions at Fixed Times

We now present an important class of conditionally monotone processes: that of the "well-reversible" diffusions at a fixed time.

1.4.3.1 The Diffusion $(X_\lambda, \lambda \geq 0; \mathbb{P}_x, x \in \mathbb{R})$

Let $\sigma : \mathbb{R}_+ \times \mathbb{R} \to \mathbb{R}$ and $b : \mathbb{R}_+ \times \mathbb{R} \to \mathbb{R}$ be two Borel measurable functions and let $(B_u, u \geq 0)$ be a standard Brownian motion starting from 0. We consider the SDE:

$$X_\lambda = x + \int_0^\lambda \sigma(s, X_s)\, dB_s + \int_0^\lambda b(s, X_s)\, ds, \ \lambda \geq 0. \tag{1.115}$$

We assume that:

(A1) For every $x \in \mathbb{R}$, this SDE admits a unique pathwise solution $(X_\lambda^{(x)}, \lambda \geq 0)$, and furthermore the mapping $x \longmapsto (X_\lambda^{(x)}, \lambda \geq 0)$ may be chosen measurable.

As a consequence of (A1), Yamada-Watanabe's theorem asserts that $(X_\lambda^{(x)}, \lambda \geq 0)$ is a strong solution of equation (1.115), and that it enjoys the strong Markov property; finally the transition kernel $P_\lambda(x, dy) = \mathbb{P}(X_\lambda^{(x)} \in dy)$ is measurable.

We now remark that, for $x \leq y$, the process $(X_\lambda^{(y)}, \lambda \geq 0)$ is stochastically greater than $(X_\lambda^{(x)}, \lambda \geq 0)$ in the following sense: for every $a \in \mathbb{R}$ and $\lambda \geq 0$,

$$\mathbb{P}\left(X_\lambda^{(y)} \geq a\right) \geq \mathbb{P}\left(X_\lambda^{(x)} \geq a\right). \tag{1.116}$$

Indeed, assuming that both $(X_\lambda^{(x)}, \lambda \geq 0)$ and $(X_\lambda^{(y)}, \lambda \geq 0)$ are defined on the same probability space, and considering the coupling time:

$$T = \inf\{\lambda \geq 0; X_\lambda^{(x)} = X_\lambda^{(y)}\}$$
$$(= +\infty \text{ if } \{\lambda \geq 0; X_\lambda^{(x)} = X_\lambda^{(y)}\} = \emptyset),$$

it is clear that, on $\{T = +\infty\}$,

$$X_\lambda^{(y)} \geq X_\lambda^{(x)} \qquad (\text{since } y \geq x)$$

while on $\{T < +\infty\}$, we have:

$$X_\lambda^{(y)} > X_\lambda^{(x)} \qquad \text{for every } \lambda \in [0, T[$$

and

$$X_\lambda^{(y)} = X_\lambda^{(x)} \qquad \text{for every } \lambda \in [T, +\infty[$$

since, as a consequence of our hypothesis (A1), and as already recalled above, (1.115) admits a unique strong Markovian solution.

On the other hand, (1.116) is equivalent to: for every bounded and increasing (resp. decreasing) function, and for every $\lambda \geq 0$:

$$x \longrightarrow \mathbb{E}_x[\phi(X_\lambda)] = \int_{\mathbb{R}} P_\lambda(x, dy)\phi(y) \text{ is increasing (resp. decreasing).} \tag{1.117}$$

Lemma 1.5. *Let $((X_\lambda)_{\lambda \geq 0}, (\mathscr{F}_\lambda)_{\lambda \geq 0}, (\mathbb{P}_x)_{x \in \mathbb{R}})$ be a Markov process in $\mathbb{R}$ which satisfies (1.116). Then, for every $n \geq 1$, every $0 \leq \lambda_1 < \cdots < \lambda_n$, every $i \in \{1, \dots, n\}$, every function $\phi : \mathbb{R}^n \to \mathbb{R}$ in $\mathscr{E}_n$, and for every $x \geq 0$,*

$$\mathbb{E}_x[\phi(X_{\lambda_1}, \dots, X_{\lambda_n}) | \mathscr{F}_{\lambda_i}] = \widetilde{\phi}_i(X_{\lambda_1}, \dots, X_{\lambda_i}), \tag{1.118}$$

where $\widetilde{\phi}_i : \mathbb{R}^i \to \mathbb{R}$ belongs to $\mathscr{E}_i$. In particular,

$$x \longrightarrow \mathbb{E}_x[\phi(X_{\lambda_1}, \dots, X_{\lambda_n})] \text{ is increasing.} \tag{1.119}$$

Proof. If $i = n$, (1.118) is obvious. If $i = n-1$, then (1.118) is satisfied since:

$$\mathbb{E}_x[\phi(X_{\lambda_1}, \dots, X_{\lambda_{n-1}}, X_{\lambda_n}) | \mathscr{F}_{\lambda_{n-1}}] = \int_{\mathbb{R}} \phi(X_{\lambda_1}, \dots, X_{\lambda_{n-1}}, y) P_{\lambda_n - \lambda_{n-1}}(X_{\lambda_{n-1}}, dy)$$

and then, for $i = n-1$, (1.118) follows immediately from (1.117). Thus, Lemma 1.5 follows by iteration of this argument. □

Observe that as a consequence of Lemma 1.5, the conditional monotonicity property (1.92) for these diffusions is equivalent to $(\widetilde{\mathrm{CM}})$.

1.4.3.2 Time-reversal at a Fixed Time

Let $x \in \mathbb{R}$ fixed. We assume that:

(A2) For every $\lambda > 0$, $\sigma(\lambda, \cdot)$ is a differentiable function and X_λ admits a $\mathscr{C}^{1,2}$ density function p on $]0, +\infty[\times \mathbb{R}$.

By setting

$$a(\lambda, y) := \sigma^2(\lambda, y) \qquad \text{for every } \lambda \geq 0 \text{ and } y \in \mathbb{R},$$

we define successively, for any fixed $\lambda_0 > 0$ and for $y \in \mathbb{R}$:

$$\begin{cases} \overline{a}^{\lambda_0}(\lambda, y) &= a(\lambda_0 - \lambda, y), \quad (0 \leq \lambda \leq \lambda_0) \\ \overline{b}^{\lambda_0}(\lambda, y) &= -b(\lambda_0 - \lambda, y) + \dfrac{1}{p(\lambda_0 - \lambda, y)} \dfrac{\partial}{\partial y}\big(a(\lambda_0 - \lambda, y)\, p(\lambda_0 - \lambda, y)\big), \\ & \quad (0 \leq \lambda < \lambda_0) \end{cases} \tag{1.120}$$

and the differential operator $L^{\lambda_0}_\lambda$, $(0 \leq \lambda < \lambda_0)$:

$$L^{\lambda_0}_\lambda f(x) = \frac{1}{2} \overline{a}^{\lambda_0}(\lambda, y) f''(y) + \overline{b}^{\lambda_0}(\lambda, y) f'(y) \quad \text{for } f \in \mathscr{C}^2_b.$$

Under some suitable conditions on a and b, U.G. Haussmann and E. Pardoux [76] (see also P.A. Meyer [129]) proved that:

(A3) The process $(\overline{X}_\lambda^{\lambda_0}, 0 \leq \lambda < \lambda_0)$ obtained by time-reversing $(X_\lambda, 0 < \lambda \leq \lambda_0)$ at time λ_0:

$$(\overline{X}_\lambda^{\lambda_0}, 0 \leq \lambda < \lambda_0) := (X_{\lambda_0 - \lambda}, 0 \leq \lambda < \lambda_0)$$

is a diffusion and there exists a Brownian motion $(\overline{B}_u, 0 \leq u \leq \lambda_0)$, independent of X_{λ_0}, such that $(\overline{X}_\lambda^{\lambda_0}, 0 \leq \lambda < \lambda_0)$ solves the SDE:

$$\begin{cases} dY_\lambda &= \overline{\sigma}^{\lambda_0}(\lambda, Y_\lambda) d\overline{B}_\lambda + \overline{b}^{\lambda_0}(\lambda, Y_\lambda) d\lambda \qquad (0 \leq \lambda < \lambda_0) \\ Y_0 &= X_{\lambda_0} \qquad (\text{with } \overline{\sigma}^{\lambda_0}(\lambda, y) = \sigma(\lambda_0 - \lambda, y)). \end{cases} \tag{1.121}$$

Note that the coefficients $\overline{b}^{\lambda_0}$ and $\overline{\sigma}^{\lambda_0}$ depend on x.

(A4) We assume furthermore that the SDE (1.121) admits a unique strong solution on $[0, \lambda_0[$; thus, this strong solution is strongly Markovian.

Note that, a priori, the solution of (1.121) is only defined on $[0, \lambda_0[$, but it can be extended on $[0, \lambda_0]$ by setting $\overline{X}_{\lambda_0}^{\lambda_0} = x$.

1.4.3.3 Our Hypotheses and the Main Result

Our goal here is not to give optimal hypotheses under which the assertions (A1)–(A4) are satisfied. We refer the reader to [76] or [131] for more details. Instead, we shall present two hypotheses (H1) and (H2), either of them implying the preceding assertions:

(H1) We assume that:

i) the functions $(\lambda, y) \longmapsto \sigma(\lambda, y)$ and $(\lambda, y) \longmapsto b(\lambda, y)$ are of $\mathscr{C}^{1,2}$ class on $]0, +\infty[\times\mathbb{R}$, locally Lipschitz continuous in y uniformly in λ, and the solution of (1.115) does not explode on $[0, \lambda_0]$,

ii) there exists $\alpha > 0$ such that:

$$a(\lambda, y) \equiv \sigma^2(\lambda, y) \geq \alpha \qquad \text{for every } y \in \mathbb{R} \text{ and } 0 \leq \lambda \leq \lambda_0,$$

and

$$\frac{\partial^2 a}{\partial y^2} \in L^\infty(]0, \lambda_0] \times \mathbb{R}_+).$$

(H2) We assume that:

i) the functions σ and b are of $\mathscr{C}^{1,2}$ class, locally Lipschitz continuous in y uniformly in λ, and the solution of (1.115) does not explode on $[0, \lambda_0]$,

ii) the functions a and b are of $\mathscr{C}^\infty$ class on $]0, +\infty[\times\mathbb{R}$ in (λ, y) and the differential operator

$$\overline{L} = \frac{\partial}{\partial \lambda} + L_\lambda$$

is hypoelliptic (see Ikeda-Watanabe [86, p.411] for the definition and properties of hypoelliptic operators), where $(L_\lambda, \lambda \geq 0)$ is the generator of the diffusion (1.115):

$$L_\lambda = \frac{1}{2}a(\lambda,\cdot)\frac{d}{dy^2} + b(\lambda,\cdot)\frac{d}{dy}. \tag{1.122}$$

Then, under either (H1) or (H2), the assertions $(A_i)_{i=1\ldots4}$ of both paragraphs 1.4.3.1 and 1.4.3.2 are satisfied, see [76]. In particular, $(\overline{X}^{\lambda_0}_\lambda, 0 \leq \lambda < \lambda_0)$ is a strong solution of equation (1.121), see P.A. Meyer [129]. Let us now give the main result of this subsection.

Theorem 1.16. *Under either (H1) or (H2), and for every $x \in \mathbb{R}$, the process $(X_\lambda, \lambda > 0)$ is conditionally monotone under $\mathbb{P}_x$.*

Proof. Let $n \in \mathbb{N}^*$ and let $\phi : \mathbb{R}^n \to \mathbb{R}$ in $\mathscr{E}_n$. For every $0 < \lambda_1 < \cdots < \lambda_n$ and every $i \in \{1, \ldots, n\}$:

$$\begin{aligned}
&\mathbb{E}_x\left[\phi(X_{\lambda_1}, \ldots, X_{\lambda_n})|X_{\lambda_i} = z\right] \\
&\quad = \mathbb{E}_x\left[\mathbb{E}_x\left[\phi(X_{\lambda_1}, \ldots, X_{\lambda_n})|\mathscr{F}_{\lambda_i}\right]|X_{\lambda_i} = z\right] \\
&\quad = \mathbb{E}_x\left[\widetilde{\phi}_i(X_{\lambda_1}, \ldots, X_{\lambda_i})|X_{\lambda_i} = z\right] \\
&\qquad \text{(by Lemma 1.5, where } \widetilde{\phi}_i : \mathbb{R}^i \to \mathbb{R} \text{ belongs to } \mathscr{E}_i \\
&\quad = \overline{\mathbb{E}}_x\left[\widetilde{\phi}_i(\overline{X}^{\lambda_i}_{\lambda_i-\lambda_1}, \ldots, \overline{X}^{\lambda_i}_0)|\overline{X}^{\lambda_i}_0 = z\right] \quad \text{(by time-reversal at } \lambda_i) \\
&\quad = \overline{\mathbb{E}}_z\left[\widetilde{\phi}_i(\overline{X}^{\lambda_i}_{\lambda_i-\lambda_1}, \ldots, \overline{X}^{\lambda_i}_{\lambda_i-\lambda_{i-1}}, z)\right]
\end{aligned}$$

and, by applying (1.119) to the reversed process $(\overline{X}^{\lambda_i}_\lambda, 0 \leq \lambda < \lambda_i)$, this last expression is a bounded function which increases with respect to z. □

Remark 1.19. Observe that we were careful to exclude the point $\lambda_1 = 0$ in Theorem 1.16, since a priori "well-reversible" diffusions can be only reversed on $]0, \lambda_0]$: $(\overline{X}^{\lambda_0}_\lambda, 0 \leq \lambda < \lambda_0) := (X_{\lambda_0 - \lambda}, 0 \leq \lambda < \lambda_0)$.

Corollary 1.9. *Let $(X_\lambda, \lambda > 0)$ the unique strong solution of (1.115), taking values in $\mathbb{R}_+$, where b and σ satisfy either (H1) or (H2). Then, for every finite positive measure μ on $]0, +\infty[$ and for every $p > 0$, the process:*

$$\left(A^{(\mu,p)}_t := \int_0^\infty e^{-t(X_\lambda)^p - h_{\lambda,p}(t)}\mu(d\lambda), t \geq 0\right) \tag{1.123}$$

is a peacock, with:

$$h_{\lambda,p}(t) = \log \mathbb{E}_x\left[\exp\left(-t(X_\lambda)^p\right)\right].$$

Proof. Let $\varepsilon > 0$ and define $\mu^{(\varepsilon)}$ to be the restriction of μ to the interval $[\varepsilon, +\infty[$: $\mu^{(\varepsilon)} := \mu_{|[\varepsilon,+\infty[}$. As $(X_\lambda, \lambda \geq \varepsilon)$ is a continuous positive process, conditions (INT1) and (INT2) are satisfied, and we may apply Theorems 1.16 and 1.12 to obtain that, for every $\psi \in \boldsymbol{C}$ and every $0 \leq s \leq t$:

$$\mathbb{E}\left[\psi(A_s^{(\mu^{(\varepsilon)})})\right] \leq \mathbb{E}\left[\psi(A_t^{(\mu^{(\varepsilon)})})\right].$$

Then, proceeding as in Point 4. of the proof of Theorem 1.12, the result follows by letting ε tend to 0. □

1.4.3.4 A Few Examples of Diffusions which are "Well-Reversible" at Fixed Times

Example 1.3 (Brownian motion with drift ν). We take $\sigma \equiv 1, b(s,y) = \nu$ and $X_\lambda = x + B_\lambda + \nu\lambda$. Then,

$$p(t,x,y) = \frac{1}{\sqrt{2\pi t}} \exp\left(-\frac{(y-(x+\nu t))^2}{2t}\right),$$

and $(\overline{X}_\lambda^{\lambda_0}, 0 \leq \lambda < \lambda_0)$ is the solution of:

$$Y_\lambda = \overline{X}_0^{\lambda_0} + \overline{B}_\lambda + \int_0^\lambda \frac{x - Y_u}{\lambda_0 - u} du$$

with $(\overline{B}_\lambda, 0 \leq \lambda < \lambda_0)$ independent from $\overline{X}_0^{\lambda_0} = X_{\lambda_0}^{(x)}$.

See Jeulin-Yor [97] for similar computations.

Example 1.4 (The Bessel process of dimension $\delta \geq 2$). We take $\sigma \equiv 1$ and $b(s,y) = \dfrac{\delta - 1}{2y}$, with $\delta = 2(\nu+1)$, $\delta \geq 2$. Then,

i) for $x > 0$:

$$p(t,x,y) = \frac{1}{t} \frac{y^{\nu+1}}{x^\nu} \exp\left(-\frac{x^2+y^2}{2t}\right) I_\nu\left(\frac{xy}{t}\right),$$

where I_ν denotes the modified Bessel function of index ν (see Lebedev [117, p.110] for the definition of I_ν), and $(\overline{X}_\lambda^{\lambda_0}, 0 \leq \lambda < \lambda_0)$ is the solution of:

$$Y_\lambda = \overline{X}_0^{\lambda_0} + \overline{B}_\lambda + \int_0^\lambda \left(\frac{1}{2Y_u} - \frac{Y_u}{\lambda_0 - u} + \frac{x}{\lambda_0 - u} \frac{I'_\nu}{I_\nu}\left(\frac{xY_u}{\lambda_0 - u}\right)\right) du,$$

ii) for $x = 0$:

$$p(t,0,y) = \frac{1}{2^\nu t^{\nu+1} \Gamma(\nu+1)} y^{2\nu+1} \exp\left(\frac{-y^2}{2t}\right),$$

and $(\overline{X}_\lambda^{\lambda_0}, 0 \leq \lambda < \lambda_0)$ is the solution of:

$$Y_\lambda = \overline{X}_0^{\lambda_0} + \overline{B}_\lambda + \int_0^\lambda \left(\frac{2\nu+1}{2Y_u} - \frac{Y_u}{\lambda_0 - u}\right) du.$$

This example has a strong resemblance with that of a Bessel process with drift, see Watanabe [169] and Pitman-Yor [144].

Example 1.5 (Squared Bessel process of dimension $\delta > 0$). We take $\sigma(s,y) = 2\sqrt{y}$ and $b \equiv \delta$. Then:
i) for $x > 0$:

$$p(t,x,y) = \frac{1}{2t}\left(\frac{y}{x}\right)^{\nu/2} \exp\left(-\frac{x+y}{2t}\right) I_\nu\left(\frac{\sqrt{xy}}{t}\right),$$

and $(\overline{X}^{\lambda_0}_\lambda, 0 \leq \lambda < \lambda_0)$ is the solution of:

$$Y_\lambda = \overline{X}^{\lambda_0}_0 + 2\int_0^\lambda \sqrt{Y_u} dB_u + 2\lambda - 2\int_0^\lambda \left(\frac{Y_u}{\lambda - u} - \frac{\sqrt{xY_u}}{\lambda_0 - u}\frac{I'_\nu}{I_\nu}\left(\frac{\sqrt{xY_u}}{\lambda_0 - u}\right)\right) du,$$

ii) for $x = 0$:

$$p(t,0,y) = \left(\frac{1}{2t}\right)^{\delta/2} \frac{1}{\Gamma(\delta/2)} y^{\frac{\delta}{2}-1} \exp\left(-\frac{y}{2t}\right),$$

and $(\overline{X}^{\lambda_0}_\lambda, 0 \leq \lambda < \lambda_0)$ is the solution of:

$$Y_\lambda = \overline{X}^{\lambda_0}_0 + 2\int_0^\lambda \sqrt{Y_u} dB_u + \delta\lambda - \int_0^\lambda \frac{2Y_u}{\lambda - u} du.$$

Note that we could also have obtained this example by squaring the results on Bessel processes.

Remark 1.20. Clearly, all the above examples entertain strong links with initial enlargements of a filtration. We refer the reader to Mansuy-Yor [125] for further examples.

1.4.4 Another Class of Markovian Peacocks

We shall introduce another set of hypotheses on the Markov process $(X_\lambda, \lambda \geq 0)$ which ensures that:

$$\left(A^{(\mu)}_t := \int_0^\infty e^{-tX_\lambda - h_\lambda(t)} \mu(d\lambda), t \geq 0\right)$$

is a peacock.

Definition 1.16 (Condition L). A right-continuous Markov process $(X_\lambda, \lambda \geq 0; \mathbb{P}_x, x \in \mathbb{R}_+)$, with values in $\mathbb{R}_+$, is said to satisfy condition L if both *i*) and *ii*) below are satisfied:

i) This process increases in the stochastic order with respect to the starting point x; in other words, for every $a \geq 0$ and $\lambda \geq 0$, and for every $0 \leq x \leq y$:

$$\mathbb{P}_y(X_\lambda \geq a) \geq \mathbb{P}_x(X_\lambda \geq a). \tag{1.124}$$

ii) The Laplace transform $\mathbb{E}_x[e^{-tX_\lambda}]$ is of the form:

$$\mathbb{E}_x[e^{-tX_\lambda}] = C_1(t,\lambda)\exp(-xC_2(t,\lambda)), \tag{1.125}$$

where C_1 and C_2 are two positive functions such that:

- for every $t > 0$ and $\lambda \geq 0$,

$$\frac{\partial}{\partial t}C_2(t,\lambda) > 0, \tag{1.126}$$

- for every $t \geq 0$ and every compact K, there exist two constants $k_K(t) > 0$ and $\widetilde{k}_K(t) < +\infty$ such that:

$$k_K(t) \leq \inf_{\lambda \in K} C_1(t,\lambda); \quad \sup_{\lambda \in K} C_2(t,\lambda) \leq \widetilde{k}_K(t). \tag{1.127}$$

Taking $x = 0$ in (1.125), we see that $C_1(\cdot,\lambda)$ is completely monotone (and hence infinitely differentiable) on $]0,+\infty[$ and continuous at 0. Consequently, $C_2(\cdot,\lambda)$ is also infinitely differentiable on $]0,+\infty[$ and continuous at 0. Moreover, we have for $t > 0$ and $\lambda \geq 0$:

$$\mathbb{E}_x\left[X_\lambda e^{-tX_\lambda}\right] = \left(-\frac{\partial}{\partial t}C_1(t,\lambda) + xC_1(t,\lambda)\frac{\partial}{\partial t}C_2(t,\lambda)\right)\exp\left(-xC_2(t,\lambda)\right)$$

and we introduce, for $t > 0$, $\lambda \geq 0$ and $y \in \mathbb{R}$:

$$\begin{cases} \alpha(t,\lambda,y) & := -\dfrac{\partial}{\partial t}C_1(t,\lambda) + yC_1(t,\lambda) \\ \beta(t,\lambda) & := C_1(t,\lambda)\dfrac{\partial}{\partial t}C_2(t,\lambda) > 0. \end{cases} \tag{1.128}$$

We can now state the main result of this subsection.

Theorem 1.17. *Let $(X_\lambda,\lambda \geq 0;\mathbb{P}_x,x \in \mathbb{R}_+)$ be a Markov process which satisfies condition (L). Then, for every $x \geq 0$ and every finite positive measure μ on $\mathbb{R}_+$,*

$$\left(A_t^{(\mu)} := \int_0^\infty e^{-tX_\lambda - h_\lambda(t)}\mu(d\lambda), t \geq 0\right)$$

is a peacock under $\mathbb{P}_x$. Here, the function h_λ is defined as:

$$h_\lambda(t) = \log\left(\mathbb{E}_x\left[e^{-tX_\lambda}\right]\right).$$

Before proving Theorem 1.17, let us give two examples of processes $(X_\lambda,\lambda \geq 0;\mathbb{P}_x,x \in \mathbb{R}_+)$ which satisfy condition (L).

Example 1.6. Let $(X_\lambda,\lambda \geq 0;\mathbb{Q}_x,x \in \mathbb{R}_+)$ be the square of a δ-dimensional Bessel process (denoted $BESQ^\delta$, $\delta \geq 0$, see [150, Chapter XI]). This process satisfies condition (L) since:

- it is stochastically increasing with respect to x; indeed, it solves a SDE which enjoys both existence and uniqueness properties, hence the strong Markov property (see paragraph 1.4.3.1),
- for every $t > 0$, we have:

$$\mathbb{Q}_x\left[e^{-tX_\lambda}\right] = \frac{1}{(1+2t\lambda)^{\frac{\delta}{2}}}\exp\left(-\frac{tx}{1+2t\lambda}\right),$$

which yields Point $ii)$ of Definition 1.16.

In particular, for $(X_t, t \geq 0)$ a squared Bessel process of dimension 0, $(A_t^{(\mu)}, t \geq 0)$ is a peacock. This case stands out of the scope of Example 1.5.

Example 1.7 (A generalization of the preceding example for $\delta = 0$). Let $(X_\lambda, \lambda \geq 0; \mathbb{P}_x, x \in \mathbb{R}_+)$ be a continuous state branching process (denoted CSBP) (see [114]).We denote by $P_\lambda(x, dy)$ the law of X_λ under $\mathbb{P}_x$, (with $x \neq 0$), and by $*$ the convolution product. Then (P_λ) satisfies:

$$P_\lambda(x, \cdot) * P_\lambda(x', \cdot) = P_\lambda(x + x', \cdot) \quad \text{for every } \lambda \geq 0, x \geq 0 \text{ and } x' \geq 0$$

which easily implies (1.124) (see [114, p.21-23]). As a consequence, one has:

$$\mathbb{E}_x\left[e^{-tX_\lambda}\right] = \exp(-xC(t, \lambda)), \tag{1.129}$$

where the function $C : \mathbb{R}_+ \times \mathbb{R}_+ \to \mathbb{R}_+$ satisfies:

- for every $\lambda \geq 0$, $C(\cdot, \lambda)$ is continuous on $\mathbb{R}_+$, differentiable on $]0, +\infty[$, and

$$\frac{\partial C}{\partial t}(t, \lambda) > 0 \quad \text{for every } t > 0,$$

- for every $t \geq 0$ and every compact K, there exists a constant $k_K(t) < \infty$ such that:

$$\sup_{\lambda \in K} C(t, \lambda) \leq k_K(t). \tag{1.130}$$

Thus, $(X_\lambda, \lambda \geq 0)$ satisfies (1.125).

Corollary 1.10. *Let $(X_\lambda, \lambda \geq 0; \mathbb{P}_x, x \in \mathbb{R}_+)$ be either a BESQ$^\delta$ or a CSBP. Then, for any finite positive measure μ on $\mathbb{R}_+$, and for every $x \geq 0$:*

$$\left(A_t^{(\mu)} := \int_0^\infty e^{-tX_\lambda - h_\lambda(t)} \mu(d\lambda), t \geq 0\right)$$

is a peacock under $\mathbb{P}_x$ with:

$$h_\lambda(t) = \log\left(\mathbb{E}_x\left[e^{-tX_\lambda}\right]\right).$$

Other peacocks which are constructed from CSBP (as well as associated martingales) are presented in Chapter 4.

Proof. Let $(X_\lambda, \lambda \geq 0)$ be a process which enjoys condition (L). Then:

1) $(-X_\lambda, \lambda \geq 0)$ being a negative process, condition (INT1) clearly holds. Moreover, by (1.127), (INT2) also holds. Thus, following the proof of Theorem 1.12, it suffices to show that $(A_t^{(\mu)}, t \geq 0)$ is a peacock when μ is a finite linear combination of Dirac measures with positive coefficients.

2) For $t \geq 0$, $a_1 \geq 0, \dots, a_n \geq 0$ and $0 \leq \lambda_1 < \cdots < \lambda_n$, we set:

$$A_t := \sum_{i=1}^n a_i e^{-tX_{\lambda_i} - h_{\lambda_i}(t)}.$$

Let $\psi \in \boldsymbol{C}$. One has:

$$\frac{\partial}{\partial t}\mathbb{E}_x[\psi(A_t)] = -\mathbb{E}_x\left[\psi'(A_t)\sum_{i=1}^n a_i e^{-tX_{\lambda_i} - h_{\lambda_i}(t)}(h'_{\lambda_i}(t) + X_{\lambda_i})\right]$$

and, we shall prove as in the proof of Theorem 1.12 that, for every $i \in \{1, \dots, n\}$, the following quantity Δ_i is negative, where:

$$\begin{aligned}\Delta_i &= \mathbb{E}_x\left[\psi'(A_t) e^{-tX_{\lambda_i} - h_{\lambda_i}(t)}(h'_{\lambda_i}(t) + X_{\lambda_i})\right] \\ &= \mathbb{E}_x\left[\psi'(A_t) e_{\lambda_i}(X_{\lambda_i})\right],\end{aligned}$$

and where we have set

$$e_{\lambda_i}(z) := e^{-tz - h_{\lambda_i}(t)}(h'_{\lambda_i}(t) + z).$$

We note that, since $\mathbb{E}\left[e^{-tX_{\lambda_i} - h_{\lambda_i}(t)}\right] = 1$:

$$\mathbb{E}_x[e_{\lambda_i}(X_{\lambda_i})] = 0. \tag{1.131}$$

Now, because the function

$$(x_1, \dots, x_n) \to \psi'\left(\sum_{j=0}^n a_j e^{-tx_j - h_{\lambda_j}(t)}\right)$$

is bounded and decreases with respect to each of its arguments, it suffices to show that: for every bounded Borel function $\phi : \mathbb{R}^n \to \mathbb{R}_+$ which decreases with respect to each of its arguments, and for every $i \in \{1, \dots, n\}$,

$$\mathbb{E}_x[\phi(X_{\lambda_1}, \dots, X_{\lambda_n}) e_{\lambda_i}(X_{\lambda_i})] \leq 0. \tag{1.132}$$

3) We now show (1.132).
a) We may suppose $i = n$. Indeed, thanks to (1.124) and to Lemma 1.5, we have, for $i < n$:

$$\begin{aligned}\mathbb{E}_x[\phi(X_{\lambda_1}, \dots, X_{\lambda_n}) e_{\lambda_i}(X_{\lambda_i})] &= \mathbb{E}_x[\mathbb{E}_x[\phi(X_{\lambda_1}, \dots, X_{\lambda_n})|\mathscr{F}_{\lambda_i}] e_{\lambda_i}(X_{\lambda_i})] \\ &= \mathbb{E}_x[\widetilde{\phi}_i(X_{\lambda_1}, \dots, X_{\lambda_i}) e_{\lambda_i}(X_{\lambda_i})],\end{aligned}$$

where $\widetilde{\phi}_i : \mathbb{R}^i \to \mathbb{R}$ is a bounded Borel function which decreases with respect to each of its arguments.

b) On the other hand, one has:

$$\begin{aligned}
&\mathbb{E}_x[\widetilde{\phi}_i(X_{\lambda_1},\dots,X_{\lambda_i})e_{\lambda_i}(X_{\lambda_i})]\\
&\quad=\mathbb{E}_x[\widetilde{\phi}_i(X_{\lambda_1},\dots,X_{\lambda_i})e^{-tX_{\lambda_i}-h_{\lambda_i}(t)}(h'_{\lambda_i}(t)+X_{\lambda_i})]\\
&\quad\leq\mathbb{E}_x[\widetilde{\phi}_i(X_{\lambda_1},\dots,X_{\lambda_{i-1}},-h'_{\lambda_i}(t))e_{\lambda_i}(X_{\lambda_i})]\\
&\qquad(\text{since } \widetilde{\phi}_i(X_{\lambda_1},\dots,X_{\lambda_i})(h'_{\lambda_i}(t)+X_{\lambda_i})\leq\widetilde{\phi}_i(X_{\lambda_1},\dots,-h'_{\lambda_i}(t))(h'_{\lambda_i}(t)+X_{\lambda_i}))\\
&\quad=\mathbb{E}_x\left[\widetilde{\widetilde{\phi}}_i(X_{\lambda_1},\dots,X_{\lambda_{i-1}})e_{\lambda_i}(X_{\lambda_i})\right],
\end{aligned}$$

where $\widetilde{\widetilde{\phi}}_i:\mathbb{R}^{i-1}\to\mathbb{R}$ is a bounded Borel function which decreases with respect to each of its arguments, and is defined by:

$$\widetilde{\widetilde{\phi}}_i(z_1,\dots,z_{i-1})=\widetilde{\phi}_i(z_1,\dots,z_{i-1},-h'_{\lambda_i}(t)). \tag{1.133}$$

c) We now end the proof of Theorem 1.17 by showing the following lemma.

Lemma 1.6. *For every $i\in\{1,\dots,n\}$ and $j\in\{0,1,\dots,i-1\}$, let $\phi:\mathbb{R}^j\to\mathbb{R}$ be a bounded Borel function which decreases with respect to each of its arguments. Then,*

$$\mathbb{E}_x[\phi(X_{\lambda_1},\dots,X_{\lambda_j})e_{\lambda_i}(X_{\lambda_i})]\leq 0. \tag{1.134}$$

In particular,

$$\mathbb{E}_x[\phi(X_{\lambda_1},\dots,X_{\lambda_{i-1}})e_{\lambda_i}(X_{\lambda_i})]\leq 0. \tag{1.135}$$

Proof. We prove this lemma by induction on j.

- For $j=0$, ϕ is constant and one has:

$$\mathbb{E}_x[\phi\, e_{\lambda_i}(X_{\lambda_i})]=\phi\,\mathbb{E}_x[e_{\lambda_i}(X_{\lambda_i})]=0 \quad (\text{from (1.131)}).$$

- On the other hand, if one assumes that (1.134) holds for $0\leq j<i-1$, then

$$\begin{aligned}
&\mathbb{E}_x[\phi(X_{\lambda_1},\dots,X_{\lambda_j},X_{\lambda_{j+1}})e_{\lambda_i}(X_{\lambda_i})]\\
&=\mathbb{E}_x[\phi(X_{\lambda_1},\dots,X_{\lambda_j},X_{\lambda_{j+1}})P_{\lambda_i-\lambda_{j+1}}e_{\lambda_i}(X_{\lambda_{j+1}})]\\
&\quad(\text{by the Markov property})\\
&=\mathbb{E}_x[\phi(X_{\lambda_1},\dots,X_{\lambda_j},X_{\lambda_{j+1}})e^{-X_{\lambda_{j+1}}C_2(t,\lambda_i-\lambda_{j+1})-h_{\lambda_i}(t)}\\
&\qquad\cdot\left(\alpha(t,\lambda_i-\lambda_{j+1},h'_{\lambda_i}(t))+X_{\lambda_{j+1}}\beta(t,\lambda_i-\lambda_{j+1})\right)]
\end{aligned} \tag{1.136}$$

$$\begin{aligned}
&\quad(\text{from (1.125) and (1.128), where } \beta>0)\\
&\leq\mathbb{E}_x\left[\phi\left(X_{\lambda_1},\dots,X_{\lambda_j},-\frac{\alpha(t,\lambda_i-\lambda_{j+1},h'_{\lambda_i}(t))}{\beta(t,\lambda_i-\lambda_{j+1})}\right)P_{\lambda_i-\lambda_{j+1}}e_{\lambda_i}(X_{\lambda_{j+1}})\right]\\
&=\mathbb{E}_x\left[\widetilde{\phi}(X_{\lambda_1},\dots,X_{\lambda_j})e_{\lambda_i}(X_{\lambda_i})\right]\leq 0 \quad (\text{by the induction hypothesis}),
\end{aligned}$$

where $\widetilde{\phi} : \mathbb{R}^j \to \mathbb{R}$ is defined by:

$$\widetilde{\phi}(z_1,\ldots,z_j) = \phi\left(z_1,\ldots,z_j, -\frac{\alpha(t,\lambda_i - \lambda_{j+1}, h'_{\lambda_i}(t))}{\beta(t,\lambda_i - \lambda_{j+1})}\right).$$

□

Notes and Comments

Theorem 1.1 (which asserts the equivalence between peacocks and 1-martingales) was obtained through the succession of works of V. Strassen ([163]), J.L. Doob ([56]) and H.G. Kellerer ([106]). The particular case of Point 1) of Theorem 1.4 (with the Brownian exponential martingale) is due to Carr, Ewald and Xiao ([43]). D. Madan drew our attention to the notion of trading cash flow (see Proposition 1.2 and Definition 1.5.) Subsection 1.2.3 is taken from [80]. Exercise 1.40 is due to J. Bertoin. The proof of the Gaussian integration by parts formula (Exercise 1.41) is taken from [165, Appendix A.6]. The notion of conditional monotonicity appears in Shaked-Shanthikumar [157, Chapter 4B, p.114-126]. The results of Section 1.4 are taken from A.M. Bogso *et al.* [33]. Other results in this chapter are new.

2

The Sheet Method

Abstract To some peacocks constructed from a Brownian motion, we associate a martingale defined with the help of the Brownian sheet. We then generalize this approach in two directions:

1) We first replace the Brownian motion (resp. the Brownian sheet) by a Lévy process (resp. a Lévy sheet).
2) We then replace the Brownian motion (resp. the Brownian sheet) by a Gaussian process (resp. a Gaussian sheet).

2.1 General Framework

Unlike in Chapter 1, which essentially contains tools to prove that a given process is a peacock, (without considering the problem of finding an associated martingale), we shall now tackle the reverse problem, i.e. we shall look for martingales associated to some given peacocks. In the present chapter, we present the *sheet method*, which we first develop in a general set-up before particularizing it to different cases (see [79]). We consider a measurable space Λ and, for every $t \geq 0$, a real-valued measurable process

$$Y_{\bullet,t} = (Y_{\lambda,t}, \lambda \in \Lambda)$$

such that:

$$\forall \lambda \in \Lambda,\ \forall t \geq 0, \quad \mathbb{E}\left[\exp(Y_{\lambda,t})\right] < +\infty.$$

Now, for any finite signed measure σ on Λ, we set:

$$\left(A_t^{(\sigma)} = \int_\Lambda \frac{\exp(Y_{\lambda,t})}{\mathbb{E}\left[\exp(Y_{\lambda,t})\right]}\sigma(d\lambda),\ t \geq 0\right)$$

and we look for conditions which ensure that $\left(A_t^{(\sigma)}, t \geq 0\right)$ is a 1-martingale, hence a peacock. This set-up encompasses those of several theorems of Chapter 1. For example:

Hirsch F., Profeta C., Roynette B., Yor M.:
Peacocks and Associated Martingales, with Explicit Constructions

- Carr, Ewald and Xiao's guiding example (1.34) is obtained by taking $\Lambda = \mathbb{R}_+$, $Y_{\lambda,t} = \nu B_{\lambda t}$ with $(B_t, t \geq 0)$ a standard Brownian motion and σ the Lebesgue measure on $[0,1]$.
- The set-up of Theorem 1.7 is obtained with: $\Lambda = \mathbb{R}_+$, $Y_{\lambda,t} = \lambda t X$ with X a r.v. and σ positive.
- The set-up of Theorem 1.8 is obtained with: $\Lambda = \mathbb{R}_+$, $Y_{\lambda,t} = \lambda X_t$ with $(X_t, t \geq 0)$ a real-valued, right-continuous process with independent increments.
- The set up of Theorem 1.11 is obtained with: Λ a metric σ-compact space, $Y_{\lambda,t} = G_{\lambda,t}$ with for each $t \geq 0$, $(G_{\lambda,t}, \lambda \in \Lambda)$ a Gaussian process.
- The set-up of Theorem 1.12 is obtained with: $\Lambda = \mathbb{R}_+$, $Y_{\lambda,t} = tX_\lambda$ with $(X_\lambda, \lambda \geq 0)$ a conditionally monotonous process.

Proposition 2.1. *Assume the existence of a measurable sheet* $(Z_{\lambda,t};\ \lambda \in \Lambda, t \geq 0)$ *such that:*

i) for every $t \geq 0$,

$$Y_{\bullet,t} \overset{(law)}{=} Z_{\bullet,t},$$

ii) for every $0 \leq s \leq t$, $Z_{\bullet,t} - Z_{\bullet,s}$ *is independent from*

$$\mathscr{Z}_s = \sigma\{Z_{\lambda,u};\ \lambda \in \Lambda, 0 \leq u \leq s\}.$$

Then the process

$$\left(M_t^{(\sigma)} := \int_\Lambda \frac{\exp(Z_{\lambda,t})}{\mathbb{E}[\exp(Z_{\lambda,t})]} \sigma(d\lambda),\ t \geq 0 \right)$$

is a $(\mathscr{Z}_t)$*-martingale and*

$$A_t^{(\sigma)} \overset{(1.d)}{=} M_t^{(\sigma)}.$$

Proof. From Point *ii*), we have for $s \leq t$:

$$\mathbb{E}[\exp(Z_{\lambda,t})] = \mathbb{E}[\exp(Z_{\lambda,s})]\mathbb{E}[\exp(Z_{\lambda,t} - Z_{\lambda,s})],$$

hence:

$$\begin{aligned}
\mathbb{E}\left[M_t^{(\sigma)} | \mathscr{Z}_s \right] &= \int_\Lambda \frac{\mathbb{E}\left[\exp(Z_{\lambda,t} - Z_{\lambda,s} + Z_{\lambda,s}) | \mathscr{Z}_s \right]}{\mathbb{E}[\exp(Z_{\lambda,t})]} \sigma(d\lambda) \\
&= \int_\Lambda \exp(Z_{\lambda,s}) \frac{\mathbb{E}[\exp(Z_{\lambda,t} - Z_{\lambda,s})]}{\mathbb{E}[\exp(Z_{\lambda,t})]} \sigma(d\lambda) \\
&= M_s^{(\sigma)}
\end{aligned}$$

and $(M_t^{(\sigma)}, t \geq 0)$ is a $(\mathscr{Z}_t)$-martingale. The equality $A_t^{(\sigma)} \overset{(1.d)}{=} M_t^{(\sigma)}$ follows directly from Point *i*). □

Exercise 2.1 (Composition of a two parameter process by an independent process). We keep the notation of Proposition 2.1. Let $(A_u, u \geq 0)$ be a process taking values in Λ, independent from $(Y_{\bullet,t}, t \geq 0)$. Let $\phi_\lambda(t)$ be defined by: $\mathbb{E}[e^{Y_{\lambda,t}}] = \exp \phi_\lambda(t)$ and let σ be a finite signed measure on $\mathbb{R}_+$.

Prove that

$$X_t^{(\sigma)} := \int_{\mathbb{R}_+} \exp\left(Y_{A_u,t} - \phi_{A_u}(t)\right) \sigma(du), \quad t \geq 0$$

is a peacock, and that

$$M_t^{(\sigma)} := \int_{\mathbb{R}_+} \exp\left(Z_{A_u,t} - \phi_{A_u}(t)\right) \sigma(du), \quad t \geq 0$$

is an associated martingale, where $(A_u, u \geq 0)$ is independent from $(Z_{\bullet,t}, t \geq 0)$. (*Hint*: $(M_t^{(\sigma)}, t \geq 0)$ is a martingale in the filtration $(\mathscr{G}_t, t \geq 0)$ defined as $\mathscr{G}_t := \sigma(A_u, u \geq 0; Z_{v,s}, v \geq 0, s \leq t)$).

Exercise 2.2 (Composition of a family of peacocks by an independent process). Let $\widehat{\Lambda}$ be a measurable space and $\mathscr{H} = \{H^\Sigma := (H_t^\Sigma, t \geq 0); \Sigma \in \widehat{\Lambda}\}$ be a measurable family of peacocks. Let $\mathscr{A}$ be a r.v. taking values in $\widehat{\Lambda}$, independent from $\mathscr{H}$.

1) Prove that $(H_t^{\mathscr{A}}, t \geq 0)$ is a peacock as soon as it is an integrable process. (*Hint*: for $\psi \in \boldsymbol{C}$, $\mathbb{E}[\psi(H_t^{\mathscr{A}})] = \int_{\widehat{\Lambda}} \mathbb{E}[\psi(H_t^\Sigma)] d\mathbb{P}_{\mathscr{A}}(\Sigma)$ where $\mathbb{P}_{\mathscr{A}}$ is the law of $\mathscr{A}$).
2) *Application*: let Λ be a measurable space and let $\mathscr{M}^+(\Lambda)$ (resp. $\mathscr{M}(\Lambda)$) the space of positive measures (resp. signed measures) which are finite on Λ. Let $\{Y_{\bullet,t} := (Y_{\lambda,t}, \lambda \in \Lambda), t \geq 0\}$ be a family of processes such that:

$$\forall (\lambda, t) \in \Lambda \times \mathbb{R}_+, \quad \mathbb{E}\left[\exp(Y_{\lambda,t})\right] = \exp(h_{\lambda,t}) < \infty$$

and that, for every $\Sigma \in \mathscr{M}^+(\Lambda)$ (resp. $\mathscr{M}(\Lambda)$)

$$\left(\int_\Lambda \exp(Y_{\lambda,t} - h_{\lambda,t}) \Sigma(d\lambda), t \geq 0\right) \quad \text{is a peacock.}$$

Let $(A_u, u \geq 0)$ be a process taking values in Λ, independent from Y. Prove that for every $\sigma \in \mathscr{M}^+(\Lambda)$ (resp. $\mathscr{M}(\Lambda)$):

$$\left(\int_\Lambda \exp(Y_{A_u,t} - h_{A_u,t}) \sigma(du), t \geq 0\right) \quad \text{is a peacock.}$$

(*Hint:* apply Question 1) with $\widehat{\Lambda} = \mathscr{M}^+(\Lambda)$ (resp. $\mathscr{M}(\Lambda)$), $H_t^\Sigma = \int_\Lambda \exp(Y_{\lambda,t} - h_{\lambda,t}) \Sigma(d\lambda)$ and $\mathscr{A}_\sigma(\omega)$ the image of σ by $A_\cdot(\omega)$, i.e.: $\int_\Lambda F(\lambda) \mathscr{A}_\sigma(\omega)(d\lambda) = \int_{\mathbb{R}_+} F(A_u(\omega)) \sigma(du)$.)

Another hint is to write, for $\psi \in \boldsymbol{C}$:

$$\mathbb{E}\left[\psi\left(\int_\Lambda \exp(Y_{A_u,t} - h_{A_u,t}) \sigma(du)\right)\right] = \mathbb{E}_A\left[\mathbb{E}_Y\left[\psi\left(\int_\Lambda \exp(Y_{A_u,t} - h_{A_u,t}) \sigma(du)\right)\right]\right]).$$

Compare with Exercise 7.16.

We shall now show that the hypotheses of Proposition 2.1 are satisfied in several frameworks.

In Section 2.2, we consider the Brownian framework and give the analogue of Proposition 2.1 (with the Brownian sheet), replacing in the definition of $(A_t^{(\sigma)}, t \geq 0)$ the term $\left(\frac{\exp(Y_{\lambda,t})}{\mathbb{E}[\exp(Y_{\lambda,t})]} = e^{B_{\lambda t} - \frac{\lambda t}{2}}\right)$ by the martingale $(h(s,B_s), s \geq 0)$, where h is a space-time harmonic function. This is the result of Baker-Yor, see Theorem 2.2.

In Section 2.3, we extend this approach, constructing at the same time a peacock and an associated martingale.

We then generalize these results to Lévy processes in Sections 2.4 and 2.5.

Finally, in Section 2.6, we give the analogue of Proposition 2.1 in a general Gaussian set-up.

2.2 The Brownian Sheet Method

2.2.1 Introduction

Let $(M_s, s \geq 0)$ be a right-continuous martingale which belongs to H^1_{loc}, i.e.: for all $t \geq 0$, $\mathbb{E}\left[\sup_{s \leq t} |M_s|\right] < \infty$. We saw in Chapter 1, Theorem 1.4 that $\left(\frac{1}{t}\int_0^t M_s ds, t \geq 0\right)$ is a ($\mathbf{F}_1$-type) peacock.

Definition 2.1 (Space-time harmonic function[1]). A function $h : \mathbb{R}_+ \times \mathbb{R} \longrightarrow \mathbb{R}$ is called a *space-time harmonic function* if:

$$(h(s,B_s), s \geq 0) \text{ is a } (\mathscr{F}_s, s \geq 0)\text{-martingale,}$$

where $(B_s, s \geq 0)$ is a standard Brownian motion and $(\mathscr{F}_s, s \geq 0)$ denotes its natural filtration.

Remark 2.1. Let h be a $\mathscr{C}^{1,2}$ function such that:

$$\int_0^t \mathbb{E}\left[\left(\frac{\partial h}{\partial x}(s,B_s)\right)^2\right] ds < \infty \quad \text{for all } t \geq 0. \tag{2.1}$$

Then h is a space-time harmonic function if and only if:

$$\forall (s,x) \in]0,+\infty[\times \mathbb{R}, \quad \frac{\partial h}{\partial s} + \frac{1}{2}\frac{\partial^2 h}{\partial x^2} = 0. \tag{2.2}$$

Such a $\mathscr{C}^{1,2}$-function h satisfying (2.1) and (2.2) will be called, in the sequel, a $\mathscr{C}^{1,2}$ space-time harmonic function.

[1] This is understood, clearly, with respect to Brownian motion.

Our aim now is to construct a martingale $(M_t^{(h)}, t \geq 0)$ associated to the peacock $\left(\frac{1}{t}\int_0^t h(s,B_s)ds, t \geq 0\right)$. We first give a few examples of space-time harmonic functions:

a) For every real ν, the function h defined by:

$$h(s,x) = \exp\left(\nu x - \frac{\nu^2}{2}s\right)$$

is a $\mathscr{C}^{1,2}$ space-time harmonic function.

b) For every integer n, the nth Hermite polynomial $H_n(s,x)$ defined by:

$$\exp\left(\lambda x - \frac{\lambda^2}{2}s\right) = \sum_{n=0}^{+\infty} \frac{\lambda^n}{n!} H_n(s,x)$$

is a $\mathscr{C}^{1,2}$ space-time harmonic function.

c) Let $(P_t, t \geq 0)$ denote the heat semi-group defined by

$$P_t f(x) = \frac{1}{\sqrt{2\pi}} \int_{-\infty}^{+\infty} f(x + \sqrt{t}y) e^{-y^2/2} dy.$$

Then, for all $a > 0$ fixed and f a Borel function on $\mathbb{R}$ such that $\mathbb{E}[|f(B_a)|] < +\infty$,

$$h(s,x) := P_{a-s}f(x) \qquad (0 \leq s \leq a,\ x \in \mathbb{R})$$

is space-time harmonic on $[0,a] \times \mathbb{R}$.

d) The function

$$h(s,x) := \frac{1}{\sqrt{1-s}} \exp\left(-\frac{x^2}{2(1-s)}\right) \qquad (0 \leq s < 1,\ x \in \mathbb{R})$$

is space-time harmonic on $[0,1[\times\mathbb{R}$. The corresponding martingale $\left(h(s,B_s) = \frac{1}{\sqrt{1-s}} \exp\left(-\frac{B_s^2}{2(1-s)}\right), s < 1\right)$ is the Girsanov density of the law of the standard Brownian bridge $(b_s, 0 \leq s < 1)$ with respect to Wiener measure.

e) From Widder [173] (see also [180], Theorem 1.3, for a probabilistic proof), every positive space-time harmonic function can be written under the form

$$h(s,x) = \int_{\mathbb{R}} e^{\nu x - \frac{\nu^2}{2}s} \sigma(d\nu)$$

where σ is a positive measure of finite total mass.

We shall also need some properties of the Brownian sheet, which we now recall.

2.2.2 Some Properties of the Brownian Sheet (see [39])

We denote by $(W_{s,t}; s \geq 0, t \geq 0)$ the standard Brownian sheet. W is defined as the centered Gaussian process, indexed by $\mathbb{R}_+ \times \mathbb{R}_+$ and with continuous paths, whose covariance is given by:

$$\mathbb{E}\left[W_{s,t} W_{s',t'}\right] = (s \wedge s')(t \wedge t').$$

Let also $(B_u, \mathscr{F}_u, u \geq 0)$ be a standard Brownian motion started from 0. Then, for every $u \geq 0$ fixed:

$$(B_{ut}, t \geq 0) \overset{(law)}{=} (W_{u,t}, t \geq 0). \tag{2.3}$$

On the other hand, we have:

$$W_{0,t} = W_{0,s} = 0$$

and for every rectangle $R = [u, v] \times [s, t]$ (with $0 \leq u \leq v$, $0 \leq s \leq t$), the "rectangular increment":

$$\Delta_R = W_{v,t} - W_{u,t} - W_{v,s} + W_{u,s} \text{ is independent from } \mathscr{G}_s \tag{2.4}$$

with:

$$\mathscr{G}_s := \sigma\{W_{u,r};\ u \geq 0, r \leq s\}$$

and satisfies the equalities in law

$$\Delta_R \overset{(law)}{=} B_{(v-u)(t-s)} \overset{(law)}{=} \sqrt{(v-u)(t-s)}\, G \tag{2.5}$$

where G is a reduced Gaussian r.v. In particular, for every $u \geq 0$ and $s < t$:

$$W_{u,t} - W_{u,s} \ \text{ is independent from } \mathscr{G}_s, \tag{2.6}$$

and the process $(W_{\cdot,t}, t \geq 0)$ is a Lévy process taking values in

$$\Omega_0 := \{\omega \in \mathscr{C}([0, +\infty[\to \mathbb{R});\ \omega_0 = 0\}.$$

Let $v \in \mathbb{R}$ and $Y_{\lambda,t} = vB_{\lambda t}$ for $\lambda \geq 0$ and $t \geq 0$. Setting

$$Z_{\lambda,t} = vW_{\lambda,t}, \qquad \lambda \geq 0, t \geq 0$$

we deduce from (2.3) and (2.6) that conditions *i)* and *ii)* in Proposition 2.1 are satisfied. Therefore, Proposition 2.1 entails:

Theorem 2.1. *Let σ be a signed finite measure on $\mathbb{R}_+$ and*

$$A_t^{(\sigma)} := \int_{\mathbb{R}_+} \exp\left(vB_{\lambda t} - \frac{v^2 \lambda t}{2}\right) \sigma(d\lambda)$$
$$M_t^{(\sigma)} := \int_{\mathbb{R}_+} \exp\left(vW_{\lambda,t} - \frac{v^2 \lambda t}{2}\right) \sigma(d\lambda).$$

Then, the process $(M_t^{(\sigma)}, t \geq 0)$ is a $(\mathscr{G}_t)$-martingale and $A_t^{(\sigma)} \overset{(1.d)}{=} M_t^{(\sigma)}$. Consequently, $(A_t^{(\sigma)}, t \geq 0)$ is a peacock.

In the following paragraph, we shall replace the function $(t,x) \longmapsto \exp\left(\nu x - \frac{\nu^2 t}{2}\right)$ by a space-time harmonic function.

2.2.3 Main Result

We now state the main result of Section 2.2.

Theorem 2.2 (D. Baker, M. Yor [12]). *Let h be a space-time harmonic function.*

$$1) \left(X_t^{(h)} := \frac{1}{t}\int_0^t h(s,B_s)ds, t \geq 0\right) \text{ is a peacock (from Theorem 1.4).} \tag{2.7}$$

$$2) \left(M_t^{(h)} := \int_0^1 h(tu, W_{u,t})du, t \geq 0\right) \text{ is a } (\mathscr{G}_t, t \geq 0) \text{ martingale.} \tag{2.8}$$

3) *This martingale is associated to* $(X_t^{(h)}, t \geq 0)$, *i.e.*

$$X_t^{(h)} \overset{(1.d)}{=} M_t^{(h)}. \tag{2.9}$$

Proof. 1) *We first prove (2.9).* We have:

$$X_t^{(h)} = \frac{1}{t}\int_0^t h(s,B_s)ds = \int_0^1 h(ut, B_{ut})du$$
(after the change of variable $s = ut$)
$$\overset{(1.d)}{=} \int_0^1 h(ut, W_{u,t})du = M_t^{(h)} \quad \text{(from (2.3))}.$$

2) *We now prove that* $(M_t^{(h)}, t \geq 0)$ *is a* $(\mathscr{G}_t, t \geq 0)$ *martingale.*

a) Since $(h(v,B_v), v \geq 0)$ is a $(\mathscr{F}_v, v \geq 0)$ martingale, with $(\mathscr{F}_v, v \geq 0)$ the natural filtration of $(B_v, v \geq 0)$, we have, for $u < v$:

$$\begin{aligned} h(u,B_u) &= \mathbb{E}\left[h(v,B_v)|\mathscr{F}_u\right] \\ &= \mathbb{E}\left[h(v, B_v - B_u + B_u)|\mathscr{F}_u\right] \\ &= \mathbb{E}\left[h(v, b + \sqrt{v-u}\,G)\right] \end{aligned}$$

with $b = B_u$ (since $B_v - B_u$ is independent from $\mathscr{F}_u$ and $B_v - B_u \overset{(law)}{=} \sqrt{v-u}\,G$ with G a reduced Gaussian r.v.). Thus:

$$h(u,b) = \mathbb{E}\left[h(v, b + \sqrt{v-u}\,G)\right] \quad \text{a.s.} \tag{2.10}$$

b) We have, for $s < t$:

$$M_t^{(h)} = \int_0^1 h(ut, W_{u,t})du = \int_0^1 h(ut, (W_{u,t} - W_{u,s}) + W_{u,s})du.$$

Then, since from (2.6), $W_{u,t} - W_{u,s}$ is independent from $\mathcal{G}_s$ and, from (2.5), $W_{u,t} - W_{u,s} \overset{(law)}{=} \sqrt{u(t-s)}G$, we have:

$$\begin{aligned}
\mathbb{E}\left[M_t^{(h)}|\mathcal{G}_s\right] &= \mathbb{E}\left[\int_0^1 h(ut,(W_{u,t}-W_{u,s})+W_{u,s})du|\mathcal{G}_s\right] \\
&= \int_0^1 \mathbb{E}\left[h(ut,(W_{u,t}-W_{u,s})+W_{u,s})|\mathcal{G}_s\right]du \\
&= \int_0^1 \mathbb{E}\left[h(ut,a+\sqrt{u(t-s)}G)\right]du \quad \text{with } a = W_{u,s} \\
&= \int_0^1 h(us,W_{u,s})du \quad \text{(from (2.10))} \\
&= M_s^{(h)}.
\end{aligned}$$

□

Remark 2.2. Note that to find a martingale associated to the peacock $(X_t^{(h)}, t \geq 0)$ defined by (2.7), we have enlarged the initial probability space (the space $\mathscr{C}(\mathbb{R}_+, \mathbb{R})$ equipped with the Wiener measure) defining the martingale $(M_t^{(h)}, t \geq 0)$ on the Brownian sheet space $\mathscr{C}(\mathbb{R}_+ \times \mathbb{R}_+, \mathbb{R})$.

Exercise 2.3 (Two-parameter peacocks and martingales). A process $(X_{s,t}; s \geq 0, t \geq 0)$ is a two-parameter peacock if:

a) for every $s,t \geq 0$, $\mathbb{E}[|X_{s,t}|] < \infty$,
b) for every convex function $\psi : \mathbb{R} \longrightarrow \mathbb{R}$, the application $(s,t) \longmapsto \mathbb{E}[\psi(X_{s,t})]$ is increasing with respect to each of its variables.

A process $(M_{s,t}; s \geq 0, t \geq 0)$ is a two-parameter martingale if:

a) for every $s,t \geq 0$, $\mathbb{E}[|M_{s,t}|] < \infty$,
b) for every $s \leq s'$, $t \leq t'$,

$$\mathbb{E}[M_{s',t'}|\mathscr{F}_{s,t}] = M_{s,t},$$

where $\mathscr{F}_{s,t} \supseteq \sigma(M_{u,v};\ u \leq s, v \leq t)$.

1) Prove that a two-parameter martingale is a two-parameter peacock.
2) Let $(B_u, u \geq 0)$ be a Brownian motion started from 0, and denote by $(L_t^x; t \geq 0, x \in \mathbb{R})$ its jointly continuous family of local times. Let also $(\beta_u, u \geq 0)$ another Brownian motion independent from $(B_u, u \geq 0)$.

i) Let $A > 0$. Prove that $(Y_{a,t} := \beta_{L_t^{A-a}};\ t \geq 0, 0 \leq a \leq A)$ is a two-parameter peacock and exhibit an associated two-parameter martingale.

(*Hint*: for x and t fixed, $L_t^x \overset{(law)}{=} (L_t^0 - |x|)^+)$, see Exercise 1.14; use this relation to obtain $Y_{a,t} \overset{(1.d)}{=} \beta_{(L_t^0+a-A)^+}$, and observe that $(a,t) \longmapsto (L_t^0 + a - A)^+$ is an increasing function of t and a.)

ii) We denote by $(\tau_l, l \geq 0)$ the right-continuous inverse of $(L_t^0, t \geq 0)$: $\tau_l := \inf\{t \geq 0; L_t^0 > l\}$. Prove that $(Z_{a,l} := \beta_{\left(L_{\tau_l}^a\right)^2};\ a \geq 0, l \geq 0)$ is a two-parameter peacock.

(*Hint*: observe that $Z_{a,l} \stackrel{(1.d)}{=} L_{\tau_l}^a \beta_1$. To prove the peacock property in a, use the second Ray-Knight theorem: $(L_{\tau_l}^a, a \geq 0)$ is a squared Bessel process of dimension 0 started at l, hence a martingale; to prove the peacock property in l, use Point 2) of Exercise 1.25.)

iii) We denote $T_b := \inf\{u \geq 0; B_u = b\}$. Let $A > 0$. Prove that $(\widetilde{Z}_{a,b} := \beta_{L_{T_b}^{A-a}}; a \leq A \leq a+b)$ is a two-parameter peacock.

(*Hint*: observe that $\widetilde{Z}_{a,b} \stackrel{(1.d)}{=} \sqrt{L_{T_b}^{A-a}}\beta_1$ and, from the first Ray-Knight theorem, $(L_{T_b}^{b-a}, 0 \leq a \leq b)$ is a squared Bessel process of dimension 2 started at 0, hence a stochastically increasing process (see [148, Chapter 7]); use then Point 2) of Exercise 1.25.)

iv) Exhibit a two-parameter martingale $(M_{a,b}; a \leq A \leq a+b)$ associated to $(\widetilde{Z}_{a,b}; a \leq A \leq a+b)$. (*Hint*: $M_{a,b} = \beta_{\inf\limits_{s \geq a+b-A} Q_s^{(4,0)}}$, where $(Q_s^{(4,0)}, s \geq 0)$ is a squared Bessel process of dimension 4 started from 0, see relation (5.2) and Question 5 *i*) in Exercise 5.7 below.)

3) A process $(M_{s,t}; s \geq 0, t \geq 0)$ is a strong two-parameter martingale if:

a) for every $s,t \geq 0$, $\mathbb{E}[|M_{s,t}|] < \infty$,

b) for every $s \leq s'$, $t \leq t'$,

$$\mathbb{E}[M_{s',t'} | \mathscr{G}_{s,t}] = M_{s,t},$$

where $\mathscr{G}_{s,t} \supseteq \sigma(M_{u,v};\ \{u \leq s, v \geq 0\} \cup \{u \geq 0, v \leq t\})$.

i) Let $(M_{s,t}; s \geq 0, t \geq 0)$ be a strong martingale. Prove that

$$X_{a,b} := \frac{1}{ab}\int_0^a ds \int_0^b dt M_{s,t}, \qquad a \geq 0, b \geq 0$$

is a two-parameter peacock. (*Hint*: prove that $(\int_0^b dt M_{s,t}, s \geq 0)$ is a martingale with respect to the filtration $\widehat{\mathscr{G}_s} := \sigma(M_{u,t};\ u \leq s, t \geq 0)$ and apply Theorem 1.4).

ii) *Application*: let $(W_{s,t}; s \geq 0, t \geq 0)$ be the standard Brownian sheet. Prove that

$$X_{a,b} := \frac{1}{ab}\int_0^a ds \int_0^b dt \exp\left(W_{s,t} - \frac{st}{2}\right), \qquad a \geq 0, b \geq 0$$

is a two-parameter peacock and exhibit an associated martingale.

(*Hint*: use the "four parameter Brownian sheet" $(W_{a,u,b,v}; a, u, b, v \geq 0)$ or the "three parameter Brownian sheet" $(W_{u,v,c}; u, v, c \geq 0)$ with $c = ab$.)

4) Let $(C_{a,b}; a, b \geq 0)$ be a two-parameter $\mathbb{R}_+$-valued process increasing in the stochastic order, i.e. $\forall a \leq b$, $a' \leq b'$, $C_{a',b'} \stackrel{(st)}{\geq} C_{a,b}$. From the two-parameter version of Kamae-Krengel's Theorem [104], there exists a process $(A_{a,b}; a, b \geq 0)$ such that:

a) $(A_{a,b}; a, b \geq 0)$ is a.s. increasing, i.e. $\forall a \leq b$, $a' \leq b'$, $A_{a',b'} \geq A_{a,b}$,

b) $\forall a, b \geq 0$, $C_{a,b} \stackrel{(law)}{=} A_{a,b}$.

Let $(X_u, u \geq 0)$ be a peacock independent from $(C_{a,b}; a,b \geq 0)$ and such that for every $a,b \geq 0$, $\mathbb{E}[|X_{C_{a,b}}|] < \infty$.

i) Show that $(X_{C_{a,b}}; a,b \geq 0)$ is a two-parameter peacock.

(*Hint*: let $(M_u, u \geq 0)$ be a martingale associated to $(X_u, u \geq 0)$; then for every $a,b \geq 0$, $X_{C_{a,b}} \overset{(law)}{=} M_{A_{a,b}}$, with M and A independent.)

ii) Prove that the results of Question 2) still hold when the Brownian motion $(\beta_u, u \geq 0)$ is replaced by a peacock $(X_u, u \geq 0)$ which satisfies some suitable integrability conditions.

2.3 Space-time Harmonic Functions for the Brownian Sheet (STHBS) and Peacocks

We shall now slightly change our point of view. Instead of looking for a martingale associated to a given peacock (as in Section 2.2), we shall now construct at the same time a peacock and an associated martingale. The previous section has shown how important is the notion of space-time harmonic function for Brownian motion. This leads us to define a similar notion for the Brownian sheet.

2.3.1 Space-Time Harmonic Functions for the Brownian Sheet

We recall that $(W_{u,t};\ u \geq 0, t \geq 0)$ denotes a standard Brownian sheet and $\mathscr{G}_t := \sigma\{W_{u,s};\ u \geq 0, s \leq t\}$.

Definition 2.2. Let $\Omega_0 = \{\omega \in \mathscr{C}(\mathbb{R}_+, \mathbb{R}); \omega_0 = 0\}$ and I be an interval of $\mathbb{R}_+$. Let $H : I \times \Omega_0 \longmapsto \mathbb{R}$. We say that H is a space-time harmonic function for the Brownian sheet (STHBS) if

$$(H(t, W_{\bullet,t}), t \in I) \text{ is a } (\mathscr{G}_t, t \in I)\text{-martingale.}$$

The notation $H(t, W_{\bullet,t})$ stands for $H(t, \omega)$ where $\omega \in \Omega_0$ is defined by $(\omega(u) = W_{u,t}, u \geq 0)$.

Here is the announced construction of a peacock and an associated $(\mathscr{G}_t, t \geq 0)$-martingale.

Theorem 2.3. *Let H be a STHBS function. Then:*

1) *$(H(t, W_{\bullet,t}), t \in I)$ is a $(\mathscr{G}_t, t \in I)$-martingale.*
2) *$(H(t, B_{\bullet t}), t \in I)$ is a peacock.*
3) $H(t, W_{\bullet,t}) \overset{(1.d)}{=} H(t, B_{\bullet t})$.

In other words, the martingale $(H(t, W_{\bullet,t}), t \in I)$ is associated to the peacock $(H(t, B_{\bullet t}), t \in I)$. The notation $H(t, B_{\bullet t})$ stands for $H(t, \omega)$ where $\omega \in \Omega_0$ is defined by $(\omega(u) = B_{ut}, u \geq 0)$.

Proof. Of course, it is sufficient to prove point 3), and this results directly from (2.3). □

Remark 2.3 (On the relationship between Theorems 2.2 and 2.3). Theorem 2.2 is in fact a particular case of Theorem 2.3. Indeed, let $h : \mathbb{R}_+ \times \mathbb{R} \longrightarrow \mathbb{R}$ be a space-time harmonic function for the Brownian motion. We know – this is Theorem 2.2 – that $\left(\int_0^1 h(ut, W_{u,t})du, t \geq 0\right)$ is a $(\mathscr{G}_t, t \geq 0)$-martingale. In other words:

$$H(t,\omega) := \int_0^1 h(ut, \omega(u))du \quad \text{is a STHBS function.}$$

From Points 2) and 3) of Theorem 2.3:

$$(H(t, B_{\cdot t})) = \int_0^1 h(ut, B_{ut})du \qquad \left(= \frac{1}{t}\int_0^t h(v, B_v)dv, t \geq 0\right)$$

is a peacock associated to the martingale $\left(H(t, W_{\bullet,t}) = \int_0^1 h(ut, W_{u,t})du, t \geq 0\right)$. This is Theorem 2.2.

Of course, Theorem 2.3 has practical interest only if we can describe the STHBS functions and give examples. To this end, we first present some general results on STHBS functions.

Theorem 2.4. *Let $h : I \times \mathbb{R} \longrightarrow \mathbb{R}$ be a space-time harmonic function for the Brownian motion. Then, $H : I \times \Omega_0 \longrightarrow \mathbb{R}$ defined by:*

$$H(t,\omega) = h(t, \omega(1))$$

is a STHBS function.

Proof. Let $0 \leq s < t$, with $s,t \in I$.

$$\begin{aligned}
\mathbb{E}\left[H(t, W_{\bullet,t})|\mathscr{G}_s\right] &= \mathbb{E}\left[h(t, W_{1,t})|\mathscr{G}_s\right] = \mathbb{E}\left[h(t, b + \sqrt{t-s}G)\right] \\
&\qquad \text{(with } b = W_{1,s} \text{ and } G \text{ a reduced Gaussian r.v.)} \\
&= h(s, W_{1,s}) \quad \text{(by 2.10)} \\
&= H(s, W_{\bullet,s}).
\end{aligned}$$

□

Theorem 2.5. *Let $H : I \times \Omega_0 \longrightarrow \mathbb{R}$ be a STHBS function. Define, for $r > 0$ and $t \in r^{-1}I$,*

$$H_r(t,\omega) = H(rt, \omega(r\bullet)).$$

Then H_r is still a STHBS function (on $r^{-1}I$).

Proof. Let $0 \le s < t$, with $s,t \in r^{-1}I$.
i) We have:

$$H(rs, W_{\bullet,rs}) = \mathbb{E}\left[H(rt, W_{\bullet,rt})|\mathscr{G}_{rs}\right] = \mathbb{E}_B\left[H(rt, \varphi + \sqrt{r(t-s)}B_{\bullet})\right]$$

with $\varphi = W_{\bullet,rs}$ and $B_{\bullet}$ a standard Brownian motion starting from 0. Thus,

$$H(rs, \varphi) = \mathbb{E}_B\left[H(rt, \varphi + \sqrt{r(t-s)}B_{\bullet})\right] \ \mathbb{P}^{(rs)}\text{-a.s.}$$

where $\mathbb{P}^{(a)}$ denotes the law of $W_{\bullet,a}$.
ii) On the other hand,

$$\begin{aligned}\mathbb{E}\left[H_r(t, W_{\bullet,t})|\mathscr{G}_s\right] &= \mathbb{E}\left[H(rt, W_{r\bullet,t})|\mathscr{G}_s\right] \\ &= \mathbb{E}_B\left[H(rt, \psi + \sqrt{r(t-s)}B_{\bullet})\right] \quad \text{with } \psi = W_{r\bullet,s}\end{aligned}$$

Indeed,

$$W_{r\bullet,t} - W_{r\bullet,s} \overset{(law)}{=} \sqrt{r(t-s)}B_{\bullet}$$

iii) Since $\mathbb{P}^{(rs)}$ is the law of $W_{\bullet,rs}$, which is the same as the law of $W_{r\bullet,s}$, and of $\sqrt{rs}B_{\bullet}$, we deduce from *i*) and *ii*):

$$\mathbb{E}\left[H_r(t, W_{\bullet,t})|\mathscr{G}_s\right] = H(rs, W_{r\bullet,s}) = H_r(s, W_{\bullet,s}) \quad \text{a.s.}$$

□

Corollary 2.1. *Let $H : \mathbb{R}_+ \times \Omega_0 \longrightarrow \mathbb{R}$ be a STHBS function and let σ be a finite signed measure on $\mathbb{R}_+$, with compact support. Then*

$$H^{(\sigma)}(t, \omega) = \int_{\mathbb{R}_+} H(rt, \omega(r\bullet))\sigma(dr)$$

is a STHBS function.

Proof. By Theorem 2.5, it suffices to notice that, if $r \in [0, a]$,

$$\mathbb{E}\left[|H(rt, W_{r\bullet,t})|\right] = \mathbb{E}\left[|H(rt, W_{\bullet,rt})|\right] \le \mathbb{E}\left[|H(at, W_{\bullet,at})|\right] < \infty,$$

since $(H(u, W_{\bullet,u}), u \ge 0)$ is a martingale. □

Remark 2.4. a) The assumption on the support of σ in Corollary 2.1 can be omitted if, for every $t \ge 0$:

$$\sup_{r \ge 0} \mathbb{E}\left[|H(rt, W_{\bullet,rt})|\right] < \infty.$$

This is satisfied, in particular, if $H \ge 0$, since then:

$$\mathbb{E}\left[|H(rt, W_{\bullet,rt})|\right] = H(0,0).$$

b) Applying successively Theorem 2.4 and Corollary 2.1 with σ the Lebesgue measure on $[0,1]$, we recover Theorem 2.2.

Exercise 2.4 (Examples of STHBS functions). 1) Let $c \in L^2(\mathbb{R}_+, du)$. Prove that the functions η^c and $\xi^c : \mathbb{R}_+ \times \Omega_0 \longrightarrow \mathbb{R}$ defined (almost surely) by:

$$\eta^c(t, B_.) := \int_0^{+\infty} c(u) dB_u$$

and

$$\xi^c(t, B_.) := \exp\left(\int_0^{+\infty} c(u) dB_u - \frac{t}{2}\int_0^{+\infty} c^2(u) du\right)$$

are STHBS functions. Deduce that $\left(\int_0^{+\infty} c\left(\frac{u}{t}\right) dB_u, t \geq 0\right)$ and $\left(\exp\left(\int_0^{+\infty} c\left(\frac{u}{t}\right) dB_u - \frac{t}{2}\int_0^{+\infty} c^2(u)\, du\right),\ t \geq 0\right)$ are peacocks.

2) Show that, for every finite signed Borel measure σ on $L^2(\mathbb{R}_+, du)$, the function $\xi^{(\sigma)}(t, B_.) := \int \xi^c(t, B_.) d\sigma(c)$ is a STHBS function.

Note that Föllmer [66] determined all the positive STHBS functions on $\mathbb{R}_+$. In particular, he showed that there exist other such functions than the above functions $\xi^{(\sigma)}$ with $\sigma \geq 0$. Compare with Widder's result on positive space-time harmonic functions for Brownian motion, presented in Point *e*) below Remark 2.1.

2.3.2 Characterization of Space-Time Harmonic Functions for the Brownian Sheet

Let us first introduce some notation. We consider the space $\Omega_0 = \{\omega \in \mathscr{C}(\mathbb{R}_+, \mathbb{R});\ \omega_0 = 0\}$ equipped with the σ-algebra $\mathscr{F}_\infty = \sigma\{\omega_u, u \geq 0\}$ and the Wiener measure $\mathbb{P}$ such that $\mathbb{P}(\omega_0 = 0) = 1$. We also define the space:

$$L^1(\mathbb{P}) = \{\Phi : \Omega_0 \longrightarrow \mathbb{R},\ \mathscr{F}_\infty - \text{measurable and } \mathbb{P}\text{-integrable}\}. \tag{2.11}$$

For an interval I of $\mathbb{R}_+$, we set:

$$\mathscr{H}_I = \{H : I \times \Omega_0 \longrightarrow \mathbb{R};\ H \text{ is a STHBS function}\}.$$

Let us note that, if $H \in \mathscr{H}_I$, $\mathbb{E}[|H(t, W_{\bullet,t})|] < +\infty$ for all $t \in I$. When $I = [0,1]$, we denote simply $\mathscr{H}_I$ by $\mathscr{H}$. We shall now prove that there exists a bijection between $L^1(\mathbb{P})$ and $\mathscr{H}$. Of course, we might, with obvious modifications, replace the interval $[0,1]$ by any compact interval $[0,a]$, with $a > 0$ (see in particular Exercise 2.7).

2.3.2.1 Definition of the Semi-Group $(\Pi_t, t \geq 0)$

The process $(W_{\bullet,t}, t \geq 0)$, being a Lévy process with values in Ω_0, is a Markov process. We denote as $(\Pi_t, t \geq 0)$ its associated semi-group. Hence, for all $\Phi : \Omega_0 \longrightarrow \mathbb{R}$, $\mathscr{F}_\infty$-measurable, bounded or positive, we have:

$$\Pi_t \Phi(\omega) = \mathbb{E}[\Phi(\omega_. + W_{\bullet,t})]. \tag{2.12}$$

In (2.12), $\omega \in \Omega_0$, $\omega_{\cdot} + W_{\cdot,t}$ denotes the element $\widetilde{\omega}$ of Ω_0 defined by $(\widetilde{\omega}_u = \omega_u + W_{u,t}, u \geq 0)$ and the expectation is taken with respect to the Brownian sheet. The Markov property reads:

$$\mathbb{E}\left[\Phi(W_{\bullet,t})|\mathscr{G}_s\right] = \Pi_{t-s}\Phi(W_{\bullet,s}) \quad (0 \leq s \leq t).$$

2.3.2.2 Bijection between $L^1(\mathbb{P})$ and $\mathscr{H}$

Theorem 2.6. 1) *The application* $\Gamma : \mathscr{H} \longrightarrow L^1(\mathbb{P})$ *defined by:*

$$\Gamma(H)(\omega) = H(1,\omega) \quad (H \in \mathscr{H}) \tag{2.13}$$

is one-to-one. Its reciprocal mapping $\Gamma^{-1} : L^1(\mathbb{P}) \longrightarrow \mathscr{H}$ *is defined by:*

$$\Gamma^{-1}(\Phi)(t,\omega) = \Pi_{1-t}\Phi(\omega) \quad (\Phi \in L^1(\mathbb{P}), 0 \leq t \leq 1). \tag{2.14}$$

2) *Let* $H \in \mathscr{H}$, *and* $\Phi = \Gamma(H)$. *Then,* $\Phi \in L^1(\mathbb{P})$ *and we have for all* $0 \leq t \leq 1$*:*

$$H(t,W_{\bullet,t}) = \mathbb{E}\left[\Phi(W_{\bullet,1})|\mathscr{G}_t\right] \tag{2.15}$$

$$= \Pi_{1-t}\Phi(W_{\bullet,t}) \tag{2.16}$$

$$= \mathbb{E}_{\widetilde{W}}\left[\Phi(W_{\bullet,t} + \widetilde{W}_{\bullet,1-t})\right] \tag{2.17}$$

$$= \mathbb{E}_{\widetilde{B}}\left[\Phi(W_{\bullet,t} + \widetilde{B}_{\bullet(1-t)})\right]. \tag{2.18}$$

3) *Furthermore, the peacock* $(H(t,B_{\cdot t}), 0 \leq t \leq 1)$ *satisfies:*

$$H(t,B_{\cdot t}) = \Pi_{1-t}\Phi(B_{\cdot t}) \tag{2.19}$$

$$= \mathbb{E}_{\widetilde{W}}\left[\Phi(B_{\cdot t} + \widetilde{W}_{\bullet,1-t})\right] \tag{2.20}$$

$$= \mathbb{E}_{\widetilde{B}}\left[\Phi(B_{\cdot t} + \widetilde{B}_{\bullet(1-t)})\right]. \tag{2.21}$$

In (2.17) (resp. (2.20)), $\widetilde{W}$ *is a copy of* W *which is independent of* W *(resp.* B*). In (2.18) (resp. (2.21)),* $\widetilde{B}$ *is a copy of* B *which is independent of* W *(resp.* B*).*

Proof. a) We prove that, if $H \in \mathscr{H}$, *then* $\Gamma(H) \in L^1(\mathbb{P})$. Let $H \in \mathscr{H}$. Since $(H(t,W_{\bullet,t}), 0 \leq t \leq 1)$ is a martingale, $\mathbb{E}\left[|H(t,W_{\bullet,t})|\right] < \infty$ for $t \in [0,1]$. In particular, since $(W_{u,1}, u \geq 0) \overset{(law)}{=} (B_u, u \geq 0)$, then $\mathbb{E}\left[|H(1,W_{\bullet,1})|\right] = \mathbb{E}\left[|H(1,B_{\cdot 1})|\right] = \mathbb{E}\left[|\Gamma(H)(B_{\cdot})|\right] < +\infty$.

b) Let $\Phi \in L^1(\mathbb{P})$ and define $\Gamma^{-1}(\Phi)$ by (2.14). We prove that $\Gamma^{-1}(\Phi) \in \mathscr{H}$. We first prove that $\mathbb{E}\left[|\Gamma^{-1}(\Phi)(t,W_{\bullet,t})|\right] < +\infty$. Indeed, we have:

$$\begin{aligned}\mathbb{E}\left[|\Gamma^{-1}(\Phi)(t,W_{\bullet,t})|\right] &= \mathbb{E}\left[|\Pi_{1-t}\Phi(W_{\bullet,t})|\right]\\ &\leq \mathbb{E}\left[\Pi_{1-t}(|\Phi|)(W_{\bullet,t})\right]\\ &= \mathbb{E}\left[\mathbb{E}\left[|\Phi(W_{\bullet,1})||\mathscr{G}_t\right]\right] \quad \text{(from the Markov property)}\\ &\leq \mathbb{E}\left[|\Phi(W_{\bullet,1})|\right] = \|\Phi\|_{L^1(\mathbb{P})}\end{aligned}$$

since $(W_{u,1}, u \geq 0) \overset{(law)}{=} (B_u, u \geq 0)$. On the other hand, it is clear, from the Markov property, that $(\Pi_{1-t}\Phi(W_{\bullet,t}), 0 \leq t \leq 1)$ is a $(\mathscr{G}_t, t \geq 0)$-martingale. Hence $\Gamma^{-1}(\Phi) \in \mathscr{H}$.

c) We prove that $\Gamma^{-1} \circ \Gamma(H) = H$ ($H \in \mathscr{H}$). We have:

$$\begin{aligned}\Gamma^{-1}(\Gamma(H))(t,W_{\bullet,t}) &= \Pi_{1-t}(H(1,\bullet))(W_{\bullet,t})\\ &= \mathbb{E}\left[H(1,W_{\bullet,1})|\mathscr{G}_t\right]\\ &= H(t,W_{\bullet,t})\end{aligned}$$

since $(H(t,W_{\bullet,t}), 0 \leq t \leq 1)$ is a martingale.

d) We prove that $\Gamma \circ \Gamma^{-1}(\Phi) = \Phi$ ($\Phi \in L^1(\mathbb{P})$). We have $\Gamma^{-1}\Phi(t,\omega) = \Pi_{1-t}\Phi(\omega)$. Thus: $\Gamma \circ \Gamma^{-1}(\Phi)(\omega) = \Pi_{1-t}\Phi(\omega)_{|t=1} = \Pi_0\Phi(\omega) = \Phi(\omega)$ since Π_0 is the identity operator. This proves Point 1) of Theorem 2.6. Points 2) and 3) of Theorem 2.6 are simply applications of the definitions and of the equality in law $(\widetilde{W}_{u,1-t}, u \geq 0) \overset{(law)}{=} (\widetilde{B}_{u(1-t)}, u \geq 0)$. □

Compare the statement of Theorem 2.6 with Exercise 2.7.

2.3.3 The Processes $\Phi^{(\mathbf{M})}$ and $\Phi^{(\mathbf{P})}$

Theorem 2.6 allows us to rewrite Theorem 2.3 in the case $I = [0,1]$.

Definition 2.3. Let $\Phi \in L^1(\Omega_0, \mathscr{F}_\infty, \mathbb{P})$ $(= L^1(\mathbb{P}))$. We denote:

$$\left(\Phi_t^{(\mathbf{M})} := \Pi_{1-t}\Phi(W_{\bullet,t}),\ 0 \leq t \leq 1\right) \tag{2.22}$$

$$\left(= \mathbb{E}\left[\Phi(W_{\bullet,1})|\mathscr{G}_t\right],\ 0 \leq t \leq 1\right) \tag{2.23}$$

and

$$\left(\Phi_t^{(\mathbf{P})} := \Pi_{1-t}\Phi(B_{\bullet t}),\ 0 \leq t \leq 1\right) \tag{2.24}$$

$$\left(= \mathbb{E}_{\widetilde{B}}\left[\Phi(B_{\bullet t} + \widetilde{B}_{\bullet(1-t)})\right],\ 0 \leq t \leq 1\right). \tag{2.25}$$

In (2.25), $\widetilde{B}$ is an independent copy of B, and $\mathbb{E}_{\widetilde{B}}$ means integration with respect to $\widetilde{B}$.

Let us note that the two processes $\left(\Phi_t^{(\mathbf{P})}, 0 \leq t \leq 1\right)$ and $\left(\Phi_t^{(\mathbf{M})}, 0 \leq t \leq 1\right)$ are not defined on the same probability space: $\Phi^{(\mathbf{M})}$ is defined on the Brownian sheet space, whereas $\Phi^{(\mathbf{P})}$ is defined on the Brownian motion space. Of course, we note $\Phi^{(\mathbf{M})}$ for martingale and $\Phi^{(\mathbf{P})}$ for peacock.

Theorem 2.7. *For all $\Phi \in L^1(\mathbb{P})$, $\left(\Phi_t^{(M)}, 0 \leq t \leq 1\right)$ is a $(\mathscr{G}_t, 0 \leq t \leq 1)$-martingale associated to the peacock $\left(\Phi_t^{(P)}, 0 \leq t \leq 1\right)$.*

In the following theorem, we gather properties of the map $\Phi \longmapsto \left(\Phi_t^{(\mathbf{P})}, 0 \leq t \leq 1\right)$.

Theorem 2.8. 1) *For every $\Phi \in L^1(\mathbb{P})$, $\Phi_0^{(P)} = \mathbb{E}[\Phi]$ and $\Phi_1^{(P)} = \Phi$. Moreover, if Φ is $\mathscr{F}_r$-measurable, then $\Phi_t^{(P)}$ is $\mathscr{F}_{rt}$-measurable.*
2) *Let $1 \leq p \leq +\infty$. For every $\Phi \in L^p(\mathbb{P})$, for every $t \in [0,1]$, $\Phi_t^{(P)} \in L^p(\mathbb{P})$ and*

$$\|\Phi_t^{(P)}\|_{L^p(\mathbb{R})} \leq \|\Phi\|_{L^p(\mathbb{R})}.$$

3) *For every $\Phi \in L^1(\mathbb{R})$ and for $t,s \in [0,1]$,*

$$\left(\Phi_t^{(P)}\right)_s^{(P)} = \Phi_{ts}^{(P)}.$$

4) *If $1 \leq p < +\infty$ and $\Phi \in L^p(\mathbb{P})$, then the map $t \in [0,1] \longmapsto \Phi_t^{(P)} \in L^p(\mathbb{P})$ is continuous in $L^p(\mathbb{P})$.*

Proof. *i*) Property 1) follows directly from (2.25).
ii) Since $(\Phi_t^{(\mathbf{P})}, 0 \leq t \leq 1)$ is a peacock, $\mathbb{E}\left[\Phi_t^{(\mathbf{P})}\right] = \mathbb{E}[\Phi]$. Let $1 \leq p < +\infty$ and $\Phi \in L^p(\mathbb{P})$. By (2.25) and Hölder's inequality:

$$|\Phi_t^{(\mathbf{P})}|^p \leq (|\Phi|^p)_t^{(\mathbf{P})}$$

and therefore

$$\|\Phi_t^{(\mathbf{P})}\|_{L^p(\mathbb{P})}^p = \mathbb{E}\left[|\Phi_t^{(\mathbf{P})}|^p\right] \leq \mathbb{E}\left[(|\Phi|^p)_t^{(\mathbf{P})}\right] = \mathbb{E}[|\Phi|^p] = \|\Phi\|_{L^p(\mathbb{R})}^p.$$

The similar result for $p = +\infty$ is obvious.
iii) Let $\Phi \in L^1(\mathbb{P})$ and $t,s \in [0,1]$. By Fubini's theorem, we obtain:

$$\begin{aligned}\left(\Phi_t^{(\mathbf{P})}\right)_s^{(\mathbf{P})} &= \mathbb{E}_{\widetilde{B}}\left[\Phi_t^{(\mathbf{P})}(B_{\bullet s} + \widetilde{B}_{\bullet(1-s)})\right] \\ &= \mathbb{E}_{\widetilde{B},\widehat{B}}\left[\Phi\left(B_{\bullet ts} + \widetilde{B}_{\bullet t(1-s)} + \widehat{B}_{\bullet(1-t)}\right)\right]\end{aligned}$$

with $\widetilde{B}$, $\widehat{B}$ two independent standard Brownian motions, which are also independent of B. Now

$$\widetilde{B}_{\bullet t(1-s)} + \widehat{B}_{\bullet(1-t)} \overset{(law)}{=} B_{\bullet(1-ts)}.$$

Therefore:

$$\left(\Phi_t^{(\mathbf{P})}\right)_s^{(\mathbf{P})} = \mathbb{E}_{\widetilde{B}}\left[\Phi(B_{\bullet ts} + \widetilde{B}_{\bullet(1-ts)})\right] = \Phi_{ts}^{(\mathbf{P})}.$$

iv) If Φ is a bounded continuous function on the Wiener space Ω_0 (equipped with the topology of the uniform convergence on compact sets), then, by (2.25) and the dominated convergence theorem, Property 4) holds. The general result follows by density, using Property 2). □

Exercise 2.5 (Explicit computation of $\Psi^{(\mathbf{P})}$ for $\Psi = \int_0^1 \Phi_s^{(\mathbf{P})} h(s)ds$). Let $\Phi \in L^1(\mathbb{P})$ and $h \in L^1([0,1], ds)$. Consider

$$\Psi = \int_0^1 \Phi_s^{(\mathbf{P})} h(s)ds.$$

Prove that:

$$\Psi_t^{(\mathbf{P})} = \frac{1}{t}\int_0^t \Phi_s^{(\mathbf{P})} h\left(\frac{s}{t}\right) ds.$$

(*Hint*: use Property 3) in Theorem 2.8.)

In particular, $\left(\frac{1}{t}\int_0^t \Phi_s^{(\mathbf{P})} h\left(\frac{s}{t}\right) ds, t \in [0,1]\right)$ is a peacock.

Exercise 2.6 (Explicit computation of $\Psi^{(\mathbf{P})}$ and $\Psi^{(\mathbf{M})}$ for $\Psi = \int_0^1 \Phi_s^{(\mathbf{P})} dB_s$). Let $\Phi \in L^2(\mathbb{P})$, $\mathscr{F}_1$-mesurable, and $\Psi = \displaystyle\int_0^1 \Phi_s^{(\mathbf{P})} dB_s$.

1) Prove that Ψ is well-defined and $\Psi \in L^2(\mathbb{P})$.

2) Show that

$$\Psi_t^{(\mathbf{P})} = \int_0^t \Phi_s^{(\mathbf{P})} dB_s.$$

Thus, in this case, the process $\left(\Psi_t^{(\mathbf{P})}, t \in [0,1]\right)$ is a Brownian martingale.

3) Let $0 \le a \le 1$. We now suppose that $\Phi(\omega) = \omega(a)$. Show that:

$$\Phi_s^{(\mathbf{P})} = B_{as}, \quad \Psi_t^{(\mathbf{P})} = \int_0^t B_{as} dB_s$$

and

$$\Psi_t^{(\mathbf{M})} = \int_0^1 W_{as,t} d_s W_{s,t}.$$

(Note that in [83, Proposition 5.2], it is proven that, if $0 < a < 1$, $\left(\Psi_t^{(\mathbf{P})}, 0 \le t \le 1\right)$ and $\left(\Psi_t^{(\mathbf{M})}, 0 \le t \le 1\right)$ do not have the same law, although they are both martingales and they have the same one-dimensional marginals.)

2.3.4 Examples

In all the following examples, Φ is a function from Ω_0 into $\mathbb{R}$, which is $\mathscr{F}_\infty$-measurable and integrable with respect to the Wiener measure $\mathbb{P}$ (such that $\mathbb{P}(\omega_0 = 0) = 1$). In other words, $\Phi \in L^1(\mathbb{P})$, see (2.11).

Example 2.1. Let $r > 0$ fixed and $f : \mathbb{R} \longrightarrow \mathbb{R}$ Borel, integrable for the centered Gaussian measure of variance r. Let Φ be defined by:

$$\Phi(\omega) := f(\omega_r).$$

Then:

$$\begin{aligned}\Phi_t^{(\mathbf{M})} &= \mathbb{E}\left[f(W_{r,1})|\mathscr{G}_t\right] \quad (0 \le t \le 1)\\ &= \mathbb{E}\left[f(W_{r,1} - W_{r,t} + W_{r,t})|\mathscr{G}_t\right]\\ &= P_{(1-t)r}f(W_{r,t})\end{aligned}$$

(from (2.22) and the relation $W_{r,1} - W_{r,t} \overset{(law)}{=} \sqrt{r(1-t)}G$, with G a reduced Gaussian r. v.). On the other hand:

$$\Phi_t^{(\mathbf{P})} = P_{(1-t)r}f(B_{rt}) \quad (0 \le t \le 1). \tag{2.26}$$

In this case, from (2.3):

$$\left(\Phi_t^{(\mathbf{M})}, 0 \le t \le 1\right) \overset{(law)}{=} \left(\Phi_t^{(\mathbf{P})}, 0 \le t \le 1\right).$$

In particular, $\left(\Phi_t^{(\mathbf{P})}, 0 \le t \le 1\right)$ is a $(\mathscr{F}_{rt}, 0 \le t \le 1)$-martingale.

Example 2.2. Let $r > 0$ be fixed and $h : [0,r] \times \mathbb{R} \longrightarrow \mathbb{R}$ be a space-time harmonic function (for Brownian motion) on $[0,r]$. We define:

$$\Phi(\omega) := h(r, \omega_r).$$

Then:

$$\begin{aligned}\Phi_t^{(\mathbf{M})} &= \mathbb{E}\left[h(r, W_{r,1})|\mathscr{G}_t\right] \quad (0 \le t \le 1)\\ &= \mathbb{E}\left[h(r, W_{r,1} - W_{r,t} + W_{r,t})|\mathscr{G}_t\right]\\ &= \mathbb{E}\left[h(r, a + \sqrt{r(1-t)}G)\right] \quad \text{with } a = W_{r,t}\\ &= h(rt, W_{r,t}) \quad (\text{from } (2.10)).\end{aligned}$$

On the other hand:

$$\Phi_t^{(\mathbf{P})} = h(rt, B_{rt}) \quad (0 \le t \le 1)$$

and, as in the previous example, $\left(\Phi_t^{(\mathbf{M})}, 0 \le t \le 1\right) \overset{(law)}{=} \left(\Phi_t^{(\mathbf{P})}, 0 \le t \le 1\right)$. Remark that, by Theorems 2.4 and 2.5 with $I = [0,r]$, the function $H(t,\omega) = h(rt, \omega_r)$ is a STHBS function on $[0,1]$. This also yields the previous result.

Example 2.3. Let $h : [0,1] \times \mathbb{R} \longrightarrow \mathbb{R}$ be a space-time harmonic function (for Brownian motion) on $[0,1]$ and let us define:

$$\Phi(\omega) := \int_0^1 h(u, \omega_u) du.$$

Then:

$$\begin{aligned}\Phi_t^{(\mathbf{M})} &= \mathbb{E}\left[\int_0^1 h(u, W_{u,1}) du | \mathscr{G}_t\right] \quad (0 \le t \le 1) \\ &= \int_0^1 h(ut, W_{u,t}) du \quad \text{(from Theorem 2.2)}\end{aligned}$$

whereas:

$$\Phi_t^{(\mathbf{P})} = \int_0^1 h(ut, B_{ut}) du.$$

This example fits with Theorem 2.2.

Example 2.4. Let $g \in L^2(\mathbb{R}_+, dx)$ and

$$\Phi(\omega) := \int_0^{+\infty} g(u) d\omega_u.$$

Then:

$$\begin{aligned}\Phi_t^{(\mathbf{M})} &= \mathbb{E}\left[\int_0^{+\infty} g(u) d_u W_{u,1} | \mathscr{G}_t\right] \quad (0 \le t \le 1) \\ &\quad \left(\text{where } \int_0^{+\infty} g(u) d_u W_{u,1} \text{ is the Wiener integral of } g \text{ with respect to } W_{\bullet,1}\right) \\ &= \int_0^{+\infty} g(u) d_u W_{u,t}. \qquad (2.27)\end{aligned}$$

On the other hand:

$$\begin{aligned}\Phi_t^{(\mathbf{P})} &= \int_0^{+\infty} g(u) d_u B_{ut} \quad (0 \le t \le 1) \\ &= \int_0^{+\infty} g\left(\frac{s}{t}\right) dB_s \quad \text{(after the change of variable } ut = s). \qquad (2.28)\end{aligned}$$

In particular, if $g(u) = u^\alpha 1_{\{u \le 1\}}$, with $\alpha > -\frac{1}{2}$, we obtain:

$$\Phi_t^{(\mathbf{P})} = \frac{1}{t^\alpha} \int_0^t s^\alpha dB_s \quad (0 \le t \le 1)$$

which is a $\mathbf{F}_2'$-type peacock (see definition in Remark 1.10), since by scaling:

$$\frac{1}{t^\alpha} \int_0^t s^\alpha dB_s \overset{(1.d)}{=} \sqrt{t} \int_0^1 u^\alpha dB_u.$$

Example 2.5 (see also Exercise 2.4). Let $c \in L^2(\mathbb{R}_+, du)$ and define:

$$\Phi(\omega) := \exp\left(\int_0^{+\infty} c(u)d\omega_u - \frac{1}{2}\int_0^{+\infty} c^2(u)du\right).$$

Then:

$$\begin{aligned}\Phi_t^{(\mathbf{M})} &= \mathbb{E}\left[\exp\left(\int_0^{+\infty} c(u)d_uW_{u,1} - \frac{1}{2}\int_0^{+\infty} c^2(u)du\right)|\mathscr{G}_t\right] \quad (0 \le t \le 1)\\ &= \exp\left(\int_0^{+\infty} c(u)d_uW_{u,t} - \frac{t}{2}\int_0^{+\infty} c^2(u)du\right).\end{aligned}$$

On the other hand:

$$\begin{aligned}\Phi_t^{(\mathbf{P})} &= \exp\left(\int_0^{+\infty} c(u)d_uB_{ut} - \frac{t}{2}\int_0^{+\infty} c^2(u)du\right) \quad (0 \le t \le 1)\\ &= \exp\left(\int_0^{+\infty} c\left(\frac{s}{t}\right)dB_s - \frac{1}{2}\int_0^{+\infty} c^2\left(\frac{s}{t}\right)ds\right).\end{aligned}$$

In the particular case $c(u) = \nu u^{\alpha} 1_{\{u \le 1\}}$ $(\alpha > -\frac{1}{2})$ we obtain:

$$\begin{aligned}\Phi_t^{(\mathbf{P})} &= \exp\left(\frac{\nu}{t^{\alpha}}\int_0^t s^{\alpha}dB_s - \frac{\nu^2 t}{2}\int_0^1 s^{2\alpha}ds\right)\\ &\overset{(1.d)}{=} \exp\left(\nu\sqrt{t}\int_0^1 s^{\alpha}dB_s - \frac{\nu^2 t}{2(2\alpha+1)}\right) \quad \text{(by scaling)}.\end{aligned}$$

Easy computations allow us to check that, for all $t \le 1$, $\mathbb{E}\left[\Phi_t^{(\mathbf{P})}\right] = 1$, an equality which agrees with the fact that an integrable peacock has a constant expectation.

Example 2.6. In Example 2.3, where the space-time harmonic function $h(t,x)$ is equal to $\exp\left(\nu x - \frac{\nu^2}{2}t\right)$ $(\nu \in \mathbb{R})$, we have shown that the $(\mathscr{G}_t, t \ge 0)$-martingale $\left(\int_0^1 du \exp\left(\nu W_{u,t} - \frac{\nu^2}{2}ut\right), t \ge 0\right)$ is associated to the F_1-type peacock $\left(\int_0^1 du \exp\left(\nu B_{ut} - \frac{\nu^2}{2}ut\right), t \ge 0\right) = \left(\frac{1}{t}\int_0^t du \exp\left(\nu B_u - \frac{\nu^2}{2}u\right), t \ge 0\right)$.
We shall now extend this example by "replacing the constant ν by a function c", i.e. by considering exponential martingales. More precisely, let $c \in L^2([0,1], dv)$ and

$$\Phi(\omega) := \int_0^1 du \exp\left(\int_0^u c(v)d\omega_v - \frac{1}{2}\int_0^u c^2(v)dv\right).$$

Then:

$$\begin{aligned}
\Phi_t^{(\mathbf{M})} &= \mathbb{E}\left[\int_0^1 du \exp\left(\int_0^u c(v) d_v W_{v,1} - \frac{1}{2}\int_0^u c^2(v)dv\right) |\mathscr{G}_t\right] \qquad (0 \le t \le 1) \\
&= \int_0^1 du \mathbb{E}\left[\exp\left(\int_0^u c(v) d_v W_{v,1} - \frac{1}{2}\int_0^u c^2(v)dv\right) |\mathscr{G}_t\right] \\
&= \int_0^1 du \exp\left(\int_0^u c(v) d_v W_{v,t} - \frac{t}{2}\int_0^u c^2(v)dv\right)
\end{aligned}$$

(see Example 2.5 for this relation). On the other hand:

$$\begin{aligned}
\Phi_t^{(\mathbf{P})} &= \int_0^1 du \exp\left(\int_0^u c(v) d_v B_{vt} - \frac{t}{2}\int_0^u c^2(v)dv\right) \\
&= \int_0^1 du \exp\left(\int_0^{ut} c\left(\frac{a}{t}\right) dB_a - \frac{1}{2}\int_0^{ut} c^2\left(\frac{a}{t}\right) da\right) \\
&\quad \text{(after the change of variable } vt = a) \\
&= \frac{1}{t}\int_0^t ds \exp\left(\int_0^s c\left(\frac{a}{t}\right) dB_a - \frac{1}{2}\int_0^s c^2\left(\frac{a}{t}\right) da\right) \qquad (2.29) \\
&\quad \text{(after the change of variable } ut = s).
\end{aligned}$$

In particular, if $c := \nu$, then:

$$\Phi^{(\mathbf{P})} = \left(\frac{1}{t}\int_0^t ds \exp\left(\nu B_s - \frac{\nu^2}{2}s\right), 0 \le t \le 1\right)$$

is a $\mathbf{F}_1$-type peacock. If $c(u) := \nu u^\alpha$ $(\alpha > -\frac{1}{2})$, we have:

$$\Phi^{(\mathbf{P})} = \left(\frac{1}{t}\int_0^t ds \exp\left(\frac{\nu}{t^\alpha}\int_0^s a^\alpha dB_a - \frac{\nu^2 s^{2\alpha+1}}{2(2\alpha+1)t^{2\alpha}}\right), 0 \le t \le 1\right).$$

2.3.5 Some Natural Questions

Let $\Phi \in L^1(\mathbb{P})$. Several questions arise naturally regarding the previous subsection:

$(Q1)$ Do the processes $(\Phi_t^{(\mathbf{M})}, 0 \le t \le 1)$ and $(\Phi_t^{(\mathbf{P})}, 0 \le t \le 1)$ admit continuous versions?

$(Q2)$ Are the processes $(\Phi_t^{(\mathbf{M})}, 0 \le t \le 1)$ and $(\Phi_t^{(\mathbf{P})}, 0 \le t \le 1)$ really different from each other? This question may be separated into two subquestions:

$(Q2a)$ For which $\Phi \in L^1(\mathbb{P})$ do we have:

$$(\Phi_t^{(\mathbf{M})}, 0 \le t \le 1) \overset{(law)}{=} (\Phi_t^{(\mathbf{P})}, 0 \le t \le 1)\ ?$$

$(Q2b)$ For which $\Phi \in L^1(\mathbb{P})$, is $(\Phi_t^{(\mathbf{P})}, 0 \le t \le 1)$ a process with bounded variation? (Below we shall see that $(\Phi_t^{(\mathbf{M})}, 0 \le t \le 1)$ admits a continuous version, and therefore is not a process with bounded variation, see Remark 1.5.)

Here are some (partial) answers to these questions:

Proposition 2.2. *Let* $\Omega_0^1 := \{\omega \in \mathscr{C}([0,1],\mathbb{R}),\ \omega_0 = 0\}$*, equipped with the uniform norm:* $\|\omega\|_\infty := \sup_{0\le s\le 1} |\omega_s|$*. Let* $\Phi : \Omega_0^1 \longrightarrow \mathbb{R}$*,* $\Phi \in L^1(\Omega_0^1, \mathbb{P})$*. Then:*

i) $(\Phi_t^{(M)}, 0 \le t \le 1)$ *admits a continuous version.*
ii) Assume that $\Phi : \Omega_0^1 \longrightarrow \mathbb{R}$ *is continuous (for the uniform norm) and that there exists* $\lambda \ge 0$ *and* $k < \frac{1}{2}$ *such that, for every* $\omega \in \Omega_0^1$*:*

$$|\Phi(\omega)| \le \lambda \exp\left(k\|\omega\|_\infty^2\right).$$

Then the peacock $(\Phi_t^{(P)}, 0 \le t \le 1)$ *admits a continuous version given by:*

$$\left(\Phi_t^{(P)} = \mathbb{E}_{\widetilde{B}}\left[\Phi(B_{t\cdot} + \sqrt{1-t}\widetilde{B}_\cdot)\right], 0 \le t \le 1\right).$$

Proof. i) Let, for $0 \le s \le t \le 1$,

$$\mathscr{G}_{s,t} = \sigma\left(W_{u,v};\ 0 \le u \le s, 0 \le v \le t\right).$$

Then, from Cairoli-Walsh [39], the two-parameter martingale $N_{s,t} := \mathbb{E}\left[\Phi(W_{\cdot,1})|\mathscr{G}_{s,t}\right]$ admits a continuous version. Now since $\Phi^{(\mathbf{M})}$ is a $(\mathscr{G}_t := \mathscr{G}_{1,t}, 0 \le t \le 1)$-martingale which is equal to $\left(\Phi_t^{(\mathbf{M})} = N_{1,t}, 0 \le t \le 1\right)$, it admits a continuous version (see also Exercise 2.9 for another proof of Point *i)* of Proposition 2.2).

ii) This point follows from the dominated convergence theorem thanks to the next lemma. □

Lemma 2.1. *One has, for* $0 \le k < 1/2$*:*

$$\frac{1}{\sqrt{1-2k}} \le \mathbb{E}\left[\exp\left(k\|B\|_\infty^2\right)\right] \le \frac{2}{\sqrt{1-2k}}.$$

Proof. The first inequality is obvious since $\|B\|_\infty^2 \ge (B_1)^2$. For the second inequality, set $S(B) = \sup_{0\le s\le 1} B_s$. We have $\|B\|_\infty = \sup(S(B), S(-B))$ and $S(B)$ and $S(-B)$ are both distributed as $|B_1|$. Therefore,

$$\begin{aligned}\mathbb{E}\left[\exp\left(k\|B\|_\infty^2\right)\right] &= \mathbb{E}\left[\sup\left(e^{kS^2(B)}, e^{kS^2(-B)}\right)\right] \\ &\le 2\mathbb{E}\left[\exp\left(kB_1^2\right)\right]\end{aligned}$$

which yields the desired result. □

Here is now an answer to Question $(Q2b)$.

Proposition 2.3. *The set of r.v.'s* $\Phi \in L^1(\mathbb{P})$ *such that* $(\Phi_t^{(P)}, 0 \le t \le 1)$ *has finite variation is dense in* $L^1(\mathbb{P})$*.*

Proof. 1) Let $\mathscr{K} := \{c : [0,+\infty[\longrightarrow \mathbb{R}, c \in \mathscr{C}^2$ with compact support in $]0,+\infty[\}$. We know (see for instance [150], p.198) that the set of r.v.'s of the form $\left\{\exp\left(\int_0^{+\infty} c(s)dB_s - \frac{1}{2}\int_0^{+\infty} c^2(s)ds\right); c \in \mathscr{K}\right\}$ is total in $L^2(\Omega_0, \mathscr{F}_\infty, \mathbb{P})$. Thus the set $\left\{\Phi^{(c)} = \exp\left(\int_0^{+\infty} c(s)d\omega_s - \frac{1}{2}\int_0^{+\infty} c^2(s)ds\right), c \in \mathscr{K}\right\}$ is total in $L^2(\Omega_0, \mathscr{F}_\infty, \mathbb{P})$. It is sufficient, to prove Proposition 2.3, to see that, for $c \in \mathscr{K}$, from Example 2.5:

$$\left(\Phi_t^{(\mathbf{P},c)}, 0 \le t \le 1\right) = \left(\exp\left(\int_0^{+\infty} c\left(\frac{v}{t}\right) dB_v - \frac{t}{2}\int_0^{+\infty} c^2(v)dv\right), 0 \le t \le 1\right) \quad (2.30)$$

has bounded variation.

2) *We now prove that $\Phi^{(P,c)}$ defined by (2.30) has bounded variation.* From (2.30), we only need to check that

$$\left(Z_t := \int_0^{+\infty} c\left(\frac{v}{t}\right) dB_v \, , \, 0 \le t \le 1\right)$$

has bounded variation (with $c \in \mathscr{C}^2$ and with compact support in $]0,+\infty[$). Thus:

$$Z_t = -\frac{1}{t}\int_0^{+\infty} c'\left(\frac{v}{t}\right) B_v dv$$

after integrating by parts, and

$$\frac{dZ_t}{dt} = \frac{1}{t}\int_0^{+\infty} \left(c'(v) + vc''(v)\right) B_{vt} dv$$

Hence: $\mathbb{E}\left[\left|\frac{dZ_t}{dt}\right|\right] = \mathscr{O}\left(\frac{1}{\sqrt{t}}\right)$ and $\int_0^1 \left|\frac{dZ_t}{dt}\right| dt < +\infty$ a.s. □

We now provide a (partial) answer to Question $Q2a$), by giving a sufficient condition for the processes $\left(\Phi_t^{(\mathbf{P})}, 0 \le t \le 1\right)$ and $\left(\Phi_t^{(\mathbf{M})}, 0 \le t \le 1\right)$ to be equal in law. Note that the Examples 2.1 and 2.2 fall precisely in this case.

Proposition 2.4. *Let $\Phi \in L^1(\mathbb{P})$, and assume that Φ belongs to the first Wiener chaos, i.e. there exists $g \in L^2(\mathbb{R}_+, du)$ such that:*

$$\Phi(\omega) = \int_0^{+\infty} g(u) d\omega_u.$$

Then:

$$(\Phi_t^{(M)}, 0 \le t \le 1) \overset{(law)}{=} (\Phi_t^{(P)}, 0 \le t \le 1)$$

if and only if, for every $a \in [0,1]$,

$$\int_0^{+\infty} g(u)g(ua)du = \int_0^{+\infty} g^2(u)du. \quad (2.31)$$

Proof. Since $(\Phi_t^{(\mathbf{M})}, 0 \leq t \leq 1)$ and $(\Phi_t^{(\mathbf{P})}, 0 \leq t \leq 1)$ are two centered Gaussian processes, they have the same law if and only if they have the same covariance, a condition which writes, from (2.27) and (2.28),

$$\int_0^{+\infty} g\left(\frac{v}{t}\right) g\left(\frac{v}{s}\right) dv = (s \wedge t) \int_0^{+\infty} g^2(v) dv.$$

Therefore, if $s \leq t$:

$$s \int_0^{+\infty} g\left(u \frac{s}{t}\right) g(u)\, du = s \int_0^{+\infty} g^2(v) dv$$

(after the change of variable $\frac{v}{s} = u$), whereas, if $s > t$:

$$t \int_0^{+\infty} g(u)\, g\left(u \frac{t}{s}\right) du = t \int_0^{+\infty} g^2(v) dv$$

(after the change of variable $\frac{v}{t} = u$). This is Proposition 2.4. □

Note that condition (2.31) is equivalent to:

$$\left(B_t^{(g)} := \frac{1}{\sqrt{\int_0^{+\infty} g^2(u) du}} \int_0^{+\infty} g\left(\frac{v}{t}\right) dB_v, t \geq 0 \right)$$

is a Brownian motion.

Exercise 2.7 (Heat semi-group and space-time harmonic functions for the Brownian motion). Let us denote by $L^1(\mathbb{R}, d\nu_a)$ the space of integrable functions for the Gaussian measure with variance a. Let $\mathscr{H}_a$ be the space of space-time harmonic functions defined on $[0,a] \times \mathbb{R}$. Prove that the mapping: $\gamma : L^1(\mathbb{R}, d\nu_a) \longrightarrow \mathscr{H}_a$ defined by $\gamma(\varphi)(t,x) = P_{a-t}\varphi(x)$ (where $(P_u, u \geq 0)$ denotes the heat semi-group) is one-to-one.

Exercise 2.8 (Martingales associated to some integrals of exponentials of squared Brownian motion). 1) Let $(W_{u,v}; u \geq 0, v \geq 0)$ denote the standard Brownian sheet and $\mathscr{G}_v = \sigma\{W_{u,s};, u \geq 0, s \leq v\}$. Prove that, for all $u > 0$,

$$\left(\frac{1}{\sqrt{1-2v}} \exp\left(-\frac{W_{u,v}^2}{(1-2v)u} \right), 0 \leq v < \frac{1}{2} \right)$$

is a $(\mathscr{G}_v, 0 \leq v < \frac{1}{2})$ martingale.

2) Deduce that, for every signed measure with finite total mass on $]0, +\infty[$,

$$\left(\frac{1}{\sqrt{1-2v}} \int_0^{+\infty} \exp\left(-\frac{v}{u(1-2v)} B_u^2 \right) \sigma(du), 0 \leq v < \frac{1}{2} \right)$$

is a peacock.

3) Deduce that, for every signed measure with finite total mass on $]0,+\infty[$,

$$\left(\sqrt{1+2t}\int_0^{+\infty} \exp\left(-\frac{t}{u}B_u^2\right)\sigma(du)\,,\,t\geq 0\right)$$

is a peacock and that

$$\left(\sqrt{1+2t}\int_0^{+\infty} \exp\left(-\frac{1+2t}{u}W^2_{u,\frac{t}{1+2t}}\right)\sigma(du)\,,\,t\geq 0\right)$$

is a martingale associated to this peacock. Compare with Theorem 1.8.

Exercise 2.9 (Another proof of the existence of a continuous version of $(\Phi_t^{(\mathbf{M})}$, $0\leq t\leq 1)$, for $\Phi\in L^2(\mathbb{P})$). 1) Let $(e_n, n\geq 0)$ be an orthonormal basis of $L^2(\mathbb{R}_+,dx)$ and set, for $n\geq 0$ and $t\in[0,1]$:

$$B_t^{(n)} := \int_0^{+\infty} e_n(u)d_uW_{u,t}$$

(where $(W_{u,t},\, u\geq 0, t\geq 0)$ denotes the standard Brownian sheet).

i) Prove that $(B^{(n)}, n\geq 0)$ is a sequence of independent Brownian motions.

ii) Prove that:

$$\mathscr{G}_t := \sigma(W_{u,s};\, u\geq 0, s\leq t) = \sigma(B_s^{(n)}, n\geq 0, s\leq t).$$

2) *i*) Let $\varphi_0,\dots,\varphi_{n-1}$ be n elements of $L^2([0,1])$ and define:

$$\mathscr{E}_n^{(\varphi)} := \exp\left(\sum_{k=0}^{n-1}\int_0^1 \varphi_k(s)dB_s^{(k)} - \frac{1}{2}\int_0^1 \varphi_k^2(s)ds\right).$$

Prove that $(\mathscr{E}_n^{(\varphi)}, \varphi\in (L^2([0,1]))^n, n\geq 0)$ is total in $L^2(\mathscr{G}_1)$ (equipped with the law of W).

ii) Prove that every $\mathscr{E}_n^{(\varphi)}$ may be written as:

$$\mathscr{E}_n^{(\varphi)} = c + \sum_{p=0}^{n-1}\int_0^1 \Phi_p(s)dB_s^{(p)}$$

where $(\Phi_p, p=0,\dots,n-1)$ is a sequence of $(\mathscr{G}_t, 0\leq t\leq 1)$ predictable processes such that:

$$\mathbb{E}\left[\sum_{p=0}^{n-1}\int_0^1 \Phi_p^2(s)ds\right] < \infty$$

iii) Prove that every $X\in L^2(\mathscr{G}_1)$ admits the following representation:

$$X = c + \sum_{n=0}^{+\infty}\int_0^1 x_n(s)dB_s^{(n)} \tag{2.32}$$

with $\mathbb{E}\left[\sum_{n=0}^{+\infty}\int_0^1 x_n^2(s)ds\right] < \infty$.

3) i) Prove that, if X admits the representation (2.32), then:

$$X_t := \mathbb{E}[X|\mathscr{G}_t] = c + \sum_{n=0}^{+\infty} \int_0^t x_n(s) dB_s^{(n)} \quad (0 \le t \le 1) \tag{2.33}$$

ii) Deduce from (2.33) and from Doob's maximal inequality that $(X_t, 0 \le t \le 1)$ admits a continuous version.

Exercise 2.10 (Computation and properties of $\Phi^{(\mathbf{P})}$ for $\Phi(\omega) = L_r^a(\omega)$). Let $a \in \mathbb{R}$ and $r \ge 0$. We define Φ by:

$$\Phi(\omega) = L_r^a(\omega)$$

where L_r^a denotes the Brownian local time at level a and time r.
1) i) Let $f : \mathbb{R} \longmapsto \mathbb{R}_+$ be a continuous function with compact support. Deduce from the occupation times density formula:

$$\int_0^r f(B_s) ds = \int_{-\infty}^{+\infty} f(a) L_r^a da$$

and from (2.26) of Example 2.1 that:

$$\int_0^r P_{(1-t)s} f(B_{ts}) ds = \int_{-\infty}^{+\infty} f(a) \Phi_t^{(\mathbf{P})} da \tag{2.34}$$

(where $(P_u, u \ge 0)$ denotes the heat semi-group).
ii) Deduce from (2.34), by identification that:

$$\Phi_t^{(\mathbf{P})} = \frac{1}{\sqrt{2\pi(1-t)}} \int_0^r \exp\left(-\frac{(B_{ts} - a)^2}{2s(1-t)}\right) \frac{ds}{\sqrt{s}}, \quad 0 \le t < 1. \tag{2.35}$$

2) i) Deduce from (2.35) that $(\Phi_t^{(\mathbf{P})}, 0 \le t < 1)$ is continuous on $[0,1[$.
ii) Prove that

$$\Phi_t^{(\mathbf{P})} = \frac{1}{\sqrt{2\pi t(1-t)}} \int_0^{tr} \exp\left(-\frac{(B_u - a)^2 t}{2u(1-t)}\right) \frac{du}{\sqrt{u}} \tag{2.36}$$

and deduce from (2.36) that $\Phi^{(\mathbf{P})}$ is of $\mathscr{C}^1$ class on $]0,1[$.

Remark 2.5. It is proven in [83] that $(\Phi_t^{(\mathbf{P})}, 0 \le t \le 1)$ is of finite variation on any interval $[0,s]$ with $0 < s < 1$, the variation being square integrable on any such interval.

Exercise 2.11 ($(\Phi_t^{(\mathbf{P})}, 0 \le t \le 1)$ and developments in Wiener chaoses). Let $\Omega_0^1 = \{\omega \in \mathscr{C}([0,1],\mathbb{R}); \omega_0 = 0\}$ equipped with the norm $\|\omega\|_\infty := \sup_{0 \le s \le 1} |\omega_s|$. Let $(\Omega_0^1, (B_t, \mathscr{F}_t)_{0 \le t \le 1}, \mathbb{P})$ be the canonical realization of a Brownian motion on $[0,1]$ started from 0. We denote by L^p the space $L^p(\Omega_0^1, \mathscr{F}_1, \mathbb{P})$ $(1 \le p \le +\infty)$ and recall that every element $\Phi \in L^2$ admits an orthogonal decomposition in Wiener chaoses; we write $(\mathscr{C}_n, n \geq 0)$ for the sequence of these chaoses:

$$\Phi = \sum_{n=0}^{+\infty} \Phi_n$$

($\Phi_n \in \mathscr{C}_n$, $\Phi_0 = \mathbb{E}[\Phi]$, this series being convergent in L^2).
1) Let $(O_h, h \geq 0)$ denote the Ornstein-Uhlenbeck semi-group defined by:

$$O_h\Phi(\omega) := \mathbb{E}_{\widetilde{B}}\left[\Phi\left(e^{-h/2}\omega_. + \sqrt{1-e^{-h}}\widetilde{B}_.\right)\right].$$

i) Show that, for all $h \geq 0$, O_h is a bounded operator, actually a contraction, on all the L^p spaces $(1 \leq p \leq +\infty)$ (*Hint*: the proof is similar to that of Property 2) in Theorem 2.8.)

ii) Verify that $(O_h, h \geq 0)$ defines a semi-group of operators, i.e.:

$$O_h \circ O_k = O_{h+k} \quad (h, k \geq 0)$$

in each L^p space $(1 \leq p \leq +\infty)$. (*Hint*: the proof is similar to that of Property 3) in Theorem 2.8.)

iii) Show that, if $1 \leq p < +\infty$ and $\Phi \in L^p$, the map $h \geq 0 \longmapsto O_h\Phi \in L^p$ is continuous. (*Hint*: the proof is similar to that of Property 3) in Theorem 2.8.) On the other hand, we recall that if $\Phi \in \mathscr{C}_n$, then:

$$O_h\Phi = e^{-\frac{nh}{2}}\Phi$$

(see for instance Nualart [135]).
2) Let, for all $t \in]0,1]$, $\sigma_t : \Omega_0^1 \longrightarrow \Omega_0^1$ defined by:

$$((\sigma_t(\omega))_u, 0 \leq u \leq 1) := \left(\frac{1}{\sqrt{t}}\omega_{ut}, 0 \leq u \leq 1\right)$$

and, for all $t \in]0,1]$, $\Sigma_t : L^2 \longrightarrow L^2$ defined by:

$$\Sigma_t\Phi(\omega) = \Phi(\sigma_t(\omega)).$$

i) Prove that for all $t \in]0,1]$, Σ_t is an isometry of L^2.

ii) Show that, for all $h \geq 0$ and $t \in]0,1]$:

$$\Sigma_t \circ O_h = O_h \circ \Sigma_t.$$

3) Let, for $t \in [0,1]$, $R_t : L^2 \longrightarrow L^2$ defined by:

$$\begin{cases} R_t\Phi := \Sigma_t \circ O_{-\log t}\Phi = O_{-\log t} \circ \Sigma_t\Phi & \text{for } t \in]0,1] \\ R_0\Phi = \mathbb{E}[\Phi] & \text{for } t = 0. \end{cases}$$

i) Show that, for all $t \in [0,1]$, R_t is a contraction of L^2 and that, if $\mathbb{E}[\Phi] = 0$, then:

$$\|R_t \Phi\|_{L^2} \leq \sqrt{t} \|\Phi\|_{L^2}.$$

ii) Show that R_1 is the identity operator, and that for $0 \leq s,t \leq 1$

$$R_s \circ R_t = R_{st}.$$

iii) Show that, for every $\Phi \in L^2$, the mapping

$$t \in [0,1] \longmapsto R_t \Phi \in L^2$$

is continuous and $(\mathscr{F}_t, 0 \leq t \leq 1)$-adapted.

4) *i*) Prove that, for every $\Phi \in L^2$:

$$(R_t \Phi(B_\cdot), 0 \leq t \leq 1) \overset{(law)}{=} \left(\Phi_t^{(\mathbf{P})}, 0 \leq t \leq 1\right).$$

(*Hint*: the definition (2.25) of $\Phi^{(\mathbf{P})}$ might be useful.) Show that this last formula is still true for $\Phi \in L^1$.

ii) Deduce then that, if $\Phi_n \in \mathscr{C}_n$:

$$\left((\Phi_n)_t^{(\mathbf{P})}, 0 < t \leq 1\right) \overset{(law)}{=} \left(t^{n/2} \Phi_n \left(\frac{1}{\sqrt{t}} B_{\cdot t}\right), 0 < t \leq 1\right).$$

In particular, recover, for $n = 1$, the results of Example 2.4.

iii) Let $\Phi \in L^2$ and $\Phi = \sum\limits_{n \geq 0} \Phi_n$ its decomposition in Wiener's chaoses. Show that, for $0 < t \leq 1$:

$$\Phi_t^{(\mathbf{P})} = \sum_{n \geq 0} t^{n/2} \Phi_n \left(\frac{1}{\sqrt{t}} B_{\cdot t}\right).$$

iv) Let $(W_{u,v}, u, v \geq 0)$ denote the standard Brownian sheet and $\mathscr{G}_t := \sigma(W_{u,v}; u \geq 0, v \leq t)$. Show that, for $0 < t \leq 1$:

$$\Phi_t^{(\mathbf{M})} = O_{-\log t} \Phi \left(\frac{1}{\sqrt{t}} W_{\bullet,t}\right) = R_t \Phi \left(W_{\bullet t^{-1},t}\right) = \sum_{n \geq 0} t^{\frac{n}{2}} \Phi_n \left(\frac{1}{\sqrt{t}} W_{\bullet,t}\right).$$

2.4 The Lévy Sheet Method

2.4.1 Introduction

Let $(\Omega = \mathscr{C}(\mathbb{R}_+, \mathbb{R}), (B_t, \mathscr{F}_t)_{t \geq 0}, \mathbb{P})$ denote the canonical realisation of a Brownian motion started from 0 and let $(\Omega' = \mathscr{C}(\mathbb{R}_+ \times \mathbb{R}_+, \mathbb{R}), (W_{s,t}, \mathscr{G}_{s,t})_{s,t \geq 0}, \mathbb{Q})$ the one of the standard Brownian sheet. In the previous sections, to some peacocks $(X_t, t \geq 0)$ defined on $(\Omega, \mathscr{F}_\infty, \mathbb{P})$, we managed to associate a martingale defined on $(\Omega', \mathbb{Q})$. To this end, we went "from Brownian motion to Brownian sheet" and used extensively the following properties:

- for all $t \geq 0$, $(B_{ut}, u \geq 0) \overset{(law)}{=} (W_{u,t}, u \geq 0)$,
- $(W_{\bullet,t}, t \geq 0)$ is a process with independent stationary increments, i.e. a Lévy process, taking values in $\mathscr{C}(\mathbb{R}_+, \mathbb{R})$.

Now, let $(L_t, t \geq 0)$ denote a Lévy process with real values. Our aim is to associate to some peacocks $(X_t = F(t, (L_u, u \geq 0)), t \geq 0)$ (functions of L) a martingale $(M_t, t \geq 0)$. To carry out this program, imitating the previous Brownian discussion, we shall go "from a Lévy process to its Lévy sheet". We start by describing what is a Lévy sheet.

2.4.2 Lévy Sheet $W^{(L)}$ Extending a Lévy Process L

2.4.2.1 The Lévy-Khintchine Representation of L

We start with a real-valued Lévy process $(L_t, t \geq 0)$ starting from 0. We denote by ψ its characteristic exponent:

$$\forall \lambda \in \mathbb{R},\ \forall t \geq 0, \quad \mathbb{E}\left[\exp(i\lambda L_t)\right] = \exp\left(-t\psi(\lambda)\right).$$

One has (*Lévy-Khintchine formula*):

$$\psi(\lambda) = \sigma^2 \frac{\lambda^2}{2} + i\gamma\lambda + \int \left(1 - e^{i\lambda x} + i\lambda x 1_{\{|x| \leq 1\}}\right) \nu(dx)$$

with $\sigma, \gamma \in \mathbb{R}$ and ν a positive measure on $\mathbb{R} \backslash \{0\}$ such that:

$$\int \frac{x^2}{1+x^2} \nu(dx) < \infty.$$

We refer e.g. to Bertoin [24] for a deep study of Lévy processes.

2.4.2.2 The Skorokhod Space

We denote by $\mathbb{D}_0$ the Skorokhod space consisting of all càdlàg functions ε from $\mathbb{R}_+$ into $\mathbb{R}$ such that $\varepsilon(0) = 0$ (we refer, for instance, to Jacod-Shiryaev [91]). The space $\mathbb{D}_0$ is equipped with the probability $\mathbb{P}$ which is the law of L. We often identify L with the coordinate process on $\mathbb{D}_0$.

We denote by $(\mathscr{F}_t, t \geq 0)$ the natural filtration of L on $(\mathbb{D}_0, \mathbb{P})$, and we set $\mathscr{F} = \mathscr{F}_\infty$. Thus, $\mathscr{F}$ is the Borel σ-field on $\mathbb{D}_0$ completed with respect to $\mathbb{P}$.

2.4.2.3 The X-integral of a Rectangle R

If $(X_{s,t}; s \geq 0, t \geq 0)$ is a real-valued two-parameter process and if

$$R =]s_1, s_2] \times]t_1, t_2], \quad s_1 \leq s_2, t_1 \leq t_2$$

is a rectangle, we set

$$\Delta_R X = X_{s_2,t_2} - X_{s_1,t_2} - X_{s_2,t_1} + X_{s_1,t_1}$$

and we denote by $|R|$ the area of R:

$$|R| = (s_2 - s_1)(t_2 - t_1).$$

2.4.2.4 Defining the Lévy Sheet $W^{(L)}$

The following results, for which we refer to Dalang-Walsh [50, Section 2] (see also [1]) are essential for our purpose.

Theorem 2.9. *There exists a real-valued two parameter process* $W^{(L)} = (W^{(L)}_{s,t}; s \geq 0, t \geq 0)$ *satisfying the following properties:*

1) *For all* $s,t \geq 0$, $W^{(L)}_{s,0} = W^{(L)}_{0,t} = 0$.
2) *Almost surely, for any* $s,t \geq 0$, $W^{(L)}_{s,\bullet}$ *and* $W^{(L)}_{\bullet,t}$ *are càdlàg functions on* $\mathbb{R}_+$.
3) *For all finite sets of disjoint rectangles* $R^1,\dots,R^n$, *the random variables* $\Delta_{R^1} W^{(L)},\dots,\Delta_{R^n} W^{(L)}$ *are independent.*
4) *For any rectangle* R,

$$\Delta_R W^{(L)} \overset{(law)}{=} L_{|R|}.$$

The process $W^{(L)}$ will be called the Lévy sheet extending L.
Let, for $t \geq 0$,

$$\mathscr{G}^{(L)}_t = \sigma\{W^{(L)}_{u,v};\ u \geq 0, 0 \leq v \leq t\}.$$

In the following theorem, we summarize some straightforward consequences of Theorem 2.9 which will be needed in the sequel.

Theorem 2.10. 1) *Let* $0 \leq t_1 \leq t_2$. *Then the process* $(W^{(L)}_{s,t_2} - W^{(L)}_{s,t_1}, s \geq 0)$ *is a Lévy process which starts from 0, independent of* $\mathscr{G}^{(L)}_{t_1}$ *and is distributed as* $(L_{(t_2-t_1)s}, s \geq 0)$. *In particular, for any fixed* $t \geq 0$, $W^{(L)}_{\bullet,t} \overset{(law)}{=} L_{t\bullet}$ *i.e.,*

$$\forall t \geq 0, \quad \left(W^{(L)}_{u,t}, u \geq 0\right) \overset{(law)}{=} (L_{tu}, u \geq 0). \tag{2.37}$$

Thus, $(W^{(L)}_{\bullet,t}, t \geq 0)$ *is a Lévy process which takes values in* $\mathbb{D}_0$, *and has the same one-dimensional marginals as* $(L_{t\bullet}, t \geq 0)$.
2) *There is the equality in law:*

$$\left(W^{(L)}_{s,t}, s,t \geq 0\right) \overset{(law)}{=} \left(W^{(L)}_{t,s}, s,t \geq 0\right).$$

Thus, 1) *may be stated with the roles of s and t exchanged.*

2.4.2.5 A Family of Peacocks and Associated Martingales

Suppose that the Lévy process L satisfies the condition:

$$\mathbb{E}\left[\exp\left(L_1\right)\right] < \infty,$$

and set

$$l = \log \mathbb{E}\left[\exp\left(L_1\right)\right].$$

We obtain easily that, for every $t \geq 0$,

$$\mathbb{E}\left[\exp\left(L_t\right)\right] = e^{tl}.$$

We then set, for $\lambda \geq 0$ and $t \geq 0$,

$$Y_{\lambda,t} = L_{\lambda t} \quad \text{and} \quad Z_{\lambda,t} = W_{\lambda,t}^{(L)}.$$

It follows from Property 1) in Theorem 2.10, that conditions i) and ii) in Proposition 2.1 are satisfied, and therefore we obtain the following theorem which extends Theorem 2.1.

Theorem 2.11. *We assume that* $\mathbb{E}\left[\exp\left(L_1\right)\right] < \infty$ *and set* $l = \log \mathbb{E}\left[\exp\left(L_1\right)\right]$. *Let* σ *be a signed finite measure on* $\mathbb{R}_+$ *and*

$$A_t^{(\sigma)} := \int_{\mathbb{R}_+} \exp\left(L_{\lambda t} - l\lambda t\right) \sigma(d\lambda),$$

$$M_t^{(\sigma)} := \int_{\mathbb{R}_+} \exp\left(W_{\lambda,t}^{(L)} - l\lambda t\right) \sigma(d\lambda).$$

Then, the process $(M_t^{(\sigma)}, t \geq 0)$ *is a* $(\mathscr{G}_t^{(L)})$*-martingale and* $A_t^{(\sigma)} \overset{(1.d)}{=} M_t^{(\sigma)}$. *Consequently,* $(A_t^{(\sigma)}, t \geq 0)$ *is a peacock.*

2.4.2.6 The Semi-Group Associated to $(W_{\bullet,t}^{(L)}, t \geq 0)$

As both semi-groups of the Lévy processes $(W_{\bullet,t}^{(L)}, t \geq 0)$ and $(L_t, t \geq 0)$ play some role in the sequel, the following definition may be helpful.

Definition 2.4. Let $\mathscr{M}_t(d\eta)$ denote the law on $\mathbb{D}_0$ of $W_{\bullet,t}^{(L)}$ ($\overset{(law)}{=} L_{t\bullet}$) and $\mu_t(dx)$ the law on $\mathbb{R}$ of L_t. The semi-groups of the Lévy processes $(W_{\bullet,t}^{(L)}, t \geq 0)$ and $(L_t, t \geq 0)$ are, respectively, given by:

$$\Pi_t F(\varepsilon) = \mathbb{E}\left[F(\varepsilon + W_{\bullet,t}^{(L)})\right] = \int_{\mathbb{D}_0} F(\varepsilon + \eta) \mathscr{M}_t(d\eta) \tag{2.38}$$

and

$$P_t f(x) = \mathbb{E}\left[f(x + L_t)\right] = \int_{\mathbb{R}} f(x+y) \mu_t(dy). \tag{2.39}$$

Remark 2.6. In the sequel, the following elementary fact shall play some important role. Let $\widetilde{L}$ be an independent copy of L. Then, for any $a, b \geq 0$,

$$L_{a\bullet} + \widetilde{L}_{b\bullet} \overset{(law)}{=} L_{(a+b)\bullet}.$$

As a consequence, the $\mathbb{D}_0$-valued r.v. $L_\bullet$ is infinitely divisible. Theorem 2.10, Point 1) then states that the Lévy sheet $W^{(L)}$ may be understood as the $\mathbb{D}_0$-valued Lévy process $(W^{(L)}_{\bullet,t}, t \geq 0)$ such that:

$$W^{(L)}_{\bullet,1} \overset{(law)}{=} L_\bullet$$

Exercise 2.12 (Wiener integrals of subordinators; peacocks and associated martingales). Let $(\alpha_t, t \geq 0)$ be a subordinator without drift. We denote by ϕ its Bernstein function and by ν its Lévy measure:

$$\mathbb{E}\left[e^{-\lambda\alpha_t}\right] = \exp(-t\phi(\lambda)) \tag{2.40}$$

$$\text{with } \phi(\lambda) = \int_0^{+\infty}(1 - e^{-\lambda x})\nu(dx) \tag{2.41}$$

$$\text{and } \int_0^{+\infty}(1 \wedge x)\nu(dx) < \infty. \tag{2.42}$$

We also assume that:

$$a := \int_0^{+\infty} x\nu(dx) < \infty. \tag{2.43}$$

Of course, (2.43) implies (2.42).
1) Prove that ϕ is differentiable on $[0, +\infty[$, that it is of C^∞ class on $]0, +\infty[$ and that:

i) $\phi'(0^+) = \int_0^{+\infty} x\nu(dx) = a$,
ii) $\mathbb{E}[\alpha_t] = at$ (in fact, $(\alpha_t - at, t \geq 0)$ is a martingale).

2) Prove that, for every λ and $x \geq 0$:

i) $(\lambda \wedge 1)\phi(x) \leq \phi(\lambda x) \leq (\lambda \vee 1)\phi(x)$,
ii) for all $n \geq 1$, there exists a constant C_n such that, for every $\lambda > 0$:

$$|\phi^{(n)}(\lambda)| \leq C_n \frac{\phi(\lambda)}{\lambda^n} \tag{2.44}$$

(one may take $C_1 = 1$ and, for $n \geq 2$, $C_n = (n-1)^{n-1}$).
3) Let $h : \mathbb{R}_+ \longrightarrow \mathbb{R}_+$ be a Borel function. Let us define:

$$A(h) := \int_0^{+\infty} h(s)d\alpha_s \quad (\leq +\infty)$$

The process $(\alpha_s, s \geq 0)$ being increasing, this integral is a classical Stieltjes integral. Prove that:

i) $\mathbb{E}[A(h)] = a\int_0^{+\infty} h(s)ds$,

ii) $\mathbb{E}\left[e^{-\lambda A(h)}\right] = \exp\left(-\int_0^{+\infty} \phi(\lambda h(s))ds\right)$ (with the usual convention $e^{-\infty} = 0$)

(one may solve the two above questions first with h a piece-wise constant function with compact support, then with h bounded with compact support, and finally obtain the general case by passing to the limit).

iii) Prove that:

$$\int_0^{+\infty} \phi(h(s))ds < \infty \Longleftrightarrow \forall\lambda > 0, \int_0^{+\infty} \phi(\lambda h(s))ds < \infty$$

iv) Prove that $A(h) < \infty$ a.s. if and only if $\int_0^{+\infty} \phi(h(s))ds < \infty$.

v) Prove that $A(h) = \infty$ a.s. if and only if $\int_0^{+\infty} \phi(h(s))ds = \infty$.

vi) Prove that $\int_0^{+\infty} h(s)ds < \infty$ implies $\int_0^{+\infty} \phi(h(s))ds < \infty$.

(*Hint*: use Points *i*) and *iv*) above, or the inequality $1 - e^{-x} \leq x$ for $x \geq 0$ which implies $\phi(\lambda) \leq a\lambda$.)

4) Let h^* be the decreasing rearrangement of h, i.e. h^* is characterized by: h^* is decreasing, and for every $a \geq 0$,

$$\text{Leb}\{x; h(x) \geq a\} = \text{Leb}\{x; h^*(x) \geq a\}$$

where Leb denotes the Lebesgue measure on $\mathbb{R}_+$.

Prove that $A(h^*) \overset{(law)}{=} A(h)$.

(*Hint:* for every $\varphi : \mathbb{R}_+ \longrightarrow \mathbb{R}_+$, increasing, $\mathscr{C}^1$ and such that $\varphi(0) = 0$ (in particular, for $\varphi = \phi$), we have

$$\begin{aligned}\int_0^{+\infty} \varphi(\lambda h(s))ds &= \int_0^{+\infty} \varphi'(t)\text{Leb}\left\{h \geq \frac{t}{\lambda}\right\}dt \\ &= \int_0^{+\infty} \varphi'(t)\text{Leb}\left\{h^* \geq \frac{t}{\lambda}\right\}dt = \int_0^{+\infty} \varphi(\lambda h^*(s))ds.)\end{aligned}$$

5) In this question, we assume that $\int_0^{+\infty} \phi(h(s))ds < \infty$ and we define for $\lambda \geq 0$:

$$\phi_h(\lambda) := \int_0^{+\infty} \phi(\lambda h(s))ds.$$

i) Prove that ϕ_h is a Bernstein function, i.e. ϕ_h is positive and continuous on $[0, +\infty[$, differentiable on $]0, +\infty[$ and ϕ_h' is completely monotone on $]0, +\infty[$. (*Hint*: use (2.44) and question 3) *iii*)).

ii) Deduce that $A(h)$ is infinitely divisible and that there exists a subordinator $(\alpha_t(h), t \geq 0)$ such that:

$$\alpha_1(h) \overset{(law)}{=} A(h).$$

(*Hint*: $(\alpha_t(h), t \geq 0)$ is the subordinator whose Laplace exponent is the Bernstein function ϕ_h.)

iii) Let $(\alpha_{s,t}; s \geq 0, t \geq 0)$ be the Lévy sheet extending the Lévy process $(\alpha_t(h), t \geq 0)$. Prove that

$$(\alpha_t(h), t \geq 0) \overset{(law)}{=} \left(\int_0^{+\infty} h(s) d_s \alpha_{s,t}, t \geq 0 \right).$$

(*Hint*: $\left(\int_0^{+\infty} h(s) d_s \alpha_{s,t}, t \geq 0\right)$ is a process with positive stationary independent increments (hence a subordinator) whose Laplace exponent is ϕ_h; see Exercise 2.19 for a similar result.)

6) *Example*. Let $(\gamma_t, t \geq 0)$ be the gamma subordinator, i.e. the subordinator characterized by:

$$\mathbb{E}\left[e^{-\lambda \gamma_t}\right] = \frac{1}{(1+\lambda)^t} = \exp(-t \log(1+\lambda)).$$

In particular, for every $t \geq 0$, γ_t is a gamma r.v. of parameter t, i.e. with density $\frac{1}{\Gamma(t)} e^{-u} u^{t-1} 1_{]0,+\infty[}(u)$, and the Bernstein function of this subordinator is $\phi(\lambda) := \log(1+\lambda)$.

i) Prove that the Lévy measure of $(\gamma_t, t \geq 0)$ is given by:

$$\nu(dx) = \frac{e^{-x}}{x} 1_{]0,+\infty[}(x) dx$$

and that $\phi'(0^+) = 1 = \int_0^{+\infty} x \nu(dx)$.

ii) Let $h : \mathbb{R}_+ \longrightarrow \mathbb{R}_+$ be a Borel function such that:

$$\int_0^{+\infty} \log(1 + h(s)) ds < \infty.$$

We set $\gamma(h) := \int_0^{+\infty} h(s) d\gamma_s$. Prove that, for every $\lambda \geq 0$,

$$\mathbb{E}\left[e^{-\lambda \gamma(h)}\right] = \exp\left(- \int_0^{+\infty} (1 - e^{-\lambda x}) \frac{dx}{x} \int_0^{+\infty} e^{-xz} \mu_h(dz) \right)$$

where μ_h is the image of the Lebesgue measure on $\mathbb{R}_+$ by the application $s \longrightarrow \frac{1}{h(s)}$. Deduce that $\gamma(h)$ is a G.G.C. (Generalized Gamma Convolution) r.v. whose Thorin measure equals μ_h (see [34] or [92]).

iii) Now, let $a \in]0,1[$ and let $X^{(a)}$ be a positive stable r.v. of index a, i.e. such that:

$$\forall \lambda \geq 0, \quad \mathbb{E}\left[e^{-\lambda X^{(a)}}\right] = e^{-\lambda^a}.$$

Prove that there is a constant C_a such that:

$$\forall \lambda \geq 0, \quad \int_0^{+\infty} \log(1 + \lambda C_a s^{-\frac{1}{a}}) ds = \lambda^a$$

and deduce that:

$$X^{(a)} \overset{(law)}{=} C_a \int_0^{+\infty} s^{-\frac{1}{a}} d\gamma_s$$

(one may check that $C_a = (B(a,1-a))^{-\frac{1}{a}} = \left(\frac{\sin(a\pi)}{\pi}\right)^{\frac{1}{a}}$).

iv) Prove that $X^{(a)}$ is infinitely divisible and that $\mathbb{E}\left[X^{(a)}\right] = +\infty$.

v) Let $(\gamma_{s,t}; s,t \geq 0)$ be the Lévy sheet extending $(\gamma_t, t \geq 0)$. Prove that:

$$\left(X_t^{(a)} := C_a \int_0^{+\infty} s^{-\frac{1}{a}} d\gamma_{s,t},\ t \geq 0\right)$$

is a subordinator such that $X_1^{(a)} = X^{(a)}$.

7) We come back to the notation of Questions 1 to 5, and we now assume that

$$\int_0^{\infty} h(s)ds < \infty.$$

Let us define, for every $t > 0$,

$$A_t(h) := \int_0^{+\infty} h\left(\frac{s}{t}\right) d\alpha_s$$

and $A_0(h) = 0$.

i) Prove that, $A_t(h) \overset{(1.d)}{=} A_t(h^*)$.

ii) Deduce that $\left(A_t(h) - ta\int_0^{+\infty} h(s)ds, t \geq 0\right)$ is a peacock.

(*Hint*: use the result of Point 1.b) of Exercise 1.34, after having noticed that the function $t \longmapsto h^*\left(\frac{s}{t}\right)$ is, for every $s \geq 0$, increasing.)

iii) Prove that $\left(\alpha_t(h) - ta\int_0^{+\infty} h(s)ds, t \geq 0\right)$ is a martingale associated to the peacock $\left(A_t(h) - ta\int_0^{+\infty} h(s)ds, t \geq 0\right)$.

Exercise 2.13 (Infinitely divisible Wald couples and peacocks). Let a and b be two positive real numbers and let c be a positive measure on $]0,+\infty[$ such that $\int_{]0,+\infty[}(x^2 \wedge x)c(dx) < \infty$. We define a function Ψ on $\{z \in \mathbb{C}; \mathrm{Re}(z) \geq 0\}$ by

$$\Psi(z) := az + b\frac{z^2}{2} + \int_{]0,+\infty[}(e^{-zx} + zx - 1)c(dx).$$

1) Let for $\lambda \in \mathbb{R}$, $\psi(\lambda) = -\Psi(\mathrm{i}\lambda)$.

i) Prove that ψ is the characteristic exponent of a Lévy process $(L_t, t \geq 0)$ without positive jumps, with Lévy measure $\check{c}(dx) := c(-dx)$.

ii) Deduce that, for every λ and $t \geq 0$, $\mathbb{E}[e^{\lambda L_t}] = \exp(t\Psi(\lambda))$.

2) Set, for $s \geq 0$, $\phi(s) = \Psi(\sqrt{2s})$.

i) Prove that ϕ is positive and continuous on $[0,+\infty[$, differentiable on $]0,+\infty[$ and for every $s > 0$:

$$\phi'(s) = b + \frac{a}{\sqrt{2s}} + \int_0^{+\infty} e^{-u\sqrt{2s}}\left(\int_u^{+\infty} xc(dx)\right)du.$$

Using the fact that $\mathbb{E}[e^{-\lambda\tau_t}] = e^{-t\sqrt{\lambda}}$ for $\lambda \geq 0$, with $(\tau_t, t \geq 0)$ the $\frac{1}{2}$-stable subordinator, prove that ϕ' is a completely monotone function on $]0,+\infty[$ and therefore, ϕ is a Bernstein function.

ii) Deduce that there exists a subordinator $(H_t, t \geq 0)$ such that, for every $\lambda \geq 0$ and $t \geq 0$,

$$\mathbb{E}\left[\exp\left(-\frac{\lambda^2}{2}H_t\right)\right] = \exp(-t\Psi(\lambda)).$$

The couple (L_1, H_1) is called an *infinitely divisible Wald couple*. We refer the reader to [153] for numerous examples of such couples.

3) *i*) Prove that for every $\lambda \geq 0$, $\left(M_s^{(\lambda)} := \exp\left(\lambda L_s - \frac{\lambda^2}{2}H_s\right), s \geq 0\right)$ (with $(L_s, s \geq 0)$ and $(H_s, s \geq 0)$ independent) is a martingale.

ii) Prove that, when $a = 0$ and $c \equiv 0$, then:

$$\left(M_s^{(\lambda)}, s \geq 0\right) \overset{(law)}{=} \left(\exp\left(\lambda B_{bs} - \frac{\lambda^2}{2}bs\right), s \geq 0\right)$$

where $(B_u, u \geq 0)$ is a standard Brownian motion started from 0.

iii) Prove that, when $b = 0$ and $c \equiv 0$, then:

$$\left(M_s^{(\lambda)}, s \geq 0\right) \overset{(law)}{=} \left(\exp\left(\lambda as - \lambda^2 \tau_{as}\right), s \geq 0\right)$$

where $(\tau_s, s \geq 0)$ is as in Point 2) *i*).

4) Let σ be a signed finite measure on $\mathbb{R}_+$ and let $\lambda \geq 0$. Define, for $t \geq 0$:

$$A_t^{(\sigma)} := \int_{\mathbb{R}_+} \exp\left(\lambda L_{ts} - \frac{\lambda^2}{2}H_{ts}\right)\sigma(ds)$$

$$M_t^{(\sigma)} := \int_{\mathbb{R}_+} \exp\left(\lambda W_{s,t}^{(L)} - \frac{\lambda^2}{2}W_{s,t}^{(H)}\right)\sigma(ds)$$

where the sheets $W^{(L)}$ and $W^{(H)}$ are assumed to be independent. Prove that $(M_t^{(\sigma)}, t \geq 0)$ is a martingale and

$$A_t^{(\sigma)} \overset{(1.d)}{=} M_t^{(\sigma)}.$$

(*Hint*: show that, for $0 \leq t_1 \leq t_2$ and $s \geq 0$:

$$\mathbb{E}\left[e^{\lambda(W_{s,t_2}^{(L)} - W_{s,t_1}^{(L)})}\right] = \exp((t_2 - t_1)s\Psi(\lambda))$$

$$\mathbb{E}\left[e^{-\frac{\lambda^2}{2}(W_{s,t_2}^{(H)} - W_{s,t_1}^{(H)})}\right] = \exp(-(t_2 - t_1)s\Psi(\lambda)).)$$

Thus, $(A_t^{(\sigma)}, t \geq 0)$ is a peacock and $(M_t^{(\sigma)}, t \geq 0)$ is an associated martingale.

5) Let σ be a finite signed measure on $\mathbb{R}_+$ and define

$$X_\lambda^{(\sigma)} := \int_0^{+\infty} e^{\lambda L_t - \frac{\lambda^2}{2}H_t}\sigma(dt), \quad \lambda \geq 0.$$

Prove that $(X_\lambda^{(\sigma)}, \lambda \geq 0)$ is a peacock.

(*Hint*: let $\theta \in C$. We have:

$$\mathbb{E}[\theta(X_\lambda^{(\sigma)})] = \mathbb{E}_H \left[\mathbb{E}_L \left[\theta\left(\int_0^{+\infty} \exp(\lambda L_t - t\Psi(\lambda)) \exp\left(-\frac{\lambda^2}{2}H_t + t\Psi(\lambda)\right) \sigma(dt)\right)\right]\right];$$

then apply Theorem 1.8 first to L, then to H, and use Fubini.)

This exercise is continued in Exercise 5.15.

Exercise 2.14 (Peacocks with respect to the dimension and sheet of squared Bessel processes). Let $\mathbb{P}_\delta$ be the law, on $\mathscr{C}(\mathbb{R}_+,\mathbb{R}_+)$, of a squared Bessel process $(Q_s^{(\delta)}, s \geq 0)$ of dimension δ ($\delta \geq 0$) started from 0.

1) Prove that there exists a $\mathscr{C}(\mathbb{R}_+,\mathbb{R}_+)$-valued Lévy process $(Z_\delta, \delta \geq 0)$ such that, for every $\delta \geq 0$, the law of Z_δ is $\mathbb{P}_\delta$.
   (*Hint*: $(\mathbb{P}_\delta, \delta \geq 0)$ satisfies: for every $\delta, \delta' \geq 0, \mathbb{P}_\delta * \mathbb{P}_{\delta'} = \mathbb{P}_{\delta+\delta'}$; see [150, Chapter XI, Th. (1.2)] and [145]).
2) Let $\lambda \geq 0$ and μ be a finite signed measure on $\mathbb{R}_+$. Let us define:

$$X_\delta^{(\lambda,\mu)} := \int_0^{+\infty} e^{-\lambda Q_s^{(\delta)} + h_{\lambda,\delta}(s)} \mu(ds), \qquad \delta \geq 0$$

   with $e^{h_{\lambda,\delta}(s)} := \dfrac{1}{\mathbb{E}[e^{-\lambda Q_s^{(\delta)}}]} = (1+2\lambda s)^{\delta/2}$. Prove that $(X_\delta^{(\lambda,\mu)}, \delta \geq 0)$ is a peacock, and find an associated martingale. (*Hint*: use Proposition 2.1.)

2.5 Space-Time Harmonic Functions for $W^{(L)}$ and Peacocks

In this section, we shall extend the results of Section 2.3, replacing the Brownian motion B (resp. the Brownian sheet W) by the Lévy process L (resp. the Lévy sheet $W^{(L)}$).

2.5.1 Space-Time Harmonic Functions for L

Definition 2.5. Let I be an interval of $\mathbb{R}_+$. A function $h(t,x)$ on $I \times \mathbb{R}$ is called a space-time harmonic function for $(L_t, t \in I)$, if the process $(h(t,L_t), t \in I)$ is a $(\mathscr{F}_t)$-martingale, or equivalently, if, for any $s,t \in I$ with $s < t$,

$$P_{t-s}h^t = h^s \quad \mu_s\text{-a.s.}$$

where, for $u \in I$, $h^u(x) = h(u,x)$ and μ_s denotes the law of L_s.

Using the expression of P_t given by (2.39), we obtain that $h : I \times \mathbb{R} \longrightarrow \mathbb{R}$ is a space-time harmonic function for $(L_t, t \in I)$ if and only if, for any $s,t \in I$, with $s < t$,

$$h(s,x) = \mathbb{E}[h(t, x+L_{t-s})] \quad \mu_s\text{-a.s.}$$

This formula extends (2.10).

Example 2.7 (Space-time harmonic functions for $(L_t, 0 \le t \le r)$). Let $r > 0$ and $f \in L^1(\mu_r)$. We define h on $[0,r] \times \mathbb{R}$ by:

$$h(t,x) = P_{r-t}f(x).$$

Then, it follows directly from Definition 2.5 that h is a space-time harmonic function for $(L_t, 0 \le t \le r)$.

Conversely, let h be a space-time harmonic function for $(L_t, 0 \le t \le r)$ and set: $f = h(r, \bullet)$. Then, for every $(t,x) \in [0,r] \times \mathbb{R}$,

$$h(t,x) = P_{r-t}f(x).$$

Example 2.8. Let $\nu \in \mathbb{R}$ such that: $\mathbb{E}[\exp(\nu L_1)] < \infty$. We set $\phi(\nu) = \log \mathbb{E}[\exp(\nu L_1)]$. Then, $h : \mathbb{R}_+ \times \mathbb{R} \longrightarrow \mathbb{R}$ defined by

$$h(t,x) = \exp(\nu x - t\phi(\nu))$$

is a space-time harmonic function for $(L_t, t \ge 0)$. Indeed, if $0 \le s \le t$, then:

$$\mathbb{E}[h(t,L_t)|\mathscr{F}_s] = \exp(\nu L_s - t\phi(\nu))\mathbb{E}[\exp(\nu L_{t-s})]$$

and we have, for any $a \ge 0$,

$$\mathbb{E}[\exp(\nu L_a)] = \exp(a\phi(\nu)).$$

Example 2.9. Let $\nu \in \mathbb{R}$ and ψ be the characteristic exponent of L (see Paragraph 2.4.2.1). Then the function $h : \mathbb{R}_+ \times \mathbb{R} \longrightarrow \mathbb{C}$ defined by:

$$h(t,x) = \exp(i\nu x + t\psi(\nu))$$

is a complex-valued space-time harmonic function for $(L_t, t \ge 0)$, i.e.: the real and imaginary parts of h are space-time harmonic functions.

2.5.2 Space-Time Harmonic Functions for $\mathbf{W}^{(L)}$

Definition 2.6. Let I be an interval of $\mathbb{R}_+$. A function $H(t,\varepsilon)$ on $I \times \mathbb{D}_0$ is called a space-time harmonic function for $\left(W^{(L)}_{\bullet,t}, t \in I\right)$ if the process $\left(H(t, W^{(L)}_{\bullet,t}), t \in I\right)$ is a $(\mathscr{G}^{(L)}_t)$-martingale, or equivalently, if for any $s,t \in I$ with $s < t$,

$$\Pi_{t-s}H^t = H^s \quad \mathscr{M}_s\text{-a.s.}$$

where, for $u \in I$, $H^u(\varepsilon) = H(u,\varepsilon)$ and $\mathscr{M}_s$ denotes the law of $W^{(L)}_{\bullet,s}$.

We have the following extension of Theorem 2.3.

Theorem 2.12. *Let H be a space-time harmonic function for $(W_{\bullet,t}, t \in I)$. Then:*

1) $\left(H(t, W^{(L)}_{\bullet,t}), t \in I\right)$ *is a* $(\mathscr{G}^{(L)}_t)$*-martingale.*
2) $(H(t, L_{\bullet t}), t \in I)$ *is a peacock.*
3) $H(t, W^{(L)}_{\bullet,t}) \overset{(1.d)}{=} H(t, L_{\bullet t})$.

Example 2.10. Let $a > 0$ and c be a bounded Borel function on $[0, a]$. The stochastic integral: $\int_0^a c(u)dL_u$ is well-defined a.s. and

$$\mathbb{E}\left[\exp\left(i\int_0^a c(u)dL_u\right)\right] = \exp\left(-\int_0^a \psi(c(u))du\right).$$

This can be seen by approximation of c by simple functions.

Define $H : \mathbb{R}_+ \times \mathbb{D}_0 \longrightarrow \mathbb{C}$ by:

$$H(t, L) = \exp\left(i\int_0^a c(u)dL_u + t\int_0^a \psi(c(u))du\right).$$

Then, if $0 \leq s \leq t$, we obtain by Theorem 2.10:

$$\begin{aligned}\mathbb{E}\left[H(t, W^{(L)}_{\bullet,t})|\mathscr{G}^{(L)}_s\right] =& \exp\left(i\int_0^a c(u)d_uW_{u,s} + t\int_0^a \psi(c(u))du\right) \\ &\times \mathbb{E}\left[\exp\left(i\int_0^a c(u)d_uL_{(t-s)u}\right)\right] \\ =& H(t, W^{(L)}_{\bullet,s}).\end{aligned}$$

(We used that the characteristic exponent of $L_{a\bullet}$ is $a\psi$.) Thus, H is a complex-valued space-time harmonic function for $(W^{(L)}_{\bullet,t}, t \geq 0)$, i.e.: the real and imaginary parts of H are space-time harmonic functions.

We may now extend Theorem 2.4, Theorem 2.5 and Corollary 2.1 to the set-up of Lévy processes. This is proposed in the following exercise.

Exercise 2.15 (Examples of space-time harmonic functions for $(W^{(L)}_{\bullet,t}, t \in I)$).
1) Let $h : I \times \mathbb{R} \longrightarrow \mathbb{R}$ be a space-time harmonic function for $(L_t, t \in I)$. Let us define:

$$H(t, \varepsilon) := h(t, \varepsilon(1)) \quad (\varepsilon \in \mathbb{D}_0). \tag{2.45}$$

Prove that H is a space-time harmonic function for $(W^{(L)}_{\bullet,t}, t \in I)$.

2) Let H be a space-time harmonic function for $(W^{(L)}_{\bullet,t}, t \in I)$. Prove that for every $r > 0$, H_r defined by:

$$H_r(t, \varepsilon) := H(rt, \varepsilon(r\bullet)) \tag{2.46}$$

is space-time harmonic for $(W^{(L)}_{\bullet,t}, t \in \frac{1}{r}I)$.

3) Let $h : I \times \mathbb{R} \longrightarrow \mathbb{R}$ be a space-time harmonic function for $(L_t, t \in I)$. Prove that for every $r > 0$, the function h_r defined by

$$h_r(t,\varepsilon) := h(rt, \varepsilon(r))$$

is space-time harmonic for $(W^{(L)}_{\bullet,t}, t \in \frac{1}{r}I)$.

4) Let H be a space-time harmonic function for $(W^{(L)}_{\bullet,t}, t \in I)$ and let σ be a finite signed measure on $\mathbb{R}_+$. Prove that, if σ has compact support or $H \geq 0$, then $H^{(\sigma)}$ defined by

$$H^{(\sigma)}(t,\varepsilon) := \int H_r(t,\varepsilon)\sigma(dr)$$

is a space-time harmonic function for $(W^{(L)}_{\bullet,t}, t \in I)$. Here, as in Question 2),

$$H_r(t,\varepsilon) = H(rt, \varepsilon(r_\bullet)).$$

5) Consider the hypotheses and notation of Example 2.8. Prove that, if σ is a finite signed measure on $\mathbb{R}_+$, then $\xi^{(\sigma)}_\nu$ defined by:

$$\xi^{(\sigma)}_\nu := \int \exp(\nu\varepsilon(r) - tr\phi(\nu))\,\sigma(dr)$$

is a space-time harmonic function for $(W^{(L)}_{\bullet,t}, t \geq 0)$. Then, using Theorem 2.12, recover Theorem 2.11.

6) We now consider the notation of Exercise 1.38. In particular, $(C_n(u,a), n \geq 0)$ denotes the sequence of Charlier polynomials.

i) Prove that, for every $n \geq 0$, $C_n(x,t)$ (resp. $C_n(t,x)$) is a space-time harmonic function for $(N_t, t \geq 0)$ (resp. $(\gamma_t, t \geq 0)$).

ii) Prove that, if σ is a finite signed measure on $\mathbb{R}_+$ with compact support, then:

$$\int C_n(\varepsilon(r), rt)\sigma(dr) \quad \left(\text{resp. } \int C_n(rt, \varepsilon(r))\sigma(dr)\right)$$

is a space-time harmonic function for $(W^{(N)}_{\bullet,t}, t \geq 0)$ (resp. $(W^{(\gamma)}_{\bullet,t}, t \geq 0)$.

iii) Prove that, if σ is as above, then

$$\left(\int C_n(N_{rt}, rt)\sigma(dr), t \geq 0\right) \quad \text{and} \quad \left(\int C_n(rt, \gamma_{rt})\sigma(dr), t \geq 0\right)$$

are peacocks.

This improves some results in Exercise 1.38, where the above result is proven for σ equal to the Lebesgue measure on $[0,1]$.

2.5.3 Definitions of the Processes $\Phi_t^{(\mathbf{M})}(W^{(L)})$ and $\Phi_t^{(\mathbf{P})}(L)$

Let $\Phi \in L^1(\mathbb{P})$ $(= L^1(\mathbb{D}_0, \mathscr{F}, \mathbb{P}))$, where $\mathbb{P}$ is the law of the given Lévy process L. We shall now be able, by imitation of Definition 2.3, to define the processes $\left(\Phi_t^{(\mathbf{M})}(W^{(L)}), t \in [0,1]\right)$ and $\left(\Phi_t^{(\mathbf{P})}(L), t \in [0,1]\right)$.

Definition 2.7. Let $\Phi \in L^1(\mathbb{P})$. We define, for $0 \le t \le 1$:

$$\Phi_t^{(\mathbf{M})}(W^{(L)}) := \mathbb{E}_{\widetilde{W}^{(L)}}\left[\Phi\left(W_{\bullet,t}^{(L)} + \widetilde{W}_{\bullet,1-t}^{(L)}\right)\right] \tag{2.47}$$

$$= \mathbb{E}_{\widetilde{L}}\left[\Phi\left(W_{\bullet,t}^{(L)} + \widetilde{L}_{\bullet(1-t)}\right)\right] \tag{2.48}$$

$$= \Pi_{1-t}\Phi\left(W_{\bullet,t}^{(L)}\right) \tag{2.49}$$

and, for $0 \le t \le 1$:

$$\Phi_t^{(\mathbf{P})}(L) := \mathbb{E}_{\widetilde{W}^{(L)}}\left[\Phi\left(L_{\bullet t} + \widetilde{W}_{\bullet,1-t}^{(L)}\right)\right] \tag{2.50}$$

$$= \mathbb{E}_{\widetilde{L}}\left[\Phi\left(L_{\bullet t} + \widetilde{L}_{\bullet(1-t)}\right)\right] \tag{2.51}$$

$$= \Pi_{1-t}\Phi\left(L_{\bullet t}\right). \tag{2.52}$$

- In (2.47), (resp. (2.50)), $\widetilde{W}^{(L)}$ is a copy of $W^{(L)}$ which is independent of $W^{(L)}$ (resp. L) and $\mathbb{E}_{\widetilde{W}^{(L)}}$ means integration with respect to $\widetilde{W}^{(L)}$.
- In (2.48), (resp. (2.51)), $\widetilde{L}$ is a copy of L independent of $W^{(L)}$ (resp. L) and $\mathbb{E}_{\widetilde{L}}$ means integration with respect to $\widetilde{L}$.
- $(\Pi_t, t \ge 0)$ denotes the semi-group of $(W_{\bullet,t}, t \ge 0)$ defined by (2.38).
- $\Phi_t^{(\mathbf{M})}(W^{(L)})$ is defined on the probability space of $W^{(L)}$ whereas $\Phi_t^{(\mathbf{P})}(L)$ is defined on the probability space of L $(\mathbb{D}_0, \mathbb{P})$.
- The equality between (2.47), (2.48) and (2.49), respectively (2.50), (2.51) and (2.52) relies merely on (2.37) (i.e. for all $t \ge 0$, $\left(W_{u,t}^{(L)}, u \ge 0\right) \overset{(law)}{=} (L_{tu}, u \ge 0)$) and on the definition of the semi-group $(\Pi_t, t \ge 0)$.

The analogue of Theorem 2.7 may be stated here as:

Theorem 2.13. *For all $\Phi \in L^1(\mathbb{P})$, $\left(\Phi_t^{(M)}(W^{(L)}), 0 \le t \le 1\right)$ is a $\left(\mathscr{G}_t^{(L)}, 0 \le t \le 1\right)$-martingale associated to the peacock $\left(\Phi_t^{(P)}(L), 0 \le t \le 1\right)$, i.e.:*

$$\Phi_t^{(P)}(L) \overset{(1.d)}{=} \Phi_t^{(M)}(W^{(L)}). \tag{2.53}$$

Proof. The relation (2.53) is obvious: it relies on (2.49) and (2.52) and on the fact that, for all $t \geq 0$, $W^{(L)}_{\bullet,t} \overset{(law)}{=} L_{\bullet t}$. Let us show that $\left(\Phi^{(\mathbf{M})}(W^{(L)})_t, 0 \leq t \leq 1\right)$ is a $\left(\mathscr{G}^{(L)}_t, 0 \leq t \leq 1\right)$-martingale. We have, for $0 \leq s \leq t \leq 1$:

$$\begin{aligned}\mathbb{E}\left[\Phi^{(\mathbf{M})}_t(W^{(L)})|\mathscr{G}^{(L)}_s\right] &= \mathbb{E}\left[\Pi_{1-t}\Phi\left(W^{(L)}_{\bullet,t}\right)|\mathscr{G}^{(L)}_s\right] \\ &= \Pi_{t-s}\Pi_{1-t}\Phi(W^{(L)}_{\bullet,s}) \quad \text{(from the Markov property)} \\ &= \Pi_{t-s+1-t}\Phi(W^{(L)}_{\bullet,s}) \quad \text{(from the semi-group property)} \\ &= \Pi_{1-s}\Phi(W^{(L)}_{\bullet,s}) = \Phi^{(\mathbf{M})}_s(W^{(L)}).\end{aligned}$$

□

Theorem 2.14. *Let us denote by $\mathscr{H}^{W^{(L)}}$ the space of space-time harmonic functions for $(W^{(L)}_{\bullet,t}, 0 \leq t \leq 1)$. Then the application $\Gamma : H \in \mathscr{H}^{W^{(L)}} \longrightarrow \Phi^H \in L^1(\mathbb{P})$ defined by:*

$$\Phi^H(\varepsilon) = H(1,\varepsilon)$$

is one-to-one. Its reciprocal mapping $\Gamma^{-1} : \Phi \in L^1(\mathbb{P}) \longrightarrow H^\Phi \in \mathscr{H}^{W^{(L)}}$ is defined by:

$$H^\Phi(t,\varepsilon) = \Pi_{1-t}\Phi(\varepsilon) \quad (0 \leq t \leq 1). \tag{2.54}$$

The proof of Theorem 2.14 is very similar to the one of Theorem 2.6 and is thus omitted. The following theorem is then an easy consequence of Theorems 2.13 and 2.14.

Theorem 2.15. *Let $\Phi \in L^1(\mathbb{P})$ and H^Φ defined by (2.54), i.e.*

$$H^\Phi(t,\varepsilon) = \Pi_{1-t}\Phi(\varepsilon) \quad (\varepsilon \in \mathbb{D}_0).$$

Then:

1) $\left(\Phi^{(M)}_t(W^{(L)}), 0 \leq t \leq 1\right) = \left(H^\Phi(t, W^{(L)}_{\bullet,t}), 0 \leq t \leq 1\right)$.
2) $\left(\Phi^{(P)}_t(L), 0 \leq t \leq 1\right) = \left(H^\Phi(t, L_{\bullet t}), 0 \leq t \leq 1\right)$.
3) *The $(\mathscr{G}^{(L)}_t, 0 \leq t \leq 1)$-martingale $\left(H^\Phi(t, W^{(L)}_{\bullet,t}), 0 \leq t \leq 1\right)$ is associated to the peacock $\left(H^\Phi(t, L_{\bullet t}), 0 \leq t \leq 1\right)$.*

The last part of this section is dedicated to illustrations and complements of Theorems 2.14 and 2.15, through a series of exercises.

Exercise 2.16. Prove that Theorem 2.8 extends to the set-up of Lévy processes.

Exercise 2.17 (Explicit computation of $\Psi^{(\mathbf{P})}(L)$ for $\Psi = \int_0^1 \Phi^{(\mathbf{P})}_s(L)h(s)ds$). Let $\Phi \in L^1(\mathbb{P})$ and $h \in L^1([0,1], ds)$. Set

$$\Psi = \int_0^1 \Phi^{(\mathbf{P})}_s(L)h(s)ds.$$

Prove that

$$\Psi_t^{(\mathbf{P})}(L) = \frac{1}{t}\int_0^t \Phi_s^{(\mathbf{P})}(L) h\left(\frac{s}{t}\right) ds.$$

Exercise 2.18 (Examples of processes $\Phi^{(\mathbf{P})}$ and $\Phi^{(\mathbf{M})}$). Let $r > 0$ and $f \in L^1(\mu_r)$, where μ_r is the law of L_r. Let us define:

$$\begin{aligned} h(t,x) &:= P_{r-t}f(x) \\ H(t,\varepsilon) &:= h(rt,\varepsilon(r)) \\ \text{and} \qquad \Phi(\varepsilon) &:= H(1,\varepsilon) = f(\varepsilon(r)). \end{aligned}$$

1) Prove that:

i) h is space-time harmonic for $(L_t, t \in [0,r])$.

ii) $\left(\Phi_t^{(\mathbf{P})}(L), 0 \le t \le 1\right) = (h(rt, L_{rt}), 0 \le t \le 1)$. Hence, $\Phi^{(\mathbf{P})}(L)$ is, in this particular set-up, a $(\mathscr{F}_{rt}, 0 \le t \le 1)$-martingale.

iii) $\left(\Phi_t^{(\mathbf{M})}(W^{(L)}), 0 \le t \le 1\right) = (h(rt, W_{r,t}), 0 \le t \le 1)$. Hence, in this particular set-up,

$$\left(\Phi_t^{(\mathbf{M})}(W^{(L)}), 0 \le t \le 1\right) \overset{(law)}{=} \left(\Phi_t^{(\mathbf{P})}(L), 0 \le t \le 1\right).$$

2) *i)* We define:

$$\begin{aligned} G(\varepsilon,t) &:= \int_0^r h(ut,\varepsilon(u))du \\ \text{and} \qquad \Phi(\varepsilon) &:= \int_0^r h(u,\varepsilon(u))du. \end{aligned}$$

Prove that G is space-time harmonic for $(W_{\bullet,t}^{(L)}, 0 \le t \le 1)$ and that $\Gamma^{-1}\Phi := H^\Phi = G$ (with the notation of Theorem 2.14).

ii) Prove that:

$$\begin{aligned} \Phi_t^{(\mathbf{M})}(W^{(L)}) &= \int_0^r h(ut, W_{u,t}^{(L)})du \quad (0 \le t \le 1) \\ \text{and} \quad \Phi_t^{(\mathbf{P})}(L) &= \int_0^r h(ut, L_{ut})du = \frac{1}{t}\int_0^{rt} h(u,L_u)du \quad (0 \le t \le 1). \end{aligned}$$

3) We assume moreover that there is $b > 0$ and $\lambda \in \mathbb{R}$ such that $\mathbb{E}\left[\exp(\lambda L_b)\right] < \infty$.

i) Prove that there exists $\phi(\lambda)$ such that for all $t \ge 0$, $\mathbb{E}\left[\exp(\lambda L_t)\right] = \exp(t\phi(\lambda))$.

ii) Let $h^{(\lambda)}(t,x) := \exp(\lambda x - t\phi(\lambda))$. Show that $h^{(\lambda)}$ is space-time harmonic for $(L_u, u \ge 0)$.

iii) Let, for u fixed, $\Phi^{(u)}$ defined by

$$\Phi^{(u)}(\varepsilon) = \exp\left(\lambda\varepsilon(u) - u\phi(\lambda)\right).$$

Prove that $\Phi_t^{(u),(\mathbf{P})}(L) = \exp\left(\lambda L_{tu} - tu\phi(\lambda)\right)$, and deduce then that $(\Phi_t^{(u),(\mathbf{P})}(L), t \ge 0)$ is a $(\mathscr{F}_{tu}, t \ge 0)$-martingale and that

$$\left(\Phi_t^{(u),(\mathbf{M})}(W^{(L)}), t \ge 0\right) \overset{(law)}{=} \left(\Phi_t^{(u),(\mathbf{P})}(L), t \ge 0\right).$$

iv) Now, let, for a fixed,

$$\Phi(\varepsilon) := \int_0^a \exp(\lambda\varepsilon(u) - u\phi(\lambda))\,du.$$

Prove that, for all $t \in [0,1]$:

$$\Phi_t^{(\mathbf{P})}(L) = \int_0^a \exp(\lambda L_{tv} - tv\phi(\lambda))\,dv.$$

Deduce then that

$$\left(\frac{1}{t}\int_0^t \exp(\lambda L_u - u\phi(\lambda))\,du, 0 \le t \le 1)\right)$$

is a peacock and that

$$\left(\int_0^1 \exp\left(\lambda W_{v,t}^{(L)} - tv\phi(\lambda)\right)dv, 0 \le t \le 1)\right)$$

is an associated $(\mathscr{G}_t^{(L)}, 0 \le t \le 1)$ martingale.

Compare this result with Theorem 2.11.

Exercise 2.19 (A Wiener integral in one parameter of a Lévy sheet is a Lévy process in the other parameter). Let $a > 0$ and c be a Borel bounded function on $[0,a]$. Prove that:

$$\left(\int_0^a c(u)d_u W_{u,t}^{(L)}, t \ge 0\right)$$

is a Lévy process whose characteristic exponent is $\psi^c(\lambda) = \displaystyle\int_0^a \psi(\lambda c(u))du$.

Exercise 2.20 (Explicit computation of $\Psi^{(\mathbf{P})}$ for $\Psi = \int_0^1 \Phi_s^{(\mathbf{P})}(L)dL_s$). In this exercise, we assume that the characteristic exponent ψ of L is given by:

$$\psi(\lambda) = \int (1 - e^{i\lambda x} + i\lambda x)\nu(dx)$$

with ν a positive measure on $\mathbb{R}\backslash\{0\}$ satisfying $0 < a := \int x^2\nu(dx) < \infty$. We denote by $\mathscr{P}_2^{(L)}$ the space of $(\mathscr{F}_s)$-predictable processes: $K = (K_s, s \ge 0)$ such that:

$$\int_0^{+\infty} \mathbb{E}\left[K_s^2\right] ds < \infty.$$

1) Prove that, for $t \ge 0$,

$$\mathbb{E}[L_t] = 0 \quad \text{and} \quad \mathbb{E}[L_t^2] = at.$$

2) Show that, if $K \in \mathscr{P}_2^{(L)}$, then the stochastic integral $\int_0^{+\infty} K_s dL_s$ is well-defined as an element of $L^2(\mathbb{P})$ and

$$\mathbb{E}\left[\left(\int_0^{+\infty} K_s dL_s\right)^2\right] = a\int_0^{+\infty} \mathbb{E}[K_s^2]ds.$$

Moreover, $\left(\int_0^t K_s dL_s, t \geq 0\right)$ is a $(\mathscr{F}_t)$-martingale. (*Hint*: approximate the process K by simple predictable processes).

3) Let $c \in L^2(\mathbb{R}_+, du)$. Prove that: $\left(\int_0^{+\infty} c(u) d_u W_{u,t}^{(L)}, t \geq 0\right)$ is a Lévy process whose characteristic exponent is $\psi^c(\lambda) := \int_0^{+\infty} \psi(\lambda c(u)) du$. Deduce that

$$\left(\int_0^{+\infty} c\left(\frac{u}{t}\right) dL_u, t \geq 0\right) \quad \text{is a peacock.}$$

4) Let $\Phi \in L^2(\mathbb{P})$ and suppose that Φ is $\mathscr{F}_1$-measurable. We set:

$$\Psi = \int_0^1 \Phi_s^{(\mathbf{P})}(L) dL_s.$$

Prove that Ψ is well-defined as an element of $L^2(\mathbb{P})$ and, for $t \in [0,1]$

$$\Psi_t^{(\mathbf{P})} = \int_0^t \Phi_s^{(\mathbf{P})} dL_s.$$

In particular, $\Psi^{(\mathbf{P})}$ is a $(\mathscr{F}_t)$-martingale.

Exercise 2.21 (A chaos decomposition formula for $\Phi^{(\mathbf{P})}$ in the Poisson case). Define $(L_t := N_t - t, t \geq 0)$ where N denotes the standard Poisson process. We recall that L enjoys the chaos decomposition property, i.e., if $\Phi \in L^2(\mathbb{P})$ there exists a sequence $(\varphi_n, n \geq 1)$ such that:

$$\Phi = \mathbb{E}[\Phi] + \sum_{n=1}^{+\infty} \int_0^{+\infty} dL_{s_1} \int_0^{s_1^-} dL_{s_2} \dots \int_0^{s_{n-1}^-} dL_{s_n} \varphi_n(s_1, \dots, s_n)$$

with

$$\sum_{n=1}^{+\infty} \int_0^{+\infty} ds_1 \int_0^{s_1} ds_2 \dots \int_0^{s_{n-1}} ds_n \varphi_n^2(s_1, \dots, s_n) < +\infty.$$

Prove that, for every $0 < t \leq 1$,

$$\Phi_t^{(\mathbf{P})} = \mathbb{E}[\Phi] + \sum_{n=1}^{+\infty} \int_0^{+\infty} dL_{s_1} \int_0^{s_1^-} dL_{s_2} \dots \int_0^{s_{n-1}^-} dL_{s_n} \varphi_n\left(\frac{s_1}{t}, \dots, \frac{s_n}{t}\right)$$

and

$$\Phi_t^{(\mathbf{M})} = \mathbb{E}[\Phi] + \sum_{n=1}^{+\infty} \int_0^{+\infty} d_{s_1} W_{s_1,t}^{(L)} \int_0^{s_1^-} d_{s_2} W_{s_2,t}^{(L)} \dots \int_0^{s_{n-1}^-} d_{s_n} W_{s_n,t}^{(L)} \varphi_n(s_1, \dots, s_n).$$

2.6 Peacocks and Gaussian Sheets

In this section, we shall apply the sheet method as defined in Section 2.1 to some Gaussian processes. To start with, we must construct a Gaussian sheet.

2.6.1 Gaussian Sheets

We first recall some of the notation introduced in Chapter 1, Subsection 1.3.5. Let Λ be a measurable space. We consider, for each $t \geq 0$, a real-valued measurable centered Gaussian process

$$G_{\bullet,t} = (G_{\lambda,t}, \lambda \in \Lambda) .$$

For $\lambda, \mu \in \Lambda$ and $t \geq 0$, we set:

$$c_{\lambda,\mu}(t) = \mathbb{E}[G_{\lambda,t} G_{\mu,t}] ,$$

and we define, for any signed finite measure σ on Λ and $t \geq 0$:

$$A_t^{(\sigma)} = \int_{\Lambda} \exp\left(G_{\lambda,t} - \frac{1}{2} c_{\lambda,\lambda}(t) \right) \sigma(d\lambda) .$$

We shall also need the following conditions:

(M_2) Λ is a separable metric space equipped with its Borel σ-field.

(C_2) The function:

$$(\lambda, \mu, t) \in \Lambda \times \Lambda \times \mathbb{R}_+ \longrightarrow c_{\lambda,\mu}(t) \in \mathbb{R}$$

is continuous.

(I_2) For every $n \geq 1$, for every $\lambda_1, \cdots, \lambda_n \in \Lambda$, the matrix function

$$t \in \mathbb{R}_+ \longrightarrow (c_{\lambda_j,\lambda_k}(t))_{1 \leq j,k \leq n} \in \mathbb{S}_n$$

is increasing with respect to the order on $\mathbb{S}_n$ induced by the convex cone $\mathbb{S}_n^+$.

Theorem 2.16. *Under (M_2), (C_2) and (I_2), there exists a measurable centered Gaussian process:*

$$(\Gamma_{\lambda,t} \,;\, \lambda \in \Lambda \,,\, t \geq 0) ,$$

such that

$$\forall (\lambda, s), (\mu, t) \in \Lambda \times \mathbb{R}_+, \qquad \mathbb{E}[\Gamma_{\lambda,s} \Gamma_{\mu,t}] = c_{\lambda,\mu}(s \wedge t) . \tag{2.55}$$

We call $(\Gamma_{\lambda,t}; \lambda \in \Lambda, t \geq 0)$ the Gaussian sheet attached to the family of Gaussian processes $(G_{\lambda,t}; \lambda \in \Lambda)_{t \geq 0}$.

Proof. We first prove that

$$[(\lambda,s),(\mu,t)] \longrightarrow c_{\lambda,\mu}(s\wedge t)$$

is a covariance on $\Lambda \times \mathbb{R}_+$.

Let $\lambda_1,\cdots,\lambda_n \in \Lambda$ and $t_1,\cdots,t_n \in \mathbb{R}_+$. We denote by u a bijection from $\{1,2,\cdots,n\}$ onto $\{1,2,\cdots,n\}$ such that, setting $s_r = t_{u(r)}$, we have:

$$s_1 \leq s_2 \leq \cdots \leq s_n \, .$$

We also set $s_0 = 0$ and we denote by v the inverse of the bijection u. Condition (I_2) ensures the existence of matricies

$$D^{(\lambda_1,\cdots,\lambda_n)}(r) \in \mathbb{S}_n^+ \quad , \quad 1 \leq r \leq n$$

such that,

$$\left[D^{(\lambda_1,\cdots,\lambda_n)}(r)\right]^2 = \frac{1}{s_r - s_{r-1}} \Big(c_{\lambda_j,\lambda_k}(s_r) - c_{\lambda_j,\lambda_k}(s_{r-1})\Big)_{1\leq j,k\leq n}$$

if $s_{r-1} < s_r$, and $D^{(\lambda_1,\cdots,\lambda_n)}(r) = 0$ if $s_{r-1} = s_r$. Let

$$B_t = (B_t^1,\cdots,B_t^n) \quad , \quad t \geq 0$$

be a standard $\mathbb{R}^n$-valued Brownian motion, independent of $G_{\cdot,0}$. We set, for $1 \leq j \leq n$,

$$Z_j = G_{\lambda_j,0} + \sum_{r=1}^{v(j)} \sum_{l=1}^{n} d_{j,l}^{(\lambda_1,\cdots,\lambda_n)}(r)\,(B_{s_r}^l - B_{s_{r-1}}^l)$$

where

$$d_{j,k}^{(\lambda_1,\cdots,\lambda_n)}(r) \quad , \quad 1 \leq j,k \leq n$$

denote the entries of the matrix $D^{(\lambda_1,\cdots,\lambda_n)}(r)$. Then,

$$\mathbb{E}[Z_j Z_k] = c_{\lambda_j,\lambda_k}(0) + \sum_{r=1}^{v(j)\wedge v(k)} (c_{\lambda_j,\lambda_k}(s_r) - c_{\lambda_j,\lambda_k}(s_{r-1})) = c_{\lambda_j,\lambda_k}(s_{v(j)\wedge v(k)}) \, .$$

Since

$$s_{v(j)\wedge v(k)} = s_{v(j)} \wedge s_{v(k)} = t_j \wedge t_k \, ,$$

we obtain

$$\mathbb{E}[Z_j Z_k] = c_{\lambda_j,\lambda_k}(t_j \wedge t_k) \, ,$$

which ensures the covariance property.

From the preceding, there exists a centered Gaussian process:

$$(\Gamma_{\lambda,t} \, ; \, \lambda \in \Lambda \, , \, t \geq 0) \, ,$$

such that

$$\mathbb{E}[\Gamma_{\lambda,s}\Gamma_{\mu,t}] = c_{\lambda,\mu}(s \wedge t) .$$

Moreover, hypotheses (M_2) and (C_2) easily entail that the Gaussian space generated by this process Γ is separable. Therefore, by [133, Corollaire 3.8, p. 44] (see also, for instance, [93, Chapter VIII]), the process admits a measurable version. □

Exercise 2.22 (Composition of a two parameter Gaussian process by an independent process). We keep the notation and hypotheses of Theorem 2.16. Let $(A_u, u \geq 0)$ be a process taking values in Λ, independent from $(G_{\bullet,t}, t \geq 0)$. Let σ be a signed finite measure on $\mathbb{R}_+$.

Prove that

$$X_t^{(\sigma)} := \int_{\mathbb{R}_+} \exp\left(G_{A_u,t} - \frac{1}{2}c_{A_u,A_u}(t)\right)\sigma(du), \quad t \geq 0$$

is a peacock, and that

$$M_t^{(\sigma)} := \int_{\mathbb{R}_+} \exp\left(\Gamma_{A_u,t} - \frac{1}{2}c_{A_u,A_u}(t)\right)\sigma(du), \quad t \geq 0$$

is an associated martingale, where Γ is independent from A. Precise the filtration with respect to which $(M_t^{(\sigma)}, t \geq 0)$ is a martingale.

One can compare this exercise with Exercise 2.1.

In some particular cases, we can give more explicit constructions, without assuming hypotheses (M_2), (C_2).

Proposition 2.5. *Assume that there exists an increasing function φ on $\mathbb{R}_+$ such that, for every $\lambda, \mu \in \Lambda$ and $t \geq 0$,*

$$c_{\lambda,\mu}(t) = \varphi(t)\, c_{\lambda,\mu}(1) .$$

Let $(G_{\bullet}^{(n)}, n \geq 0)$ be a sequence of independent copies of $G_{\bullet,1}$, and let $(e_n, n \geq 0)$ be a Hilbert basis of $L^2(\mathbb{R}_+)$. We set

$$\Gamma_{\lambda,t} = \sum_{n=0}^{\infty}\left(\int_0^{\varphi(t)} e_n(s)ds\right) G_{\lambda}^{(n)} .$$

Then,

$$(\Gamma_{\lambda,t} \quad ; \quad \lambda \in \Lambda \, , \, t \geq 0) \, ,$$

is a measurable centered Gaussian process such that (2.55) is satisfied.

Proof. Since the function φ is increasing, the result follows from Parseval's identity. □

Proposition 2.6. *Let $g : \Lambda \times \mathbb{R}_+ \longrightarrow \mathbb{R}$ be a measurable function such that, for every $\lambda \in \Lambda$, $g(\lambda, \centerdot) \in L^2(\mathbb{R}_+)$. We suppose that*

$$G_{\lambda,t} = t \int_0^\infty g(\lambda, s) dB_s$$

where (B_s) is a standard Brownian motion. We denote by $(W_{s,t}\ ;\ s,t \geq 0)$ the Brownian sheet and we set:

$$\Gamma_{\lambda,t} = \int_0^\infty g(\lambda, u)\ \mathrm{d}_u W_{u,t^2}\ .$$

Then,

$$(\Gamma_{\lambda,t}\ ;\ \lambda \in \Lambda\ ,\ t \geq 0)\ ,$$

is a measurable centered Gaussian process such that (2.55) is satisfied.

Proof. We have

$$c_{\lambda,\mu}(t \wedge s) = (t \wedge s)^2 \int_0^\infty g(\lambda, u) g(\mu, u) du$$

and the result follows. □

Proposition 2.7. *Let $h : \Lambda \times \mathbb{R}_+ \longrightarrow \mathbb{R}$ be a measurable function such that, for every $\lambda \in \Lambda$, $h(\lambda, \centerdot) \in L^2_{loc}(\mathbb{R}_+)$. We suppose that*

$$G_{\lambda,t} = \int_0^t h(\lambda, s) dB_s$$

where (B_s) is a standard Brownian motion. We set:

$$\Gamma_{\lambda,t} = G_{\lambda,t}\ .$$

Then,

$$(\Gamma_{\lambda,t}\ ;\ \lambda \in \Lambda\ ,\ t \geq 0)\ ,$$

is a measurable centered Gaussian process such that (2.55) is satisfied.

The proof is straightforward.

The following proposition states the properties of Γ which are essential in the sequel.

Proposition 2.8. *Let $(\Gamma_{\lambda,t}\ ;\ \lambda \in \Lambda, t \geq 0)$ be a measurable centered Gaussian process such that (2.55) holds. We set, for $t \geq 0$,*

$$\mathscr{G}_t = \sigma\{\Gamma_{\lambda,s}\ ;\ \lambda \in \Lambda\ , 0 \leq s \leq t\}\ .$$

Then,

1) *For $0 \leq s \leq t$, the process $(\Gamma_{\lambda,t} - \Gamma_{\lambda,s}, \lambda \in \Lambda)$ is independent of the σ-field $\mathscr{G}_s$.*
2) *For any $t \geq 0$,*

$$\Gamma_{\centerdot,t} \overset{(law)}{=} G_{\centerdot,t}\ .$$

The proof is straightforward.

2.6.2 Application to Peacocks

Looking at Theorem 2.16 and Proposition 2.8, we see that, in a Gaussian setting, the hypotheses of Proposition 2.1 are satisfied. Therefore, we can state the following result:

Proposition 2.9. *Assume there exists a measurable centered Gaussian process:*

$$(\Gamma_{\lambda,t}\,;\,\lambda\in\Lambda\,,\,t\geq 0)\,,$$

such that (2.55) is satisfied. We set, for $t\geq 0$,

$$\mathscr{G}_t=\sigma\{\Gamma_{\lambda,s}\,;\,\lambda\in\Lambda\,,0\leq s\leq t\}\,.$$

Let σ be a signed finite measure on Λ. We set, for $t\geq 0$,

$$M_t^{(\sigma)}=\int_\Lambda \exp\left(\Gamma_{\lambda,t}-\frac{1}{2}c_{\lambda,\lambda}(t)\right)\,\sigma(\mathrm{d}\lambda)\,.$$

Then $(M_t^{(\sigma)},t\geq 0)$ is a $(\mathscr{G}_t)$-martingale and, for each $t\geq 0$,

$$M_t^{(\sigma)}\overset{(law)}{=}A_t^{(\sigma)}\,.$$

In particular, $(A_t^{(\sigma)},t\geq 0)$ is a peacock.

Theorem 2.16, Proposition 2.5, Proposition 2.6 and Proposition 2.7 give conditions which entail the hypothesis of the above proposition. In particular, Theorem 2.16 and Proposition 2.9 yield another proof (with slightly different hypotheses) of Theorem 1.11 under Condition (I_2).

Notes and Comments

Theorem 2.2 is due to Baker and Yor [12], who introduced the Brownian sheet to exhibit a martingale associated to the guiding peacock. Subsections 2.3, 2.4 and 2.5 come from [83] and [84], where the interested reader may find many complements. Many properties of the Brownian sheet and of the Lévy sheet extending a Lévy process are described in Cairoli-Walsh [39] and Dalang-Walsh [50]. Subsection 2.6 about the Gaussian sheet is taken from [81] and [79].

3

The Time Reversal Method

Abstract We associate to some $\mathbf{F}_2$-type peacocks (see Definition 1.8) a martingale by using a time reversal method. Then, quite similarly as in Chapter 2, where we exhibit some peacocks and associated martingales defined from the Brownian sheet, we construct, thanks to time reversal arguments, a new family of peacocks and associated martingales.

3.1 Martingales Associated to a $\mathbf{F}_2$-type Peacock in a Discrete Setting

Let $(Z_1, Z_2, \dots, Z_n, \dots)$ be a sequence of i.i.d real r.v's, and $(\mathscr{F}_n := \sigma(Z_1, Z_2, \dots, Z_n))$ the associated natural filtration. Let $(Y_n, n \geq 1)$ be a centered $(\mathscr{F}_n, n \geq 1)$-martingale and

$$\left(X_n := \sum_{i=1}^{n} Y_i,\ n \geq 1 \right).$$

From Point 2) of Theorem 1.6, we know that $(X_n, n \geq 0)$ is a peacock. Therefore, are we able to find an associated martingale?

First observe that, since Y_n is $\mathscr{F}_n$-measurable, there exists, for every $n \geq 1$, a Borel function $f_n : \mathbb{R}^n \longrightarrow \mathbb{R}$ such that:

$$Y_n = f_n(Z_1, Z_2, \dots, Z_n) \quad \text{a.s.}$$

Thus:

$$X_n = \sum_{j=1}^{n} f_j(Z_1, Z_2, \dots, Z_j).$$

We now define, for every $n \geq 1$:

$$\begin{aligned} M_n &:= f_1(Z_n) + f_2(Z_n, Z_{n-1}) + \dots + f_n(Z_n, Z_{n-1}, \dots, Z_1) \\ &= \sum_{j=1}^{n} f_j(Z_n, Z_{n-1}, \dots, Z_{n-j+1}). \end{aligned} \tag{3.1}$$

Hirsch F., Profeta C., Roynette B., Yor M.:
Peacocks and Associated Martingales, with Explicit Constructions

Let us note that M_n was obtained from X_n by "reversing the time", i.e. by replacing in the definition of X_n the n-tuple $(Z_1, Z_2, \dots, Z_n)$ by $(Z_n, Z_{n-1}, \dots, Z_1)$. Hence, we have:

$$X_n \overset{(1.d)}{=} M_n. \tag{3.2}$$

Theorem 3.1. *Under the preceding notation and hypothesis, and with the extra assumption that, for every $n \geq 1$, f_n is a symmetric function of its arguments, we have:*

i) $(M_n, n \geq 1)$ *is a* $\mathscr{F}_n$*-martingale,*

ii) $(M_n, n \geq 1)$ *is associated to the peacock* $(X_n, n \geq 1)$*:* $X_n \overset{(1.d)}{=} M_n$.

Proof. Point *ii)* is an immediate consequence of (3.2).
1) We first observe, since Y_n is a centered martingale, that for every $n \geq 1$ and every reals $x_1, x_2, \dots, x_{n-1}$:

$$\mathbb{E}\left[f_n(x_1, x_2, \dots, x_{n-1}, Z)\right] = f_{n-1}(x_1, x_2, \dots, x_{n-1}) \tag{3.3}$$

$$\text{and } \mathbb{E}[f_1(Z)] = 0 \tag{3.4}$$

where Z is a r.v. with the same law as the Z_i's, and (3.3) is true $\nu^{\otimes(n-1)}$ a.s., with ν the common law of the Z_i's.

2) We now prove that $(M_n, n \geq 1)$ is a $(\mathscr{F}_n, n \geq 1)$ martingale. We have, from (3.1):

$$\begin{aligned}
\mathbb{E}[M_n | \mathscr{F}_{n-1}] =& \mathbb{E}[f_1(Z_n) | \mathscr{F}_{n-1}] + \mathbb{E}[f_2(Z_n, Z_{n-1}) | \mathscr{F}_{n-1}] + \dots \\
& + \mathbb{E}[f_n(Z_n, Z_{n-1}, \dots, Z_1) | \mathscr{F}_{n-1}] \\
=& 0 + \mathbb{E}[f_2(Z_n, Z_{n-1}) | \mathscr{F}_{n-1}] + \dots + \mathbb{E}[f_n(Z_n, Z_{n-1}, \dots, Z_1) | \mathscr{F}_{n-1}] \\
& \quad \text{(from (3.4))} \\
=& \mathbb{E}[f_2(Z_{n-1}, Z_n) | \mathscr{F}_{n-1}] + \dots + \mathbb{E}[f_n(Z_1, Z_2, \dots, Z_{n-1}, Z_n) | \mathscr{F}_{n-1}] \\
& \quad \text{(since } f_j,\ j = 2, \dots, n \text{ is a symmetric function of its arguments)} \\
=& f_1(Z_{n-1}) + f_2(Z_{n-2}, Z_{n-1}) + \dots + f_{n-1}(Z_1, Z_2, \dots, Z_{n-1}) \\
& \quad \text{(from (3.3))} \\
=& f_1(Z_{n-1}) + f_2(Z_{n-1}, Z_{n-2}) + \dots + f_{n-1}(Z_{n-1}, \dots, Z_2, Z_1) \\
=& M_{n-1}.
\end{aligned}$$

□

This theorem can be generalized in the set-up of exchangeable variables, see [77].

Exercise 3.1. Let $(Z_k, k \in \mathbb{N}^*)$ be a sequence of i.i.d. centered r.v.'s, and let, for every $i \geq 0$:

$$X_n^{(i)} := \sum_{k=0}^{n-i} (n - k - i + 1) Z_{k+1} Z_{k+2} \dots Z_{k+i}, \qquad n \geq i.$$

Prove that $(X_n^{(i)}, n \geq i)$ is a peacock, and exhibit an associated martingale.

(*Hint*: prove that $(M_n^{(i)} := \sum_{k=0}^{n-i}(n-k-i+1)Z_{n-k-i+1}Z_{n-k-i+2}\dots Z_{n-k}, n \geq i)$ is a martingale, and verify, using a time-reversal argument that, for every $n \geq i$, $M_n^{(i)} \overset{(law)}{=} X_n^{(i)}$.)

3.2 Martingales Associated to a $\mathbf{F}_2$-type Peacock in a Brownian Setting

3.2.1 The One-Dimensional Case

Let us denote as $(\Omega, (B_t, t \geq 0), (\mathscr{F}_t, t \geq 0), \mathbb{P})$ the canonical realization of the standard Brownian motion started at 0. Let $\varphi : \mathbb{R}_+ \times \mathbb{R} \longrightarrow \mathbb{R}$ be a $\mathscr{C}^{1,2}$ function such that:

$$\int_0^t \mathbb{E}\left[\left(\frac{\partial \varphi}{\partial x}(s, B_s)\right)^2\right] ds < \infty \qquad \text{for all } t \geq 0,$$

and

$$\forall (s,x) \in]0,+\infty[\times\mathbb{R}, \quad \frac{\partial \varphi}{\partial s} + \frac{1}{2}\frac{\partial^2 \varphi}{\partial x^2} = 0.$$

From Remark 2.1, φ is a space-time harmonic function. We assume furthermore that the martingale $(\varphi(s, B_s), s \geq 0)$ is centered. (This is equivalent to $\varphi(0,0) = 0$). Then, from Point 2) of Theorem 1.4,

$$\left(X_t^{(\varphi)} := \int_0^t \varphi(s, B_s) ds, t \geq 0\right)$$

is a $\mathbf{F}_2$-type peacock. (See Definition 1.8). Examples of space-time harmonic functions are given in Section 2.2.

Theorem 3.2. *Define:*

$$\left(M_t^{(\varphi)} := \int_0^t \varphi(t-s,\, B_t - B_s) ds, t \geq 0\right).$$

Then:

1) $\left(M_t^{(\varphi)}, t \geq 0\right)$ *is a martingale.*
2) $\left(M_t^{(\varphi)}, t \geq 0\right)$ *is associated to* $\left(X_t^{(\varphi)}, t \geq 0\right)$*, i.e.:*

$$M_t^{(\varphi)} \overset{(1.d)}{=} X_t^{(\varphi)}.$$

Remark 3.1. 1) Unlike with the Brownian sheet method of Chapter 2, the martingale $\left(M_t^{(\varphi)}, t \geq 0\right)$ is defined on the same probability space as the original Brownian motion $(B_t, t \geq 0)$, and is a martingale in its natural filtration.

2) $\left(M_t^{(\varphi)}, t \geq 0\right)$ is obtained from $\left(X_t^{(\varphi)}, t \geq 0\right)$ by time-reversing the original Brownian motion at time t. Indeed, the time reversal of $(B_u, u \leq t)$ at time t is the process $(B_t - B_{t-u}, 0 \leq u \leq t)$ and we have:

$$M_t^{(\varphi)} = \int_0^t \varphi(t-s, B_t - B_s)ds = \int_0^t \varphi(u, B_t - B_{t-u})du$$

(after the change of variable $u = t - s$). On the other hand, since $(B_u, 0 \leq u \leq t) \overset{(law)}{=} (B_t - B_{t-u}, 0 \leq u \leq t)$, Point 2) of Theorem 3.2 is clear.

Proof. From the previous Remark, it is sufficient to prove Point 1). Itô's formula, applied between the times s and t $(s < t)$, yields:

$$\begin{aligned}\varphi(t-s, B_t - B_s) =& \varphi(s-s, B_s - B_s) + \int_s^t \frac{\partial \varphi}{\partial x}(u-s, B_u - B_s)dB_u \\ &+ \int_s^t \left(\frac{\partial \varphi}{\partial u} + \frac{1}{2}\frac{\partial^2 \varphi}{\partial x^2}\right)(u-s, B_u - B_s)du \\ =& \int_s^t \frac{\partial \varphi}{\partial x}(u-s, B_u - B_s)dB_u \end{aligned} \tag{3.5}$$

since $(\varphi(s, B_s), s \geq 0)$ is a centered martingale, hence $\varphi(0,0) = 0$, and

$$\frac{\partial \varphi}{\partial s} + \frac{1}{2}\frac{\partial^2 \varphi}{\partial x^2} = 0.$$

Therefore, we have:

$$\begin{aligned}M_t^{(\varphi)} &= \int_0^t \varphi(t-s, B_t - B_s)ds \\ &= \int_0^t ds \int_s^t \frac{\partial \varphi}{\partial x}(u-s, B_u - B_s)dB_u \\ &= \int_0^t dB_u \int_0^u \frac{\partial \varphi}{\partial x}(u-s, B_u - B_s)ds.\end{aligned}$$

The process $\left(\int_0^u \frac{\partial \varphi}{\partial x}(u-s, B_u - B_s)ds, u \geq 0\right)$ being previsible, $(M_t^{(\varphi)}, t \geq 0)$ is a $(\mathscr{F}_t, t \geq 0)$-martingale. □

3.2.2 Hermite Polynomials and Peacocks

Let, for any integer n, $H_n(s,x)$ denote the nth Hermite polynomial defined via the generating function expansion:

$$e^{\lambda x - \frac{\lambda^2}{2}s} = \sum_{n=0}^{+\infty} \frac{\lambda^n}{n!} H_n(s,x) \qquad (s \geq 0,\ x \in \mathbb{R}).$$

We recall that $H_0(s,x) = 1$, $H_1(s,x) = x$, $H_2(s,x) = x^2 - s$ and

$$\frac{\partial H_n}{\partial x}(s,x) = nH_{n-1}(s,x).$$

Since H_n is a space-time harmonic function, $(H_n(s,B_s), s \geq 0)$ is a martingale. From Theorem 3.2, $(M_t^{(n)} := \int_0^t H_n(t-s, B_t - B_s)ds, t \geq 0)$ is a martingale associated to the peacock $(X_t^{(n)} := \int_0^t H_n(s,B_s)ds, t \geq 0)$. In this particular case, the martingale $(M_t^{(n)}, t \geq 0)$ has a particularly nice form.

Theorem 3.3. *There is the formula:*

$$M_t^{(n)} = n! \int_0^t dB_{u_1} \int_0^{u_1} dB_{u_2} \dots \int_0^{u_{n-2}} dB_{u_{n-1}} \int_0^{u_{n-1}} u_n dB_{u_n}, \quad t \geq 0 \tag{3.6}$$

Compare with the classical formula:

$$H_n(t,B_t) = \int_0^t dB_{u_1} \int_0^{u_1} dB_{u_2} \dots \int_0^{u_{n-2}} dB_{u_{n-1}} \int_0^{u_{n-1}} dB_{u_n}.$$

Proof. We have:

$$\begin{aligned} M_t^{(n)} &= \int_0^t H_n(t-u_2, B_t - B_{u_2})du_2 \\ &= \int_0^t du_2 \int_{u_2}^t \frac{\partial H_n}{\partial x}(u_1 - u_2, B_{u_1} - B_{u_2})dB_{u_1} \\ &\quad \text{(from Itô's formula, since } H_n \text{ is space-time harmonic, see (3.5))} \\ &= n\int_0^t du_2 \int_{u_2}^t H_{n-1}(u_1 - u_2, B_{u_1} - B_{u_2})dB_{u_1} \\ &= n\int_0^t dB_{u_1} \int_0^{u_1} H_{n-1}(u_1 - u_2, B_{u_1} - B_{u_2})du_2 \quad \text{(applying Fubini).} \end{aligned}$$

Iterating this procedure, we obtain:

$$M_t^{(n)} = n(n-1)\dots 3 \int_0^t dB_{u_1} \int_0^{u_1} dB_{u_2} \dots 2 \int_0^{u_{n-1}} H_1(u_{n-1} - u_n, B_{u_{n-1}} - B_{u_n})du_n \tag{3.7}$$

But, since $H_1(s,x) = x$,

$$\begin{aligned} \int_0^{u_{n-1}} H_1(u_{n-1} - u_n, B_{u_{n-1}} - B_{u_n})du_n &= \int_0^{u_{n-1}} (B_{u_{n-1}} - B_{u_n})du_n \\ &= \int_0^{u_{n-1}} du_n \int_{u_n}^{u_{n-1}} dB_v \\ &= \int_0^{u_{n-1}} dB_v \int_0^v du_n \quad \text{(applying Fubini)} \\ &= \int_0^{u_{n-1}} u_n dB_{u_n}. \end{aligned} \tag{3.8}$$

Finally, gathering (3.7) and (3.8), we obtain (3.6). □

Remark 3.2 (Back to the Brownian sheet). 1) Thanks to time reversal, we have obtained a martingale $(M_t^{(n)}, t \geq 0)$ given by (3.6) associated to the peacock $(X_t^{(n)} := \int_0^t H_n(s, B_s) ds, t \geq 0)$.

2) On the other hand, consider the peacock:

$$\left(\frac{1}{t} \int_0^t H_n(s, B_s) ds = \int_0^1 H_n(ut, B_{ut}) du, t \geq 0 \right). \tag{3.9}$$

$a)$ Now, replacing in (3.9) the process $(B_{ut}, t \geq 0)$ by the process (with the same law) $(W_{u,t}, t \geq 0)$, where $(W_{u,t},\ u, t \geq 0)$ denotes the Brownian sheet, we obtain:

$$\frac{1}{t} \int_0^t H_n(s, B_s) ds \overset{(1.d)}{=} \int_0^1 H_n(ut, W_{u,t}) du =: M_t^{(1,n)}$$

and the process $(M_t^{(1,n)}, t \geq 0)$ is a $(\mathscr{G}_t, t \geq 0)$-martingale (with $\mathscr{G}_t = \sigma(W_{u,s};\ u \geq 0, s \leq t))$ associated to the peacock $\left(\frac{1}{t} \int_0^t H_n(s, B_s) ds, t \geq 0\right)$ (see Chapter 2, Theorem 2.2).

$b)$ But we also have from Theorem 3.3:

$$\begin{aligned}
&\frac{1}{t} \int_0^t H_n(s, B_s) ds \\
&\quad \overset{(1.d)}{=} \frac{1}{t} \int_0^t H_n(t - s, B_t - B_s) ds \\
&\quad = \frac{n!}{t} \int_0^t dB_{u_1} \int_0^{u_1} dB_{u_2} \dots \int_0^{u_{n_2}} dB_{u_{n-1}} \int_0^{u_{n-1}} u_n dB_{u_n} \\
&\quad = n! \int_0^1 dB_{s_1 t} \int_0^{s_1} dB_{s_2 t} \dots \int_0^{s_{n_2}} dB_{s_{n-1} t} \int_0^{s_{n-1}} s_n dB_{s_n t} \\
&\quad \text{(after the changes of variables } u_1 = s_1 t,\ u_2 = s_2 t,\ \dots,\ u_n = s_n t) \\
&\quad \overset{(1.d)}{=} n! \int_0^1 d_{s_1} W_{s_1,t} \int_0^{s_1} d_{s_2} W_{s_2,t} \dots \int_0^{s_{n_2}} d_{s_{n-1}} W_{s_{n-1},t} \int_0^{s_{n-1}} s_n d_{s_n} W_{s_n,t} \\
&\quad \text{(since } (B_{st}, s \geq 0) \overset{(law)}{=} (W_{s,t}, s \geq 0)) \\
&\quad =: M_t^{(2,n)}.
\end{aligned}$$

It is easily shown that $(M_t^{(2,n)}, t \geq 0)$ is a $(\mathscr{G}_t)$-martingale. (We omit the proof). Thus, the two martingales

$$\left(M_t^{(1,n)} = \int_0^1 H_n(ut, W_{u,t}) du, t \geq 0 \right)$$

and

$$\left(M_t^{(2,n)} = n! \int_0^1 d_{s_1} W_{s_1,t} \dots \int_0^{s_{n_2}} d_{s_{n-1}} W_{s_{n-1},t} \int_0^{s_{n-1}} s_n d_{s_n} W_{s_n,t}\ ,\ t \geq 0 \right)$$

are associated to the peacock $\left(\frac{1}{t} \int_0^t H_n(s, B_s) ds, t \geq 0\right)$.

c) A natural question is: are these two martingales equal in law, i.e. do we have:

$$\left(M_t^{(1,n)}, t \geq 0\right) \stackrel{(law)}{=} \left(M_t^{(2,n)}, t \geq 0\right) ?$$

The answer to this question is positive. Indeed, applying n times Itô's formula, we have:

$$\begin{aligned} M_t^{(1,n)} &= \int_0^1 H_n(ut, W_{u,t}) du \\ &= n! \int_0^1 (1-u_1) d_{u_1} W_{u_1,t} \int_0^{u_1} d_{u_2} W_{u_2,t} \ldots \int_0^{u_{n-1}} d_{u_n} W_{u_n,t}. \end{aligned}$$

On the other hand, replacing in the definition of $(M_t^{(2,n)}, t \geq 0)$ the process $(W_{s,t}, 0 \leq s \leq 1, t \geq 0)$ by the process with the same law $(\widehat{W}_{s,t}, 0 \leq s \leq 1, t \geq 0) := (W_{1,t} - W_{1-s,t}, 0 \leq s \leq 1, t \geq 0)$, we obtain:

$$\begin{aligned} (M_t^{(2,n)}, t \geq 0) &\stackrel{(law)}{=} \left(n! \int_0^1 d_{s_1} \widehat{W}_{s_1,t} \int_0^{s_1} d_{s_2} \widehat{W}_{s_2,t} \ldots \int_0^{s_{n-1}} s_n d_{s_n} \widehat{W}_{s_n,t}, t \geq 0\right) \\ &= (M_t^{(1,n)}, t \geq 0). \end{aligned}$$

We see that the two processes $(M_t^{(1,n)}, t \geq 0)$ and $(M_t^{(2,n)}, t \geq 0)$ have the same law, which confirms a posteriori that $(M_t^{(2,n)}, t \geq 0)$ is a martingale. Thus, there is the commutative diagram:

$$\begin{array}{ccc} \dfrac{1}{t} \displaystyle\int_0^t H_n(s, B_s) ds & \xrightarrow{\text{time reversal}} & \dfrac{n!}{t} \displaystyle\int_0^t dB_{u_1} \int_0^{u_1} dB_{u_2} \ldots \int_0^{u_{n-1}} u_n dB_{u_n} \\ \Big\downarrow \text{Brownian sheet method} & & \Big\downarrow \text{Brownian sheet method} \\ \displaystyle\int_0^1 H_n(ut, W_{u,t}) du & \xrightarrow{\text{time reversal}} & n! \displaystyle\int_0^1 d_{s_1} W_{s_1,t} \int_0^{s_1} d_{s_2} W_{s_2,t} \ldots \int_0^{s_{n-1}} s_n d_{s_n} W_{s_n,t}. \end{array}$$

The first line of this diagram consists of two peacocks whose one-dimensional marginals are equal in law, and the second line gives associated martingales which are equal in law.

Exercise 3.2 **($\exp(\xi_t) \int_0^t \exp(-\xi_s) d\beta_s$ is a Markovian peacock).** Let $(\xi_t, t \geq 0)$ be a Lévy process such that, for every $u > 0$, $\mathbb{E}\left[e^{2\xi_u}\right] < \infty$. We define:

$$\Pi_t := \exp(\xi_t) \int_0^t \exp(-\xi_s) d\beta_s, \quad (t \geq 0)$$

where $(\beta_s, s \geq 0)$ is a Brownian motion independent from $(\xi_t, t \geq 0)$.
1) Prove that there exists a Brownian motion $(\widetilde{\beta}_t, t \geq 0)$ independent from $(\xi_t, t \geq 0)$ such that:

$$\Pi_t \stackrel{(1.d)}{=} \int_0^t \exp(\xi_u) d\beta_u \stackrel{(1.d)}{=} \widetilde{\beta}_{\int_0^t \exp(2\xi_u) du}.$$

2) Deduce that $(\Pi_t, t \geq 0)$ is a peacock.
3) Prove that this peacock is (homogeneous) Markovian.

Exercise 3.3 ("Doob-Meyer" decomposition for some peacocks). Let $(B_u, u \geq 0)$ be a Brownian motion started at 0.

1) Let $(F(t,(B_u, 0 \leq u \leq t)), t \geq 0)$ be an increasing predictable process. Prove that $(F(t,(B_t - B_{t-u}, 0 \leq u \leq t)), t \geq 0)$ is a stochastically increasing process (see Exercise 1.25).

2) Let $h : \mathbb{R}_+ \times \mathbb{R} \longrightarrow \mathbb{R}$ be a space-time harmonic function for the Brownian motion such that $h(0,0) = 0$, and such that there exists two continuous functions $C_1, C_2 : \mathbb{R} \longrightarrow \mathbb{R}_+$ such that:

$$\sup_{0 \leq s \leq t} |h(s,x)| \leq C_1(t) \exp\left(C_2(t)|x|\right).$$

We consider the peacock $\left(X_t^{(h)} := \int_0^t h(s, B_s) ds, t \geq 0\right)$ (see Theorem 1.4). Prove that there exists a stochastically increasing process $(Y_t^{(h)}, t \geq 0)$ such that $\left((X_t^{(h)})^2 - Y_t^{(h)}, t \geq 0\right)$ is a peacock.
(*Hint*: write the Doob-Meyer decomposition of the submartingale $\left((M_t^{(h)})^2 := \left(\int_0^t h(t-s, B_t - B_{t-s}) ds\right)^2, t \geq 0\right)$ and use time reversal.)

3) Let $\psi : \mathbb{R} \longrightarrow \mathbb{R}$ be a convex, continuous and sublinear function. Prove that there exists a stochastically increasing process $(Y_t^{(\psi,h)}, t \geq 0)$ such that $\left(\psi(X_t^{(h)}) - Y_t^{(\psi,h)}, t \geq 0\right)$ is a peacock.

Exercise 3.4 (A nilpotent system of SDE's). Let β and γ be two independent Brownian motions started from 0. Let $n \in \mathbb{N}^*$, $(\lambda_0, \lambda_1, \dots, \lambda_n, \mu_0, \mu_1, \dots, \mu_n)$ $2n+2$ reals, all different from 0. We define $\delta_i := \mu_i \mu_{i-1}$ for $i = 1, \dots, n$ and $\delta_0 = 0$. We consider the $(n+1)$-dimensional SDE:

$$Z_t^{(i)} = \lambda_i \gamma_t + \mu_i \int_0^t Z_s^{(i-1)} d\beta_s + \frac{\delta_i}{2} \int_0^t Z_s^{(i-2)} ds \quad i = 0, \dots, n \tag{3.10}$$

(with $Z^{(-1)} = Z^{(-2)} = 0$).

i) Prove that, for every $i = 0, \dots, n$, the process $(Z_t^{(i)}, t \geq 0)$ is a peacock.
(*Hint*: let $c_0 = \lambda_0$ and for $1 \leq i \leq n$, $c_i = \lambda_0 \mu_1 \dots \mu_i$. Define the polynomial Q of degree n by $Q^{(i)}(0) = \dfrac{\lambda_{n-i}}{c_{n-i}}$, $0 \leq i \leq n$, where $Q^{(i)}$ denotes the ith derivative of Q. Set for $0 \leq i \leq n$:

$$X_t^{(i)} := c_i \int_0^t Q^{(n-i)}(\beta_t - \beta_u) d\gamma_u.$$

Prove that $(X^{(i)}, i = 0, \dots, n)$ is solution of (3.10) and apply a time-reversal argument.)

ii) Show that one can replace in the previous question the Brownian motion $(\gamma_t, t \geq 0)$ by an integrable and centered Lévy process.

3.2.3 The n-Dimensional Case, $n \geq 1$

Let $(\Omega, (B_t^{(n)}, t \geq 0), (\mathscr{F}_t^{(n)}, t \geq 0), \mathbb{P})$ be the canonical realization of the standard n-dimensional Brownian motion started at 0. In this context, Theorem 3.2 admits a version which we state without proof.

Theorem 3.4. 1) *Let $\varphi : \mathbb{R}_+ \times \mathbb{R}^n \longrightarrow \mathbb{R}$ be a $\mathscr{C}^{1,2}$ function such that $\varphi(0,0) = 0$ and, for $(t,x) \in]0, +\infty[\times \mathbb{R}^n$,*

$$\frac{\partial \varphi}{\partial t} + \frac{1}{2}\Delta \varphi = 0.$$

We also assume that, for every $t > 0$:

$$\int_0^t \mathbb{E}\left[|\nabla \varphi(s, B_s^{(n)})|^2\right] ds < +\infty.$$

As a consequence, $\left(\varphi(s, B_s^{(n)}), s \geq 0\right)$ is a centered $(\mathscr{F}_s^{(n)})$-martingale. Then:

$$\left(M_t^{(\varphi)} := \int_0^t \varphi(t-s, B_t^{(n)} - B_s^{(n)}) ds \right.$$
$$\left. = \int_0^t dB_u^{(n)} \centerdot \int_0^u \nabla \varphi(u-s, B_u^{(n)} - B_s^{(n)}) ds, \; t \geq 0 \right)$$

is a martingale associated to the $\mathbf{F}_2$-type peacock:

$$\left(X_t^{(\varphi)} := \int_0^t \varphi(s, B_s^{(n)}) ds, \; t \geq 0 \right).$$

2) *Let $\varphi : \mathbb{R}^n \longrightarrow \mathbb{R}$ be a harmonic function (i.e. satisfying $\Delta \varphi = 0$) such that, for every $t > 0$,*

$$\int_0^t \mathbb{E}\left[|\nabla \varphi(B_s^{(n)})|^2\right] ds < +\infty.$$

Then:

$$\left(M_t^{(\varphi)} := \int_0^t \left(\varphi(B_t^{(n)} - B_s^{(n)}) - \varphi(0)\right) ds \right.$$
$$\left. = \int_0^t dB_u^{(n)} \centerdot \int_0^u \nabla \varphi(B_u^{(n)} - B_s^{(n)}) ds, \; t \geq 0 \right)$$

is a martingale associated to the $\mathbf{F}_2$-type peacock:

$$\left(X_t^{(\varphi)} := \int_0^t \left(\varphi(B_s^{(n)}) - \varphi(0)\right) ds, \; t \geq 0 \right).$$

3) $(n = 2)$ *Let φ be a holomorphic function from $\mathbb{C}$ to $\mathbb{C}$ such that:*

$$\int_0^t \mathbb{E}\left[|\varphi'(B_s^{(2)})|^2\right] ds < +\infty$$

with $(B_s^{(2)} := B_s + i\widetilde{B}_s, s \geq 0)$*, B and $\widetilde{B}$ being two independent 1-dimensional Brownian motions. Then:*

$$\left(M_t^{(\varphi)} := \int_0^t \left(\varphi(B_s^{(2)}) - \varphi(0)\right) ds \right.$$
$$\left. = \int_0^t dB_u^{(2)} \int_0^u \varphi'(B_u^{(2)} - B_s^{(2)}) ds,\ t \geq 0\right)$$

is a ($\mathbb{C}$-valued) martingale associated to the ($\mathbb{C}$-valued) peacock:

$$\left(X_t^{(\varphi)} := \int_0^t \left(\varphi(B_s^{(2)}) - \varphi(0)\right) ds,\ t \geq 0\right)$$

($(X_t, t \geq 0)$ is a $\mathbb{C}$-valued peacock if for every convex function $\psi : \mathbb{C} \longrightarrow \mathbb{R}$, $t \longmapsto \mathbb{E}[\psi(X_t)]$ is increasing).

Exercise 3.5 (A F_1-type peacock defined from a n-dimensional squared Bessel process). For $n \geq 1$, let $(R_t^{(n)}, t \geq 0)$ denote a n-dimensional Bessel process started from 0. Prove that the martingale:

$$\left(2\int_0^t dB_u^{(n)} \bullet \int_0^u (B_u^{(n)} - B_s^{(n)}) ds, t \geq 0\right)$$

is associated to the peacock $\left(t^2 \int_0^1 \left((R_s^{(n)})^2 - ns\right) ds, t \geq 0\right)$. (*Hint*: apply Theorem 3.4 with $\varphi(t,x) = |x|^2 - nt$, use $(R_t^{(n)}, t \geq 0) \overset{(law)}{=} (|B_t^{(n)}|, t \geq 0)$ and the scaling property of $(B_t^{(n)}, t \geq 0)$).

3.3 Harmonic Functions in $\mathbb{R}^n \backslash \{0\}$ and Peacocks

In this section, we encounter particular instances of the following situation: let $(M_t, t \geq 0)$ denote a *strict* local martingale, i.e. a local martingale which is not a martingale, e.g. its expectation over time is not constant. Thus, $\int_0^t ds (M_s - M_0)$ is not a peacock, and we may ask which adapted process $\alpha_t^{(M)}$ we might substract from $\int_0^t ds (M_s - M_0)$ so that the resulting process:

$$\left(\int_0^t ds (M_s - M_0) - \alpha_t^{(M)},\ t \geq 0\right)$$

is a peacock.

To start with, here are two well-known examples of strict local martingales (see [150, Chapter V, Exercise (2.13) p.194], [137] or Exercise 1.9) which we shall study in the following Subsections 3.3.1 and 3.3.2.

- Let $y \in \mathbb{R}^2$, $y \neq 0$ and $(B_u, u \geq 0)$ be a 2-dimensional Brownian motion started from 0. Then, the process

$$\left(X_s^{(2)} = \log|B_s - y| - \log|y|, s \geq 0\right) \quad \text{is a strict local martingale.}$$

- Let $y \in \mathbb{R}^3$, $y \neq 0$ and $(B_s, s \geq 0)$ be a 3-dimensional Brownian motion started at 0. Then, the process

$$\left(X_s^{(3)} = \frac{1}{|B_s - y|} - \frac{1}{|y|}, s \geq 0\right) \quad \text{is a strict local martingale.}$$

Those processes are both local martingales (from Itô's formula), but are not martingales since their expectations are not constant.

Exercise 3.6 (Examples of integrals of strict local martingales which are not peacocks). In this exercise, we shall prove that the two processes $\left(\int_0^t X_s^{(2)} ds, t \geq 0\right)$ and $\left(\int_0^t X_s^{(3)} ds, t \geq 0\right)$ are not peacocks, by showing that their expectations depend on t. This implies in particular from Theorem 1.4 that both processes $(X_s^{(i)}, s \geq 0)_{i=2,3}$ are not martingales.

Let $(R_s, s \geq 0;\ \mathbb{P}_a^{(\nu)})$ denote a Bessel process of index ν (i.e. of dimension $2(\nu+1)$) started from $a > 0$.
1) Let $\nu > 0$. Prove that:

$$\mathbb{E}_a^{(\nu)}\left[\frac{1}{R_s^{2\nu}} - \frac{1}{a^{2\nu}}\right] = \mathbb{E}_a^{(-\nu)}\left[\frac{R_{s\wedge T_0}^{2\nu}}{a^{2\nu}}\left(\frac{1}{R_s^{2\nu}} - \frac{1}{a^{2\nu}}\right)\right].$$

(*Hint*: use the absolute continuity formula: $\mathbb{P}_{a|\mathscr{F}_s}^{(\nu)} = \frac{R_s^{2\nu}}{a^{2\nu}} \cdot \mathbb{P}_{a|\mathscr{F}_s\cap\{s<T_0\}}^{(-\nu)}$ where $T_0 := \inf\{s \geq 0;\ R_s = 0\}$.) Deduce that:

$$\begin{aligned}\int_0^t \left(\mathbb{E}_a^{(\nu)}\left[\frac{1}{R_s^{2\nu}}\right] - \frac{1}{a^{2\nu}}\right) ds &= -\frac{1}{a^{2\nu}} \int_0^t \mathbb{P}_a^{(-\nu)}(T_0 < s) ds \\ &= -\frac{1}{a^{2\nu}} \mathbb{E}_a^{(-\nu)}\left[(t - T_0)^+\right]. \qquad (3.11)\end{aligned}$$

2) We recall that, under $\mathbb{P}_a^{(-\nu)}$, there is the equality in law:

$$T_0 \overset{(law)}{=} \frac{a^2}{2\gamma_\nu}$$

where γ_ν is a gamma r.v. with parameter ν (see Getoor [68] or Yor [180]). Prove that:

$$\lim_{\nu \downarrow 0} \frac{1}{a^{2\nu}} \frac{1}{\nu} \mathbb{E}_a^{(-\nu)}\left[(t - T_0)^+\right] = \int_0^{+\infty} e^{-u}\left(t - \frac{a^2}{2u}\right)^+ \frac{du}{u}.$$

(We recall that $\nu\Gamma(\nu)=\Gamma(1+\nu)\xrightarrow[\nu\to 0]{} 1$ and that the density of γ_ν is given by $f_{\gamma_\nu}(u)=\frac{1}{\Gamma(\nu)}u^{\nu-1}e^{-u}1_{\{u\geq 0\}}$.)

3) Prove that:

$$\frac{1}{\nu}\int_0^t \mathbb{E}_a^{(\nu)}\left[\frac{1}{R_s^{2\nu}}-\frac{1}{a^{2\nu}}\right]ds=\frac{1}{\nu}\int_0^t \mathbb{E}_a^{(\nu)}\left[\exp(-2\nu\log(R_s))-\exp(-2\nu\log(a))\right]ds$$

$$\xrightarrow[\nu\to 0]{} -2\int_0^t \mathbb{E}_a^{(0)}\left[\log R_s-\log a\right]ds.$$

4) Let $y\in\mathbb{R}^2$, $y\neq 0$ and $(B_u,u\geq 0)$ be a 2-dimensional Brownian motion started from 0. Deduce from the preceding questions that:

$$\int_0^t \mathbb{E}\left[\log\left(\frac{|B_s-y|}{|y|}\right)\right]ds=-\frac{1}{2}\int_0^{+\infty} e^{-u}\left(t-\frac{|y|^2}{2u}\right)^+\frac{du}{u}.$$

(*Hint*: $(|B_s-y|,s\geq 0)$ is a 2-dimensional Bessel process starting at $|y|$.)

5) Let $(B_s,s\geq 0)$ be a 3-dimensional Brownian motion started at 0. Prove similarly that:

$$\int_0^t\left(\mathbb{E}\left[\frac{1}{|B_s-y|}\right]-\frac{1}{|y|}\right)ds=-\frac{1}{|y|}\int_0^t\frac{(t-u)}{\sqrt{2\pi u^3}}\exp\left(-\frac{|y|^2}{2u}\right)du.$$

3.3.1 $n = 2$

Let $(B_s,s\geq 0)$ be a 2-dimensional Brownian motion started at 0. From Exercise 3.6, the process :

$$\left(\int_0^t(\log|B_s-y|-\log|y|)\,ds,t\geq 0\right)$$

is not a peacock. However, we shall see that modulo some suitable renormalization (involving intersection local times), this process will become a peacock. To this end, we first recall the definition and some properties of intersection local times (see [113]).

Proposition 3.1. *a) There exists a process $(\alpha(y,t),\ y\in\mathbb{R}^2\backslash\{0\},t\geq 0)$, which is jointly continuous on $\mathbb{R}^2\backslash\{0\}\times\mathbb{R}_+$ such that the following density of occupation formula holds, for every $f:\mathbb{R}^2\longrightarrow\mathbb{R}_+$ Borel:*

$$\int_0^t ds\int_0^s du f(B_u-B_s)=\int_{\mathbb{R}^2}f(y)\alpha(y,t)dy. \tag{3.12}$$

b) $\pi\alpha(y,t)+t\log|y|$ admits an a.s. limit as $y\to 0$:

$$\lim_{y\to 0}(\pi\alpha(y,t)+t\log|y|)=:\widetilde{\alpha}(0,t). \tag{3.13}$$

c) The intersection local time remains unchanged if we replace $(B_u, u \leq t)$ by its time-reversed process at time t, i.e. $(B_t - B_{t-u}, 0 \leq u \leq t)$.

We shall now be able to answer our former question by exhibiting an adapted process $(\alpha_t^{(M)}, t \geq 0)$ such that $\left(\int_0^t (\log|B_s - y| - \log|y|)\, ds - \alpha_t^{(M)}, t \geq 0\right)$ is a peacock.

Theorem 3.5 ($n = 2$). 1) *Case $y \neq 0$.*

a) The process $\left(X_t^{(y)} := \int_0^t (\log|B_s - y| - \log|y|)\, ds - \pi\alpha(y,t),\ t \geq 0\right)$ *is a peacock.*

b) The process $\left(M_t^{(y)} := \int_0^t dB_v \centerdot \int_0^v \frac{B_v - B_s - y}{|B_v - B_s - y|^2} ds,\ t \geq 0\right)$ *is a martingale associated to the peacock $(X_t^{(y)}, t \geq 0)$.*

2) *Case $y = 0$.*

a) The process $\left(X_t^{(0)} := \int_0^t \log|B_s| ds - \pi\widetilde{\alpha}(0,t),\ t \geq 0\right)$ *is a peacock.*

b) The process $\left(M_t^{(0)} := \int_0^t dB_v \centerdot \int_0^v \frac{B_v - B_s}{|B_v - B_s|^2} dv,\ t \geq 0\right)$ *is a martingale associated to the peacock $(X_t^{(0)}, t \geq 0)$.*

c) The processes $\left(\int_0^t (\log|B_s| - \mathbb{E}[\log|B_s|]) ds, t \geq 0\right)$ *and $(\widetilde{\alpha}(0,t) - \mathbb{E}[\widetilde{\alpha}(0,t)], t \geq 0)$ are two peacocks.*

In this statement, and in the following, $\centerdot$ stands for the usual scalar product in $\mathbb{R}^2$.

Proof. We only sketch the proof, following [178, p.341], where the Tanaka-Rosen formula (3.16) is established.

We first prove Point 1). Let $(f_n, n \geq 1)$ be a sequence of regular functions from $\mathbb{R}^2$ to $\mathbb{R}$ such that:

$$f_n(z) = \log|z| \quad \text{for } |z| \geq \frac{1}{n}.$$

From Itô's formula, applied between times s and t $(s < t)$, we have, with $y \neq 0$:

$$f_n(B_t - B_s - y) = f_n(B_s - B_s - y) + \int_s^t \nabla f_n(B_v - B_s - y) \centerdot dB_v + \frac{1}{2}\int_0^t \Delta f_n(B_v - B_s - y) dv. \quad (3.14)$$

Thus, integrating both sides of (3.14) with respect to ds:

$$\int_0^t (f_n(B_t - B_s - y) - f_n(-y))\, ds = \int_0^t ds \int_s^t \nabla f_n(B_v - B_s - y) \centerdot dB_v + \frac{1}{2}\int_0^t ds \int_s^t \Delta f_n(B_v - B_s - y) dv$$

$$= \int_0^t dB_v \cdot \int_0^v \nabla f_n(B_v - B_s - y) ds + \frac{1}{2} \int_{\mathbb{R}^2} \Delta f_n(z-y) \alpha(z,t) dz \tag{3.15}$$

(where dz denotes here Lebesgue measure on $\mathbb{R}^2$)

from Fubini and the density of occupation formula (3.12). Letting $n \to +\infty$ in (3.15) and using:

i) for $z \neq 0$:

$$f_n(z) \xrightarrow[n\to+\infty]{} \log|z| \text{ , and } \nabla f_n(z) \xrightarrow[n\to+\infty]{} \frac{z}{|z|^2},$$

ii) $\Delta f_n(z)dz \xrightarrow[n\to+\infty]{} 2\pi\delta_0(dz)$

we obtain the Tanaka-Rosen formula:

$$\int_0^t (\log|B_t - B_s - y| - \log|y|) ds = \int_0^t dB_v \cdot \int_0^v \frac{B_v - B_s - y}{|B_v - B_s - y|^2} ds + \pi\alpha(y,t). \tag{3.16}$$

All preceding limits are in probability, and can be properly justified, see [178].

We now observe that $\left(\int_0^t dB_v \cdot \int_0^v \frac{B_v - B_s - y}{|B_v - B_s - y|^2} ds, t \geq 0 \right)$ is a martingale since the r.v. $\int_0^v \frac{ds}{|B_v - B_s - y|} \overset{(law)}{=} \int_0^v \frac{ds}{|B_s - y|}$ belongs to all L^p spaces, $\frac{1}{2}\int_0^v \frac{ds}{|B_s - y|}$ being the bounded variation term of the semimartingale $(|B_v - y|, v \geq 0)$. It then suffices, to prove Point 1) of Theorem 3.5, to observe that, by time reversal:

$$\int_0^t (\log|B_t - B_s - y| - \log|y|) ds - \pi\alpha(y,t)$$
$$\overset{(1.d)}{=} \int_0^t (\log|B_s - y| - \log|y|) ds - \pi\alpha(y,t) \tag{3.17}$$

since $\alpha(y,t)$ is left unchanged by time reversal. Points 2)a) and b) of Theorem 3.5 are then obtained by letting $y \to 0$ in (3.16), (3.17) and by using (3.13).

We now prove Point 2)c) of Theorem 3.5. We have, since $B_0 = 0$:

$$\int_0^t (\log|B_s| - \mathbb{E}[\log|B_s|]) ds = t \int_0^1 (\log|B_{tu}| - \mathbb{E}[\log|B_{tu}|]) du$$
$$\overset{(1.d)}{=} t \int_0^1 \left(\log(\sqrt{t}|B_u|) - \mathbb{E}\left[\log(\sqrt{t}|B_u|)\right]\right) du$$

(by scaling)

$$= t \int_0^1 (\log|B_u| - \mathbb{E}[\log|B_u|]) du.$$

Thus, $\left(\int_0^t (\log|B_s| - \mathbb{E}[\log|B_s|]) ds, t \geq 0 \right)$ is a peacock (from Proposition 1.4), since $\int_0^1 (\log|B_u| - \mathbb{E}[\log|B_u|]) du$ is a centered r.v. We refer the reader to Chapter 7 for the construction of martingales associated to such peacocks (of the form $(tX, t \geq 0)$ with $X = \int_0^1 (\log|B_u| - \mathbb{E}[\log|B_u|])\, du$ which is a centered r.v.).

From Points 2)a) and b) of Theorem 3.5:

$$\int_0^t dB_v \boldsymbol{\cdot} \int_0^v \frac{B_v - B_s}{|B_v - B_s|^2} ds \overset{(1.d)}{=} \int_0^t \log|B_s| ds - \pi\widetilde{\alpha}(0,t)$$

we easily deduce, by scaling, that:

$$\pi\widetilde{\alpha}(0,t) \overset{(1.d)}{=} t\log(\sqrt{t}) + t\pi\widetilde{\alpha}(0,1).$$

Thus:

$$\begin{aligned}\pi\widetilde{\alpha}(0,t) - \mathbb{E}[\pi\widetilde{\alpha}(0,t)] &\overset{(1.d)}{=} t\log(\sqrt{t}) + t\pi\widetilde{\alpha}(0,1) - t\log(\sqrt{t}) - t\mathbb{E}[\pi\widetilde{\alpha}(0,1)] \\ &= t\,(\pi\widetilde{\alpha}(0,1) - \mathbb{E}[\pi\widetilde{\alpha}(0,1)])\end{aligned}$$

and $(\widetilde{\alpha}(0,t) - \mathbb{E}[\widetilde{\alpha}(0,t)], t \geq 0)$ is a peacock (from Proposition 1.4). □

3.3.2 $n = 3$

Let $(B_s, s \geq 0)$ be a 3-dimensional Brownian motion started at 0. From Exercise 3.6, the process

$$\left(\int_0^t \left(\frac{1}{|B_s - y|} - \frac{1}{|y|}\right) ds, t \geq 0\right)$$

is not a peacock.

Nevertheless, let us recall that, as in dimension 2, there exists for $y \neq 0$ an intersection local time process $(\alpha(y,t), t \geq 0)$ such that, for every Borel and positive function f, we have:

$$\int_0^t ds \int_0^s du f(B_u - B_s) = \int_{\mathbb{R}^3} f(y)\alpha(y,t) dy.$$

However, for $n = 3$, there is no analogue of (3.13), that is: there is no analogue of Varadhan's renormalization[1]. Nevertheless, for $y \neq 0$, we shall prove that, as in dimension 2, the process $\left(\int_0^t \left(\frac{1}{|B_s - y|} - \frac{1}{|y|}\right) ds, t \geq 0\right)$ becomes a peacock after a suitable renormalization.

Theorem 3.6 ($n = 3$). 1) *Case $y \neq 0$.*

a) The process $\left(\int_0^t \left(\frac{1}{|B_s - y|} - \frac{1}{|y|}\right) ds + 2\pi\alpha(y,t), t \geq 0\right)$ *is a peacock.*

b) $\left(-\int_0^t dB_u \boldsymbol{\cdot} \int_0^u \frac{B_u - B_s - y}{|B_u - B_s - y|^2} ds, t \geq 0\right)$ *is a martingale associated to this peacock.*

[1] This question has a strong link with Westwater's studies [170, 171, 172] and Symanzik's program.

2) *Case* $y = 0$. *The process* $\left(\int_0^t \left(\frac{1}{|B_s|} - \mathbb{E}\left[\frac{1}{|B_s|}\right]\right) ds, t \geq 0\right)$ *is a peacock. More precisely, we have* $\int_0^t \left(\frac{1}{|B_s|} - \mathbb{E}\left[\frac{1}{|B_s|}\right]\right) ds \overset{(1.d)}{=} \sqrt{t}X$ *where* $X = \int_0^1 \left(\frac{1}{|B_s|} - \mathbb{E}\left[\frac{1}{|B_s|}\right]\right) ds$ *is a centered r.v.*

Remark 3.3. 1) The proof of this Theorem is very similar to that of Theorem 3.5, so we refer the reader to Yor [178, p.347]. Note that this time, the fundamental relation is $\Delta\left(\frac{1}{|z|}\right) = -4\pi\delta_0$.
2) Note that there is no analogue of Varadhan's renormalization in dimension 3. In fact, there is instead a kind of central limit theorem, see [178].
3) It might be interesting to extend Theorems 3.5 and 3.6 to other functions than $z \longmapsto \log|z|$ $(n = 2)$ and $z \longmapsto \frac{1}{|z|}$ $(n = 3)$. For example, one may like to study the case of the meromorphic function $z \longmapsto \frac{1}{z}$ for $n = 2$, although this does not seem easy.

3.4 Martingales Associated to a F_2-type Peacock in a Lévy Process Framework

In this section, we shall generalize Theorem 3.2, by replacing Brownian motion by a Lévy process.

Theorem 3.7. *Let* $(L_s, \mathscr{F}_s, s \geq 0)$ *be a Lévy process started at 0 and* $\varphi : \mathbb{R}_+ \times \mathbb{R} \longrightarrow \mathbb{R}$ *be a space-time harmonic function for this process, i.e. such that:*

$$(\varphi(s, L_s)), s \geq 0) \quad \text{is a } (\mathscr{F}_s, s \geq 0)\text{-martingale.}$$

We assume furthermore that this martingale belongs to H^1_{loc} *and is centered which is equivalent to* $\phi(0,0) = 0$. *Then:*

i) $\left(X_t^{(\varphi)} := \int_0^t \varphi(s, L_s) ds,\ t \geq 0\right)$ *is a* $\boldsymbol{F}_2$*-type peacock (see Theorem 1.4 and Definition 1.8).*

ii) We define

$$\left(M_t^{(\varphi)} := \int_0^t \varphi(t - s, L_t - L_s) ds,\ t \geq 0\right).$$

Then $(M_t^{(\varphi)}, t \geq 0)$ *is a* $(\mathscr{F}_t, t \geq 0)$*-martingale associated to* $(X_t^{(\varphi)}, t \geq 0)$*:*

$$M_t^{(\varphi)} \overset{(1.d)}{=} X_t^{(\varphi)}.$$

Proof. 1) *We first prove that* $(M_t^{(\varphi)}, t \geq 0)$ *is a martingale.* We have, for $s \leq t$:

$$\begin{aligned}\mathbb{E}\left[M_t^{(\varphi)}|\mathscr{F}_s\right] &= \mathbb{E}\left[\int_0^t \varphi(t-u, L_t - L_u)du|\mathscr{F}_s\right] \\ &= \int_0^s \mathbb{E}\left[\varphi(t-u, L_t - L_u)|\mathscr{F}_s\right]du + \int_s^t \mathbb{E}\left[\varphi(t-u, L_t - L_u)|\mathscr{F}_s\right]du.\end{aligned} \tag{3.18}$$

But, for $s \leq u \leq t$:

$$\begin{aligned}\mathbb{E}\left[\varphi(t-u, L_t - L_u)|\mathscr{F}_s\right] &= \mathbb{E}\left[\varphi(t-u, L_t - L_u)\right] \\ &\qquad \text{(since } L_t - L_u \text{ is independent from } \mathscr{F}_s\text{)} \\ &= \mathbb{E}\left[\varphi(t-u, L_{t-u})\right] = 0,\end{aligned} \tag{3.19}$$

since $L_t - L_u \overset{(law)}{=} L_{t-u}$. For $u \leq s \leq t$:

$$\begin{aligned}&\mathbb{E}\left[\varphi(t-u, L_t - L_u)|\mathscr{F}_s\right] \\ &= \mathbb{E}\left[\varphi(t-u, L_t - L_s + L_s - L_u)|\mathscr{F}_s\right] \\ &= \mathbb{E}\left[\varphi(t-u, a + L_t - L_s)\right] \quad (\text{with } a = L_s - L_u) \\ &(\text{since } L_s - L_u \text{ is } \mathscr{F}_s\text{-measurable and } L_t - L_s \text{ is independent from } \mathscr{F}_s) \\ &= \mathbb{E}\left[\varphi(t-u, a + L_{t-s})\right] \quad (\text{since } L_t - L_s \overset{(law)}{=} L_{t-s}) \\ &= \varphi(s-u, L_s - L_u) \quad (\text{from the hypothesis on } \varphi).\end{aligned} \tag{3.20}$$

Gathering (3.18), (3.19) and (3.20), we finally obtain:

$$\mathbb{E}\left[M_t^{(\varphi)}|\mathscr{F}_s\right] = \int_0^s \varphi(s-u, L_s - L_u)du = M_s^{(\varphi)}.$$

We now prove that $M_t^{(\varphi)} \overset{(1.d)}{=} X_t^{(\varphi)}$. We have, for fixed $t \geq 0$:

$$\begin{aligned}M_t^{(\varphi)} &= \int_0^t \varphi(t-u, L_t - L_u)du \\ &\overset{(law)}{=} \int_0^t \varphi(t-u, L_{t-u})du \\ &\quad \text{(since, by time reversal, } (L_u, 0 \leq u \leq t) \overset{(law)}{=} (L_t - L_{(t-u)^-}, 0 \leq u \leq t) \\ &\quad \text{and } L_v = L_{v^-}\ dv \times \mathbb{P} \text{ a.s.)} \\ &= \int_0^t \varphi(s, L_s)ds \quad (\text{after the change of variable } t-u=s).\end{aligned}$$

□

Note that the above proof yields an alternative proof of Theorem 3.2.

3.5 Peacocks and Martingales Indexed by Some Predictable Processes

In the Brownian set-up of Section 3.2, we associated to some $\mathbf{F}_2$-type peacock a martingale using the time reversal method. We shall now construct at the same time a peacock with bounded variation and an associated martingale. In some way, this approach is the same as the one of Section 2.3, except that "the time reversal method now replaces the use of the Brownian sheet".

3.5.1 Notation and Definition

Let $(\Omega := \{\omega \in \mathscr{C}([0,+\infty[,\mathbb{R}); \omega(0)=0\}, (B_t, t\geq 0), (\mathscr{F}_t, t\geq 0), \mathbb{P})$ denote the canonical realisation of the standard Brownian motion started at 0, and $\mathscr{F}_\infty = \bigvee_{t\geq 0} \mathscr{F}_t$. For every $t \geq 0$, we denote:

$$\Omega^{(t)} := \{\omega \in \mathscr{C}([0,t],\mathbb{R}); \omega(0)=0\}.$$

Definition 3.1. 1) For every $t\geq 0$, let $R_t : \Omega^{(t)} \longrightarrow \Omega^{(t)}$ be the time reversal operator at time t defined by:

$$((R_t(\omega))_u, 0\leq u\leq t) := (\omega_t - \omega_{t-u}, 0\leq u\leq t). \tag{3.21}$$

2) For every $s\geq 0$, let $T_s : \Omega \longrightarrow \Omega$ be the translation operator defined by:

$$((T_s(\omega))_u, u\geq 0) := (\omega_{s+u} - \omega_s, u\geq 0). \tag{3.22}$$

It is easily shown that:

Proposition 3.2. 1) *For every* $0\leq a\leq b$, $R_a \circ R_b = T_{b-a}$ *on* $\Omega^{(b)}$
2) *For every* $a,b\geq 0$, $T_a\circ T_b = T_{a+b}$.

Definition 3.2. We denote by $\mathscr{P}_2^{(\mathrm{loc})}$ the set of predictable processes $(h(u,(B_a, a\leq u)), u\geq 0)$ such that:

$$\forall t\geq 0, \quad \mathbb{E}\left[\int_0^t h^2(u,(B_a,a\leq u))du\right] < +\infty. \tag{3.23}$$

With each $h\in \mathscr{P}_2^{(\mathrm{loc})}$, we define a martingale $(N_v^{(h)}, v\geq 0)$ by:

$$\left(N_v^{(h)} := \int_0^v h(u,(B_a,a\leq u))dB_u, v\geq 0\right). \tag{3.24}$$

Of course, for every $v\geq 0$, we have:

$$\mathbb{E}\left[N_v^{(h)}\right] = 0 \quad \text{and} \quad \mathbb{E}\left[(N_v^{(h)})^2\right] = \mathbb{E}\left[\int_0^v h^2(u,(B_a,a\leq u))du\right].$$

3.5.2 Main Result

We shall now construct two applications:

$$\begin{cases} X^{(\mathrm{P})}: & \mathscr{P}_2^{(\mathrm{loc})} \longrightarrow \begin{Bmatrix} \text{peacocks with bounded variation on} \\ \text{any compact interval} \end{Bmatrix} \\ X^{(\mathrm{M})}: & \mathscr{P}_2^{(\mathrm{loc})} \longrightarrow \{ \text{centered martingales in } L^2 \} \end{cases}$$

such that: $X_t^{(\mathrm{P})}(h) \overset{(1.d)}{=} X_t^{(\mathrm{M})}(h)$.

Theorem 3.8. *We define, for every* $h \in \mathscr{P}_2^{(loc)}$*:*

$$\left(X_t^{(P)}(h) := \int_0^t N_v^{(h)} \circ R_v(B_.)dv, t \geq 0\right) \tag{3.25}$$

$$\left(X_t^{(M)}(h) := \int_0^t N_v^{(h)} \circ R_v \circ R_t(B_.)dv, t \geq 0\right) \tag{3.26}$$

(in these formulae, $B_.$ stands for the element of $\omega \in \Omega$ defined by $(\omega_u = B_u, u \geq 0)$). Then:

1) $(X_t^{(M)}(h), t \geq 0)$ *is a* $(\mathscr{F}_t, t \geq 0)$*-martingale, centered and square-integrable. Moreover, we may write:*

$$X_t^{(M)}(h) = \int_0^t N_v^{(h)} \circ T_{t-v}(B_.)dv \quad \textit{(from Proposition 3.2)} \tag{3.27}$$

$$= \int_0^t dB_r \left(\int_0^r h(r-s, (\widehat{B}_u^{(s)}, u \leq r-s))ds\right) \tag{3.28}$$

where $\left(\widehat{B}_u^{(s)}, u \geq 0\right)$ *is the Brownian motion* $(B_{s+u} - B_s, u \geq 0)$.

2) $\left(X_t^{(P)}(h), t \geq 0\right)$ *is a peacock with bounded variation on any interval* $[0,a]$, $a > 0$.

3) $\left(X_t^{(M)}(h), t \geq 0\right)$ *is associated to* $\left(X_t^{(P)}(h), t \geq 0\right)$, *i.e.*

$$X_t^{(M)}(h) \overset{(1.d)}{=} X_t^{(P)}(h).$$

Remark 3.4. 1) This theorem is, in some sense, analogous to Theorem 2.7 where we have replaced the Brownian sheet method by the time reversal method.
2) A difference between the Brownian sheet method and the time reversal method is that the latter does not necessitate any enlargement of the initial probability space: $(X_t^{(\mathrm{P})}(h), t \geq 0)$ and $(X_t^{(\mathrm{M})}(h), t \geq 0)$ are defined on the same Wiener space.

Before proving Theorem 3.8, we start by giving a few examples.

Example 3.1 (This example shows that Theorem 3.2 is just a particular case of Theorem 3.8). Let $\varphi : \mathbb{R}_+ \times \mathbb{R} \longrightarrow \mathbb{R}$ be as in paragraph 3.2.1. From Itô's formula:

$$\varphi(s, B_s) = \int_0^s \frac{\partial \varphi}{\partial x}(u, B_u) dB_u. \tag{3.29}$$

Thus, the martingale $(\varphi(v, B_v), v \geq 0)$ is the martingale $(N_v^{(h)}, v \geq 0)$ defined by (3.24) with $h(u, (B_a, a \leq u)) = \dfrac{\partial \varphi}{\partial x}(u, B_u)$. Applying formulae (3.25) and (3.26) in this situation, we obtain:

$$X_t^{(\mathrm{P})}(h) = \int_0^t N_v^{(h)} dv = \int_0^t \varphi(v, B_v) dv$$

since $[R_v(B_\centerdot)]_v = B_v$, and from (3.28):

$$\begin{aligned}
X_t^{(\mathrm{M})}(h) &= \int_0^t dB_r \int_0^r h(r - v, (\widehat{B}_u^{(v)}, u \leq r - v)) dv \\
&= \int_0^t dB_r \int_0^r \frac{\partial \varphi}{\partial x}(r - v, B_r - B_v) dv \quad \text{(from (3.29))} \\
&= \int_0^t dv \int_v^t \frac{\partial \varphi}{\partial x}(r - v, B_r - B_v) dB_r \\
&= \int_0^t dv \varphi(t - v, B_t - B_v) \quad \text{(from Itô's formula)}.
\end{aligned}$$

Hence, in this particular case, we recover Theorem 3.2.

Example 3.2. Let $f : \mathbb{R}_+ \longrightarrow \mathbb{R}$ belong to $L^2_{\mathrm{loc}}(\mathbb{R}_+, du)$. We define:

$$N_v^{(h)} := \int_0^v f(u) dB_u.$$

In this situation, the predictable process $h(u, (B_a, a \leq u)) = (f(u), u \geq 0)$ does not depend on $(B_u, u \geq 0)$. From (3.25) and (3.28), we have:

$$X_t^{(\mathrm{P})}(h) = \int_0^t dv N_v^{(h)} \circ R_v(B_\centerdot) = \int_0^t dv \int_0^v f(v - s) dB_s = \int_0^t dB_r \int_0^{t-r} f(v) dv$$

and

$$X_t^{(\mathrm{M})}(h) = \int_0^t dB_r \int_0^r f(r - u) du = \int_0^t dB_r \int_0^r f(v) dv.$$

Example 3.3. Let $f : \mathbb{R}_+ \longrightarrow \mathbb{R}$ belong to $L^2_{\mathrm{loc}}(\mathbb{R}_+, du)$. We define:

$${}^{(f)}N_v := \exp\left(\int_0^v f(u) dB_u - \frac{1}{2} \int_0^v f^2(u) du\right) - 1$$

$({}^{(f)}N_v, v \geq 0)$ is a centered martingale which may be written, thanks to Itô's formula:

$$N_v^{(h)} = \int_0^v h(u, (B_a, a \leq u)) dB_u$$

with $h(u,(B_a,a\leq u)) = f(u)\exp\left(\int_0^u f(v)dB_v - \frac{1}{2}\int_0^u f^2(v)dv\right)$. Thus, from (3.28) and (3.25), we have:

$$\begin{aligned}
X_t^{(\mathrm{M})}(h) &= \int_0^t dB_r \int_0^r ds f(r-s)\exp\left(\int_0^{r-s} f(k)d\widehat{B}_k^{(s)} - \frac{1}{2}\int_0^{r-s} f^2(k)dk\right)\\
&= \int_0^t dB_r \int_0^r ds f(r-s)\exp\left(\int_s^r f(v-s)dB_v - \frac{1}{2}\int_s^r f^2(v-s)dv\right)
\end{aligned}$$

after the change of variable $k = v - s$, and:

$$X_t^{(\mathrm{P})}(h) = \int_0^t dv \left\{\exp\left(\int_0^v f(v-u)dB_u - \frac{1}{2}\int_0^v f^2(v-u)du\right) - 1\right\}.$$

In the particular case where f is constant, i.e. $f\equiv\lambda$, we obtain:

$$X_t^{(\mathrm{P})}(h) = \int_0^t dv\left(e^{\lambda B_v - \frac{\lambda^2}{2}v} - 1\right)$$

and

$$\begin{aligned}
X_t^{(\mathrm{M})}(h) &= \lambda\int_0^t dB_r \int_0^r ds \exp\left(\lambda(B_r - B_s) - \frac{\lambda^2}{2}(r-s)\right)\\
&= \int_0^t \left(e^{\lambda(B_t-B_s)-\frac{\lambda^2}{2}(t-s)} - 1\right)ds \quad \text{(from Itô's formula)}.
\end{aligned}$$

This is a particular case of Theorem 3.2 (or of Example 3.1) with $\varphi(s,B_s) = e^{\lambda B_s - \frac{\lambda^2}{2}s} - 1$.

Example 3.4. Let $f:\mathbb{R}\longrightarrow\mathbb{R}_+$ be a $\mathscr{C}^2$ function such that, for every x real, $f(x)>0$ and $\frac{f''(x)}{f(x)} = q(x)\geq 0$. Let:

$$N_v := f(B_v)\exp\left(-\frac{1}{2}\int_0^v q(B_u)du\right) - f(0).$$

Under suitable hypotheses, (see for instance [152]), $(N_v, v\geq 0)$ is a martingale which may be written:

$$N_v = \int_0^v h(u,(B_a,a\leq u))dB_u$$

with $h(u,(B_a,a\leq u)) = f'(B_u)\exp\left(-\frac{1}{2}\int_0^u q(B_v)dv\right)$. Thus, applying formulae (3.28) and (3.25), we obtain:

$$X_t^{(\mathrm{P})}(h) = \int_0^t N_v\circ R_v dv = \int_0^t dv\left(f(B_v)\exp\left(-\frac{1}{2}\int_0^v q(B_v - B_u)du\right) - f(0)\right)$$

(one might want to compare this peacock to the one given by Theorem 1.4, namely $\widetilde{X}_t = \int_0^t dv\left(f(B_v)\exp\left(-\frac{1}{2}\int_0^v q(B_u)du\right) - f(0)\right)$) and:

$$X_t^{(M)}(h) = \int_0^t dB_r \int_0^r ds f'(B_r - B_s)\exp\left(-\frac{1}{2}\int_s^r q(B_v - B_s)dv\right)$$
$$= \int_0^t dv\left(f(B_t - B_{t-v})\exp\left(-\frac{1}{2}\int_0^v q(B_{t-u} - B_{t-v})du\right) - f(0)\right)$$

from (3.26). In the particular case $f(x) := \frac{1}{\gamma} + |x|$ with $\gamma > 0$, then $q = 2\gamma\delta_0$ (with δ_0 the Dirac measure at 0), we obtain (although f is not $\mathscr{C}^2$):

$$X_t^{(P)}(h) = \int_0^t dv\left(\left(\frac{1}{\gamma} + |B_v|\right)e^{-\gamma L_v^{B_v}} - \frac{1}{\gamma}\right)$$

where $(L_t^a;\, t \geq 0, a \in \mathbb{R})$ denotes the bicontinuous family of local times of B, and, from (3.26):

$$X_t^{(M)}(h) = \int_0^t dv\left(\left(\frac{1}{\gamma} + |B_t - B_v|\right)e^{-\gamma(L_t^{B_v} - L_v^{B_v})} - \frac{1}{\gamma}\right).$$

Proof. Point 3) of Theorem 3.8 is clear. Indeed, from (3.25) and (3.26), $X_t^{(M)}(h) = X_t^{(P)}(h) \circ R_t$ and R_t preserves the law of $(B_u, 0 \leq u \leq t)$. Point 2) is an easy consequence of the definition (3.25) of $(X_t^{(P)}(h), t \geq 0)$, since $(N_v^{(h)} \circ R_v(B_.), v \geq 0)$ is a continuous process.

Thus, to prove Theorem 3.8, it suffices to prove (3.28). We have, from (3.27):

$$X_t^{(M)}(h) = \int_0^t N_v^{(h)} \circ T_{t-v}(B_.)dv$$
$$= \int_0^t dv \int_0^v h(s, (B_{t-v+u} - B_{t-v}, u \leq s))d_sB_{t-v+s}$$

(from the Definitions (3.22) and (3.24))

$$= \int_0^t dv \int_{t-v}^t h(r - t + v, (B_{t-v+u} - B_{t-v}, u \leq r - t + v))dB_r \tag{3.30}$$

(after the change of variable $t - v + s = r$)

$$= \int_0^t ds \int_s^t h(r - s, (\widehat{B}_u^{(s)}, u \leq r - s))dB_r$$

(after the change of variable $v = t - s$,

and with $(\widehat{B}_u^{(s)}, u \geq 0) = (B_{s+u} - B_s, u \geq 0)$)

$$= \int_0^t dB_r \int_0^r h(r - s, (\widehat{B}_u^{(s)}, u \leq r - s))ds \quad \text{(applying Fubini).}$$

This proves Theorem 3.8, since

$$\left(\widehat{h}(r, (B_a, a \leq r)) := \int_0^r h(r - s, (\widehat{B}_u^{(s)}, u \leq r - s))ds,\ r \geq 0\right)$$

is a predictable process satisfying, for every $t \geq 0$:

$$\int_0^t \mathbb{E}\left[\widehat{h}^2(r,(B_a,a \leq r))\right] dr < +\infty.$$

Indeed:

$$\begin{aligned}
\int_0^t \mathbb{E}\left[\widehat{h}^2(r,(B_a,a \leq r))\right] dr &= \int_0^t dr \mathbb{E}\left[\left(\int_0^r h(r-s,(\widehat{B}_u^{(s)},u \leq r-s))ds\right)^2\right] \\
&\leq \int_0^t rdr \int_0^r \mathbb{E}\left[h^2(r-s,(\widehat{B}_u^{(s)},u \leq r-s))\right] ds \quad (3.31) \\
&\quad \text{(from Cauchy-Schwarz's inequality)} \\
&\leq \int_0^t rdr \int_0^r \mathbb{E}\left[h^2(r-s,(B_u,u \leq r-s))\right] ds \\
&\quad \left(\text{since } (\widehat{B}_u^{(s)},u \geq 0) \overset{(law)}{=} (B_u,u \geq 0)\right) \\
&= \int_0^t rdr \int_0^r \mathbb{E}\left[h^2(v,(B_u,u \leq v))\right] dv \\
&\leq \int_0^t \mathbb{E}\left[h^2(v,(B_u,u \leq v))\right] dv \int_0^t rdr \\
&= \frac{t^2}{2} \int_0^t \mathbb{E}\left[h^2(v,(B_u,u \leq v))\right] dv < \infty \quad \text{(from (3.23))}.
\end{aligned}$$

□

3.6 The Time Reversal Method in a More General Set-up

We recall that $\mathscr{P}_2^{(\mathrm{loc})}$ denotes the set of predictable processes $(h(u,(B_a,a \leq u)),u \geq 0)$ such that:

$$\forall t \geq 0, \quad \mathbb{E}\left[\int_0^t h^2(u,(B_a,a \leq u))du\right] < \infty.$$

We denote by $\mathscr{P}_b^{(\mathrm{loc})}$ the set of predictable processes which are bounded on each interval $[0,a]$, $a > 0$, and we define, for $h \in \mathscr{P}_2^{(\mathrm{loc})}$:

$$\left(N_v^{(h)} := \int_0^v h(u,(B_a,a \leq u))dB_u,\ v \geq 0\right)$$

and, for $\theta \in \mathscr{P}_b^{(\mathrm{loc})}$:

$$\begin{aligned}
X_t^{(\mathrm{M})}(h,\theta) &:= \int_0^t \theta(t-v,(B_a,a \leq t-v))N_v^{(h)} \circ T_{t-v}(B_.)dv \quad (3.32) \\
&= \int_0^t \theta(t-v,(B_a,a \leq t-v))N_v^{(h)} \circ R_v \circ R_t(B_.)dv \quad (3.33)
\end{aligned}$$

and

$$X_t^{(\mathrm{P})}(h,\theta) := \int_0^t \left\{\theta(t-v,(B_a, a \le t-v))N_v^{(h)} \circ T_{t-v}\right\} \circ R_t(B_\cdot)dv. \tag{3.34}$$

It is clear, from its definition, that $(X_t^{(\mathrm{P})}(h,\theta), t \ge 0)$ is a process with bounded variation on any interval $[0,a]$, $a > 0$. Note that in (3.32) (resp. in (3.33)), T_{t-v} (resp. $R_v \circ R_t$) does not operate on θ. Of course, the definitions of $X_t^{(\mathrm{M})}(h,\theta)$ and $X_t^{(\mathrm{P})}(h,\theta)$ extend the definitions of $X_t^{(\mathrm{M})}(h)$ and $X_t^{(\mathrm{P})}(h)$, since, by making $\theta = 1$ in (3.32) and (3.34), we have $X_t^{(\mathrm{M})}(h,1) = X_t^{(\mathrm{M})}(h)$ and $X_t^{(\mathrm{P})}(h,1) = X_t^{(\mathrm{P})}(h)$ (since $R_t \circ R_t = Id$).
We denote by $\mathscr{M}_2^{(\mathrm{loc})}$ the set of centered martingales $(M_t, t \ge 0)$ of the form:

$$\left(M_t := \int_0^t h(u,(B_a, a \le u))dB_u,\ t \ge 0\right)$$

with $h \in \mathscr{P}_2^{(\mathrm{loc})}$.

Theorem 3.9 (Generalization of Theorem 3.8). *Let $h \in \mathscr{P}_2^{(loc)}$ and $\theta \in \mathscr{P}_b^{(loc)}$. Then:*

1) $(X_t^{(M)}(h,\theta), t \ge 0)$ is a $(\mathscr{F}_t, t \ge 0)$-martingale belonging to $\mathscr{M}_2^{(loc)}$ which may be written:

$$X_t^{(M)}(h,\theta) = \int_0^t dB_r \int_0^r h(r-s,(\widehat{B}_u^{(s)}, u \le r-s))\theta(s,(B_a, a \le s))ds \tag{3.35}$$

with $(\widehat{B}_u^{(s)}, u \ge 0) = (B_{s+u} - B_s, u \ge 0)$.

2) The set of martingales $\left\{(X_t^{(M)}(h,\theta), t \ge 0);\ h \in \mathscr{P}_2^{(loc)}, \theta \in \mathscr{P}_b^{(loc)}\right\}$ is total in $\mathscr{M}_2^{(loc)}$.

3) $(X_t^{(P)}(h,\theta), t \ge 0)$ is a peacock associated to $(X_t^{(M)}(h,\theta), t \ge 0)$, i.e.

$$(X_t^{(P)}(h,\theta), t \ge 0) \overset{(1.d)}{=} (X_t^{(M)}(h,\theta), t \ge 0).$$

Proof. Once again, Point 3) is clear since $(X_t^{(\mathrm{M})}(h,\theta), t \ge 0) = (X_t^{(\mathrm{P})}(h,\theta), t \ge 0) \circ R_t$ and R_t preserves the law of $(B_u, 0 \le u \le t)$.

We first prove (3.35). To this end, we shall take up the computations we made in the proof of Theorem 3.8 to show (3.28). (3.30) gives:

$$X_t^{(\mathrm{M})}(h,\theta) = \int_0^t dv\theta(t-v,(B_a, a \le t-v))\ldots$$
$$\int_{t-v}^t h(r-t+v,(\widehat{B}_u^{(t-v)}, u \le r-t+v))dB_r$$
$$\text{with } (\widehat{B}_u^{(t-v)}, u \ge 0) = (B_{t-v+u} - B_{t-v}, u \ge 0)$$

$$= \int_0^t ds\theta(s,(B_a,a\le s))\int_s^t h(r-s,(\widehat{B}_u^{(s)},u\le r-s))dB_r$$

(after the change of variable $t-v=s$)

$$= \int_0^t dB_r\int_0^r \theta(s,(B_a,a\le s))h(r-s,(\widehat{B}_u^{(s)},u\le r-s))ds \qquad (3.36)$$

from Fubini, and since $\theta(s,(B_a,a\le s)$ is $\mathscr{F}_s$-measurable whereas $\int_s^t h(r-s,(\widehat{B}_u^{(s)},u\le r-s))dB_r$ is $\sigma(B_v-B_s,s\le v\le t)$-measurable.

We now prove that $(X_t^{(M)}(h,\theta),t\ge 0)\in \mathscr{M}_2^{(loc)}$.

$$\begin{aligned}
&\mathbb{E}\left[\left(X_t^{(\mathrm{M})}(h,\theta)\right)^2\right]\\
&= \int_0^t dr\mathbb{E}\left[\left(\int_0^r h(r-s,(\widehat{B}_u^{(s)},u\le r-s))\theta(s,(B_a,a\le s))ds\right)^2\right]\\
&\le \int_0^t rdr\mathbb{E}\left[\int_0^r h^2(r-s,(\widehat{B}_u^{(s)},u\le r-s))\theta^2(s,(B_a,a\le s))ds\right]\\
&\le C\int_0^t rdr\mathbb{E}\left[\int_0^r h^2(r-s,(\widehat{B}_u^{(s)},u\le r-s))ds\right]
\end{aligned}$$

since θ is a bounded process, and we are back to (3.31). Since $h\in\mathscr{P}_2^{(\mathrm{loc})}$, this implies that $\mathbb{E}\left[\left(X_t^{(\mathrm{M})}(h,\theta)\right)^2\right]<\infty$, i.e. $(X_t^{(\mathrm{M})}(h,\theta),t\ge 0)\in\mathscr{M}_2^{(\mathrm{loc})}$.

We now prove Point 2) of Theorem 3.9. Let us define for $h\in\mathscr{P}_2^{(\mathrm{loc})}$, $\theta\in\mathscr{P}_b^{(\mathrm{loc})}$ and $A>0$, the predictable process $(\widehat{h}^{(\theta)}(u,(B_a,a\le u)),u\ge 0)$ by:

$$\widehat{h}^{(\theta)}(u,(B_a,a\le u)) = 1_{\{u\le A\}}\int_0^u \theta(s,(B_a,a\le s))h(u-s,(\widehat{B}_a^{(s)},a\le u-s))ds.$$

From (3.36), to obtain Point 3) of Theorem 3.9, it is sufficient to prove that, for $\nu\in\mathscr{P}_2^{(\mathrm{loc})}$, the relation: $\forall h\in\mathscr{P}_2^{(\mathrm{loc})},\forall\theta\in\mathscr{P}_b^{(\mathrm{loc})},\forall A>0$,

$$\mathbb{E}\left[\int_0^A \nu(u,(B_a,a\le u))\widehat{h}^{(\theta)}(u,(B_a,a\le u))du\right]=0 \qquad (3.37)$$

implies $\nu\equiv 0$ $ds\times\mathbb{P}$ a.e. We now write (3.37) by replacing $\theta(s,(B_a,a\le s))$ by $f(s)\theta(s,(B_a,a\le s))$ and $h(s,(B_a,a\le s))$ by $g(s)h(s,(B_a,a\le s))$ where f and g are two Borel bounded deterministic functions. We obtain:

$$\begin{aligned}
0=\mathbb{E}\Bigg[\int_0^A & du\nu(u,(B_a,a\le u))\\
&\int_0^u dsf(s)\theta(s,(B_a,a\le s))g(u-s)h(u-s,(\widehat{B}_a^{(s)},a\le u-s))\Bigg].
\end{aligned}$$

Since this equality is satisfied for every f, g and A, this implies:

$$\mathbb{E}\left[\nu(u,(B_a,a\le u))\theta(s,(B_a,a\le s))h(u-s,(\widehat{B}_a^{(s)},a\le u-s))\right]=0\;duds\,\text{a.e.} \quad (3.38)$$

Let u fixed such that (3.38) is true a.e., with $s\le u$. The monotone class theorem implies that:

$$\text{for every positive r.v. } \zeta_u,\ \mathscr{F}_u\text{-measurable,} \quad \mathbb{E}\left[\nu(u,(B_a,a\le u))\zeta_u\right]=0. \quad (3.39)$$

Note that, in (3.38), $\theta(s,(B_a,a\le s))$ is $\mathscr{F}_s$-measurable whereas $h(u-s,(\widehat{B}_a^{(s)},a\le u-s))$ is $\mathscr{F}_{s,u}$-measurable with $\mathscr{F}_{s,u}=\sigma(B_r-B_s,s\le r\le u)$. Of course, since $\nu(u,(B_a,a\le u))$ is $\mathscr{F}_u$-mesurable, (3.39) implies $\nu(u,(B_a,a\le u))=0$ $\mathbb{P}$ a.s., i.e. $\nu(u,(B_a,a\le u))=0$ $du\times\mathbb{P}$ a.e. This is Point 2) of Theorem 3.9. □

Notes and Comments

The results of Section 3.3 are inspired from [179], [178] and [113]. Other results in this chapter are new.

4

The Time Inversion Method

Abstract Denote by $(\Lambda_t, t \geq 0)$ an integrable Lévy process, i.e. for any $t \geq 0$, $\mathbb{E}[|\Lambda_t|] < \infty$. Then, $(t\Lambda_{(1/t)}, t > 0)$ is a martingale in its natural filtration. Martingales of this type appear as being naturally associated to $\mathbf{F}_1$-type peacocks or peacocks defined from squared Bessel processes of dimension 0, or, more generally stable CSBP with index $\gamma \in]1,2]$. We then generalize the preceding results of this chapter in Theorem 4.5, through a more abstract approach. Finally, we give examples of applications of that theorem.

4.1 Martingale Constructed from an Integrable Lévy Process by Time Inversion

We start by recalling some classical results which we shall use in the sequel.

Proposition 4.1. *Let $(\Lambda_t, t \geq 0)$ be an integrable Lévy process. Then, the process*

$$\widetilde{\Lambda}_t := \begin{cases} t\Lambda_{(1/t)} & \text{for } t > 0, \\ \mathbb{E}[\Lambda_1] & \text{for } t = 0, \end{cases} \tag{4.1}$$

is a martingale in its natural filtration.

Proof. See Exercise 4.1. □

Proposition 4.2. *Let $(Z_s, s \geq 0)$ be a "good" Markov process, taking real values, recurrent, started from 0, and admitting a bicontinuous family of local times.*
1) *Let $(\tau_l, l \geq 0)$ denote the right-continuous inverse of its local time at 0:*

$$\tau_l := \inf\{s \geq 0;\ L_s^0 > l\}.$$

Then, $(\tau_l, l \geq 0)$ is a subordinator.
2) *Let $\varphi : \mathbb{R} \longrightarrow \mathbb{R}_+$ be a Borel function, locally bounded. Then:*

$$\left(\Lambda_l^{(\varphi)} := \int_0^{\tau_l} \varphi(Z_s)ds, l \geq 0\right) \text{ is a subordinator.}$$

Hirsch F., Profeta C., Roynette B., Yor M.:
Peacocks and Associated Martingales, with Explicit Constructions

This proposition is classical; Point 2 follows from the strong Markov property and the fact that $Z_{\tau_l} = 0$.

Exercise 4.1 (($t\Lambda_{(1/t)}, t \geq 0$) is a martingale in its natural filtration). 1) Let $X_1, X_2, \dots, X_n, \dots$ denote a sequence of i.i.d. integrable r.v's. Define:

$$S_0 = 0, \qquad S_n = \sum_{i=1}^{n} X_i \qquad (n \geq 1).$$

i) Prove that, for every $1 \leq i \leq n$, $\mathbb{E}[X_i | S_n] = \frac{S_n}{n}$.

ii) Let $\mathscr{F}^+_{n+1} := \sigma(S_{n+1}, S_{n+2}, \dots)$. Prove that, for every $n \geq 1$:

$$\mathbb{E}\left[\frac{S_n}{n} | \mathscr{F}^+_{n+1}\right] = \frac{S_{n+1}}{n+1}.$$

2) Let $(\Lambda_t, t \geq 0)$ be an integrable Lévy process.

i) Prove that $\mathbb{E}[\Lambda_t] = t\mathbb{E}[\Lambda_1]$.

ii) Let $\mathscr{F}^+_t := \sigma(\Lambda_s; s \geq t)$. Prove that $\left(\frac{1}{t}\Lambda_t, t \geq 0\right)$ is a $\mathscr{F}^+_t$-inverse martingale, i.e, for every $s \leq t$:

$$\mathbb{E}\left[\frac{\Lambda_s}{s} | \mathscr{F}^+_t\right] = \frac{\Lambda_t}{t}.$$

(*Hint*: use discretization and 1) (see [90]).)

iii) Let $(\widetilde{\Lambda}_t, t \geq 0)$ be defined by (4.1). Prove that $(\widetilde{\Lambda}_t, t \geq 0)$ is a martingale with respect to the filtration $\left(\mathscr{G}_t := \sigma\left\{\Lambda_u, u \geq \frac{1}{t}\right\}, t > 0\right)$, which is the natural filtration of $(\widetilde{\Lambda}_t, t \geq 0)$.

(Note that Λ being a càdlàg process, the martingale $\widetilde{\Lambda}$ as defined by (4.1) is left-continuous (and right-limited). To get a right-continuous process, one should take $\widetilde{\Lambda}_t := t\Lambda_{\left(\frac{1}{t}\right)^-}$.)

iv) Identify $(\widetilde{\Lambda}_t, t \geq 0)$ when Λ_t is a Brownian motion with drift.

3) *An alternative proof by Fourier transform.* Let $(\Lambda_t, t \geq 0)$ be an integrable Lévy process and let ψ its characteristic exponent:

$$\mathbb{E}\left[\exp(i\lambda\Lambda_t)\right] = \exp(-t\psi(\lambda)).$$

i) Prove that ψ is differentiable and, for every $t \geq 0$ and $\lambda \in \mathbb{R}$,

$$\mathbb{E}\left[\Lambda_t \exp(i\lambda\Lambda_t)\right] = it\psi'(\lambda)\exp(-t\psi(\lambda)).$$

ii) Let $0 < s < t$. Show that, for every $\lambda \in \mathbb{R}$:

$$\mathbb{E}\left[\frac{1}{s}\Lambda_s \exp(i\lambda\Lambda_t)\right] = \mathbb{E}\left[\frac{1}{t}\Lambda_t \exp(i\lambda\Lambda_t)\right].$$

iii) Deduce finally that, for $0 < s < t$:

$$\mathbb{E}\left[\frac{\Lambda_s}{s} | \mathscr{F}^+_t\right] = \frac{\Lambda_t}{t}.$$

4.2 Feller's Process and $\mathbf{F}_1$-type Peacocks

4.2.1 Feller's Process and Second Ray-Knight Theorem

Let $(Q_s^{(r,0)}, s \geq 0)$ denote a squared Bessel process with dimension 0, starting from $r > 0$. This process, which is also called Feller's diffusion, is the unique solution of the SDE:

$$Y_t = r + 2\int_0^t \sqrt{Y_s} dB_s$$

where $(B_s, s \geq 0)$ is a Brownian motion. Thus, $(Q_s^{(r,0)}, s \geq 0)$ is a martingale, and for all $\alpha \geq 0$, from Theorem 1.4:

$$\left(X_t^{(\alpha,r)} := \frac{1}{t^{\alpha+1}} \int_0^t Q_s^{(r,0)} s^\alpha ds, t \geq 0\right) \tag{4.2}$$

is a $\mathbf{F}_1$-type peacock. The aim of this section is to find an associated martingale to $(X_t^{(\alpha,r)}, t \geq 0)$. For this purpose, we recall the second Ray-Knight theorem for Brownian local times.

Theorem 4.1 (Knight [109], Ray [149]). *Let $(B_s, s \geq 0)$ be a Brownian motion started from 0 and denote $(L_t^a;\ t \geq 0, a \in \mathbb{R})$ the bicontinuous family of its local times. Let $\left(\tau_l(B) := \inf\{u \geq 0; L_u^0 > l\}, l \geq 0\right)$ the right-continuous inverse of $(L_u^0, u \geq 0)$. Then for all $r > 0$:*

$$\left(L^x_{\tau_r(B)}, x \geq 0\right) \overset{(law)}{=} \left(Q_x^{(r,0)}, x \geq 0\right).$$

Let us define the process $\left(\Lambda_l^{(\alpha)}, l \geq 0\right)$ by:

$$\left(\Lambda_l^{(\alpha)} := \int_0^{\tau_l(B)} (B_v)^\alpha 1_{[0,1]}(B_v) dv, l \geq 0\right).$$

From Proposition 4.2 (applied with $(Z_s = B_s, s \geq 0)$ and $\varphi(x) = x^\alpha 1_{[0,1]}(x)$), the process $\left(\Lambda_l^{(\alpha)}, l \geq 0\right)$ is a subordinator (the Laplace exponent of which we shall compute in Subsection 4.2.3 below). Hence, from Proposition 4.1, the process

$$\left(M_t^{(\alpha,r)} := t\Lambda_{\frac{r}{t}}^{(\alpha)}, t \geq 0\right)$$

is a martingale.

4.2.2 A Martingale Associated to Some Integrals of Feller's Process

Theorem 4.2. *The martingale $(M_t^{(\alpha,r)},\ t \geq 0)$ is associated to the peacock $(X_t^{(\alpha,r)}, t \geq 0)$, defined by (4.2).*

Proof. We have:

$$
\begin{aligned}
(X_t^{(\alpha,r)}, t \geq 0) &= \left(\frac{1}{t^{\alpha+1}} \int_0^t Q_s^{(r,0)} s^\alpha ds, t \geq 0\right) \\
&\overset{(law)}{=} \left(\frac{1}{t^{\alpha+1}} \int_0^t L^s_{\tau_r(B)} s^\alpha ds, t \geq 0\right) \\
&\quad \text{(from Ray-Knight's Theorem)} \\
&= \left(\frac{1}{t^{\alpha+1}} \int_0^{\tau_r(B)} (B_u)^\alpha 1_{[0,t]}(B_u) du, t \geq 0\right) \qquad (4.3) \\
&\quad \text{(from the density of occupation times formula).}
\end{aligned}
$$

From the scaling property of Brownian motion, denoting $(B_u^{(t)}, u \geq 0) := (tB_{\frac{u}{t^2}}, u \geq 0)$, there is the identity $(B_u^{(t)}, u \geq 0) \overset{(law)}{=} (B_u, u \geq 0)$. Replacing in (4.3), we obtain, this time for fixed t:

$$
\begin{aligned}
X_t^{(\alpha,r)} &\overset{(1.d)}{=} \frac{1}{t^{\alpha+1}} \int_0^{\tau_r(B^{(t)})} (B_u^{(t)})^\alpha 1_{[0,t]}(B_u^{(t)}) du \\
&= \frac{1}{t} \int_0^{\tau_r(B^{(t)})} (B_{\frac{u}{t^2}})^\alpha 1_{[0,1]}(B_{\frac{u}{t^2}}) du \\
&= t \int_0^{\frac{\tau_r(B^{(t)})}{t^2}} (B_v)^\alpha 1_{[0,1]}(B_v) dv \\
&\quad \text{(after the change of variable } \frac{u}{t^2} = v) \\
&= t \int_0^{\tau_{\frac{r}{t}}(B)} (B_v)^\alpha 1_{[0,1]}(B_v) dv \\
&\quad \text{(since we easily deduce from the density of occupation times} \\
&\quad \text{formula that } \tau_r(B^{(t)}) = t^2 \tau_{\frac{r}{t}}(B)) \\
&= M_t^{(\alpha,r)}.
\end{aligned}
$$

□

4.2.3 Explicit Computations of Bernstein Functions

As a complement to Theorem 4.2, we shall compute explicitely the Bernstein function associated to $(\Lambda_t^{(\alpha)}, t \geq 0)$, i.e. we shall compute $\phi^{(\alpha)}$ such that, for all $\lambda \geq 0$ and $t \geq 0$:

$$\mathbb{E}\left[e^{-\lambda \Lambda_t^{(\alpha)}}\right] = \exp\left(-t\phi^{(\alpha)}(\lambda)\right).$$

Proposition 4.3. 1) *For all $\alpha \geq 0$, we have:*

$$\phi^{(\alpha)}(\lambda) = \frac{c_\theta}{2}(2\lambda)^\theta \frac{I_{1-\theta}}{I_{\theta-1}}\left(2\theta\sqrt{2\lambda}\right) \tag{4.4}$$

where I_γ denotes the modified Bessel function of index γ (c.f. [117, p. 108]) with $\theta = \dfrac{1}{2+\alpha}$ *and* $c_\theta = \dfrac{\pi\theta^{2\theta}}{\theta((\Gamma(\theta))^2\sin(\pi\theta))}$.
2) *For $\alpha = 0$ the previous formula simplifies:*

$$\phi^{(0)}(\lambda) = \frac{1}{2}\sqrt{2\lambda}\tanh(\sqrt{2\lambda}). \tag{4.5}$$

Proof. We have, from Theorem 4.2:

$$\mathbb{E}\left[e^{-\lambda t \Lambda_{1/t}^{(\alpha)}}\right] = \mathbb{E}\left[\exp\left(-\frac{\lambda}{t^{1+\alpha}}\int_0^t Q_s^{(1,0)} s^\alpha ds\right)\right].$$

Hence, since for every $t \geq 0$, by scaling:

$$\left(Q_s^{(1,0)}, s \geq 0\right) = \left(t Q_{s/t}^{(1/t,0)}, s \geq 0\right),$$

we obtain, for $r = 1$,

$$\begin{aligned}\mathbb{E}\left[e^{-\lambda t \Lambda_{1/t}^{(\alpha)}}\right] &= \mathbb{E}\left[\exp\left(-\frac{\lambda}{t^{1+\alpha}}\int_0^t t Q_{s/t}^{(1/t,0)} s^\alpha ds\right)\right] \\ &= \mathbb{E}\left[\exp\left(-\lambda t\int_0^1 Q_u^{(1/t,0)} u^\alpha du\right)\right].\end{aligned}$$

The latter expression has already been computed by J. Pitman and M. Yor [145](see also [69] and [110]). This leads to:

$$\mathbb{E}\left[e^{-\lambda t \Lambda_{1/t}^{(\alpha)}}\right] = \exp\left(-\frac{c_\theta}{2t}(2\lambda t)^\theta \frac{I_{1-\theta}}{I_{\theta-1}}\left(2\theta\sqrt{2\lambda t}\right)\right)$$

which replacing λt by λ gives:

$$\mathbb{E}\left[e^{-\lambda \Lambda_{1/t}^{(\alpha)}}\right] = \exp\left(-\frac{c_\theta}{2t}(2\lambda)^\theta \frac{I_{1-\theta}}{I_{\theta-1}}\left(2\theta\sqrt{2\lambda}\right)\right).$$

This is the announced result upon replacing $\frac{1}{t}$ by t. Finally, to obtain (4.5), we shall take $\alpha = 0$ in (4.4), using the fact that $c_{\frac{1}{2}} = 1$, $I_{1/2}(z) = \sqrt{\dfrac{2}{\pi z}}\sinh(z)$ and $I_{-1/2}(z) = \sqrt{\dfrac{2}{\pi z}}\cosh(z)$ (see [117, p. 112]), or more simply, using the Cameron-Martin type formula (see [150, Chapter XI]):

$$\mathbb{E}\left[\exp\left(-\lambda\int_0^1 Q_s^{(r,0)} ds\right)\right] = \exp\left(-\frac{r}{2}\sqrt{2\lambda}\tanh\sqrt{2\lambda}\right).$$

□

Remark 4.1. Kellerer's Theorem 1.1 asserts the existence of a martingale associated to the peacock $(X_t^{(\alpha,r)}, t \geq 0)$. In fact, as indicated in the comments following this theorem, this martingale may be chosen Markovian. We now verify that this is indeed the case here.

Proposition 4.4. *Let $(\Lambda_t, t \geq 0)$ be an integrable Lévy process. Then, the martingale $(\widetilde{\Lambda}_t := t\Lambda_{(1/t)}, t \geq 0)$ is an (inhomogeneous) Markov process.*

Proof. We denote, for $u \geq 0$, $\mathscr{F}_u^+ = \sigma(\Lambda_b, b \geq u)$. Since $(\Lambda_a, a \geq 0)$ is a Markov process, for every $v < u$, there exists a Borel kernel $Q_{v,u}$ such that, for every Borel bounded function f:

$$\mathbb{E}[f(\Lambda_v)|\mathscr{F}_u^+] = \mathbb{E}[f(\Lambda_v)|\Lambda_u] = Q_{v,u}f(\Lambda_u)$$

(with $Q_{v,u}f(x) = \int Q_{v,u}(x,dy)f(y)$). Now, let $s < t$ and $\widetilde{\mathscr{F}}_s := \sigma(\widetilde{\Lambda}_a, a < s) = \mathscr{F}_{1/s}^+$. We have then:

$$\begin{aligned}\mathbb{E}\left[f(\widetilde{\Lambda}_t)|\widetilde{\mathscr{F}}_s\right] &= \mathbb{E}\left[f(t\Lambda_{(1/t)})|\Lambda_{(1/s)}\right] \\ &= \int Q_{\frac{1}{t},\frac{1}{s}}(\Lambda_{(1/s)}, dy) f(ty) \\ &= \left[Q_{\frac{1}{t},\frac{1}{s}} f(t\,\bullet)\right]\left(\frac{\widetilde{\Lambda}_s}{s}\right)\end{aligned}$$

which proves Proposition 4.4. □

Exercise 4.2 (Explicit computation of the infinitesimal generator of $(t\Lambda_{(1/t)}, t > 0)$). Let $(\Lambda_u, u \geq 0)$ be an integrable subordinator with Lévy measure ν. We assume that Λ_t admits a density $p(t,\bullet)$ with respect to Lebesgue measure, and that p is smooth and strictly positive on $]0,+\infty[\times]0,+\infty[$. We denote $(\widetilde{\Lambda}_t := t\Lambda_{(1/t)}, t \geq 0)$.

1) *i*) Prove that the conditional density of $\Lambda_{(1/t)}$ with respect to $s\Lambda_{(1/s)} = y$ $(s < t)$ is given by:

$$\frac{p\left(\frac{1}{t},x\right)p\left(\frac{1}{s}-\frac{1}{t},\frac{y}{s}-x\right)}{p\left(\frac{1}{s},\frac{y}{s}\right)} 1_{\{0 \leq x \leq \frac{y}{s}\}}.$$

ii) Deduce that, for f a bounded Borel function:

$$\begin{aligned}\mathbb{E}\left[f(\widetilde{\Lambda}_t)|\widetilde{\Lambda}_s = y\right] &= \mathbb{E}\left[f(t\Lambda_{(1/t)})|s\Lambda_{\frac{1}{s}} = y\right] \\ &= \frac{1}{p\left(\frac{1}{s},\frac{y}{s}\right)} \int_0^{\frac{y}{s}} f(tx)p\left(\frac{1}{t},x\right)p\left(\frac{1}{s}-\frac{1}{t},\frac{y}{s}-x\right)dx. \qquad (4.6)\end{aligned}$$

2) *i*) We now assume that f is smooth. Prove that the derivative with respect to t of the RHS of (4.6) equals:

$$\frac{\partial}{\partial t}\left\{\frac{1}{p\left(\frac{1}{s},\frac{y}{s}\right)} \int_0^{\frac{y}{s}} (f(tx)-f(y))p\left(\frac{1}{t},x\right)p\left(\frac{1}{s}-\frac{1}{t},\frac{y}{s}-x\right)dx\right\}.$$

ii) Deduce that the derivative with respect to t of (4.6) equals, at $t = s$:

$$\frac{y}{s} f'(y) + \frac{1}{s^2 p\left(\frac{1}{s}, \frac{y}{s}\right)} \widehat{L}\left((f(s\centerdot) - f(y))\, p\left(\frac{1}{s}, \centerdot\right)\right)\left(\frac{y}{s}\right)$$

where $\widehat{L}$ is defined by

$$\widehat{L}g(y) = \lim_{t\to 0} \frac{1}{t} \mathbb{E}\left[(g(y - \Lambda_t) - g(y))\, 1_{\{\Lambda_t < y\}}\right].$$

3) *i*) Prove that

$$\widehat{L}g(y) = \int_0^y (g(y-z) - g(y))\, \nu(dz).$$

(*Hint*: use the fact that, on $]0, +\infty[$, $\lim\limits_{t\to 0} \frac{1}{t} p(t,z)dz = \nu(dz)$, i.e. the family of measures $(\frac{1}{t} p(t,z)dz, t > 0)$ converges weakly on $]0, +\infty[$ towards ν when $t \to 0$.)

ii) Let $\widetilde{L} = (\widetilde{L}_s, s > 0)$ be the infinitesimal generator of the Markov process $(\widetilde{\Lambda}_t := t\Lambda_{\frac{1}{t}}, t \geq 0)$. Prove that, for f regular:

$$\widetilde{L}f(s,y) = \frac{yf'(y)}{s} + \frac{1}{s^2 p\left(\frac{1}{s}, \frac{y}{s}\right)} \int_0^{\frac{y}{s}} (f(y - sx) - f(y))\, p\left(\frac{1}{s}, \frac{y}{s} - x\right) \nu(dx).$$

4) Our aim is now to show that, denoting $f_*(y) := y$

$$\widetilde{L}f_*(s,y) = 0. \tag{4.7}$$

i) Prove that $(\Lambda_t, t \geq 0)$ is integrable if and only if $\int_0^{+\infty} x\nu(dx) < \infty$.

ii) Prove that (4.7) is equivalent to: for every $t, z \geq 0$,

$$t\int_0^z p(t, z-x)x\nu(dx) = zp(t,z). \tag{4.8}$$

iii) Prove that (4.8) is equivalent to: for every $t, \lambda \geq 0$,

$$\left(t\int_0^{+\infty} xe^{-\lambda x}\nu(dx)\right) \mathbb{E}\left[e^{-\lambda\Lambda_t}\right] = \mathbb{E}\left[\Lambda_t e^{-\lambda\Lambda_t}\right]$$

and deduce that (4.7) holds.

iv) Show that the above results yield a new proof that $(\widetilde{\Lambda}_t := t\Lambda_{\frac{1}{t}}, t \geq 0)$ is a martingale.

5) When $(\Lambda_t, t \geq 0)$ is the standard gamma process, i.e. $p(t,x) = \dfrac{1}{\Gamma(t)} e^{-x} x^{t-1} 1_{\{x \geq 0\}}$ and $\nu(dx) = \dfrac{e^{-x}}{x} dx$, prove that:

$$\widetilde{L}f(s,y) = \frac{yf'(y)}{s} + \frac{1}{s^2}\int_0^1 (f(yz) - f(y))\, \frac{z^{\frac{1}{s}-1}}{1-z} dz.$$

We refer to [148, Problem 2.1, p. 60–63] for more detailed computations when $(\Lambda_t, t \geq 0)$ is a gamma process.

4.3 Continuous Space Branching Processes (CSBP) and F_1-type Peacocks

We now extend to continuous space branching processes (CSBP), which are stable with index $\gamma \in]1,2]$, what we have done in Section 4.2 with respect to the 0-dimensional squared Bessel process. To this end, we need to describe such a process and its realization with the help of a Ray-Knight type theorem. This is the purpose of the following subsection, in which we use the results and the notation from Le Gall [114] and Duquesne-Le Gall [61].

4.3.1 Stable CSBP with Index γ and Ray-Knight Theorems

A process $(Q_t, t \geq 0)$ is said to be a stable CSBP of index $\gamma \in]1,2]$ if it is a Feller Markov process taking values in $\mathbb{R}_+$, such that its transition semi-group $(\mathbf{Q}_t, t \geq 0)$ satisfies the branching property:

$$\mathbf{Q}_t(x, \cdot) * \mathbf{Q}_t(x', \cdot) = \mathbf{Q}_t(x + x', \cdot) \qquad (x, x' \geq 0).$$

Consequently, there is the formula:

$$\int_0^{+\infty} e^{-\lambda y} \mathbf{Q}_t(x, dy) = \mathbb{E}_x \left[e^{-\lambda Q_t} \right] = \exp(-x u_t(\lambda)) \tag{4.9}$$

where, in the case of the stable CSBP with index γ, $(u_t(\lambda), t \geq 0, \lambda \geq 0)$ is the unique solution of:

$$\begin{cases} \dfrac{\partial u_t}{\partial t}(\lambda) & = -c(u_t(\lambda))^{\gamma} \quad \text{(where } c \text{ is a positive parameter)} \\ u_0(\lambda) & = \lambda. \end{cases} \tag{4.10}$$

Of course, (4.10) may be solved explicitly and we obtain:

$$u_t(\lambda) = \frac{\lambda}{(1 + ct\varepsilon\lambda^{\varepsilon})^{1/\varepsilon}}, \qquad \text{where } \varepsilon = \gamma - 1.$$

Under these conditions, for every $x \geq 0$, the process $(Q_t, t \geq 0)$ is a martingale with respect to the probability $\mathbb{P}_x$ such that $\mathbb{P}_x(Q_0 = x) = 1$. Indeed, since $(Q_t, t \geq 0)$ is Markovian, it is sufficient to see that, for every $t \geq 0$ and $x, y \geq 0$:

$$\int_0^{+\infty} \mathbf{Q}_t(x, dy) y = x.$$

Now, differentiating (4.9), we obtain:

$$\frac{\partial}{\partial \lambda} \int_0^{+\infty} e^{-\lambda y} \mathbf{Q}_t(x, dy)_{|\lambda=0} = -\int_0^{+\infty} \mathbf{Q}_t(x, dy) y = -x \frac{\partial u_t}{\partial \lambda}(\lambda)_{|\lambda=0}.$$

On the other hand

$$\frac{\partial u_t}{\partial \lambda}(\lambda)_{|\lambda=0} = \lim_{\lambda \to 0} \frac{u_t(\lambda)}{\lambda} = \lim_{\lambda \to 0} \frac{1}{(1+ct\varepsilon\lambda^{\varepsilon})^{1/\varepsilon}} = 1.$$

In terms of branching processes, the fact that $(Q_t, t \geq 0)$ is a martingale is equivalent to the property that we are working here in the "critical case" framework.

As for the squared Bessel process of dimension 0, the stable CSBP of index γ can be realized with the help of the local times of another process: this is an extension of Ray-Knight's Theorem, due to J.F. Le Gall [114], which we now recall:

a) Let $(Y_t, t \geq 0)$ be the Lévy process with characteristic exponent $\psi(\lambda) = c\lambda^{\gamma}$ ($c > 0, \gamma \in]1,2]$), i.e. such that:

$$\mathbb{E}\left[e^{-\lambda Y_t}\right] = \exp(-ct\lambda^{\gamma}) \quad (t, \lambda \geq 0).$$

This process is spectrally positive, i.e. its Lévy measure does not charge $\mathbb{R}_-$, or equivalently, it has only positive jumps.

b) Let $S_t := \sup_{s \leq t} Y_s$ and $I_t := \inf_{s \leq t} Y_s$. Both processes $(S_t - Y_t, t \geq 0)$ and $(Y_t - I_t, t \geq 0)$ are Markovian. Since 0 is regular for either process, they both admit a local time at 0 (see [37] or [32, Chapter IV]).

c) Let, for $t > 0$ fixed, $(\widehat{Y}_s^{(t)}, 0 \leq s \leq t)$ be the time reversal process at time t of the process $(Y_s, 0 \leq s \leq t)$:

$$(\widehat{Y}_s^{(t)}, 0 \leq s \leq t) := (Y_t - Y_{(t-s)^-}, 0 \leq s \leq t)$$

with, by convention, $Y_{0^-} = 0$. Since the process $(\widehat{Y}_s^{(t)}, 0 \leq s \leq t)$ is distributed as $(Y_s, 0 \leq s \leq t)$, the process $(\widehat{S}_u^{(t)} - \widehat{Y}_u^{(t)}, 0 \leq u \leq t)$ also admits a local time at 0, which we denote $(\widehat{L}_u^0, 0 \leq u \leq t)$ (For clarity, we do not use further superscript t).

Definition 4.1. The process $(H_t := \widehat{L}_t^0, t \geq 0)$ is called the height process associated to the Lévy process $(Y_u, u \geq 0)$.

Here is the statement of Le Gall's extension of the second Ray-Knight theorem.

Theorem 4.3 (Le Gall [114]). 1) *The process $(H_t, t \geq 0)$ admits a family of local times $(L_t^a(H), t \geq 0, a \geq 0)$ such that, for every Borel function $h : \mathbb{R}_+ \longrightarrow \mathbb{R}_+$, the density of occupation formula*

$$\int_0^t h(H_s)ds = \int_0^{+\infty} h(a) L_t^a(H) da \quad a.s. \tag{4.11}$$

holds.

2) *For all $r \geq 0$, there is the identity:*

$$\begin{aligned}\tau_r(H) &:= \inf\{s \geq 0; L_s^0(H) > r\}\\ &= \inf\{s \geq 0; Y_s < -r\} = \inf\{s \geq 0; I_s < -r\}.\end{aligned}$$

3) *For fixed $r > 0$, the process $(L_{\tau_r}^a(H), a \geq 0)$ is a stable CSBP, with index γ, started from r.*

Remark 4.2. When $\gamma = 2$ (and $c = 1$), the process $(Y_t, t \geq 0)$ is a standard Brownian motion. In this situation, it is not difficult to see that $H_t = S_t - Y_t$, $t \geq 0$, or from Lévy's equivalence theorem:

$$(H_t, t \geq 0) \overset{(law)}{=} (|B_t|, t \geq 0),$$

where $(B_t, t \geq 0)$ is a Brownian motion started from 0.The second Ray-Knight theorem asserts that

$$(L^a_{\tau_r(H)}, a \geq 0) \overset{(law)}{=} (L^a_{\tau_r}(|B|), a \geq 0) \overset{(law)}{=} (Q_a^{(r,0)}, a \geq 0).$$

Thus, the classical Ray-Knight theorem corresponds to the particular case $\gamma = 2$ (and $c = 1$) in Theorem 4.3.

4.3.2 A Martingale Associated to the Peacock $(X_t^{(\gamma,\alpha,r)}, t \geq 0)$

We denote by $(Q_t^{(\gamma,r)}, t \geq 0)$ the stable CSBP of index γ started from r. Since, from Subsection 4.3.1, this process is a martingale, then for all $\alpha \geq 0$, the process:

$$\left(X_t^{(\gamma,\alpha,r)} := \frac{1}{t^{1+\alpha}} \int_0^t s^\alpha Q_s^{(\gamma,r)} ds, t \geq 0\right)$$

is a $\mathbf{F}_1$-type peacock, as Theorem 1.4 asserts. The aim of this section is to find a martingale associated to this peacock.

Theorem 4.4. 1) *For every $\gamma \in]1,2]$ and $\alpha \geq 0$, the process:*

$$\left(\Lambda_l^{(\gamma,\alpha)} := \int_0^{\tau_l(H)} (H_v)^\alpha 1_{[0,1]}(H_v) dv, l \geq 0\right)$$

is an integrable subordinator. Thus, from Proposition 4.1, the process, with $\varepsilon = \gamma - 1$:

$$\left(t^\varepsilon \int_0^{\tau_{\frac{r}{t^\varepsilon}}(H)} (H_v)^\alpha 1_{[0,1]}(H_v) dv = t^\varepsilon \Lambda_{\frac{r}{t^\varepsilon}}^{(\gamma,\alpha)}, t \geq 0\right)$$

is a martingale.
2) *This martingale is associated to the peacock $(X_t^{(\gamma,\alpha,r)}, t \geq 0)$, i.e.*

$$X_t^{(\gamma,\alpha,r)} \overset{(1.d)}{=} t^\varepsilon \Lambda_{\frac{r}{t^\varepsilon}}^{(\gamma,\alpha)}. \tag{4.12}$$

Proof. 1) *Scaling property*. Since the Lévy process $(Y_u, u \geq 0)$ satisfies:

$$\mathbb{E}\left[e^{-\lambda Y_u}\right] = e^{-uc\lambda^\gamma}$$

it enjoys the following scaling property:

$$\text{for any } t > 0, \quad (tY_{\frac{u}{t^\gamma}}, u \geq 0) \overset{(law)}{=} (Y_u, u \geq 0).$$

The height process $(H_t, t \geq 0)$ associated to Y inherits this scaling property:

$$\text{for any } t > 0, \quad (tH_{\frac{u}{t^\gamma}}, u \geq 0) \overset{(law)}{=} (H_u, u \geq 0),$$

and, from the density of occupation formula (4.11), we easily deduce that, for any $t > 0$:

$$\left(L^{at}_{\tau_r}(H), a \geq 0\right) \overset{(law)}{=} \left(t^{\gamma-1} L^a_{\tau_{\frac{r}{t^{\gamma-1}}}}(H), a \geq 0\right) \tag{4.13}$$

where

$$\tau_l(H) := \inf\{u \geq 0; L^0_u(H) > l\}.$$

2) *We now show formula (4.12).* We have:

$$\begin{aligned}
(X^{(\gamma,\alpha,r)}_t, t \geq 0) &= \left(\frac{1}{t^{\alpha+1}} \int_0^t Q^{(\gamma,r)}_s s^\alpha ds, t \geq 0\right) \\
&\overset{(law)}{=} \left(\frac{1}{t^{\alpha+1}} \int_0^t L^a_{\tau_r}(H) a^\alpha da, t \geq 0\right) \\
&\qquad \text{(from Theorem 4.3)} \\
&= \left(\int_0^1 L^{bt}_{\tau_r}(H) b^\alpha db, t \geq 0\right).
\end{aligned} \tag{4.14}$$

Thus, this time for fixed t only, we deduce from (4.13) that:

$$\begin{aligned}
X^{(\gamma,\alpha,r)}_t &\overset{(1.d)}{=} t^{\gamma-1} \int_0^1 L^b_{\tau_{\frac{r}{t^{\gamma-1}}}(H)}(H) b^\alpha db \\
&= t^{\gamma-1} \int_0^{\tau_{\frac{r}{t^{\gamma-1}}}(H)} (H_v)^\alpha 1_{[0,1]}(H_v) dv \\
&\qquad \text{(from the density of occupation formula (4.11))} \\
&= t^\varepsilon \Lambda^{(\gamma,\alpha)}_{\frac{r}{t^\varepsilon}}.
\end{aligned} \tag{4.15}$$

3) *We now show that $(\Lambda^{(\gamma,\alpha)}_t, t \geq 0)$ is a subordinator.* If the process $(H_u, u \geq 0)$ were Markovian, this property would follow from Proposition 4.2. But although $(H_u, u \geq 0)$ is not Markovian, $(\Lambda^{(\gamma,\alpha)}_t, t \geq 0)$ is nonetheless a subordinator since:

- for every $l \geq 0$, $H_{\tau_l(H)} = 0$,
- for every $t \geq 0$, H_t only depends on the values taken by the process $(Y_u - I_u, u \geq 0)$ during the excursion above 0 which straddles time t, i.e. it only depends on the values of $Y_u - I_u$ for $u \in [g_t, d_t]$, with $g_t := \sup\{s \leq t; Y_u - I_u = 0\}$ and $d_t := \inf\{s > t; Y_u - I_u = 0\}$ (see [61, p. 566]).

4) *It remains to prove that* $(\Lambda_t^{(\gamma,\alpha)}, t \geq 0)$ *is integrable.* This follows from (4.13), since, from this relation written for $r = 1$, we obtain:

$$\begin{aligned}\mathbb{E}\left[\Lambda_t^{(\gamma,\alpha)}\right] &= \mathbb{E}\left[\int_0^{\tau_t(H)} (H_v)^\alpha 1_{[0,1]}(H_v)dv\right] \\ &= t^{1+\frac{\alpha+1}{\varepsilon}}\mathbb{E}\left[\int_0^{\frac{1}{t^{1/\varepsilon}}} a^\alpha L_{\tau_1}^a(H)da\right] \quad (\text{with } \varepsilon = \gamma - 1) \\ &\quad \text{(from the density of occupation formula)} \\ &= t^{1+\frac{\alpha+1}{\varepsilon}} \int_0^{\frac{1}{t^{1/\varepsilon}}} a^\alpha da < \infty\end{aligned}$$

because, from Point 3) of Theorem 4.3 and Subsection 4.3.1,

$$\mathbb{E}\left[L_{\tau_1}^a(H)\right] = \mathbb{E}[Q_a^{(\gamma,1)}] = 1.$$

□

Exercise 4.3 (Le Gall's extension of the second Ray-Knight Theorem in a discrete set-up). Consider a Galton-Watson forest such that the number of offsprings of each individual has expectation 1 (this is the critical case). Let us draw the following picture to represent that forest (see [114, p.15])

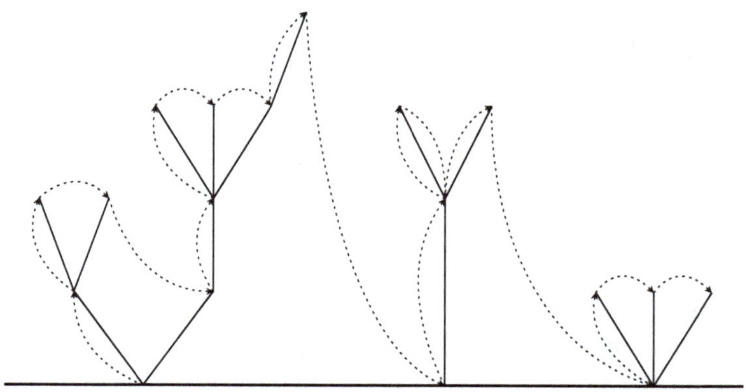

Fig. 4.1. Numbering of the Galton-Watson forest

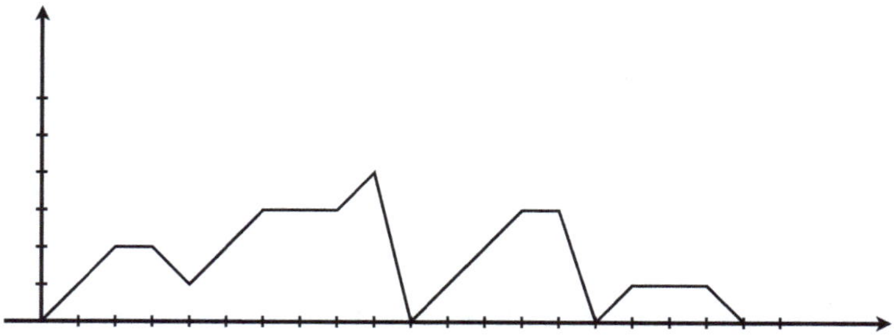

Fig. 4.2. Graph of the height process $(H_n, n \geq 0)$

The values of this forest are numbered following the order indicated on the Fig. 4.1, and we denote by H_n the generation to which the nth vertex belongs (see the graph of H_n above).

$(H_n, n \geq 0)$ is the height process. Define $L_n^a(H)$ the local time of H, at level a, by:

$$L_n^a(H) := \#\{p \leq n; H_p = a\} \qquad (a, n \in \mathbb{N}).$$

1) Show the occupation formula:

$$\sum_{a=0}^{\infty} h(a) L_n^a(H) = \sum_{k=0}^{n} h(H_k)$$

for every $h : \mathbb{N} \longrightarrow \mathbb{R}_+$ and every $n \in \mathbb{N}$.

2) Let $\tau_r(H) := \inf\{k \geq 0; L_k^0(H) > r\}$. Prove that $\tau_r(H)$ is equal to the number of vertices which are featured in the union of the r first trees of the forest, plus 1.

4.4 A Scaling Property Extension of the Previous Results

Clearly, the preceding Sections 4.2 and 4.3 have brought out the importance of the scaling property. In the light of this remark, we now generalize the above Theorems 4.2 and 4.4.

4.4.1 Hypothesis and Notation

• Let $(Z_{x,t}, x \geq 0, t \geq 0)$ be a measurable and integrable process indexed by $\mathbb{R}_+ \times \mathbb{R}_+$. Consider the subprocess $(X_x, x \geq 0)$:

$$X_x := Z_{x,1}, \quad x \geq 0, \tag{4.16}$$

and assume that $\sup\limits_{x \geq 0} \mathbb{E}[|Z_{x,1}|] < \infty$. We make two further hypotheses:

(H_1) For every $t \geq 0$, $(X_{xt}, x \geq 0) \overset{(law)}{=} (Z_{x,t}, x \geq 0)$.
(H_2) For every $y \geq 0$, $(Z_{y,t}, t \geq 0)$ is a martingale in the filtration $(\mathcal{G}_t := \sigma(Z_{u,s}; u \geq 0, s \leq t), t \geq 0)$.

Observe that, from (H_1) and (H_2):

$$\mathbb{E}[X_{xt}] = \mathbb{E}[Z_{x,t}] = \mathbb{E}[Z_{x,1}] = \mathbb{E}[X_x]. \tag{4.17}$$

In particular, the expectation of X_x does not depend on x. Note also that under (H_1) and (H_2), if Z is positive, then $\sup_{x \geq 0} \mathbb{E}[|Z_{x,1}|] < \infty$. Indeed, in this case, $\mathbb{E}[Z_{x,1}] = \mathbb{E}[X_0]$.

• We now consider $\varphi : \mathbb{R}_+ \longrightarrow \mathbb{R}$ Borel and integrable, and define the two processes:

$$X_t^{(\varphi)} := \frac{1}{t} \int_0^{+\infty} \varphi\left(\frac{x}{t}\right) X_x \, dx, \quad t \geq 0 \tag{4.18}$$

and

$$M_t^{(\varphi)} := \int_0^{+\infty} \varphi(y) Z_{y,t} \, dy, \quad t \geq 0. \tag{4.19}$$

As a consequence of the integrability of φ, the two processes are themselves integrable. Here is the announced extension of Theorems 4.2 and 4.4.

4.4.2 The Time Inversion Method in a More General Set-up

Theorem 4.5. *Under hypotheses H_1) and H_2) and for every integrable function φ:*

i) $(M_t^{(\varphi)}, t \geq 0)$ is a $(\mathcal{G}_t, t \geq 0)$ martingale,
ii) this martingale is associated to the peacock $(X_t^{(\varphi)}, t \geq 0)$.

Proof. It is clear from (H_2) and (4.17) that $(M_t^{(\varphi)}, t \geq 0)$ is a $(\mathcal{G}_t, t \geq 0)$ martingale. Let us prove Point *ii*). We have:

$$X_t^{(\varphi)} := \frac{1}{t} \int_0^{+\infty} \varphi\left(\frac{x}{t}\right) X_x dx = \int_0^{+\infty} \varphi(y) X_{yt} dy$$
$$\overset{(1.d)}{=} \int_0^{+\infty} \varphi(y) Z_{y,t} dy = M_t^{(\varphi)} \quad (\text{from } H_1)).$$

□

Of course, we must explain in which way Theorem 4.5 extends Theorems 4.2 and 4.4! This is the purpose of the following examples.

Example 4.1. *a*) With the notation of Section 4.3, taking $r = 1$ to simplify, we consider:

$$Z_{x,t} := t^{\varepsilon} L^x_{\tau_{1/t^{\varepsilon}}}(H) \qquad (\varepsilon = \gamma - 1),$$

where $(H_t, t \geq 0)$ is the height process of the Lévy process $(Y_t, t \geq 0)$ whose characteristic exponent is $c\lambda^\gamma$ with $\gamma \in]1,2]$, i.e. such that $\mathbb{E}\left[e^{-\lambda Y_t}\right] = \exp(-ct\lambda^\gamma)$. Then, Theorem 4.2 corresponds to the case $\gamma = 2$ and Theorem 4.4 to the general situation, $c > 0$, $\gamma \in]1,2]$. Let us define, following (4.16):

$$X_x := Z_{x,1} = L^x_{\tau_1}(H), \quad x \geq 0.$$

Clearly, for every $x \geq 0$ fixed:

$$Z_{x,t} = t^\varepsilon L^x_{\tau_{1/t^\varepsilon}}(H), \quad t \geq 0$$

is a martingale in its natural filtration, since $(L^x_{\tau_l}(H), l \geq 0)$ is a Lévy process (and even a subordinator) from Propositions 4.1 and 4.2. It is then easy to see that $(Z_{x,t}, t \geq 0)$ remains a martingale in the larger filtration $(\mathcal{G}_t, t \geq 0)$, thus, the hypothesis $H_2)$ is satisfied. On the other hand, from (4.13):

$$\begin{aligned}
(X_{xt}, t \geq 0) &= \left(L^{xt}_{\tau_1}(H), x \geq 0\right) \\
&\overset{(law)}{=} \left(t^\varepsilon L^x_{\tau_{1/t^\varepsilon}}(H), x \geq 0\right) \quad \text{(from (4.13))} \\
&= (Z_{x,t}, x \geq 0).
\end{aligned}$$

Thus, for the present choice $\left(Z_{x,t} = t^\varepsilon L^x_{\tau_{1/t^\varepsilon}}(H), x, t \geq 0\right)$, both hypotheses (H_1) and (H_2) are satisfied.

b) We now choose, for $\alpha \geq 0$:

$$\varphi_\alpha(x) := x^\alpha 1_{[0,1]}(x).$$

With this choice and the notation of (4.18) and (4.19):

$$\begin{aligned}
\left(X_t^{(\varphi_\alpha)} = \frac{1}{t}\int_0^{+\infty} \frac{x^\alpha}{t^\alpha} 1_{\{\frac{x}{t} \leq 1\}} X_x dx, t \geq 0\right) &= \left(\frac{1}{t^{1+\alpha}}\int_0^{+\infty} L^x_{\tau_1} x^\alpha dx, t \geq 0\right) \\
&= \left(\frac{1}{t^{1+\alpha}}\int_0^{t} Q_x^{(\gamma,1)} x^\alpha dx, t \geq 0\right) \\
&= (X_t^{(\gamma,\alpha,1)}, t \geq 0)
\end{aligned}$$

whereas:

$$\begin{aligned}
\left(M_t^{(\varphi_\alpha)} := \int_0^{+\infty} y^\alpha 1_{[0,1]}(y) Z_{y,t} dy, t \geq 0\right) &= \left(t^\varepsilon \int_0^1 y^\alpha L^y_{\tau_{1/t^\varepsilon}}(H) dy, t \geq 0\right) \\
&= \left(t^\varepsilon \Lambda^{(\gamma,\alpha)}_{1/t^\varepsilon}, t \geq 0\right) \quad \text{(from (4.15))}.
\end{aligned}$$

Thus, with this choice of φ, Theorem 4.5 is precisely Theorem 4.4.

c) Note that, when φ is no longer of the form $\varphi(x) = x^\alpha 1_{[0,1]}(x)$, Theorem 4.5 still applies although the peacock $(X_t^{(\varphi)}, t \geq 0)$ is no longer, in general, of the form $\left(\frac{1}{\alpha(t)}\int_0^t N_s d\alpha(s), t \geq 0\right)$ for a martingale $(N_s, s \geq 0)$, i.e. it is not a $\mathbf{F}_1$-type peacock.

Example 4.2 (Back to the Brownian sheet). Hypotheses H_1) and H_2) above and the contents of Chapter 2 of this Monograph bring naturally the careful reader to think about the Brownian sheet. Indeed, with the notation of Chapter 2, let $(W_{u,t}; u \geq 0, t \geq 0)$ a standard Brownian sheet. Let $h : \mathbb{R}_+ \times \mathbb{R} \longrightarrow \mathbb{R}$ be a space-time harmonic function for Brownian motion (i.e. such that $(h(s, B_s), s \geq 0)$ is a Brownian martingale). Define:

$$Z_{x,t} := h(xt, W_{x,t}); \quad x, t \geq 0$$

and note that:

$$(X_x = h(x, W_{x,1}), x \geq 0) \overset{(law)}{=} (h(x, B_x), x \geq 0).$$

Properties of the Brownian sheet (see Subsection 2.2.2) imply that both hypotheses H_1) and H_2) are satisfied. Thus, Theorem 4.5 implies that, for every integrable function φ, the process:

$$M_t^{(\varphi)} := \int_0^{+\infty} \varphi(y) h(yt, W_{y,t}) dy, \quad t \geq 0$$

is a martingale associated to the peacock:

$$X_t^{(\varphi)} := \frac{1}{t} \int_0^{+\infty} \varphi\left(\frac{x}{t}\right) h(x, B_x) dx, \quad t \geq 0.$$

Now, Theorem 2.2 corresponds to the particular case $\varphi(x) = 1_{[0,1]}(x)$.

Example 4.3 (Symmetric stable processes). Let $(Y_t, t \geq 0)$ be a symmetric stable Lévy process with index $\beta \in]1,2]$. Thus:

$$\mathbb{E}\left[e^{i\lambda Y_t}\right] = \exp\left(-t|\lambda|^\beta\right) \quad (t \geq 0, \lambda \in \mathbb{R}) \tag{4.20}$$

(see [24]). This process admits a continuous family of local times $(L_t^a, t \geq 0, a \in \mathbb{R})$ (see [37]) such that, for every bounded Borel function h:

$$\int_0^t h(Y_s) ds = \int_{\mathbb{R}} L_t^a h(a) da. \tag{4.21}$$

From (4.20), we deduce:

$$\text{for any } t > 0, \quad (Y_u, u \geq 0) \overset{(law)}{=} \left(tY_{\frac{u}{t^\beta}}, u \geq 0\right)$$

and, from the density of occupation formula (4.21):

$$\left(L_{\tau_1}^{xt}, x \geq 0\right) \overset{(law)}{=} \left(t^{\beta-1} L^x_{\tau_{\frac{1}{t^{\beta-1}}}}, x \geq 0\right), \tag{4.22}$$

with $\tau_l := \inf\{u \geq 0; L_u^0 > l\}$. We further define:

$$Z_{x,t} = t^\varepsilon L^x_{\tau_{1/t^\varepsilon}}, \quad x \geq 0, t \geq 0 \quad (\text{with } \varepsilon = \beta - 1)$$

and

$$X_x = Z_{x,1} = L^x_{\tau_1}, \quad x \geq 0.$$

The processes $(Z_{x,t}, x \geq 0, t \geq 0)$ and $(X_x, x \geq 0)$ satisfy both hypotheses $H_1)$ and $H_2)$:

- $H_1)$ is a consequence of the scaling property (4.22).
- From Proposition 4.1, $(tL^x_{\tau_{1/t}}, t \geq 0)$ is a martingale in its natural filtration since, for every $x \geq 0$, $(L^x_{\tau_l}, l \geq 0)$ is a subordinator. Now, as in Example 4.1, $(tL^x_{\tau_{1/t}}, t \geq 0)$ remains a martingale in the larger filtration $(\mathscr{G}_t, t \geq 0)$ so that $H_2)$ is also satisfied.

We can therefore apply Theorem 4.6, which becomes in this set-up:

Theorem 4.6. *Let $(Y_t, t \geq 0)$ be a symmetric stable Lévy process of index $\beta \in]1,2]$, and let $(L^a_t, t \geq 0, a \geq 0)$ the family of its local times. Let $(\tau_l, l \geq 0)$ the right-continuous inverse of $(L^0_t, t \geq 0)$. Then, for every integrable Borel function $\varphi : \mathbb{R} \longrightarrow \mathbb{R}$:*

$$M^{(\varphi)}_t := t^{\beta-1} \int_0^{+\infty} \varphi(y) L^y_{\tau_{\frac{1}{t^{\beta-1}}}}(Y)dy, \quad t \geq 0$$

is a martingale associated to the peacock

$$X^{(\varphi)}_t := \frac{1}{t} \int_0^{+\infty} \varphi\left(\frac{x}{t}\right) L^x_{\tau_1}(Y)dx, \quad t \geq 0.$$

When we choose for φ the function $\varphi_\alpha(x) = x^\alpha 1_{[0,1]}(x)$, we then have:

$$\begin{aligned} X^{(\varphi_\alpha)}_t &= \frac{1}{t} \int_0^{+\infty} \left(\frac{x}{t}\right)^\alpha 1_{[0,1]}\left(\frac{x}{t}\right) L^x_{\tau_1}(Y)dx \\ &= \frac{1}{t^{1+\alpha}} \int_0^t x^\alpha L^x_{\tau_1}(Y)dx \end{aligned}$$

and the process $(X^{(\varphi_\alpha)}_t, t \geq 0)$ is a peacock.

Remark 4.3. It is proven in [63, Theorem 11] that the following relation holds:

$$\left(L^x_{\tau_1} + \frac{1}{2}\eta^2_x, x \geq 0\right) \overset{(law)}{=} \left(\frac{1}{2}\left(\eta_x + \sqrt{2}\right)^2, x \geq 0\right) \tag{4.23}$$

where, on the left hand side, $(\eta_x, x \geq 0)$ is a fractional Brownian motion with Hurst index $H = \frac{\beta-1}{2}$ (i.e. $\mathbb{E}[\eta_x\eta_y] = \frac{1}{2}\left(x^{2H} + y^{2H} - |x-y|^{2H}\right)$) independent from $(L^x_{\tau_1}, x \geq 0)$. We then deduce from (4.23) that $\mathbb{E}[L^x_{\tau_1}] = 1$, hence $\mathbb{E}[L^x_{\tau_1}]$ does not depend on x. Nevertheless, for $\beta < 2$, $(L^x_{\tau_1}, x \geq 0)$ is not a martingale since its paths are Hölder continuous with index $\frac{\beta-1}{2} - \varepsilon$ (for every $\varepsilon > 0$), and therefore are not of squared bounded variation.

Notes and Comments

Proposition 4.1 is due to J. Jacod and P. Protter [90] and Proposition 4.2 is classical. The computations of Proposition 4.3 are due to J. Pitman and M. Yor [145] (see also R.K. Getoor and M.J. Sharpe [69] and F. Knight [110]). Results about CSBP are due to J.F Le Gall [114]. All other results of this chapter are new.

5

The Sato Process Method

Abstract We study various peacocks defined from self-decomposable laws. We construct associated martingales from Sato processes or Sato sheets.

5.1 Introduction

This chapter is devoted to the study of consequences of self-decomposability assumptions for the construction of martingales associated to certain peacocks.

In Section 5.2, we recall the main general results which we shall use in the sequel, concerning the notions of self-decomposability and Sato process.

In Section 5.3, we present, in particular, two classes of peacocks for which an associated martingale can be defined from a Sato process:

- The first one is the class $\mathbf{S}_1$ consisting of the processes:

$$V_t := Y_t - t^H \, \mathbb{E}[Y_1], \quad t \geq 0$$

  where $H > 0$ and $Y = (Y_t, t \geq 0)$ satisfies:

  a) Y_1 is an integrable self-decomposable random variable,

  b) $Y_t \overset{(1.d)}{=} t^H \, Y_1$.

  Clearly, $\mathbf{S}_1 \subset \mathbf{F}'_2$.

- The second one is the class $\mathbf{S}_2$ consisting of the processes:

$$\left(\int \exp(t\,vX - \phi(t\,v)) d\sigma(v) \,,\, t \geq 0 \right)$$

  where X is a self-decomposable random variable such that:

$$\forall v \geq 0 \qquad \mathbb{E}[\exp(vX)] < \infty,$$

  $\phi(v) = \log(\mathbb{E}[\exp(vX)])$ and σ is a signed finite measure on $\mathbb{R}_+$.

Hirsch F., Profeta C., Roynette B., Yor M.:
Peacocks and Associated Martingales, with Explicit Constructions

Section 5.4 is devoted to a precise description of martingales associated to a particular class of processes in $\mathbf{S}_1$, namely those for which we have

$$Y_1 = \int_{]0,+\infty[} R_N^2(s)d\mu(s)$$

where μ is a positive measure on $]0,+\infty[$ and $(R_N(s), s \geq 0)$ is a Bessel process of dimension $N > 0$ starting from 0. For certain measures μ, an associated martingale may be described from a process of hitting times related to a perturbed Bessel process.

Finally, we introduce in Section 5.5 the Sato sheet attached to a self-decomposable r.v. We prove in particular that, if L is a Lévy process such that L_1 is self-decomposable and, for every $r \geq 0$, $\mathbb{E}[\exp(rL_1)] < \infty$, and if σ is a signed finite measure on $\mathbb{R}_+$, then the process

$$\left(\int \exp(tL_\lambda - \lambda\,\phi_L(t))\,d\sigma(\lambda)\ ,\ t \geq 0\right)$$

(where $\phi_L(r) = \log(\mathbb{E}[\exp(rL_1)])$) is a peacock and an associated martingale may be described from a Sato sheet extending the Lévy process L.

5.2 Self-Decomposability and Sato Processes

We recall, in this section, some general facts concerning the notion of self-decomposability. We refer the reader, for background, complements and references, to Sato [155, Chapter 3].

Definition 5.1 (Self-decomposability). A random variable X is said to be *self-decomposable* if, for each u with $0 < u < 1$, there is the equality in law:

$$X \overset{(law)}{=} uX + \widehat{X}_u$$

for some variable $\widehat{X}_u$ independent of X.

Definition 5.2 (Sato process).

i) An *additive process* $(U_t,\ t \geq 0)$ is a stochastically continuous process with càdlàg paths, independent increments, and satisfying: $U_0 = 0$.

ii) An additive process (U_t) which is H*-self-similar* for some $H > 0$, meaning that, for each $c > 0$, $(U_{ct},\ t \geq 0) \overset{(law)}{=} (c^H U_t,\ t \geq 0)$, will be called a *Sato process* or, more precisely, a H*-Sato process*.

The following theorem, for which we refer to Sato's book [155, Chapter 3, Sections 16–17], gives characterizations of the self-decomposability property.

Theorem 5.1. *Let X be a real valued random variable. Then, X is self-decomposable if and only if one of the following equivalent properties is satisfied:*

1) *X is infinitely divisible and its Lévy measure is* $\frac{h(x)}{|x|}\mathrm{d}x$ *with* h *increasing on* $]-\infty,0[$ *and decreasing on* $]0,+\infty[$.

2) *There exists a Lévy process* $(C_s,\ s\geq 0)$ *such that*

$$X \overset{(law)}{=} \int_0^\infty \mathrm{e}^{-s} dC_s.$$

3) *For any (or some)* $H>0$, *there exists a* H*-Sato process* $(U_t,\ t\geq 0)$ *such that* $X \overset{(law)}{=} U_1$.

Remark 5.1. a) In Theorem 5.1, 2), the Lévy process $C=(C_s,\ s\geq 0)$ is uniquely determined in law by X. It will be called the *background driving Lévy process (*BDLP*)* of X, and denoted by C^X.

b) In Theorem 5.1, 3), the H-Sato process $U=(U_t,\ t\geq 0)$ is uniquely determined in law by X. It will be said to be *attached* to X and will be denoted by $(S_t^{H,X}, t\geq 0)$, or simply by $(S_t^X, t\geq 0)$ if H is fixed equal to 1. Obviously,

$$(S_t^{H,X},\ t\geq 0) \overset{(law)}{=} (S_{t^H}^X,\ t\geq 0)\,.$$

c) We note that, if $X\geq 0$, then the function h in Theorem 5.1, 1), vanishes on $]-\infty,0[$, C^X is a subordinator and $S^{H,X}$ is an increasing process.

The relation between $(C_s^X, s\geq 0)$ and $(S_t^X,\ t\geq 0)$ was made precise by Jeanblanc-Pitman-Yor [94, Theorem 1]:

Theorem 5.2. *Let* X *be a self-decomposable random variable.*

1) *One has:*

$$(C_s^X,\ s\geq 0) \overset{(law)}{=} \left(\int_{\mathrm{e}^{-s}}^1 t^{-1} dS_t^X,\ s\geq 0\right).$$

2) *We extend the definition of* $(C_s^X,\ s\geq 0)$ *to* $s\leq 0$ *by taking, for* $(C_s^X,\ s\leq 0)$, *an independent copy of* $(-C_{-s}^X,\ s\leq 0)$. *Then,* $(C_s^X,\ s\in\mathbb{R})$ *is a Lévy process indexed by* $\mathbb{R}$, *and*

$$(S_t^X,\ t\geq 0) \overset{(law)}{=} \left(\int_{-\log t}^{+\infty} \mathrm{e}^{-s} dC_s^X,\ t\geq 0\right).$$

3) *The equality in law stated in 1) extends in the following manner:*

$$(C_s^X,\ s\in\mathbb{R}) \overset{(law)}{=} \left(\int_{\mathrm{e}^{-s}}^1 t^{-1} dS_t^X,\ s\in\mathbb{R}\right)$$

where, if $s\leq 0$,

$$\int_{\mathrm{e}^{-s}}^1 t^{-1} dS_t^X \quad \textit{is set for} \quad -\int_1^{\mathrm{e}^{-s}} t^{-1} dS_t^X.$$

Exercise 5.1 (One-dimensional marginals of Sato processes are self-decomposable). Let $(U_t, t \geq 0)$ be a H-Sato process. Prove that, for every $a \geq 0$, U_a is self-decomposable. (*Hint*: $(U_{ta}, t \geq 0)$ is a H-Sato process attached to U_a.)

Exercise 5.2 (Increments of Sato processes are infinitely divisible). Let X be a self-decomposable random variable, and denote by ψ its characteristic exponent. Show that, if $0 \leq s \leq t$, then $S_t^X - S_s^X$ is an infinitely divisible random variable whose characteristic exponent $\psi_{s,t}$ is given by:

$$\psi_{s,t}(\lambda) = \psi(t\lambda) - \psi(s\lambda).$$

Exercise 5.3 (Sato processes with stationary increments are stable processes).
1) Let X be a positive self-decomposable r.v. such that:

$$\mathbb{E}[e^{-\lambda X}] = \exp\left(-\lambda a - \int_0^{+\infty} (1 - e^{-\lambda x}) \frac{h(x)}{x} dx\right)$$

with $a \geq 0$, h positive, decreasing and such that $\int_0^{+\infty} (1 \wedge \frac{1}{x}) h(x) dx < \infty$. Let $(S_t^X, t \geq 0)$ be the H-Sato process attached to X. Prove that $(S_t^X, t \geq 0)$ is a subordinator if and only if:

- either $H = 1$ and $h \equiv 0$; X is constant,
- or $H > 1$, $a = 0$ and $h(x) = \dfrac{c}{x^{1/H}}$ with c a positive constant; X is a positive stable r.v. of index $\frac{1}{H}$.

2) Let X be a self-decomposable r.v. whose characteristic exponent ψ is given by:

$$\psi(\lambda) = ia\lambda + \sigma^2 \frac{\lambda^2}{2} + \int_{-\infty}^{+\infty} \left(1 - e^{i\lambda x} + i\lambda x 1_{\{|x| \leq 1\}}\right) \frac{h(x)}{|x|} dx,$$

with h positive, increasing on $]-\infty, 0[$, decreasing on $]0, +\infty[$, and such that: $\int_{-\infty}^{+\infty} (|x| \wedge \frac{1}{|x|}) h(x) dx < \infty$. Let $(S_t^X, t \geq 0)$ be the H-Sato process attached to X. Prove that $(S_t^X, t \geq 0)$ is a Lévy process if and only if:

- either $H = \frac{1}{2}$, $a = 0$, σ arbitrary, and $h \equiv 0$; $X \overset{(law)}{=} \sigma G$ with G a centered reduced Gaussian r.v.,
- or $H = 1$, a is arbitrary, $\sigma = 0$, and $h(x) = \dfrac{c}{|x|}$, $c \geq 0$,
- or $H > \frac{1}{2}$, $H \neq 1$, $\sigma = 0$, $h(x) = \dfrac{c_1}{x^{1/H}} 1_{\{x>0\}} + \dfrac{c_2}{|x|^{1/H}} 1_{\{x<0\}}$ with $c_1, c_2 \geq 0$, and $a = \dfrac{H}{1-H}(c_1 - c_2)$.

This corresponds to the following characteristic functions ψ:

- $H = \frac{1}{2}$: $\psi(\lambda) = \frac{\sigma^2}{2} \lambda^2$,
- $H = 1$: $\psi(\lambda) = ia\lambda + C|\lambda|$, $C > 0$,

- $H > \frac{1}{2}, H \neq 1$:

$$\psi(\lambda) = \begin{cases} C\lambda^{1/H} & \text{if } \lambda \geq 0, \\ \overline{C}(-\lambda)^{1/H} & \text{if } \lambda \leq 0, \end{cases}$$

with $C = d\left(1 - i\frac{c_1 - c_2}{c_1 + c_2}\tan\left(\frac{\pi}{2H}\right)\right)$, where d is a positive constant.

In conclusion: $(S_t^X, t \geq 0)$, the H-Sato process attached to X, is a Lévy process if and only if X is a $\frac{1}{H}$-stable r.v. (See Bertoin [24] or Lukacs [121] for a definition of stable r.v.'s)

Exercise 5.4 (Examples of a 1-Sato process). Let X be a centered and self-decomposable r.v. and $(M_t, t \geq 0)$ be the 1-Sato process attached to X. $(M_t, t \geq 0)$ is a martingale.

1) Define for $\alpha > -1$:

$$\left(M_t^{(\alpha)} := \int_0^t u^\alpha dM_u \,,\, t \geq 0\right).$$

Prove that $(M_t^{(\alpha)}, t \geq 0)$ is a martingale.

2) Prove that $M_1^{(\alpha)} = \displaystyle\int_0^1 u^\alpha dM_u$ is a centered self-decomposable r.v.
(*Hint*: for $0 < c < 1$, applying the scaling property of $(M_t, t \geq 0)$:
$\int_0^1 u^\alpha dM_u = \int_0^c u^\alpha dM_u + \int_c^1 u^\alpha dM_u \overset{(law)}{=} c^{\alpha+1}\int_0^1 u^\alpha dM_u + \int_c^1 u^\alpha d\widetilde{M}_u$ where $\widetilde{M}_u = M_u - M_c$, and $\int_0^1 u^\alpha dM_u$ and $\int_c^1 u^\alpha d\widetilde{M}_u$ are independent.)

3) Find a 1-Sato process $(N_t^{(\alpha)}, t \geq 0)$ attached to $\displaystyle\int_0^1 u^\alpha dM_u$. (*Hint*: $N_t^{(\alpha)} = \displaystyle\int_0^{t^{\frac{1}{\alpha+1}}} u^\alpha dM_u$).

Exercise 5.5 (Another characterization of self-decomposability). 1) Let X be an infinitely divisible r.v. and let ψ its characteristic exponent.

i) Prove that if X is self-decomposable, then ψ is differentiable on $\mathbb{R}\backslash\{0\}$ and the characteristic exponent of $\widetilde{\psi}$ of the Lévy process C^X is given, for $\lambda \neq 0$, by

$$\widetilde{\psi}(\lambda) = \lambda\psi'(\lambda).$$

ii) Assume conversely that ψ is differentiable on $\mathbb{R}\backslash\{0\}$ and that the function $\widetilde{\psi}$ defined by:

$$\widetilde{\psi}(\lambda) = \lambda\psi'(\lambda) \ \text{ for } \lambda \neq 0, \qquad \widetilde{\psi}(0) = 0$$

is the characteristic exponent of a Lévy process. Prove then that X is self-decomposable. (*Hint*: prove that Property 2) in Theorem 5.1 is satisfied.)

2) We recall that a Bernstein function F is a positive continuous function on $[0, +\infty[$ such that $F(0) = 0$, which is differentiable on $]0, +\infty[$, and whose derivative F' is

completely monotonous on $]0,+\infty[$. Thus, Bernstein functions are Laplace exponents of subordinators.
Let X be an infinitely divisible, positive r.v. and let ϕ be its Laplace exponent.

i) Prove that if X is self-decomposable, then the Laplace exponent $\widetilde{\phi}$ of the subordinator C^X is given, for $s > 0$, by:

$$\widetilde{\phi}(s) = s\phi'(s).$$

ii) Assume conversely that the function $\widetilde{\phi}$ defined by:

$$\widetilde{\phi}(s) = s\phi'(s) \text{ for } s > 0, \qquad \widetilde{\phi}(0) = 0$$

is a Berstein function. Prove that X is self-decomposable.

Exercise 5.6 (Examples of BDLP's).

1) Let $(L_t,\, t \geq 0)$ be a Lévy process. Prove that, if L_1 is self-decomposable, then L_a is self-decomposable for every $a \geq 0$, and

$$(C_s^{L_a},\, s \geq 0) \overset{(law)}{=} (C_{as}^{L_1},\, s \geq 0).$$

2) Let B be the standard Brownian motion starting from 0. Show that B_1 is self-decomposable and

$$(C_s^{B_1},\, s \geq 0) \overset{(law)}{=} (B_{2s},\, s \geq 0).$$

3) More generally, let $0 < \alpha \leq 2$ and let $X^{(\alpha)}$ be the symmetric stable process of index α, whose characteristic exponent is: $\psi(\lambda) = |\lambda|^\alpha$. Show that $X_1^{(\alpha)}$ is self-decomposable and

$$(C_s^{X_1^{(\alpha)}},\, s \geq 0) \overset{(law)}{=} (X_{\alpha s}^{(\alpha)},\, s \geq 0).$$

4) Let $0 < \alpha \leq 1$ and let $\tau^{(\alpha)}$ be the stable subordinator of index α, whose Laplace exponent is: $\phi(\lambda) = \lambda^\alpha$. Show that $\tau_1^{(\alpha)}$ is self-decomposable and

$$(C_s^{\tau_1^{(\alpha)}},\, s \geq 0) \overset{(law)}{=} (\tau_{\alpha s}^{(\alpha)},\, s \geq 0).$$

5) Let $(\gamma_t,\, t \geq 0)$ be the gamma subordinator whose Laplace exponent is: $\phi(\lambda) = \log(1+\lambda)$. Prove that γ_1 is self-decomposable and C^{γ_1} is the compound Poisson process whose Lévy measure is:

$$1_{]0,+\infty[}(x)\mathrm{e}^{-x}\,dx.$$

Long Exercise 5.7 (About last passage times of some diffusions). Let $(X_t, t \geq 0)$ be a regular diffusion started from 0 and taking values in $\mathbb{R}_+$, with speed measure $m(dx) = \rho(x)dx$ such that $m([0,+\infty[) = +\infty$ and scale function s which we assume to be strictly increasing and of $\mathscr{C}^2$ class on $]0,+\infty[$. We also suppose that

$$\lim_{x\to 0} m([0,x])s(x) > -\infty$$

and that 0 is an entrance boundary, and $+\infty$ a natural boundary (see [35, p.14]). We denote $q(t,\centerdot)$ the density of the r.v. X_t with respect to the speed measure.

Let $(\overline{X}_t, t \geq 0)$ be the new diffusion started from 0 and taking values in $\mathbb{R}_+$ obtained from $(X_t, t \geq 0)$ by Biane's transformation (see [148, p.222]), i.e. the speed measure $\overline{m}$ and scale function $\overline{s}$ of $\overline{X}$ are respectively given by:

$$\begin{cases} \overline{m}(dx) = \overline{\rho}(x)dx, \quad \text{with } \overline{\rho}(x) = (m([0,x)])^2 s'(x), \\ \overline{s}(x) = -\dfrac{1}{m([0,x])}. \end{cases} \tag{5.1}$$

This transformation was introduced by Biane [29] to generalize a celebrated identity from Ciesielski and Taylor [47], originally obtained for Bessel processes. We denote $(\overline{P}_t, t \geq 0)$ the semi-group associated to the process $(\overline{X}_t, t \geq 0)$ and $\overline{q}(t,\centerdot)$ the density of the r.v. $\overline{X}_t$ with respect to the speed measure $\overline{m}$.

1) *i*) Prove that $(\overline{X}_t, t \geq 0)$ never comes back to 0. (*Hint*: for $0 < a < y$, $(-\overline{s}(\overline{X}_{t\wedge T_a}), t \geq 0)$, with $\overline{T}_a := \inf\{u \geq 0, \overline{X}_t = a\}$, is a bounded martingale under $\overline{\mathbb{P}}_y$.)

ii) Prove that $(\overline{X}_t, t \geq 0)$ is transient. (*Hint*: $(-\overline{s}(\overline{X}_t), t \geq 0)$ is a positive and continuous local martingale, hence a supermartingale which converges a.s.)

We recall the following results (see [148, p. 222]):

a) The process $(X_t, t \geq 0)$ is associated to $(\inf_{s\geq t} \overline{X}_s, t \geq 0)$, i.e.:

$$X_t \overset{(1.d)}{=} \inf_{s \geq t} \overline{X}_s. \tag{5.2}$$

b) There is the identity between transition densities:

$$q(t,x) = \int_x^{+\infty} \overline{q}(t,y) m([0,y]) s'(y) dy \quad (x \geq 0, t > 0) \tag{5.3}$$

2) We define, for every $a \geq 0$, $\overline{G}_a := \sup\{u \geq 0; \overline{X}_u = a\}$.

i) Prove that $\overline{G}_0 = 0$ a.s.

ii) Prove that $\overline{\mathbb{P}}(\overline{G}_a < t) = \mathbb{P}(X_t > a) = \int_a^{+\infty} q(t,y) m(dy)$.

iii) Recover Pitman-Yor's result [144]:

$$\overline{\mathbb{P}}(\overline{G}_a \in dt) = -\frac{\overline{q}(t,a)}{\overline{s}(a)} dt.$$

(*Hint*: use the relation $\dfrac{\partial}{\partial t} q(t,y) = \dfrac{\partial^2}{\partial m \partial s} q(t,y)$.)

3) We define the semi-group $(\widehat{P}_t, t \geq 0)$ by

$$\widehat{P}_t f(x) = \frac{1}{\overline{s}(x)} \overline{P}_t(f\overline{s})(x).$$

In other words, $\widehat{P}$ is the Doob $\overline{s}$-transform of $\overline{P}$. Let $(\widehat{X}_t, t \geq 0)$ be the diffusion associated to the semi-group $(\widehat{P}_t, t \geq 0)$.

i) Prove that the speed measure $\widehat{m}$ and the scale function $\widehat{s}$ of the process $(\widehat{X},t \geq 0)$ are respectively given by:

$$\begin{cases} \widehat{m}(dx) = \widehat{\rho}(x)dx, & \text{with } \widehat{\rho}(x) = s'(x) \\ \widehat{s}(x) = m([0,x]). \end{cases} \tag{5.4}$$

ii) Let $\overline{U}$ be the potential measure of the process $\overline{X}$, i.e.

$$\int_0^{+\infty} f(y)\overline{U}(dy) := \int_0^{+\infty} \mathbb{E}[f(\overline{X}_t)]dt = \int_0^{+\infty} dt \int_0^{+\infty} f(y)\overline{q}(t,y)\overline{m}(dy).$$

Prove that $\overline{U}(dy) = -\overline{s}(y)\overline{m}(dy)$.

iii) Prove that the semi-groups $(\overline{P}_t,t \geq 0)$ and $(\widehat{P}_t,t \geq 0)$ are in duality with respect to the measure $\overline{U}$:

$$\int_0^{+\infty} \overline{P}_t f(y)\, g(y)\overline{U}(dy) := \int_0^{+\infty} f(y)\, \widehat{P}_t g(y)\overline{U}(dy).$$

iv) Let $a > 0$. Prove that the process $(\overline{X}_{\overline{G}_a - t}, 0 \leq t \leq \overline{G}_a)$ equals in law the process $(\widehat{X}_t,t \geq 0)$ started from a and stopped when it hits 0.
(*Hint*: use Nagasawa's Theorem about time-reversal of a diffusion at a last passage time, see [150, Theorem 4.5, p. 315].)

v) Deduce that the process $(\overline{G}_a,a \geq 0)$ has independent increments. (*Hint*: use the strong Markov property of $\widehat{X}$.)

4) We now assume that the process $(\overline{X}_t,t \geq 0)$ is $\frac{1}{H}$ $(H > 0)$ auto-similar, i.e., for every $c > 0$:

$$(\overline{X}_{tc},t \geq 0) \overset{(law)}{=} (c^{1/H}\overline{X}_t,t \geq 0).$$

i) Prove that $(\overline{G}_a,a \geq 0)$ is an increasing H-Sato process.

ii) Deduce that for every $a \geq 0$, $\overline{G}_a$ is self-decomposable.
(*Hint*: see Exercise 5.1.)

We shall now particularize the preceding results when $(X_t,t \geq 0)$ is a Bessel process $(R_t^{(\nu)},t \geq 0)$ with index $\nu > -1$ (i.e. with dimension $d = 2(\nu+1) > 0$). Recall that:

- if $\nu > 0$: $s^{(\nu)}(x) = -x^{-2\nu}$ and $\rho^{(\nu)}(x) = \dfrac{x^{2\nu+1}}{\nu}$,
- if $\nu = 0$: $s^{(0)}(x) = 2\log x$ and $\rho^{(0)}(x) = x$,
- if $\nu < 0$: $s^{(\nu)}(x) = x^{-2\nu}$ and $\rho^{(\nu)}(x) = -\dfrac{x^{2\nu+1}}{\nu}$.

5) *i*) Prove that the process $(\overline{X}_t,t \geq 0)$ is the Bessel process $(R_t^{(\nu+1)},t \geq 0)$ of index $\nu+1$ (i.e. with dimension $\overline{d} = d+2$) started at 0.

ii) Prove that $(\overline{X}_{\overline{G}_a - t}, 0 \leq t \leq \overline{G}_a)$ is the Bessel process $(R_t^{(-\nu-1)},t \geq 0)$ of index $-\nu-1$ (i.e. with dimension $4-\overline{d}$) started from a and stopped at 0. (*Hint*: use Question 3), or [174, 175]. Note that, although $4-\overline{d}$ might be negative, there is no difficulty in defining this process as the square root of a squared Bessel process of dimension $4-\overline{d}$, since this process is stopped when it hits 0; observe also that $\overline{d}$ cannot be equal to 2, since $d > 0$.)

iii) Use Point *ii*) of Question 2) to recover Getoor's result [68]:

$$G_a^{(\nu+1)} \overset{(law)}{=} \frac{a^2}{2\gamma_{\nu+1}} \quad (\nu > -1)$$

where $G_a^{(\nu+1)} := \sup\{u \geq 0; R_u^{(\nu+1)} = a\}$ and $\gamma_{\nu+1}$ is a gamma r.v. of parameter $\nu+1$:

$$\mathbb{P}(\gamma_{\nu+1} \in dt) = \frac{1}{\Gamma(\nu+1)} e^{-t} t^{\nu} 1_{\{t \geq 0\}} dt.$$

(*Hint*: use the scaling property of $R^{(\nu)}$ and the fact that $(R_1^{(\nu)})^2 \overset{(law)}{=} 2\gamma_{\nu+1}$.)

iv) Let $(R_t^{(\nu)}, t \geq 0)$ be a Bessel process of index ν, with $-1 < \nu < 0$, started from $y > 0$, and let $T^{(y)} = \inf\{u \geq 0; R_u^{(\nu)} = 0\}$. Recover Kent's result [107]:

$$T^{(y)} \overset{(law)}{=} \frac{y^2}{2\gamma_{|\nu|}}.$$

(*Hint*: prove by time-reversal that $T^{(y)} = G_y^{(-\nu)}$.)

v) Prove that for every $\lambda, a \geq 0$:

$$\mathbb{E}\left[\exp\left(-\frac{\lambda^2}{2} G_a^{(\nu)}\right)\right] = \frac{2}{\Gamma(\nu)} K_\nu(a\lambda) \left(\frac{a\lambda}{2}\right)^{\nu}$$

where K_ν denote the MacDonald function of index ν, see [117, p.110]. (*Hint*: use the formulae $K_\mu(z) = \frac{1}{2}\left(\frac{z}{2}\right)^{\mu} \int_0^{+\infty} e^{-\frac{z^2}{4t} - t} \frac{dt}{t^{\mu+1}}$ and $K_\mu = K_{-\mu}$, see [117, p.119]).

vi) Deduce the existence of a decreasing function $h^{(\nu)} : \mathbb{R}_+ \longrightarrow \mathbb{R}_+$ such that $\int_0^{+\infty} (x \wedge 1) \frac{h^{(\nu)}(x)}{x} dx < \infty$ which satisfies:

$$\frac{2}{\Gamma(\nu)} K_\nu(\lambda) \left(\frac{\lambda}{2}\right)^{\nu} = \exp\left(-\int_0^{+\infty} \left(1 - e^{-\frac{\lambda^2 x}{2}}\right) \frac{h^{(\nu)}(x)}{x} dx\right).$$

(*Hint*: use the fact that the r.v. $G_1^{(\nu)}$, with $\nu > 0$, is self-decomposable.) We refer the reader to Pitman-Yor [144] or Ismail-Kelker [87] for more information about this function $h^{(\nu)}$.

vii) Prove that $\displaystyle\int_0^{+\infty} e^{-\frac{\lambda^2 x}{2}} h^{(\nu)}(x) dx = \frac{K_{\nu-1}(\lambda)}{\lambda K_\nu(\lambda)}$.

(*Hint*: use the relation $\dfrac{K_\nu'(\lambda)}{K_\nu(\lambda)} = -\dfrac{K_{\nu-1}(\lambda)}{K_\nu(\lambda)} - \dfrac{\nu}{\lambda}$, see [117, p.110].)

viii) Let, for $\nu > 1$, $f^{(\nu-1)}$ be the density, under $\mathbb{P}_1^{(\nu-1)}$ of the r.v.

$$\int_0^{G_1^{(\nu-1)}} 1_{\{R_s^{(\nu-1)} > 1\}} ds.$$

We recall (cf. [144, p.339]) that $f^{(\nu-1)}(x) = 4(\nu-1)\eta^{(\nu)}(x)$, $(\nu > 1, x \geq 0)$ where the function $\eta^{(\nu)}$ satisfies to:

$$\int_0^{+\infty} x e^{-\lambda x} \eta^{(\nu)}(x) dx = \frac{K_{\nu-1}(\sqrt{\lambda})}{2\sqrt{\lambda} K_\nu(\sqrt{\lambda})} \quad (\lambda > 0, \nu > 1).$$

Deduce that $h^{(\nu)}(x) = \frac{1}{4(\nu-1)} f^{(\nu-1)}\left(\frac{x}{2}\right)$

ix) Prove that the BDLP of the positive and self-decomposable r.v. $G_1^{(\nu)}$ is a subordinator $(C_s^{(\nu)}, s \geq 0)$ whose Lévy measure $c^{(\nu)}(dx)$ satisfies to: $c^{(\nu)}(]x, +\infty[) = h^{(\nu)}(x)$, $(x \geq 0, \nu > 1)$. We define

$$\begin{cases} \widetilde{G}_a := \displaystyle\int_{-\log a}^{+\infty} e^{-2s} dC_{2s}^{(\nu,-)} & \text{if } 0 \leq a \leq 1 \\ \widetilde{G}_a := \widetilde{G}_1 + \displaystyle\int_0^{\log a} e^{2s} dC_{2s}^{(\nu,+)} & \text{if } a \geq 1 \end{cases}$$

where $(C_s^{(\nu,-)}, s \geq 0)$ and $(C_s^{(\nu,+)}, s \geq 0)$ are two independent copies of the subordinator $(C_s^{(\nu)}, s \geq 0)$. Prove that:

$$(G_a^{(\nu)}, a \geq 0) \overset{(law)}{=} (\widetilde{G}_a, a \geq 0).$$

Exercise 5.8 (Are one-dimensional marginals of squared Bessel processes self-decomposable?). Let $Y_u^{N,x}$ be the value at time $u > 0$ of the squared Bessel process of dimension $N \geq 0$, starting from $x \geq 0$. We recall (see [150, Chapter XI, §1]) the following formula:

$$\mathbb{E}\left[\exp(-t Y_u^{N,x})\right] = (1 + 2tu)^{-N/2} \exp\left(-\frac{x^2 t}{1 + 2tu}\right).$$

1) Prove that $Y_u^{N,x}$ is infinitely divisible and that its Lévy measure admits on $]0, +\infty[$ the density:

$$\varphi(y) = \left(N + \frac{x^2}{2u^2} y\right) \frac{1}{2y} \mathrm{e}^{-y/2u}.$$

2) Deduce from this that the random variable $Y_u^{N,x}$ is self-decomposable if and only if $x^2 \leq Nu$.

5.3 Sato Processes and Peacocks

We now present two classes of peacocks for which an associated martingale can be defined from a Sato process.

Theorem 5.3. *Let $H > 0$. Suppose that $Y = (Y_t, t \geq 0)$ satisfies:*

a) Y_1 is an integrable self-decomposable random variable.

b) $Y_t \overset{(1.d)}{=} t^H Y_1$.

Then the process

$$V_t := Y_t - t^H \mathbb{E}[Y_1], \quad t \geq 0$$

is a peacock, and an associated martingale is

$$M_t := S_t^{H,Y_1} - t^H \mathbb{E}[Y_1], \quad t \geq 0.$$

Proof. Since S^{H,Y_1} is a H-self-similar process, we have:

$$V_t \overset{(1.d)}{=} t^H V_1 \overset{(1.d)}{=} t^H M_1 \overset{(1.d)}{=} M_t \,.$$

Since S^{H,Y_1} is an additive process, M also is an additive process which is centered. Consequently, M is a martingale. □

Theorem 5.4. *Let* X *be a self-decomposable random variable. We denote by* ψ *its characteristic exponent, that is:*

$$\forall \lambda \in \mathbb{R}, \quad \mathbb{E}[\exp(\mathrm{i}\lambda X)] = \exp(-\psi(\lambda))\,.$$

Let σ *be a signed finite measure on* $\mathbb{R}$ *such that:*

$$\forall t \geq 0, \quad \int |\exp(\psi(t\lambda))| d|\sigma|(\lambda) < \infty.$$

We set, for every $t \geq 0$,

$$\begin{aligned} A_t^{X,\sigma} &= \int \exp[\mathrm{i}t\lambda X + \psi(t\lambda)]\, \sigma(\mathrm{d}\lambda), \\ M_t^{X,\sigma} &= \int \exp[\mathrm{i}\lambda S_t^X + \psi(t\lambda)]\, \sigma(\mathrm{d}\lambda), \\ \text{and } \mathscr{G}_t^X &= \sigma\{S_s^X;\, s \leq t\}. \end{aligned}$$

Then $(M_t^{X,\sigma})$ *is a* $\mathbb{C}$*-valued* $(\mathscr{G}_t^X)$*-martingale and*

$$A_t^{X,\sigma} \overset{(1.d)}{=} M_t^{X,\sigma}.$$

Proof. Since S^X is 1-self-similar, $tX \overset{(1.d)}{=} S_t^X$, and therefore,

$$A_t^{X,\sigma} \overset{(1.d)}{=} M_t^{X,\sigma}.$$

On the other hand, since S^X is an additive process, for every $\lambda \in \mathbb{R}$ and $0 \leq s \leq t$,

$$\begin{aligned} &\mathbb{E}\left[\exp(\mathrm{i}\lambda\,(S_t^X - S_s^X))\right] = \exp(-(\psi(t\lambda) - \psi(s\lambda))) \\ \text{and} \quad &\mathbb{E}\left[\exp(\mathrm{i}\lambda\, S_t^X + \psi(t\lambda)) \mid \mathscr{G}_s^X\right] = \exp(\mathrm{i}\lambda\, S_s^X + \psi(s\lambda)). \end{aligned}$$

The desired result follows. □

As a consequence of the previous theorem and Jensen's inequality, we obtain that $A^{X,\sigma}$ is a $\mathbb{C}$-valued peacock, which means that for every convex function $H : \mathbb{C} \longrightarrow \mathbb{R}$, the function:

$$t \geq 0 \longrightarrow \mathbb{E}[H(A_t^{X,\sigma})] \in]-\infty, +\infty]$$

is increasing. We give below a real valued version of this result.

Theorem 5.5. *Let X be a self-decomposable random variable. We assume that:*

$$\forall v \geq 0, \quad \mathbb{E}[\exp(vX)] < \infty.$$

We set:

$$\mathbb{E}[\exp(vX)] = \exp[\phi(v)] .$$

Let σ be a signed finite measure on $\mathbb{R}_+$. Then,

$$\left(\int \exp(t\,vX - \phi(t\,v)) d\sigma(v) \; , \; t \geq 0 \right)$$

is a peacock, and an associated martingale is:

$$\left(\int \exp(vS_t^X - \phi(tv)) d\sigma(v) \; , \; t \geq 0 \right) .$$

Theorem 5.5 above should be compared with Theorem 1.7, which is valid without the self-decomposability assumption, but only for positive measures σ.

Exercise 5.9 (Remarkable polynomials built from a self-decomposable r.v.). Let X be a real valued random variable such that $\mathbb{E}[|X|^p] < \infty$ for some integer $p \geq 1$. Then, the function

$$\lambda \in \mathbb{R} \longrightarrow \Psi(\lambda) := \mathbb{E}[\exp(\mathrm{i}\lambda X)]$$

is of $\mathscr{C}^p$-class; furthermore, as a characteristic function, it does not vanish in some neighborhood of 0.

1) Show that a sequence $(P_r^X;\, 0 \leq r \leq p)$ of polynomials is defined by induction by:

$$P_0^X = 1 \qquad \text{and}$$

$$\forall 1 \leq r \leq p, \quad (P_r^X)' = r P_{r-1}^X \text{ and } \mathbb{E}[P_r^X(X)] = 0 .$$

2) Prove that, for $0 \leq r \leq p$ and $x \in \mathbb{R}$,

$$P_r^X(x) = (-\mathrm{i})^r \frac{\partial^r}{\partial \lambda^r} \left[\frac{\exp(\mathrm{i}\lambda x)}{\Psi(\lambda)} \right] \bigg|_{\lambda=0} .$$

3) We now assume moreover that X is self-decomposable. Prove that, for $0 \leq r \leq p$, $(t^r P_r^X(X),\, t \geq 0)$ is a peacock, and an associated martingale is:

$$\left(t^r P_r^X \left(t^{-1} S_t^X\right) , \, t \geq 0 \right) .$$

(*Hint:* use the previous question and the fact that, by Theorem 5.4,

$$\left(\frac{\exp(\mathrm{i}\,\mu\, S_t^X)}{\Psi(t\mu)},\, t\geq 0\right)$$

is a $(\mathscr{G}_t^X)$-martingale.)

4) Suppose that $X = B_a$ where B_a denotes a centered real valued Gaussian random variable with variance a. Prove that

$$\forall r\geq 0 \qquad P_r^{B_a}(x) = a^{r/2}\, h_r\left(\frac{x}{\sqrt{a}}\right)$$

with (h_r) the sequence of *Hermite polynomials* defined by:

$$h_r(x) = (-1)^r \mathrm{e}^{x^2/2}\,\frac{d^r}{dx^r}\left(\mathrm{e}^{-x^2/2}\right).$$

Prove that a martingale associated to the peacock:

$$(t^r\, h_r(B_1)\,,\, t\geq 0)$$

is:

$$\left(t^r\, h_r\left(\frac{B_{t^2}}{t}\right),\, t\geq 0\right).$$

5) Suppose now that $X = \gamma_a$ where γ_a denotes a gamma variable with expectation a. Prove that

$$\forall r\geq 0 \qquad P_r^{\gamma_a}(x) = (-1)^r\, C_r(a,x)\,,$$

where $(C_r(a,x),\, r\geq 0)$ denotes the sequence of *Charlier polynomials* defined by

$$C_r(a,x) = \left.\frac{\partial^r}{\partial\lambda^r}\left[(1+\lambda)^a \exp(-\lambda\, x)\right]\right|_{\lambda=0}$$

(see for instance Schoutens [156] and Exercise 1.38).

We refer the reader to Barrieu-Schoutens [15] for related computations.

5.4 Quadratic Functionals of Bessel Processes $(R_N(t),\, t\geq 0)$

In this subsection, we shall illustrate Theorem 5.3 for random variables which are quadratic functionals of Bessel processes.

5.4.1 The Process $(R_N^2(t),\, t\geq 0)$

In the sequel, we denote by $(R_N(t),\, t\geq 0)$ the Bessel process of dimension $N>0$, starting from 0.

5.4.1.1 Self-Decomposability of $R_N^2(1)$

As is well-known (see, for instance, Revuz-Yor [150, Chapter XI]) one has

$$\mathbb{E}[\exp(-\lambda R_N^2(1))] = (1+2\lambda)^{-N/2}.$$

In other words,

$$R_N^2(1) \overset{(law)}{=} 2\gamma_{N/2}$$

where, for $a > 0$, γ_a denotes a gamma random variable of index a. Now, the classical Frullani's formula yields:

$$\frac{N}{2}\log(1+2\lambda) = \frac{N}{2}\int_0^{\infty}(1-\mathrm{e}^{-\lambda t})\frac{\mathrm{e}^{-t/2}}{t}dt.$$

Then, $R_N^2(1)$ satisfies the property 1) in Theorem 5.1 with

$$h(x) = \frac{N}{2}1_{]0,+\infty[}(x)\,\mathrm{e}^{-x/2}$$

and it is therefore self-decomposable (see also Exercise 5.8).

The process R_N^2 is 1-self-similar and $\mathbb{E}[R_N^2(1)] = N$. By Theorem 5.3, the process

$$V_t^N := R_N^2(t) - tN, \quad t \geq 0$$

is a peacock, and an associated martingale is

$$M_t^N := U_t^N - tN, \quad t \geq 0$$

where (U_t^N) denotes the 1-Sato process attached to $R_N^2(1)$ by Theorem 5.1.

We remark that, in this case, the process (V_t^N) itself is a *continuous martingale* and therefore obviously a peacock. In the following paragraphs, we give two expressions for the process (U_t^N). As we will see, this process is purely discontinuous with finite variation; consequently, the martingales (V_t^N) and (M_t^N), which have the same one-dimensional marginals, do not have the same law.

5.4.1.2 Expression of (U_t^N) from a Compound Poisson Process

We denote by $(\Pi_s, s \geq 0)$ the compound Poisson process with Lévy measure:

$$1_{]0,+\infty[}(t)\,\mathrm{e}^{-t}\,\mathrm{d}t.$$

This process allows to compute the distributions of a number of perpetuities

$$\int_0^{\infty}\mathrm{e}^{-\Lambda_s}d\Pi_s$$

where (Λ_s) is a particular Lévy process, independent of Π; see, e.g., Nilsen-Paulsen [134]. In the case $\Lambda_s = rs$, the following result seems to go back at least to Harrison [75].

Proposition 5.1. *The BDLP* (C_s^N) *of the self-decomposable random variable* $R_N^2(1)$, *in the sense of Remark 5.1, is:*

$$C_s^N = 2\,\Pi_{Ns/2}, \quad s \geq 0.$$

Proof. $a)$ First, recall that for a subordinator $(\tau_s,\, s \geq 0)$ and $f : \mathbb{R}_+ \longrightarrow \mathbb{R}_+$ Borel, there is the formula:

$$\mathbb{E}\left[\exp\left(-\int_0^\infty f(s)d\tau_s\right)\right] = \exp\left(-\int_0^\infty \phi(f(s))ds\right),$$

where ϕ is the Lévy exponent of $(\tau_s,\, s \geq 0)$. Consequently, a slight amplification of this formula is:

$$\mathbb{E}\left[\exp\left(-\mu\int_0^\infty f(s)d\tau_{As}\right)\right] = \exp\left(-\int_0^\infty \phi(\mu f(u/A))du\right)$$

for every $\mu, A > 0$.

$b)$ We set $C_s^N = 2\,\Pi_{Ns/2}$. Then, as a consequence of the previous formula with $\mu = 2, f(s) = \lambda\,\mathrm{e}^{-s}, A = N/2$, we get:

$$\mathbb{E}\left[\exp\left(-\lambda\int_0^\infty \mathrm{e}^{-s}dC_s^N\right)\right] = \exp\left(-\int_0^\infty \phi(2\lambda\,\mathrm{e}^{-\frac{2}{N}u})du\right)$$

with, for $x > 0$,

$$\phi(x) = \int_0^\infty (1-\mathrm{e}^{-tx})\,\mathrm{e}^{-t}dt = \frac{x}{1+x}.$$

$c)$ We obtain by change of variable:

$$\begin{aligned}\exp\left(-\int_0^\infty \phi(2\lambda\,\mathrm{e}^{-\frac{2}{N}u})du\right) &= \exp\left(-\int_0^\infty \frac{2\lambda\,\mathrm{e}^{-\frac{2}{N}u}}{1+2\lambda\,\mathrm{e}^{-\frac{2}{N}u}}du\right)\\ &= \exp\left(-\frac{N}{2}\int_0^{2\lambda}\frac{1}{1+x}dx\right)\\ &= (1+2\lambda)^{-N/2}.\end{aligned}$$

Consequently,

$$\mathbb{E}\left[\exp\left(-\lambda\int_0^\infty \mathrm{e}^{-s}dC_s^N\right)\right] = (1+2\lambda)^{-N/2},$$

which proves the result. □

By application of Theorem 5.2 we get:

Corollary 5.1. *One has:*

$$(U_t^N,\, t \geq 0) \overset{(law)}{=} \left(2\int_{-\frac{N}{2}\log t}^\infty \mathrm{e}^{-2s/N}d\Pi_s,\, t \geq 0\right)$$

where $(\Pi_s,\, s \leq 0)$ *denotes an independent copy of* $(-\Pi_{-s},\, s \leq 0)$.

5.4.1.3 Expression of (U_t^N) from the Local Time of a Perturbed Bessel Process

There is by now a wide literature on perturbed Bessel processes, a notion originally introduced by Le Gall-Yor [115, 116], and then studied by Chaumont-Doney [44], Doney-Warren-Yor [54]. We also refer the interested reader to Doney-Zhang [55].

We first introduce the perturbed Bessel process $(R_{1,\alpha}(t),\, t \geq 0)$ starting from 0, for $\alpha < 1$, as the nonnegative continuous strong solution of the equation

$$R_{1,\alpha}(t) = B_t + \frac{1}{2} L_t(R_{1,\alpha}) + \alpha M_t(R_{1,\alpha}) \tag{5.5}$$

where $L_t(R_{1,\alpha})$ is the semimartingale local time of $R_{1,\alpha}$ in 0 at time t, and

$$M_t(R_{1,\alpha}) = \sup_{0 \leq s \leq t} R_{1,\alpha}(s),$$

(B_t) denoting a standard linear Brownian motion starting from 0. (The strong solution property has been established in Chaumont-Doney [44].)

It is clear that the process $R_{1,0}$ is nothing else but the Bessel process R_1 (reflected Brownian motion).

We also denote by $T_t(R_{1,\alpha})$ the hitting time:

$$T_t(R_{1,\alpha}) = \inf\{s;\ R_{1,\alpha}(s) > t\}.$$

We set $L_{T_t}(R_{1,\alpha})$ for $L_{T_t(R_{1,\alpha})}(R_{1,\alpha})$.

Finally, in the sequel, we set

$$\alpha_N = 1 - \frac{N}{2}.$$

Proposition 5.2. *For any $\alpha < 1$, the process $(L_{T_t}(R_{1,\alpha}),\, t \geq 0)$ is a 1-Sato process, and we have*

$$(U_t^N,\, t \geq 0) \overset{(law)}{=} (L_{T_t}(R_{1,\alpha_N}),\, t \geq 0).$$

Proof. By the uniqueness in law of the solution to the equation (5.5), the process $R_{1,\alpha}$ is $(1/2)$-self-similar. As a consequence, the process $(L_{T_t}(R_{1,\alpha}),\, t \geq 0)$ is 1-self-similar.

On the other hand, the pair $(R_{1,\alpha}, M(R_{1,\alpha}))$ is a strong Markov process (see Doney-Warren-Yor [54, p. 239]). As

$$R_{1,\alpha}(u) = M_u(R_{1,\alpha}) = t \quad \text{if} \quad u = T_t(R_{1,\alpha}),$$

the fact that $(L_{T_t}(R_{1,\alpha}),\, t \geq 0)$ is an additive process follows from the strong Markov property.

Finally, we need to prove:

$$R_N^2(1) \overset{(law)}{=} L_{T_1}(R_{1,\alpha_N}).$$

For the remainder of the proof, we denote R_{1,α_N} by R, and $L_t(R_{1,\alpha_N})$, $T_t(R_{1,\alpha_N})$, $M_t(R_{1,\alpha_N}) \cdots$ are simply denoted respectively by L_t, T_t, M_t $\cdots$ As a particular case of the "balayage formula" (Yor [177], Revuz-Yor [150, VI.4]) we deduce from equation (5.5), that:

$$R_t \exp(-\lambda L_t) = \int_0^t \exp(-\lambda L_s) dR_s$$
$$= \int_0^t \exp(-\lambda L_s) dB_s + \frac{1-\exp(-\lambda L_t)}{2\lambda} + \alpha_N \int_0^t \exp(-\lambda L_s) dM_s.$$

Hence,

$$\exp(-\lambda L_t)(1+2\lambda R_t) = 1 + 2\lambda \int_0^t \exp(-\lambda L_s) dB_s + 2\lambda\, \alpha_N \int_0^t \exp(-\lambda L_s) dM_s.$$

From this formula, we learn that the martingale

$$\left(\int_0^{u \wedge T_t} \exp(-\lambda L_s) dB_s,\ u \geq 0\right)$$

is bounded; hence, by applying the optional stopping theorem, we get:

$$\mathbb{E}[\exp(-\lambda L_{T_t})](1+2\lambda t) = 1 + 2\lambda\, \alpha_N \mathbb{E}\left[\int_0^{T_t} \exp(-\lambda L_s) dM_s\right]$$
$$= 1 + 2\lambda\, \alpha_N \int_0^t \mathbb{E}[\exp(-\lambda L_{T_u})] du,$$

by time-changing. Setting

$$\varphi_\lambda(t) = \mathbb{E}[\exp(-\lambda L_{T_t})],$$

we obtain:

$$\varphi_\lambda(t) = \frac{1}{1+2\lambda t} + \frac{2\lambda\, \alpha_N}{1+2\lambda t} \int_0^t \varphi_\lambda(u) du.$$

Consequently

$$\varphi_\lambda(t) = (1+2\lambda t)^{-N/2}.$$

Therefore,

$$\mathbb{E}[\exp(-\lambda L_{T_1})] = (1+2\lambda)^{-N/2} = \mathbb{E}[\exp(-\lambda R_N^2(1))],$$

which proves the desired result. □

5.4.2 The Process $\left(\int_0^t R_N^2(s) ds,\ t \geq 0\right)$

5.4.2.1 A Class of Sato Processes

Let $(\ell_t,\, t \geq 0)$ be the local time in 0 of a linear Brownian motion $(B_t,\, t \geq 0)$ starting from 0. We denote, as usual, by $(\tau_t,\, t \geq 0)$ the inverse of this local time:

$$\tau_t = \inf\{s \geq 0;\ \ell_s > t\}.$$

Proposition 5.3. *Let $f(x,u)$ be a Borel function on $\mathbb{R}_+ \times \mathbb{R}_+$ such that:*

$$\forall t > 0, \quad \int\int_{\mathbb{R}_+ \times [0,t]} |f(x,u)| dx du < \infty. \tag{5.6}$$

Then the process $A^{(f)}$ defined by:

$$A_t^{(f)} = \int_0^{\tau_t} f(|B_s|, \ell_s) ds, \quad t \geq 0$$

is an integrable additive process. Furthermore,

$$\mathbb{E}[A_t^{(f)}] = 2 \int\int_{\mathbb{R}_+ \times [0,t]} f(x,u) dx du.$$

Proof. Assume first that f is nonnegative. Then,

$$A_t^{(f)} = \sum_{0 \leq u \leq t} \int_{\tau_{u-}}^{\tau_u} f(|B_s|, u) ds.$$

The Brownian excursions theory (Revuz-Yor [150, Chapter XII, Proposition 1.10]) yields:

$$\mathbb{E}[A_t^{(f)}] = \int_0^t \mathrm{d}u \int n(\mathrm{d}\varepsilon) \int_0^{V(\varepsilon)} ds\, f(|\varepsilon_s|, u)$$

where n denotes the Itô measure of Brownian excursions and $V(\varepsilon)$ denotes the life time of the excursion ε. The entrance law under n is given by:

$$n(\varepsilon_s \in \mathrm{d}x;\, s < V(\varepsilon)) = (2\pi s^3)^{-1/2} |x| \exp(-x^2/(2s)) dx.$$

Therefore:

$$\mathbb{E}[A_t^{(f)}] = 2 \int_0^t \mathrm{d}u \int_0^\infty dx\, f(x,u).$$

The additivity of the process $A^{(f)}$ follows easily from the fact that, for any $t \geq 0$, $(B_{\tau_t + s},\, s \geq 0)$ is a Brownian motion starting from 0, which is independent of $\mathscr{B}_{\tau_t}$ (where $(\mathscr{B}_u)$ is the natural filtration of B). □

Corollary 5.2. *We assume that f is a Borel function on $\mathbb{R}_+ \times \mathbb{R}_+$ satisfying (5.6) and which is m-homogeneous for $m > -2$, meaning that*

$$\forall a > 0,\ \forall (x,u) \in \mathbb{R}_+ \times \mathbb{R}_+, \quad f(ax, au) = a^m f(x,u).$$

Then the process $A^{(f)}$ is a $(m+2)$-Sato process.

Proof. This is a direct consequence of the scaling property of Brownian motion. □

5.4.2.2 A Particular Case

Let $N > 0$. We denote by $A^{(N)}$ the process $A^{(f)}$ with

$$f(x,u) = \frac{N^2}{4} 1_{(x \leq \frac{2}{N} u)}.$$

By Proposition 5.3, $(A_t^{(N)})$ is an integrable process and

$$\mathbb{E}[A_t^{(N)}] = \frac{N t^2}{2}.$$

We now consider the process Y_N defined by

$$Y_N(t) = \int_0^t R_N^2(s) ds, \quad t \geq 0.$$

Theorem 5.6. *The process $A^{(N)}$ is a 2-Sato process and*

$$Y_N(t) \overset{(1.d)}{=} A_t^{(N)}.$$

Proof. It is a direct consequence of Corollary 5.2 that $A^{(N)}$ is a 2-Sato process. By Le Gall-Yor [115] (see also Mansuy-Yor [126, Theorem 3.4, p.38]), the following extension of the Ray-Knight theorem holds:

For any $u > 0$,

$$(L_{\tau_u}^{a-(2u/N)}, 0 \leq a \leq (2u/N)) \overset{(law)}{=} (R_N^2(a), 0 \leq a \leq (2u/N))$$

where L_t^x denotes the local time of the semimartingale $(|B_s| - \frac{2}{N} \ell_s, s \geq 0)$ in x at time t.

We remark that

$$s \in [0, \tau_t] \Longrightarrow |B_s| - \frac{2}{N} \ell_s \geq -\frac{2t}{N}.$$

Therefore, the occupation times formula entails:

$$A_t^{(N)} = \frac{N^2}{4} \int_{-2t/N}^0 L_{\tau_t}^x dx = \frac{N^2}{4} \int_0^{2t/N} L_{\tau_t}^{x-(2t/N)} dx.$$

Thus, by the above mentioned extension of the Ray-Knight theorem,

$$A_t^{(N)} \overset{(1.d)}{=} \frac{N^2}{4} \int_0^{2t/N} R_N^2(s) ds.$$

The scaling property of R_N also yields the identity in law:

$$A_t^{(N)} \overset{(1.d)}{=} \int_0^t R_N^2(s) ds,$$

and the result follows from the definition of Y_N. □

We may now apply Theorem 5.3 to get:

Corollary 5.3. *The process V_N defined by:*

$$V_N(t) = Y_N(t) - \frac{Nt^2}{2}, \quad t \geq 0$$

is a peacock and an associated martingale is M_N defined by:

$$M_N(t) = A_t^{(N)} - \frac{Nt^2}{2}, \quad t \geq 0.$$

Moreover, M_N is a centered 2-Sato process.

5.4.2.3 Representation of $A^{(N)}$ as a Process of Hitting Times

Theorem 5.7. *The process $A^{(N)}$ is identical in law to the process*

$$T_t(R_{1,\alpha_N}), \quad t \geq 0$$

where R_{1,α_N} denotes the perturbed Bessel process defined in Paragraph 5.4.1.3 and

$$T_t(R_{1,\alpha_N}) = \inf\{s;\ R_{1,\alpha_N}(s) > t\}.$$

The proof can be found in Le Gall-Yor [116]. Nevertheless, for the convenience of the reader, we give again the proof below. A more general result, based on Doney-Warren-Yor [54], shall also be stated in the next subsection.

Proof. In this proof, we adopt the following notation: (B_t) still denotes a standard linear Brownian motion starting from 0, $S_t = \sup_{0 \leq s \leq t} B_s$ and $\sigma_t = \inf\{s;\ B_s > t\}$. Moreover, for $a < 1$ and $t \geq 0$, we set:

$$H_t^a := (B_t - aS_t), \quad X_t^a := \int_0^t 1_{\{B_s > aS_s\}} ds \quad \text{and} \quad Z_t^a := \inf\{s;\ X_s^a > t\}. \qquad \square$$

Lemma 5.1. *Let $a < 1$. Then*

$$M_t(H^a) := \sup_{0 \leq s \leq t} H_s^a = (1-a)\,S_t.$$

Consequently, $-aS_t = \alpha M_t(H^a)$, with $\alpha = -a/(1-a)$.

Proof. Since $a < 1$, we have, for $0 \leq s \leq t$,

$$(B_s - aS_s) \leq (1-a)\,S_s \leq (1-a)\,S_t.$$

Moreover, there exists $s_t \in [0,t]$ such that $B_{s_t} = S_t$ and therefore $S_{s_t} = S_t$. Hence, $B_{s_t} - aS_{s_t} = (1-a)\,S_t$. $\square$

Lemma 5.2. *Let $a < 1$ and recall $\alpha = -a/(1-a)$. We set:*

$$D_t^a = H^a_{Z_t^a},\; t \geq 0.$$

Then the processes D^a and $R_{1,\alpha}$ are identical in law.

Proof. $a)$ Since Z_t^a is a time of increase of the process

$$X_s^a = \int_0^s 1_{\{H_u^a > 0\}} du,\; s \geq 0,$$

we get: $D_t^a = H^a_{Z_t^a} \geq 0$. Moreover, since the process $(H^a)^+$ is obviously Z^a-continuous, the process D^a is continuous.
$b)$ By Tanaka's formula,

$$(H_t^a)^+ = \int_0^t 1_{\{H_s^a > 0\}} d(B_s - a S_s) + \frac{1}{2} L_t(H^a)$$

where $L_t(H^a)$ denotes the local time of the semimartingale H^a in 0 at time t. If $s > 0$ belongs to the support of dS_s, then $B_s = S_s$ and, since $a < 1$, $B_s - aS_s > 0$. Therefore,

$$(H_t^a)^+ = \int_0^t 1_{\{B_s - aS_s > 0\}} dB_s - a S_t + \frac{1}{2} L_t(H^a).$$

By Lemma 5.1, $-aS_t = \alpha M_t(H^a)$. Consequently,

$$D_t^a = \int_0^{Z_t^a} 1_{\{B_s - aS_s > 0\}} dB_s + \frac{1}{2} L_{Z_t^a}(H^a) + \alpha M_{Z_t^a}(H^a).$$

By the Dubins-Schwarz theorem, the process

$$\int_0^{Z_t^a} 1_{\{B_s - aS_s > 0\}} dB_s, \quad t \geq 0$$

is a Brownian motion. On the other hand, it is easy to see that

$$L_{Z_t^a}(H^a) = L_t(D^a) \quad \text{and} \quad M_{Z_t^a}(H^a) = M_t(D^a).$$

Therefore, the process D^a is a continuous and nonnegative solution to equation (5.5). □

To conclude the proof of Theorem 5.7, we observe that by Lévy's equivalence theorem ([150, Theorem VI.2.3]), the process $A^{(N)}$ is identical in law to the process

$$\frac{N^2}{4} \int_0^{\sigma_t} 1_{\{B_s > (1-\frac{2}{N}) S_s\}} ds, \quad t \geq 0.$$

By the scaling property of B, the above process has the same law as

$$\int_0^{\sigma_{Nt/2}} 1_{\{B_s > (1-\frac{2}{N}) S_s\}} ds = X^{1-\frac{2}{N}}_{\sigma_{Nt/2}}, \quad t \geq 0.$$

Now,

$$X_{\sigma_{Nt/2}}^{1-\frac{2}{N}} = \inf\left\{X_u^{1-\frac{2}{N}};\ S_u > \frac{Nt}{2}\right\}$$

and, by Lemma 5.1,

$$S_u = \frac{N}{2} M_u(H^{1-\frac{2}{N}}).$$

Thus,

$$\begin{aligned} X_{\sigma_{Nt/2}}^{1-\frac{2}{N}} &= \inf\{X_u^{1-\frac{2}{N}};\ M_u(H^{1-\frac{2}{N}}) > t\} \\ &= \inf\{v;\ M_v(D^{1-\frac{2}{N}}) > t\} = \inf\{v;\ D_v^{1-\frac{2}{N}} > t\}. \end{aligned}$$

The result then follows from Lemma 5.2. □

Corollary 5.4. *The process*

$$T_t(R_{1,\alpha_N}),\quad t \geq 0$$

is a 2-Sato process and

$$\int_0^t R_N^2(s)ds \overset{(1.d)}{=} T_t(R_{1,\alpha_N}).$$

5.4.3 The Process $\left(\frac{1}{K^2}\int_0^t s^{\frac{2(1-K)}{K}} R_N^2(s)ds,\ t \geq 0\right)$

In this subsection we extend Corollary 5.4. We fix two positive real numbers N and K. We first recall some important results on general perturbed Bessel processes $R_{K,\alpha}$ with $\alpha < 1$.

5.4.3.1 Perturbed Bessel Processes

We follow, in this paragraph, Doney-Warren-Yor [54]. We first recall the definition of the process $(R_{K,\alpha}(t), t \geq 0)$ with $K > 0$ and $\alpha < 1$.
The case $K = 1$ was already introduced in Paragraph 5.4.1.3. For $K > 1$, $R_{K,\alpha}$ is defined as a continuous nonnegative solution to

$$R_t = B_t + \frac{K-1}{2}\int_0^t \frac{1}{R_s} ds + \alpha M_t(R), \tag{5.7}$$

and, for $0 < K < 1$, $R_{K,\alpha}$ is defined as the square root of a continuous nonnegative solution to

$$X_t = 2\int_0^t \sqrt{X_s} dB_s + Kt + \alpha M_t(X). \tag{5.8}$$

We note that, for any $K > 0$, $(R_{K,0}(t)) \overset{(law)}{=} (R_K(t))$. As in the case $K = 1$, for any $K > 0$, the pair $(R_{K,\alpha}, M(R_{K,\alpha}))$ is strong Markov.

We denote, as before,

$$T_t(R_{K,\alpha}) = \inf\{s;\ R_{K,\alpha}(s) > t\}.$$

The following theorem, due to Doney-Warren-Yor [54, Theorem 5.2, p. 246] is an extension of the Ciesielski-Taylor theorem and of the Ray-Knight theorem.

Theorem 5.8. *The following identities in law hold.*

1)
$$\int_0^\infty 1_{(R_{K+2,\alpha}(s)\leq 1)} ds \overset{(law)}{=} T_1(R_{K,\alpha}).$$

2)
$$(L_\infty^a(R_{K+2,\alpha}),\, a \geq 0) \overset{(law)}{=} (\frac{a^{1-K}}{K} R_{2(1-\alpha)}(a^K),\, a \geq 0).$$

5.4.3.2 Identification of the Sato Process Attached to $Y_{N,K}$

We denote, for $N > 0$ and $K > 0$, by $Y_{N,K}$ the process:

$$Y_{N,K}(t) = \frac{1}{K^2} \int_0^t s^{\frac{2(1-K)}{K}} R_N^2(s) ds, \quad t \geq 0.$$

We also recall the notation:

$$\alpha_N = 1 - \frac{N}{2}.$$

Theorem 5.9. *The process*

$$T_{t^{1/K}}(R_{K,\alpha_N}), \quad t \geq 0$$

is a $(2/K)$*-Sato process and*

$$Y_{N,K}(t) \overset{(1.d)}{=} T_{t^{1/K}}(R_{K,\alpha_N}).$$

Proof. In the following proof, we denote R_{K,α_N} simply by R, and we set T_t and M_t for, respectively, $T_t(R)$ and $M_t(R)$.

The first part of the statement follows from the $(1/2)$-self-similarity of R and from the strong Markovianity of (R,M), taking into account that, for any $t \geq 0$,

$$R_{T_t} = M_{T_t} = t.$$

By occupation times formula, we deduce from 1) in Theorem 5.8,

$$\int_0^1 L_\infty^x(R_{K+2,\alpha_N}) dx \overset{(law)}{=} T_1.$$

Using then 2) in Theorem 5.8, we obtain:

$$\int_0^1 L_\infty^x(R_{K+2,\alpha_N}) dx \overset{(law)}{=} \int_0^1 \frac{x^{1-K}}{K} R_N^2(x^K) dx.$$

By change of variable, the last integral is equal to $Y_{N,K}(1)$, and hence,

$$Y_{N,K}(1) \overset{(law)}{=} T_1.$$

The final result now follows by self-similarity. □

Corollary 5.5. *The process*

$$V_{N,K}(t) := Y_{N,K}(t) - \frac{N}{2K} t^{2/K}, \quad t \geq 0$$

is a peacock, and an associated martingale is

$$M_{N,K}(t) := T_{t^{1/K}}(R_{K,\alpha_N}) - \frac{N}{2K} t^{2/K}, \quad t \geq 0,$$

which is a centered $(2/K)$-Sato process.

Finally, we have proven, in particular, that for any $\rho > -2$ and any $N > 0$, the random variable

$$\int_0^1 s^\rho R_N^2(s) ds$$

is self-decomposable. This result will be generalized and made precise in the next subsection, using completely different arguments.

5.4.4 The Quadratic Random Variables $\int R_N^2(s) d\mu(s)$

In this subsection, we consider a fixed measure μ on $\mathbb{R}_+^* =]0, +\infty[$ such that:

$$\int_{\mathbb{R}_+^*} s d\mu(s) < \infty.$$

5.4.4.1 Spectral Study of an Operator

We associate with μ an operator $K^{(\mu)}$ on $E = L^2(\mu)$ defined by

$$\forall f \in E, \quad K^{(\mu)} f(t) = \int_{\mathbb{R}_+^*} f(s)\, (t \wedge s) d\mu(s)$$

where $\wedge$ denotes the infimum.

Lemma 5.3. *The operator $K^{(\mu)}$ is a nonnegative symmetric Hilbert-Schmidt operator.*

Proof. As a consequence of the obvious inequality:

$$(t \wedge s)^2 \leq t s,$$

we get

$$\int\int_{(\mathbb{R}_+^*)^2} (t \wedge s)^2 d\mu(t) d\mu(s) \leq \left(\int_{\mathbb{R}_+^*} s d\mu(s) \right)^2,$$

and therefore $K^{(\mu)}$ is a Hilbert-Schmidt operator.

On the other hand, denoting by $(\bullet,\bullet)_E$ the scalar product in E, we have:

$$(K^{(\mu)} f, g)_E = \mathbb{E}\left[\int f(t)\, B_t d\mu(t) \int g(s)\, B_s d\mu(s)\right]$$

where B is a standard Brownian motion starting from 0. This entails that $K^{(\mu)}$ is non-negative symmetric. □

Lemma 5.4. *Let $\lambda \in \mathbb{R}$. Then λ is an eigenvalue of $K^{(\mu)}$ if and only if $\lambda > 0$ and there exists $f \in L^2(\mu)$, $f \neq 0$, such that:*

i) $\lambda f'' + f \cdot \mu = 0$ *in the (Schwartz) distribution sense on* $\mathbb{R}_+^*$, (5.9)

ii) f admits a representative which is absolutely continuous on $\mathbb{R}_+$, f' admits a representative which is right-continuous on $\mathbb{R}_+^*$,
(In the sequel, f and f' respectively always denote such representatives.)

iii)
$$f(0) = 0 \quad \textit{and} \quad \lim_{t \to \infty} f'(t) = 0.$$

Proof. Let $f \in L^2(\mu)$ and $g = K^{(\mu)} f$. We have, for μ-a.e. $t > 0$,

$$g(t) = \int_0^t du \int_{]u,+\infty[} f(s) d\mu(s). \tag{5.10}$$

Thus g admits a representative (still denoted by g) which is absolutely continuous on $\mathbb{R}_+$ and $g(0) = 0$. Moreover, g' admits a representative which is right-continuous on $\mathbb{R}_+^*$ and is given by:

$$g'(t) = \int_{]t,+\infty[} f(s) d\mu(s). \tag{5.11}$$

In particular

$$|g'(t)| \leq t^{-1/2} \left[\int_{]t,+\infty[} f^2(s) d\mu(s) \int_{]t,+\infty[} u d\mu(u)\right]^{1/2}. \tag{5.12}$$

Hence:

$$\lim_{t \to \infty} g'(t) = 0.$$

Besides, (5.11) entails:

$$g'' + f \cdot \mu = 0 \quad \text{in the distribution sense on } \mathbb{R}_+^*.$$

Consequently, 0 is not an eigenvalue of $K^{(\mu)}$ and the "only if" part is proven.

Conversely, let $f \in L^2(\mu)$, $f \neq 0$, and $\lambda > 0$ such that properties *i*), *ii*), *iii*) hold. Then

$$\lambda f'(t) = \int_{]t,\infty[} f(s) d\mu(s).$$

Hence

$$\lambda f(t) = \int_0^t \mathrm{d}u \int_{]u,\infty[} f(s) d\mu(s) = K^{(\mu)} f(t),$$

which proves the "if" part. □

We note that, since 0 is not an eigenvalue of $K^{(\mu)}$, $K^{(\mu)}$ is actually a *positive definite* symmetric operator. On the other hand, by the previous proof, the functions $f \in L^2(\mu)$, $f \neq 0$, satisfying properties *i*), *ii*), *iii*) in the statement of Lemma 5.4, are the eigenfunctions of the operator $K^{(\mu)}$ corresponding to the eigenvalue $\lambda > 0$.

Lemma 5.5. *Let f be an eigenfunction of $K^{(\mu)}$. Then,*

$$|f(t)| = o(t^{1/2}) \quad and \quad |f'(t)| = o(t^{-1/2})$$

when t tends to ∞.

Proof. This is a direct consequence of (5.12). □

Lemma 5.6. *Let f_1 and f_2 be eigenfunctions of $K^{(\mu)}$ with respect to the same eigenvalue. Then,*

$$\forall t > 0, \quad f_1'(t) f_2(t) - f_1(t) f_2'(t) = 0.$$

Proof. By (5.9),

$$(f_1' f_2 - f_1 f_2')' = 0 \quad \text{in the sense of distributions on } \mathbb{R}_+^*.$$

By right-continuity, there exists $C \in \mathbb{R}$ such that:

$$\forall t > 0, \quad f_1'(t) f_2(t) - f_1(t) f_2'(t) = C.$$

Letting t tend to $+\infty$, we deduce from Lemma 5.5 that $C = 0$. □

Lemma 5.7. *Let f be a solution of (5.9) with $\lambda > 0$, and let $a > 0$. We assume as previously that f (resp. f') denotes the representative which is absolutely continuous (resp. right-continuous) on $\mathbb{R}_+^*$. If $f(a) = f'(a) = 0$, then, for any $t \geq a$, $f(t) = 0$.*

Proof. This lemma is quite classical if the measure μ admits a continuous density with respect to the Lebesgue measure (see, for instance, [48]). The proof may be easily adapted to this more general case. □

We are now able to state the main result of this subsection.

Theorem 5.10. *The operator $K^{(\mu)}$ is a positive definite symmetric compact operator whose all eigenvalues are simple, i.e. the dimension of each eigenspace is 1.*

Proof. It only remains to prove that the eigenvalues are simple. For this purpose, let $\lambda > 0$ be an eigenvalue and let f_1 and f_2 be eigenfunctions with respect to this eigenvalue. Let $a > 0$ with $\mu(\{a\}) = 0$. By Lemma 5.6,

$$f_1'(a)\, f_2(a) - f_1(a)\, f_2'(a) = 0.$$

Hence, there exist c_1 and c_2 with $c_1^2 + c_2^2 > 0$ such that, setting $f = c_1 f_1 + c_2 f_2$, we have

$$f(a) = f'(a) = 0.$$

By Lemma 5.7, $f(t) = 0$ for any $t \geq a$. But, since $\mu(\{a\}) = 0$, f' is also left-continuous at a. Then, we may reason on $]0,a]$ as on $[a,+\infty[$ and therefore we also have $f(t) = 0$ for $0 < t \leq a$. Finally,

$$c_1 f_1 + c_2 f_2 = 0,$$

which proves the result. □

In the following, we denote by $\lambda_1 > \lambda_2 > \cdots$ the decreasing (possibly finite) sequence of the eigenvalues of $K^{(\mu)}$. Of course, this sequence depends on μ, which we omit in the notation. The following corollary plays an essential role in the sequel.

Corollary 5.6. *There exists a Hilbert basis $(f_n)_{n \geq 1}$ in $L^2(\mu)$ such that:*

$$\forall n \geq 1 \quad K^{(\mu)} f_n = \lambda_n f_n.$$

Since $K^{(\mu)}$ is Hilbert-Schmidt,

$$\sum_{n \geq 1} \lambda_n^2 < \infty.$$

It will be shown in Paragraph 5.4.4.4 (see Theorem 5.11) that actually

$$\sum_{n \geq 1} \lambda_n < \infty,$$

i.e. $K^{(\mu)}$ is trace-class.

We now consider two particular types of measures μ.

5.4.4.2 $\mu = \sum_{j=1}^{n} a_j \delta_{t_j}$

Let $a_1, \cdots, a_n$ be positive real numbers and $0 < t_1 < \cdots < t_n$. We denote by δ_t the Dirac measure at t and we consider, in this paragraph,

$$\mu = \sum_{j=1}^{n} a_j\, \delta_{t_j}.$$

By the previous study, the sequence of eigenvalues of $K^{(\mu)}$ is finite if and only if the space $L^2(\mu)$ is finite dimensional, that is if μ is of the above form. In this case, the eigenvalues of $K^{(\mu)}$ are the eigenvalues of the matrix $(m_{i,j})_{1\le i,j\le n}$ with

$$m_{i,j} = \sqrt{a_i a_j}\, t_{i\wedge j}.$$

In particular, by the previous study, such a matrix has n distinct eigenvalues, which are > 0.

5.4.4.3 $\mu = Ct^\rho\, 1_{]0,1]}(t)\,\mathrm{d}t$

In this paragraph, we consider

$$\mu = Ct^\rho\, 1_{]0,1]}(t)\,\mathrm{d}t$$

with $C > 0$ and $\rho > -2$. By Lemma 5.4, the eigenfunctions f of $K^{(\mu)}$ associated with $\lambda > 0$ are characterized by:

$$\lambda f''(x) + Cx^\rho f(x) = 0 \quad \text{on} \quad]0,1[, \tag{5.13}$$

$$f(0) = 0, \quad f'(1) = 0.$$

We set $\sigma = (\rho+2)^{-1}$ and $\nu = \sigma - 1$. For $a > -1$, we recall the definition of the Bessel function J_a:

$$J_a(x) = \sum_{k=0}^{\infty} \frac{(-1)^k (x/2)^{a+2k}}{k!\,\Gamma(a+k+1)}.$$

Then, the only function f satisfying (5.13) and $f(0) = 0$ is, up to a multiplicative constant,

$$f(x) = x^{1/2} J_\sigma\left(2\sigma\sqrt{\frac{C}{\lambda}}\, x^{1/2\sigma}\right).$$

We deduce from the equality, which is valid for $a > 1$,

$$aJ_a(x) + xJ_a'(x) = xJ_{a-1}(x)$$

that $f'(1) = 0$ if and only if

$$J_\nu\left(2\sigma\sqrt{\frac{C}{\lambda}}\right) = 0.$$

Denote by $(j_{\nu,k},\ k \ge 1)$ the sequence of the positive zeros of J_ν. Then the sequence $(\lambda_k,\ k \ge 1)$ of eigenvalues of $K^{(\mu)}$ is given by:

$$2\sigma\sqrt{\frac{C}{\lambda_k}} = j_{\nu,k}, \quad k \ge 1$$

or, since $\sigma = \nu + 1$,

$$\lambda_k = 4C(\nu+1)^2 j_{\nu,k}^{-2}, \quad k \geq 1.$$

Particular case. Suppose $\rho = 0$. Then $\nu = -1/2$ and

$$J_\nu(x) = J_{-1/2}(x) = \left(\frac{2}{\pi x}\right)^{1/2} \cos(x).$$

Hence,

$$\lambda_k = 4C\pi^{-2}(2k-1)^{-2}, \quad k \geq 1.$$

5.4.4.4 Representation of $\int B_s^2 d\mu(s)$

We again consider the general setting defined in Paragraph 5.4.4.1, the notation of which we keep.

In this paragraph, we study the random variable

$$Y_1^{(\mu)} := \int B_s^2 d\mu(s).$$

The use of the operator $K^{(\mu)}$ and of its spectral decomposition in the type of study we develop below, is called the Karhunen-Loeve decomposition method. It has a long history which goes back at least to Kac-Siegert [101, 100]. We also refer to the recent paper [51] and to the references therein.

Theorem 5.11. *The eigenvalues* $(\lambda_k,\, k \geq 1)$ *of the operator* $K^{(\mu)}$ *satisfy*

$$\sum_{k\geq 1} \lambda_k = \int_{\mathbb{R}_+^*} t\,d\mu(t) \; (< \infty, \textit{ by hypothesis}).$$

Moreover, there exists a sequence $(\Gamma_n,\, n \geq 1)$ *of independent normal variables such that:*

$$Y_1^{(\mu)} \overset{(law)}{=} \sum_{n\geq 1} \lambda_n \Gamma_n^2.$$

Proof. We deduce from Corollary 5.6, by the Bessel-Parseval equality, that:

$$Y_1^{(\mu)} = \sum_{n\geq 1} \left(\int B_s f_n(s) d\mu(s)\right)^2 \quad \text{a.s.}$$

Taking expectations, we get

$$\int_{\mathbb{R}_+^*} t\,d\mu(t) = \sum_{n\geq 1} (K^{(\mu)} f_n, f_n)_E = \sum_{n\geq 1} \lambda_n.$$

We set, for $n \geq 1$,

$$\Gamma_n = \frac{1}{\sqrt{\lambda_n}} \int B_s f_n(s) d\mu(s).$$

Then $(\Gamma_n,\ n \geq 1)$ is a Gaussian sequence and

$$\mathbb{E}[\Gamma_n \Gamma_m] = \frac{1}{\sqrt{\lambda_n \lambda_m}} (K^{(\mu)} f_n, f_m)_E = \delta_{n,m}$$

where $\delta_{n,m}$ denotes Kronecker's symbol. Hence, the result follows. □

Corollary 5.7. *The Laplace transform of* $Y_1^{(\mu)}$ *is*

$$F_1^{(\mu)}(t) = \prod_{n\geq 1} (1 + 2t\lambda_n)^{-1/2}.$$

Proof. This is a direct consequence of the previous theorem, taking into account that, if Γ is a normal variable, then

$$\Gamma^2 \overset{(law)}{=} 2\gamma_{1/2}.$$

□

5.4.4.5 Representation of $\int R_N^2(s) d\mu(s)$

We now consider the random variable

$$Y_N^{(\mu)} := \int R_N^2(s) d\mu(s).$$

A number of explicit computations of the Laplace transforms of these variables are found in Pitman-Yor ([145, Section 2]).

Theorem 5.12. *There exists a sequence* $(\Theta_{N,n},\ n \geq 1)$ *of independent variables with, for any* $n \geq 1$,

$$\Theta_{N,n} \overset{(law)}{=} R_N^2(1) \overset{(law)}{=} 2\gamma_{N/2}$$

such that

$$Y_N^{(\mu)} \overset{(law)}{=} \sum_{n\geq 1} \lambda_n \Theta_{N,n}. \tag{5.14}$$

Moreover, the Laplace transform of $Y_N^{(\mu)}$ *is*

$$F_N^{(\mu)}(t) = \prod_{n\geq 1} (1 + 2t\lambda_n)^{-N/2}. \tag{5.15}$$

Proof. It is clear, for instance from Revuz-Yor [150, Chapter XI, Theorem 1.7], that

$$F_N^{(\mu)}(t) = [F_1^{(\mu)}(t)]^N.$$

Therefore, by Corollary 5.7, formula (5.15) holds. Formula (5.14) then follows directly by the injectivity of the Laplace transform. □

Corollary 5.8. *The random variable $Y_N^{(\mu)}$ is self-decomposable. The function h, which is decreasing on $]0,+\infty[$ and associated with $Y_N^{(\mu)}$ in Theorem 5.1, is*

$$h(x) = \frac{N}{2} \sum_{n \geq 1} \exp\left(-\frac{1}{2\lambda_n} x\right).$$

Proof. We saw in Subsection 5.4.1.1 that $R_N^2(1)$ satisfies the property 1) in Theorem 5.1 with

$$h(x) = \frac{N}{2} 1_{]0,+\infty[}(x) \, \mathrm{e}^{-x/2}$$

and it is therefore self-decomposable. Using then the representation (5.14) of $Y_N^{(\mu)}$, we obtain the desired result. □

As a consequence, following Bondesson [34], we see that $Y_N^{(\mu)}$ is a generalized gamma convolution (GGC) whose Thorin measure is the discrete measure:

$$\frac{N}{2} \sum_{n \geq 1} \delta_{1/2\lambda_n}.$$

Particular case. We consider here, as in Subsection 5.4.3, the particular case:

$$\mu = \frac{1}{K^2} t^{\frac{2(1-K)}{K}} 1_{]0,1]}(t) \, \mathrm{d}t.$$

Then, $Y_N^{(\mu)}$ is the random variable $Y_{N,K}(1)$ studied in Subsection 5.4.3. As a consequence of Paragraph 5.4.4.3 with

$$C = \frac{1}{K^2} \quad \text{and} \quad \rho = \frac{2}{K} - 2,$$

we have

$$\lambda_k = j_{\nu,k}^{-2}, \quad k \geq 1$$

with $\nu = \frac{K}{2} - 1$. Moreover, by Theorem 5.9,

$$Y_2^{(\mu)} \overset{(law)}{=} T_1(R_K).$$

It is known (see Kent [107] and, for instance, Borodin-Salminen [35, formula 2.0.1, p. 387]) that

$$\mathbb{E}[\exp(-t\,T_1(R_K))] = \frac{2^{-\nu}}{\Gamma(\nu+1)} \frac{(\sqrt{2t})^{\nu}}{I_\nu(\sqrt{2t})}$$

where I_ν denotes the modified Bessel function:

$$I_\nu(x) = \sum_{k=0}^{\infty} \frac{(x/2)^{\nu+2k}}{k!\,\Gamma(\nu+k+1)}.$$

We set:

$$\widehat{I}_\nu(x) = \sum_{k=0}^{\infty} \frac{(x/2)^{2k}}{k!\,\Gamma(\nu+k+1)}.$$

Therefore, by formula (5.15) in the case $N=2$, we recover the following representation:

$$\widehat{I}_\nu(x) = \frac{1}{\Gamma(\nu+1)} \prod_{k\geq 1} \left(1+\frac{x^2}{j_{\nu,k}^2}\right).$$

In particular ($\nu = -1/2$),

$$\cosh(x) = \prod_{k\geq 1} \left(1+\frac{4x^2}{\pi^2\,(2k-1)^2}\right).$$

Likewise we obtain, for $\nu = 1/2$,

$$\frac{\sinh(x)}{x} = \prod_{k\geq 1} \left(1+\frac{x^2}{\pi^2\,k^2}\right).$$

5.4.4.6 Sato Process Attached to $Y_N^{(\mu)}$

Theorem 5.13. *Let (U_t^N) be the 1-Sato process attached to $R_N^2(1)$ (cf. Subsection 5.4.1). Then, the 1-Sato process attached to $Y_N^{(\mu)}$ is $(U_t^{(N,\mu)})$ defined by:*

$$U_t^{(N,\mu)} = \sum_{n\geq 1} \lambda_n\, U_t^{N,n}, \quad t \geq 0$$

where $((U_t^{N,n}),\ n \geq 1)$ denotes a sequence of independent processes such that, for $n \geq 1$,

$$(U_t^{N,n}) \overset{(law)}{=} (U_t^N).$$

Proof. This is a direct consequence of Theorem 5.12. □

Corollary 5.9. *The process*

$$V_t^{(N,\mu)} := \int_{\mathbb{R}_+^*} (R_N^2(t\,s) - N\,t\,s)\,d\mu(s), \quad t \geq 0$$

is a peacock and an associated martingale is

$$M_t^{(N,\mu)} := U_t^{(N,\mu)} - N t \int_{\mathbb{R}_+^*} s\,d\mu(s), \quad t \geq 0.$$

The above martingale $(M_t^{(N,\mu)})$ is purely discontinuous. We also may associate to the peacock $(V_t^{(N,\mu)})$ a continuous martingale, as we now state.

Theorem 5.14. *A continuous martingale associated to the peacock* $(V_t^{(N,\mu)})$ *is*

$$\sum_{n\geq 1} \lambda_n \left((R_N^{(n)})^2(t) - Nt\right), \quad t \geq 0$$

where $((R_N^{(n)}(t)),\ n \geq 1)$ *denotes a sequence of independent processes such that, for* $n \geq 1$,

$$(R_N^{(n)}(t)) \overset{(law)}{=} (R_N(t)).$$

Proof. This is again a direct consequence of Theorem 5.12. □

We can also explicit the relation between $U^{(N,\mu)}$ and $U^{(N',\mu)}$. Let $C^{(N,\mu)}$ (resp. $C^{(N',\mu)}$) be the BDLP of $Y_N^{(\mu)}$ (resp. $Y_{N'}^{(\mu)}$). We see, by Laplace transform, that

$$(C_s^{(N',\mu)},\ s \geq 0) \overset{(law)}{=} (C_{N's/N}^{(N,\mu)},\ s \geq 0). \tag{5.16}$$

Then, using the relations between the processes U and C given in Theorem 5.2, we obtain:

Proposition 5.4. *We have:*

$$(U_t^{(N',\mu)},\ t \geq 0) \overset{(law)}{=} \left(\int_0^{t^{N'/N}} s^{\frac{N-N'}{N'}} dU_s^{(N,\mu)},\ t \geq 0\right).$$

Proof. By Theorem 5.2,

$$(U_t^{(N',\mu)},\ t \geq 0) \overset{(law)}{=} \left(\int_{-\log t}^{\infty} e^{-s} dC_s^{(N',\mu)},\ t \geq 0\right).$$

Therefore, we obtain by (5.16),

$$(U_t^{(N',\mu)},\ t \geq 0) \overset{(law)}{=} \left(\int_{-\log t}^{\infty} e^{-s} dC_{N's/N}^{(N,\mu)},\ t \geq 0\right)$$

or, after the change of variable: $N's/N = u$,

$$(U_t^{(N',\mu)},\ t \geq 0) \overset{(law)}{=} \left(\int_{-\log t^{N'/N}}^{\infty} e^{-Nu/N'} dC_u^{(N,\mu)},\ t \geq 0\right).$$

Then Theorem 5.2 yields:

$$(U_t^{(N',\mu)},\ t \geq 0) \overset{(law)}{=} \left(\int_{-\log t^{N'/N}}^{\infty} e^{\frac{N'-N}{N'}u} dU_{e^{-u}}^{(N,\mu)},\ t \geq 0\right)$$

and, after the change of variable: $e^{-u} = s$,

$$(U_t^{(N',\mu)},\ t \geq 0) \overset{(law)}{=} \left(\int_0^{t^{N'/N}} s^{\frac{N-N'}{N'}} dU_s^{(N,\mu)},\ t \geq 0\right),$$

which is the desired result. □

Corollary 5.10. *For $N > 0$ and $K > 0$, we set, with the notation of Subsection 5.4.3,*

$$T_t^{N,K} = T_t(R_{K,\alpha_N}), \quad t \geq 0.$$

Then, for $N > 0$, $N' > 0$ and $K > 0$,

$$(T_t^{N',K}, t \geq 0) \overset{(law)}{=} \left(\int_0^{t^{N'/N}} s^{2\frac{N-N'}{N'}} dT_s^{N,K}, t \geq 0 \right).$$

Proof. By Theorem 5.9, $(T_{t^{1/2}}^{N,K})$ is the 1-Sato process attached to $Y_N^{(\mu)}$ defined from

$$\mu = \frac{1}{K^2} t^{\frac{2(1-K)}{K}} 1_{]0,1]}(t)\, \mathrm{d}t.$$

□

It is well-known, and goes back to Shiga-Watanabe [160], that $R_N^2(\centerdot)$, considered as a random variable taking values in $C(\mathbb{R}_+, \mathbb{R}_+)$ is infinitely divisible. Furthermore, we have proven the self-decomposability of the quadratic functionals: $\int_{]0,\infty[} R_N^2(s) d\mu(s)$. It then seems natural to wonder about the self-decomposability of $R_N^2(\centerdot)$, but this property is ruled out, as the 2-dimensional vectors: $(R_N^2(t_1), R_N^2(t_1+t_2))$ are not self-decomposable. The aim of the following exercise is to prove this negative result.

Exercise 5.10 (Two-dimensional marginals of squared Bessel processes are not self-decomposable). Consider, for $N > 0$, $t_1, t_2 > 0$, the $\mathbb{R}^2$-valued random variable

$$Y := (R_N^2(t_1), R_N^2(t_1+t_2)).$$

For such an $\mathbb{R}^2$-valued random variable, we can also define the notion of self-decomposability as in Section 5.2. Theorem 5.1, suitably modified, is still valid.

1) Prove that:

$$\mathbb{E}[\exp(-\lambda_1 R_N^2(t_1) - \lambda_2 R_N^2(t_1+t_2))] = [P(\lambda)]^{-N/2}$$

with

$$P(\lambda) = 1 + 2\lambda_1 t_1 + 2\lambda_2(t_1+t_2) + 4\lambda_1\lambda_2 t_1 t_2.$$

2) Suppose that Y is self-decomposable. Then, by the two-dimensional extension of Theorem 5.1, we have:

$$\log(P(\lambda)) = \int\int_{(\mathbb{R}_+^*)^2} (1 - \exp(-\lambda_1 x_1 - \lambda_2 x_2)) \frac{H(x_1,x_2)}{x_1^2 + x_2^2} dx_1 dx_2$$

with H a decreasing function on each half line with origin $(0,0)$.
Taking the derivative with respect to λ_1, show that:

$$\frac{2t_1(1+2\lambda_2 t_2)}{P(\lambda)} = \int\int_{(\mathbb{R}_+^*)^2} \exp(-\lambda_1 x_1 - \lambda_2 x_2) \frac{x_1 H(x_1,x_2)}{x_1^2 + x_2^2} dx_1 dx_2 .$$

Letting then λ_2 tend to ∞, show that:

$$\frac{2t_1t_2}{t_1+t_2+2\lambda_1 t_1 t_2}=0,$$

which yields a contradiction.

Conclude that the $\mathbb{R}^2$-valued random variable Y is not self-decomposable.

5.5 The Sato Sheet Method

5.5.1 The Sato Sheet Attached to a Self-Decomposable Random Variable

In this subsection, we fix a real valued self-decomposable random variable X, and we keep the notation of Section 5.2. In particular, S^X denotes the Sato process attached to X and C^X denotes the BDLP of X. We now shall extend the Sato process S^X into a Sato sheet $\widehat{S}^X$.

In Subsection 2.4.2.4 of Chapter 2, we defined a Lévy sheet $W^{(L)}$ extending a Lévy process L. We now define $(\widehat{C}^X_{s,t};\ s\geq 0, t\geq 0)$ as the Lévy sheet extending the Lévy process C^X. Thus, $\widehat{C}^X = W^{(C^X)}$. We then extend the definition of $(\widehat{C}^X_{s,t})$ to $s\leq 0\,, t\geq 0$, by taking, for $(\widehat{C}^X_{s,t}\ ;\ s\leq 0\,, t\geq 0)$, an independent copy of $(-\widehat{C}^X_{-s,t}\ ;\ s\leq 0\,, t\geq 0)$. Denoting by $\mathbb{D}_0$ the Skorohod space consisting of all càdlàg functions from $\mathbb{R}_+$ into $\mathbb{R}$, vanishing at 0, we have the following extension of Theorem 2.4.2:

Proposition 5.5. *The $\mathbb{D}_0$-valued process $(\widehat{C}^X_{s,\bullet}\ ,\ s\in\mathbb{R})$ has independent increments and, for every $s_1,s_2\in\mathbb{R}$ such that $s_1\leq s_2$, the process*

$$(\widehat{C}^X_{s_2,t}-\widehat{C}^X_{s_1,t}\ ,\ t\geq 0)$$

is a Lévy process starting from 0 and distributed as

$$(C^X_{(s_2-s_1)t}\ ,\ t\geq 0)\,.$$

Hence, $(\widehat{C}^X_{s,\bullet}\ ,\ s\in\mathbb{R})$ is a $\mathbb{D}_0$-valued Lévy process indexed by $\mathbb{R}$.

Corollary 5.11. *For any $a\in\mathbb{R}$, the $\mathbb{D}_0$-valued processes*

$$(\widehat{C}^X_{s+a,\bullet}-\widehat{C}^X_{a,\bullet}\ ,\ s\in\mathbb{R})\quad \textit{and}\quad (\widehat{C}^X_{s,\bullet}\ ,\ s\in\mathbb{R})$$

have the same law.

According to Theorem 5.2, we are led to the following definition.

Definition 5.3. We define, for $\lambda\geq 0$, and $t\geq 0$:

$$\widehat{S}^X_{\lambda,t}=\int_{-\log t}^{+\infty} e^{-s}\,\mathrm{d}_s\widehat{C}^X_{s,\lambda}\,.$$

We call $\widehat{S}^X$, the *Sato sheet attached to X*.

We gather, in the next statement, several important properties of the sheet $(\widehat{S}^X_{\lambda,t};\ \lambda \geq 0, t \geq 0)$.

Theorem 5.15. *The sheet $\widehat{S}^X$ satisfies the following:*

i) the process $(\widehat{S}^X_{\bullet,t},\ t \geq 0)$ has independent increments,
ii) for any fixed $c \geq 0$,

$$(\widehat{S}^X_{\bullet,ct},\ t \geq 0) \overset{(law)}{=} (c\widehat{S}^X_{\bullet,t},\ t \geq 0),$$

(In particular, for any fixed $t \geq 0$, $\widehat{S}^X_{\bullet,t} \overset{(law)}{=} t\widehat{S}^X_{\bullet,1}$.)
iii) for any fixed $t \geq 0$, $\widehat{S}^X_{\bullet,t}$ is a Lévy process,
iv) $\widehat{S}^X_{1,\bullet} \overset{(law)}{=} S^X_{\bullet}$, and, in particular, $\widehat{S}^X_{1,1} \overset{(law)}{=} X$.

Proof. Since, by Proposition 5.5, the $\mathbb{D}_0$-valued process $(\widehat{C}^X_{s,\bullet},\ s \in \mathbb{R})$ has independent increments, we deduce directly from Definition 5.3 that the $\mathbb{D}_0$-valued process $(\widehat{S}^X_{\bullet,t},\ t \geq 0)$ also has independent increments.

Let $c > 0$. Then, by the change of variable: $s = u - \log c$, we obtain:

$$\widehat{S}^X_{\lambda,ct} = c \int_{-\log t}^{+\infty} e^{-u}\, d_u \left(\widehat{C}^X_{u-\log c,\lambda} - \widehat{C}^X_{-\log c,\lambda}\right)$$

and therefore, Property ii) follows from Corollary 5.11.

By Theorem 2.4.2, we know that the process $(\widehat{C}^X_{\bullet,t},\ t \geq 0)$ also has stationary independent increments. This entails easily Property iii).

By Theorem 2.4.2, $\widehat{C}^X_{\bullet,1} \overset{(law)}{=} C^X_{\bullet}$. Therefore, Property iv) follows from Definition 5.3 and Theorem 5.2. □

Remark 5.2. According to Properties *i)* and *ii)* in Theorem 5.15, the process $(\widehat{S}^X_{\bullet,t},\ t \geq 0)$ could be called a $\mathbb{D}_0$-valued 1-Sato process.

Exercise 5.11. Let $a > 0$ and B_a be a real valued, centered, Gaussian r.v. with variance a. Show that:

$$\left(\widehat{S}^{B_a}_{\lambda,t};\ \lambda \geq 0, t \geq 0\right) \overset{(law)}{=} (W_{a\lambda,t^2};\ \lambda \geq 0, t \geq 0)$$

where W denotes the standard Brownian sheet.

5.5.2 The Sato Sheet Method

In this subsection, we consider a real valued Lévy process $L = (L_t,\ t \geq 0)$ such that the random variable L_1 is self-decomposable. This implies in particular that L is self-decomposable as a process, see Exercise 5.13. We denote by ψ the characteristic exponent of L. The Sato sheet $\widehat{S}^{L_1}$ will be denoted by $\widehat{S}^{(L)}$. We define the filtration $(\mathscr{G}^{(L)}_t,\ t \geq 0)$ as the filtration of the process $(\widehat{S}^{(L)}_{\bullet,t},\ t \geq 0)$. In other words,

$$\mathscr{G}^{(L)}_t = \sigma\{\widehat{S}^{(L)}_{\lambda,s};\ \lambda \geq 0, 0 \leq s \leq t\}.$$

We first give two useful propositions:

Proposition 5.6. *For every $t \geq 0$,*

$$\widehat{S}^{(L)}_{\bullet,t} \overset{(law)}{=} t\,L_{\bullet}.$$

Proof. By Property ii) in Theorem 5.15,

$$\widehat{S}^{(L)}_{\bullet,t} \overset{(law)}{=} t\,\widehat{S}^{(L)}_{\bullet,1}.$$

Now, by Property iii), $\widehat{S}^{(L)}_{\bullet,1}$ is a Lévy process, and, by Property iv), this Lévy process coincides with L at time 1. This yields the desired result. □

According to Proposition 5.6, $\left(\widehat{S}^{(L)}_{\lambda,t};\ \lambda \geq 0, t \geq 0\right)$ will be called the *Sato sheet extending the self-decomposable Lévy process L.*

Proposition 5.7. *Let $h : \mathbb{R}_+ \longrightarrow \mathbb{R}$ be a bounded Borel function with compact support, and let*

$$X_h = \int_0^{\infty} h(s) dL_s.$$

Then, X_h is a self-decomposable r.v. to which the following Sato process is attached:

$$S^{X_h}_t = \int_0^{\infty} h(s)\, d_s \widehat{S}^{(L)}_{s,t}, \quad t \geq 0.$$

Moreover, the characteristic exponent of X_h is:

$$\psi_h(\lambda) = \int_0^{\infty} \psi(\lambda\, h(s)) ds.$$

Proof. By Properties i) and ii) in Theorem 5.15, the process:

$$\int_0^{\infty} h(s)\, \mathrm{d}_s \widehat{S}^{(L)}_{s,t}, \quad t \geq 0$$

is a Sato process. Besides, by Proposition 5.6,

$$\int_0^{\infty} h(s)\, \mathrm{d}_s \widehat{S}^{(L)}_{s,t} \overset{(1.d)}{=} t X_h.$$

This proves the first part of the proposition.

The second part is classical and can be easily obtained by approximating the function h by simple functions. □

Theorem 5.16. *Let h, X_h, ψ_h and S^{X_h} be as in Proposition 5.7. We set, for every $t \geq 0$,*

$$A^h_t = \exp(\mathrm{i} t X_h + \psi_h(t)) \quad \textit{and} \quad M^h_t = \exp(\mathrm{i} S^{X_h}_t + \psi_h(t)).$$

Then (M^h_t) is a $\mathbb{C}$-valued $(\mathscr{G}^{(L)}_t)$-martingale and

$$A^h_t \overset{(1.d)}{=} M^h_t.$$

In particular, A^h is a $\mathbb{C}$-valued peacock.

Proof. By Proposition 5.7, we see, using Property *i*) in Theorem 5.15, that, if $0 \le s \le t$, then $S_t^{X_h} - S_s^{X_h}$ is independent of $\mathscr{G}_s^{(L)}$. This leads easily to the result. □

The important fact to notice, in the above theorem, is that the involved filtration, namely $(\mathscr{G}_t^{(L)})$, does not depend on h. This allows to consider integrals, with respect to h, of processes A^h, as in the following corollary.

Corollary 5.12. *Let σ be a signed finite measure on $\mathbb{R}_+$ such that:*

$$\forall r > 0, \quad \int \exp(r\lambda) d|\sigma|(\lambda) < \infty .$$

Then,

$$\left(\int \exp\left(\mathrm{i}\widehat{S}_{\lambda,t}^{(L)} + \lambda\, \psi(t)\right) d\sigma(\lambda)\, ,\, t \ge 0\right)$$

is a $\mathbb{C}$-valued $(\mathscr{G}_t^{(L)})$-martingale associated to the $\mathbb{C}$-valued peacock:

$$\left(\int \exp\left[\mathrm{i}t L_\lambda + \lambda\, \psi(t)\right] d\sigma(\lambda)\, ,\, t \ge 0\right).$$

The previous results admit real valued versions under some exponential integrability condition.

Proposition 5.8. *Assume that, for every $r \ge 0$,*

$$\mathbb{E}[\exp(rL_1)] < \infty,$$

and set $\phi_L(r) = \log(\mathbb{E}[\exp(rL_1)])$. Then the following properties hold.

1) *Let h be a nonnegative bounded Borel function on $\mathbb{R}_+$ with compact support. Then,*

$$\left(\exp\left(t\int_0^{+\infty} h(s)\, dL_s - \int_0^{+\infty} \phi_L(t\, h(s))ds\right), t \ge 0\right)$$

is a peacock, an associated $(\mathscr{G}_t^{(L)})$-martingale being:

$$\left(\exp\left(\int_0^{+\infty} h(s)\, d_s\widehat{S}_{s,t}^{(L)} - \int_0^{+\infty} \phi_L(t\, h(s))ds\right), t \ge 0\right).$$

2) *Let σ be a signed finite measure on $\mathbb{R}_+$. Then,*

$$\left(\int \exp\left(t L_\lambda - \lambda\, \phi_L(t)\right) d\sigma(\lambda)\, ,\, t \ge 0\right)$$

is a peacock and an associated martingale is:

$$\left(\int \exp\left(\widehat{S}_{\lambda,t}^{(L)} - \lambda\, \phi_L(t)\right) d\sigma(\lambda)\, ,\, t \ge 0\right).$$

Remark that Point 2) in the previous proposition precises Theorem 1.8, in the case where the considered process is a self-decomposable Lévy process.

Exercise 5.12 (BDLP of the Wiener integral of a Lévy process). With the assumption and notation of Proposition 5.7, prove that the BDLP of X_h is:

$$C_s^{X_h} = \int_0^\infty h(u)\, d_u \widehat{C}_{s,u}^{L_1}, \quad s \geq 0.$$

Exercise 5.13 (A class of Lévy processes which are self-decomposable in $\mathbb{D}_0$). Let $L = (L_s, \geq 0)$ be a Lévy process such that the r.v. L_1 is self-decomposable. We set, for $0 \leq t \leq 1$,

$$L_{\bullet}^{(t)} = \widehat{S}_{\bullet,1}^{(L)} - \widehat{S}_{\bullet,t}^{(L)}.$$

1) Prove that $L^{(t)}$ is a Lévy process whose characteristic exponent $\psi^{(t)}$ is given by:

$$\psi^{(t)}(\lambda) = \psi(\lambda) - \psi(t\lambda).$$

2) Prove that

$$L \overset{(law)}{=} tL + \widetilde{L}^{(t)}$$

where, in the RHS, $\widetilde{L}^{(t)}$ is a copy of $L^{(t)}$ which is independent of L. (In this sense, L is self-decomposable *as a Lévy process*.)

3) We denote by $\mathbb{P}$ (resp. $\mathbb{P}^{(t)}$) the law of L (resp. $L^{(t)}$) on $\mathbb{D}_0$. For $\Phi \in L^1(\mathbb{P})$, we denote by $\Phi^{(\mathbf{P})} = (\Phi_t^{(\mathbf{P})}, 0 \leq t \leq 1)$ the process defined by:

$$\Phi_t^{(\mathbf{P})}(L) = \mathbb{E}_{\mathbb{P}^{(t)}}[\Phi(tL + \bullet)], \quad 0 \leq t \leq 1.$$

Prove that $\Phi^{(\mathbf{P})}$ is a peacock admitting as associated $(\mathscr{G}_t^{(L)})$-martingale:

$$\Phi_t^{(\mathbf{M})} = \mathbb{E}_{\mathbb{P}^{(t)}}[\Phi(\widehat{S}_{\bullet,t}^{(L)} + \bullet)], \quad 0 \leq t \leq 1.$$

4) We assume in this question that L is the standard Brownian motion starting from 0. Show that, for $0 < t \leq 1$,

$$\Phi_t^{(\mathbf{P})} = O_{-2\log t}\Phi \quad \text{and} \quad \Phi_t^{(\mathbf{M})} = \Phi_t^{(\mathbf{P})}(t^{-1}W_{\bullet,t^2})$$

where O denotes the Ornstein-Uhlenbeck semi-group on Wiener space (see Exercise 2.11) and W denotes the standard Brownian sheet.

Exercise 5.14 (Double integrals of exponentials of Lévy processes and associated martingales). Assume that, for every $r \geq 0$,

$$\mathbb{E}[\exp(rL_1)] < \infty$$

and set $\phi_L(r) = \log(\mathbb{E}[\exp(rL_1)])$. Let μ be a signed finite measure on $\mathbb{R}_+ \times \mathbb{R}_+$. Prove that

$$\left(\int \int \exp(t\,uL_v - v\,\phi_L(ut))\; \mu(du,dv)\;,\, t \geq 0 \right)$$

is a peacock and give an associated martingale. Compare with Exercise 1.39.

Exercise 5.15 (Self-decomposable Wald couples). We consider here the framework of Exercise 2.13 on infinitely divisible Wald couples, and we keep the same notation and hypotheses.

1) Recall that the Laplace exponent ϕ of H_1 is given by:

$$\phi(s) = \alpha\sqrt{2s} + \beta s + \int_{]0,+\infty[} \left(e^{-x\sqrt{2s}} + x\sqrt{2s} - 1 \right) c(dx).$$

i) Prove that the function $k(u) := \sqrt{u}(1 - e^{-\sqrt{u}})$, $u \geq 0$ is a Bernstein function. (*Hint*: for $u \geq 0$, $k'(u) = \frac{1}{2}\left(e^{-\sqrt{u}} + \int_0^1 e^{-s\sqrt{u}} ds \right)$ and, for every $a \geq 0$, $u \longmapsto e^{-a\sqrt{u}}$ is completely monotonous.)

ii) Let $\widetilde{\phi}(s) = s\phi'(s)$ for $s > 0$, and $\widetilde{\phi}(0) = 0$. Prove that, for $s \geq 0$,

$$\widetilde{\phi}(s) = \frac{\alpha}{\sqrt{2}}\sqrt{s} + \beta s + \frac{1}{2}\int_{]0,+\infty[} k(2x^2 s) c(dx),$$

and deduce from *i*) that $\widetilde{\phi}$ is a Bernstein function.

iii) Prove that H_1 is self-decomposable. (*Hint*: use Point 2) in Exercise 5.5).

2) We assume in the sequel that $c(dx) = c(x)dx$ with $x \longmapsto xc(x)$ decreasing on $\mathbb{R}_+^*$.

i) Prove that L_1 is self-decomposable.

The couple (L_1, H_1) is then called a *self-decomposable Wald couple*.

ii) Prove that, for $t \geq 0$, $(S_t^{L_1}, S_{t^2}^{H_1})$ is a self-decomposable Wald couple, corresponding to the function: $\Psi_t(\lambda) = \Psi(t\lambda)$.

iii) Let σ be a signed finite measure on $\mathbb{R}_+$ and let $\lambda \geq 0$. Define, for $t \geq 0$,

$$\widetilde{A}_t^{(\sigma)} := \int_{\mathbb{R}_+} \exp\left(\lambda t L_s - \frac{\lambda^2 t^2}{2} H_s \right) \sigma(ds)$$

$$\widetilde{M}_t^{(\sigma)} := \int_{\mathbb{R}_+} \exp\left(\lambda \widehat{S}_{s,t}^{(L)} - \frac{\lambda^2}{2} \widehat{S}_{s,t^2}^{(H)} \right) \sigma(ds)$$

with $\widehat{S}^{(L)}$ and $\widehat{S}^{(H)}$ assumed to be independent. Prove that $(\widetilde{M}_t^{(\sigma)}, t \geq 0)$ is a martingale and that $\widetilde{A}_t^{(\sigma)} \overset{(1.d)}{=} \widetilde{M}_t^{(\sigma)}$.

(*Hint*: show that, for $0 \leq t_1 \leq t_2$ and $s \geq 0$,

$$\mathbb{E}\left[e^{\lambda(\widehat{S}_{s,t_2}^{(L)} - \widehat{S}_{s,t_1}^{(L)})} \right] = \exp(s(\Psi(\lambda t_2) - \Psi(\lambda t_1)))$$

$$\mathbb{E}\left[e^{-\frac{\lambda^2}{2}(\widehat{S}_{s,t_2^2}^{(H)} - \widehat{S}_{s,t_1^2}^{(H)})} \right] = \exp(-s(\Psi(\lambda t_2) - \Psi(\lambda t_1))).)$$

Thus, $(\widetilde{A}_t^{(\sigma)}, t \geq 0)$ is a peacock and $(\widetilde{M}_t^{(\sigma)}, t \geq 0)$ is an associated martingale.
See also Question 5) of Exercise 2.13 for a direct proof of the peacock property of $(\widetilde{A}_t^{(\sigma)}, t \geq 0)$.

Notes and Comments

All notions relative to self-decomposable r.v.'s and Sato processes are taken from Sato's book [155, Chapter 3]. Section 5.4 essentially takes up the content of [82]. Sections 5.3 and 5.5 originate from [80]. Theorem 5.2 (description of the BDLP of X) comes from Jeanblanc-Pitman-Yor [94]. The notion of perturbed Bessel processes, which appears in Section 5.4 stems from Le Gall-Yor [115, 116], and was studied by Chaumont-Doney [44], Doney-Warren-Yor [54] (see also Doney-Zhang [55]). Note that the self-decomposability property has also been used in Madan-Yor [123, Theorem 4, Theorem 5] in a very different manner, to construct martingales with the one-dimensional marginals of $(tX, t \geq 0)$.

6

The Stochastic Differential Equation Method

Abstract To certain peacocks $(X_t, t \geq 0)$, we associate martingales $(M_t, t \geq 0)$ which solve stochastic differential equations (SDE's) of the form $(Z_t = \int_0^t \sigma(s, Z_s) dB_s, t \geq 0)$.

6.1 M. Pierre's Uniqueness Theorem for a Fokker-Planck Equation

The following uniqueness theorem is due to M. Pierre [140]. We present a detailed proof for the convenience of the reader. This proof lies somewhat out of the main line of development of our book and may be omitted in a first reading.

Theorem 6.1. *Let $U =]0, +\infty[\times \mathbb{R}$ and $a : U \longrightarrow \mathbb{R}_+$ be a Borel function satisfying the following hypothesis:*

(H) *For any $0 < t < T$ and $R > 0$, there exist real numbers $\varepsilon(t, T, R) > 0$ and $m(T, R) > 0$ such that:*

$$\forall (s,x) \in [t,T] \times [-R,R], \qquad a(s,x) \geq \varepsilon(t,T,R),$$

and

$$\forall (s,x) \in]0,T] \times [-R,R], \qquad a(s,x) \leq m(T,R).$$

Let μ be a probability measure on $\mathbb{R}$ such that $\int |x| \mu(dx) < +\infty$. Then, there exists at most one family of probability measures $(p(t,dx), t \geq 0)$ such that:

(FP_1) $t \geq 0 \longmapsto p(t,dx)$ *is weakly continuous.*
(FP_2) $p(0,dx) = \mu(dx)$ *and*

$$\frac{\partial p}{\partial t} - \frac{\partial^2}{\partial x^2}(ap) = 0 \quad \text{in } \mathscr{D}'(U)$$

(i.e. in the sense of Schwartz distributions in the open set U).

Hirsch F., Profeta C., Roynette B., Yor M.:
Peacocks and Associated Martingales, with Explicit Constructions

Proof. We first consider a family of probability measures $(p(t,dx),t \geq 0)$ satisfying conditions (FP_1) and (FP_2). □

Lemma 6.1. *For any $\varphi \in \mathscr{D}(\mathbb{R})$ (i.e. a $\mathscr{C}^\infty$-function with compact support in $\mathbb{R}$) and every $t \geq 0$,*

$$\int \varphi(x)p(t,dx) = \int \varphi(x)\mu(dx) + \iint_{]0,t[\times\mathbb{R}} \varphi''(x)a(s,x)p(s,dx)ds. \tag{6.1}$$

Proof. By (FP_2), for any $\alpha \in \mathscr{D}(]0,+\infty[)$ and $\varphi \in \mathscr{D}(\mathbb{R})$:

$$-\iint \alpha'(s)\varphi(x)p(s,dx)ds = \iint \alpha(s)\varphi''(x)a(s,x)p(s,dx)ds.$$

Applying this equality to a sequence (α_n) which suitably converges to $1_{]0,t[}$, and using (FP_1), we get the equality (6.1). □

We may notice that, conversely, (6.1) entails (FP_1) and (FP_2). We set

$$P(t,dx) := \int_0^t a(s,x)p(s,dx)ds.$$

Then, $(P(t,dx),t \geq 0)$ is an increasing family of positive measures, which is vaguely continuous and $P(0,dx) = 0$.

Lemma 6.2. *1) For every $t \geq 0$, $P(t,dx)$ admits a continuous density with respect to the Lebesgue measure. We denote by $P(t,x)$ this density.*
2) The function P is continuous on $[0,+\infty[\times\mathbb{R}$, increasing with respect to t, and for every $T \geq 0$ and every $(t,x) \in [0,T]\times\mathbb{R}$:

$$0 \leq P(t,x) \leq P(t,0) + \int |y|\mu(dy). \tag{6.2}$$

3) For every $t \geq 0$, the function $P(t,\bullet)$ admits a right derivative at any $x \in \mathbb{R}$, and

$$\frac{\partial P}{\partial x}(t,x) = \int_{]x,+\infty[} \mu(du) - \int_{]x,+\infty[} p(t,du) \tag{6.3}$$

where $\dfrac{\partial P}{\partial x}$ denotes the right derivative. In particular,

$$\left|\frac{\partial P}{\partial x}(t,x)\right| \leq 1. \tag{6.4}$$

Proof. $i)$ We set:

$$M(x) := \begin{cases} -\displaystyle\int_0^{+\infty} (u\wedge x)\mu(du) & \text{if } x \geq 0 \\ \displaystyle\int_{-\infty}^0 (u\vee x)\mu(du) - x & \text{if } x \leq 0. \end{cases}$$

Then M is a convex function whose right derivative is given by :

$$M'(x) = -\int_{]x,+\infty[} \mu(du).$$

On the other hand, we have by (6.1),

$$\frac{\partial^2 P}{\partial x^2}(t,dx) = p(t,dx) - \mu(dx) \quad \text{in } \mathscr{D}'(\mathbb{R}). \tag{6.5}$$

Since $\frac{\partial^2 M}{\partial x^2} = \mu$, it follows that

$$\frac{\partial^2}{\partial x^2}(P(t,\centerdot) + M) = p(t,dx).$$

This shows that $P(t,\centerdot)$ may be represented as the difference of two convex functions, which implies Point 1).

ii) By the previous step, $P(t,\centerdot)+M$ is a convex function. Using then the vague continuity and the increase of $P(t,\centerdot)$ with respect to t, we obtain easily that P is continuous on $\mathbb{R}_+ \times \mathbb{R}$ and increasing with respect to t. Moreover, $P(t,\centerdot)$ admits a right derivative at any x, denoted by $\frac{\partial P}{\partial x}(t,x)$, which is right-continuous.

iii) Integrating (6.5), we get:

$$\frac{\partial P}{\partial x}(t,x) = \int_{]x,+\infty[} \mu(du) - \int_{]x,+\infty[} p(t,du) + C(t).$$

As a consequence, $\lim_{|x|\to+\infty} \frac{\partial P}{\partial x}(t,x) = C(t)$. Since $P \geq 0$, we have $C(t) = 0$. Consequently,

$$\frac{\partial P}{\partial x}(t,x) = \int_{]x,+\infty[} \mu(du) - \int_{]x,+\infty[} p(t,du)$$

and Point 3) is proven.

iv) Finally, integrating (6.3), we obtain:

$$P(t,x) = -\int_0^{+\infty} (u\wedge x)p(t,du) + \int_0^{+\infty} (u\wedge x)\mu(du) + P(t,0) \quad \text{if } x \geq 0 \tag{6.6}$$

$$= \int_{-\infty}^0 (u\vee x)p(t,du) - \int_{-\infty}^0 (u\vee x)\mu(du) + P(t,0) \quad \text{if } x \leq 0 \tag{6.7}$$

and (6.2) follows easily. □

Lemma 6.3. *There exists $p \in L^2_{loc}(U)$ such that, for almost every $t \geq 0$:*

$$p(t,dx) = p(t,x)dx.$$

Proof. We fix $\alpha \in \mathscr{D}(]0,+\infty[)$ and $\xi \in \mathscr{D}(\mathbb{R})$ and we assume $\alpha \geq 0$ and $\xi \geq 0$. There exist $0 < t_1 < t_2$ and $R > 0$ such that the support of α is contained in $[t_1, t_2]$ and the support of ξ is contained in $[-R, R]$. We set:

$$\varepsilon := \varepsilon(t_1, t_2, R) \quad \text{and} \quad m := m(t_2, R).$$

We denote by $\widetilde{P}$ the function:

$$\widetilde{P}(t,x) := \xi(x)\left(\alpha(t)P(t,x) - \int_0^t \alpha'(s)P(s,x)ds\right). \tag{6.8}$$

We have, integrating by parts:

$$\widetilde{P}(t,x)dx = \xi(x)\int_0^t \alpha(s)a(s,x)p(s,dx)ds. \tag{6.9}$$

Differentiating (6.9) with respect to t yields:

$$\frac{\partial \widetilde{P}}{\partial t} = \xi \alpha a p \quad \text{in } \mathscr{D}'(U).$$

Differentiating (6.8) with respect to x leads to:

$$\frac{\partial^2 \widetilde{P}}{\partial x^2} = \xi \alpha p - \xi \int_0^t \alpha'(s)p(s,dx)ds + \varphi \quad \text{in } \mathscr{D}'(U),$$

for some $\varphi \in L^\infty(U)$, by (6.4).

On the other hand,

$$\xi \int_0^t |\alpha'(s)|p(s,dx)ds \leq \frac{\|\alpha'\|_\infty}{\varepsilon}\xi P(t_2,dx).$$

Therefore, $\xi \int_0^t |\alpha'(s)|p(s,dx)ds \in L^\infty(U)$. Consequently,

$$\frac{\partial \widetilde{P}}{\partial t} \leq m\xi\alpha p \leq m\frac{\partial^2 \widetilde{P}}{\partial x^2} + \phi \quad \text{with } \phi \in L^\infty(U),$$

or

$$\frac{\partial \widetilde{P}}{\partial t} - m\frac{\partial^2 \widetilde{P}}{\partial x^2} \leq \phi, \quad \phi \in L^\infty(U). \tag{6.10}$$

Let $(\rho_n, n \geq 0)$ be a standard regularizing sequence on $\mathbb{R}$ and $\widetilde{P}_n = \widetilde{P} * \rho_n$ where $*$ denotes the convolution on $\mathbb{R}$. We obtain from (6.10):

$$\frac{\partial \widetilde{P}_n}{\partial t} - m\frac{\partial^2 \widetilde{P}_n}{\partial x^2} \leq \phi_n, \quad \text{with } \phi_n = \phi * \rho_n. \tag{6.11}$$

Since $\dfrac{\partial \widetilde{P}_n}{\partial t} = \alpha\left[(\xi a p) * \rho_n\right] \geq 0$, (6.11) entails:

$$\iint \left(\frac{\partial \widetilde{P}_n}{\partial t}\right)^2 \leq \iint \phi_n \frac{\partial \widetilde{P}_n}{\partial t} + m\iint \frac{\partial^2 \widetilde{P}_n}{\partial x^2}\frac{\partial \widetilde{P}_n}{\partial t}. \tag{6.12}$$

By integration by parts,

$$\iint \frac{\partial^2 \widetilde{P}_n}{\partial x^2} \frac{\partial \widetilde{P}_n}{\partial t} = -\frac{1}{2} \iint \frac{\partial}{\partial t} \left(\frac{\partial \widetilde{P}_n}{\partial x} \right)^2 = -\frac{1}{2} \int_{\mathbb{R}} \left(\frac{\partial \widetilde{P}_n}{\partial x}(t_2, x) \right)^2 dx.$$

Therefore,

$$\iint \left(\frac{\partial \widetilde{P}_n}{\partial t} \right)^2 \leq \iint \phi_n \left(\frac{\partial \widetilde{P}_n}{\partial t} \right) \leq \frac{1}{2} \left(\iint_{[t_1,t_2]\times[-R-1,R+1]} \phi_n^2 + \iint \left(\frac{\partial \widetilde{P}_n}{\partial t} \right)^2 \right)$$

and

$$\iint \left(\frac{\partial \widetilde{P}_n}{\partial t} \right)^2 \leq \iint_{[t_1,t_2]\times[-R-1,R+1]} \phi_n^2 \leq C$$

with C independent from n. Thus the sequence $(\alpha[(\xi ap) * \rho_n], n \geq 0)$ is bounded in L^2, therefore, by weak compactness, $\alpha \xi ap \in L^2$, and $\alpha \xi p \leq \frac{1}{\varepsilon} \alpha \xi ap$ also belongs to L^2.
Finally, for any $\alpha \in \mathscr{D}(]0, +\infty[)$ and $\xi \in \mathscr{D}(\mathbb{R})$ $\alpha \xi p \in L^2$, or $p \in L^2_{\text{loc}}(U)$. □

We now consider another family of probability measures $(\widehat{p}(t, dx), t \geq 0)$ satisfying (FP_1) and (FP_2) and we set

$$\widehat{P}(t,x)dx = \int_0^t a(s,x)\widehat{p}(s,dx)ds.$$

Let $q = p - \widehat{p}$ and $Q = P - \widehat{P}$. We have

$$\frac{\partial Q}{\partial t} = aq \quad \text{and} \quad \frac{\partial^2 Q}{\partial x^2} = q \quad \text{(by (6.5))}.$$

Thus,

$$\frac{\partial Q}{\partial t} - a \frac{\partial^2 Q}{\partial x^2} = 0 \quad \text{in } \mathscr{D}'(U). \tag{6.13}$$

On the other hand, $q \in L^2_{\text{loc}}$. Let $0 < t_1 < t_2$, $R > 0$ and $V = [t_1, t_2] \times [-R, R]$. Then,

$$\iint_V aq^2 = \iint_V \frac{\partial Q}{\partial t} \frac{\partial^2 Q}{\partial x^2}.$$

Let, as before, $(\rho_n, n \geq 0)$ be a regularizing sequence, and $Q_n = Q * \rho_n$.

$$\iint_V (aq * \rho_n)(q * \rho_n) = \iint_V \frac{\partial Q_n}{\partial t} \frac{\partial^2 Q_n}{\partial x^2},$$

and, by integration by parts,

$$\iint_V \frac{\partial Q_n}{\partial t} \frac{\partial^2 Q_n}{\partial x^2} = \int_{t_1}^{t_2} dt \left(\left[\frac{\partial Q_n}{\partial t} \frac{\partial Q_n}{\partial x} \right]_{x=-R}^{x=R} - \frac{1}{2} \frac{\partial}{\partial t} \int_{-R}^{R} \left(\frac{\partial Q_n}{\partial x} \right)^2 dx \right)$$

$$\leq \int_{t_1}^{t_2} \left[(aq * \rho_n) \frac{\partial Q_n}{\partial x} \right]_{-R}^{R} dt + \frac{1}{2} \int_{-R}^{R} \left(\frac{\partial Q_n}{\partial x}(t_1, x) \right)^2 dx.$$

Now, the following estimates hold:

$$\left| \frac{\partial Q_n}{\partial x}(t,R) \right| \leq \sup \left\{ \left| \frac{\partial Q}{\partial x}(t,x) \right| \; ; \; x \in [R-1, R+1] \right\}$$

and

$$\int_{t_1}^{t_2} |(aq * \rho_n)(R)| \, dt \leq \sup \left\{ P(t_2,x) + \widehat{P}(t_2,x) \; ; \; x \in \mathbb{R} \right\} \leq C_{t_2} \quad \text{(by (6.2))}.$$

Letting n tend to $+\infty$, we obtain, since $q \in L^2_{\text{loc}}(U)$,

$$\begin{aligned} \iint_V aq^2 \leq C_{t_2} \sup & \left\{ \left| \frac{\partial Q}{\partial x}(t,x) \right| \; ; \; t \in]0,t_2], x \in [R-1, R+1] \right\} \\ & + C_{t_2} \sup \left\{ \left| \frac{\partial Q}{\partial x}(t,x) \right| \; ; \; t \in]0,t_2], x \in [-R-1, -R+1] \right\} \\ & + \frac{1}{2} \int_{-R}^{R} \left(\frac{\partial Q}{\partial x}(t_1,x) \right)^2 dx. \end{aligned} \tag{6.14}$$

By (6.3), $\lim\limits_{t_1 \to 0} \frac{\partial Q}{\partial x}(t_1,x) = 0$ a.s. and, by (6.4), $\left| \frac{\partial Q}{\partial x}(t_1,x) \right| \leq 2$. Therefore,

$$\lim_{t_1 \to 0} \int_{-R}^{R} \left(\frac{\partial Q}{\partial x}(t_1,x) \right)^2 dx = 0.$$

We have:

$$\frac{\partial Q}{\partial x}(t,x) = \int_{]x,+\infty[} \widehat{p}(t,du) - \int_{]x,+\infty[} p(t,du)$$

and

$$\begin{aligned} \sup & \left\{ \int_{]x,+\infty[} p(t,du) \; ; \; t \in]0,t_2], x \in [R-1, R+1] \right\} \\ & \leq \sup \left\{ \int_{]R-1,+\infty[} p(t,du) \; ; \; t \in]0,t_2] \right\} \\ & \leq \sup \left\{ \int \theta(x+2-R) p(t,dx) \; ; \; t \in]0,t_2] \right\} \end{aligned}$$

with θ an increasing, continuous and positive function such that $\theta(x) = 0$ for $x \leq 0$ and $\theta(x) = 1$ for $x \geq 1$. For any $R > 0$,

$$t \in [0,t_2] \longmapsto \int \theta(x+2-R) p(t,dx)$$

is continuous by (FP_1), and, for any $t \in [0,t_2]$,

$$R \longmapsto \int \theta(x+2-R)p(t,dx)$$

is decreasing and converges to 0 when R tends to $+\infty$. Therefore, by Dini's lemma,

$$\lim_{R\to+\infty} \sup_{t\in[0,t_2]} \int \theta(x+2-R)p(t,dx) = 0.$$

We have the same result for $\widehat{p}$ instead of p, and therefore, the first term in the RHS of (6.14) tends to 0 when R tends to $+\infty$. The second term may be treated similarly. Finally, letting first t_1 tend to 0, and then R tend to $+\infty$ in (6.14), we obtain:

$$\iint_{]0,t_2[\times\mathbb{R}} aq^2\, dt\, dx = 0,$$

hence, for almost every $t > 0$, $p(t,dx) = \widehat{p}(t,dx)$. By (FP_1), this equality actually holds for every $t \geq 0$. This ends the proof of Theorem 6.1. □

Remark 6.1. It follows from (6.6) and (6.7) that, under the hypotheses of Theorem 6.1, if $(p(t,dx), t \geq 0)$ satisfies (FP_1) and (FP_2) then, for every $t \geq 0$:

$$\sup_{s\leq t} \int_{-\infty}^{+\infty} |u| p(s,du) < \infty.$$

6.2 The Fokker-Planck Equation Method

Let $U =]0,+\infty[\times\mathbb{R}$ and $\overline{U} = \mathbb{R}_+ \times \mathbb{R}$. We consider in this section a continuous function σ on $\overline{U}$ such that $\sigma(t,x) > 0$ for every $(t,x) \in U$. The following general result holds.

Theorem 6.2. *Let $a \in \mathbb{R}$.*
1) *The stochastic differential equation*

$$Z_t = a + \int_0^t \sigma(s,Z_s)dB_s \tag{6.15}$$

admits a weak non-exploding solution $(Y_t, t \geq 0)$, unique in law, and which enjoys the strong Markov property.

2) *Let $p(t,dx)$ be the law of Y_t. Then, $(p(t,dx), t \geq 0)$ is the unique family of probability measures such that:*

$$t \geq 0 \longmapsto p(t,dx) \text{ is weakly continuous} \tag{6.16}$$

$$\text{and} \quad \begin{cases} p(0,dx) = \delta_a(dx), \\ \dfrac{\partial p}{\partial t} - \dfrac{1}{2}\dfrac{\partial^2}{\partial x^2}(\sigma^2 p) = 0 \\ \text{in the sense of distributions on } U. \end{cases} \tag{6.17}$$

We shall say that $(p(t,dx),t \geq 0)$ solves the Fokker-Planck equation (6.17) with initial condition δ_a.

Proof. i) Existence of a weak non-exploding solution.
Since σ is continuous on $\overline{U}$, the SDE (6.15) admits a weak solution $(Y_t, 0 \leq t < e)$ up to an explosion time e, and

$$\lim_{t \to e} |Y_t| = +\infty$$

almost surely on $\{e < +\infty\}$ (see for example [86, Chapter IV, Theorem 2.3]). By an easy extension of the Dambis Dubins-Schwarz theorem [150, Chapter V, Theorem 1.6], there exists a Brownian motion β such that:

$$Y_t = a + \beta_{\int_0^t \sigma^2(s,Y_s)ds} \quad \text{on } \{t < e\}.$$

Since $\lim_{t \to e} |Y_t| = \infty$ on $\{e < +\infty\}$ and

$$\overline{\lim_{u \longrightarrow +\infty}} \beta_u = +\infty, \qquad \underline{\lim_{u \longrightarrow +\infty}} \beta_u = -\infty,$$

we conclude that $\mathbb{P}(e < +\infty) = 0$.

ii) Uniqueness of the 1-dimensional marginals.
By Itô's formula, for every function f of $\mathscr{C}^2$ class and with compact support:

$$\mathbb{E}[f(Y_t)] = f(a) + \frac{1}{2} \int_0^t \mathbb{E}[f''(Y_s)\sigma^2(s,Y_s)]ds$$

and therefore, denoting by $p(t,dx)$ the law of Y_t,

$$\int f(x)p(t,dx) = f(a) + \frac{1}{2} \int_0^t ds \int f''(x)\sigma^2(s,x)p(s,dx).$$

This entails directly that $(p(t,dx),t \geq 0)$ satisfies (6.16) and (6.17).

The uniqueness property stated in Point 2) of Theorem 6.2 is then a direct consequence of M. Pierre's theorem [140], the statement and the proof of which are presented in Section 6.1.

iii) Uniqueness in law.
Let $(X_t, t \geq 0)$ and $(Y_t, t \geq 0)$ be two solutions of equation (6.15). Let, for every $t \geq 0$, μ_t (resp. ν_t) the law of X_t (resp. Y_t). From Point *ii*) above, for every $t \geq 0$, $\mu_t = \nu_t$.

Let, for $n \in \mathbb{N}^*$ and $0 < t_1 < \ldots < t_n$, $\mu_{t_1,\ldots,t_n}$ (resp. $\nu_{t_1,\ldots,t_n}$) be the law of the n-tuple $(X_{t_1},\ldots,X_{t_n})$ (resp. $(Y_{t_1},\ldots,Y_{t_n})$). We shall prove by induction that $\mu_{t_1,\ldots,t_n} = \nu_{t_1,\ldots,t_n}$.

Assume that $\mu_{t_1,\ldots,t_n} = \nu_{t_1,\ldots,t_n}$ and $t_n < t_{n+1}$. Let $f : \mathbb{R}^n \longrightarrow \mathbb{R}_+$, $g : \mathbb{R} \longrightarrow \mathbb{R}_+$ be two functions of $\mathscr{C}^\infty$ class with compact support. We have, applying Itô's formula between t_n and t_{n+1}:

$$\begin{aligned}\mathbb{E}[f(X_{t_1},\ldots,X_{t_n})g(X_{t_{n+1}})] &= \mathbb{E}[f(X_{t_1},\ldots,X_{t_n})g(X_{t_n})] \\ &\quad + \frac{1}{2} \int_{t_n}^{t_{n+1}} \mathbb{E}[f(X_{t_1},\ldots,X_{t_n})g''(X_s)\sigma^2(s,X_s)]ds,\end{aligned}$$

i.e.

$$\iint f(x_1,\dots,x_n)g(y)\mu_{t_1,\dots,t_n,t_{n+1}}(dx_1,\dots,dx_n,dy)$$
$$= \mathbb{E}[f(X_{t_1},\dots,X_{t_n})g(X_{t_n})] \tag{6.18}$$
$$+\frac{1}{2}\int_{t_n}^{t_{n+1}} ds \iint f(x_1,\dots,x_n)g''(y)\sigma^2(s,y)\mu_{t_1,\dots,t_n,s}(dx_1,\dots,dx_n,dy).$$

Let us define, for every $s > t_n$, the positive measure $\overline{\mu}^f_{t_1,\dots,t_n,s}$ on $\mathbb{R}$ by:

$$\overline{\mu}^f_{t_1,\dots,t_n,s}(dy) := \int_{\mathbb{R}^n} f(x_1,\dots,x_n)\mu_{t_1,\dots,t_n,s}(dx_1,\dots,dx_n,dy)$$

and, for $s = t_n$:

$$\overline{\mu}^f_{t_1,\dots,t_n,t_n}(dy) := \int_{\mathbb{R}^{n-1}} f(x_1,\dots,x_{n-1},y)\mu_{t_1,\dots,t_n}(dx_1,\dots,dx_{n-1},dy).$$

(6.18) reads then:

$$\int_{\mathbb{R}} g(y)\overline{\mu}^f_{t_1,\dots,t_n,t_{n+1}}(dy) = \int_{\mathbb{R}} g(y)\overline{\mu}^f_{t_1,\dots,t_n,t_n}(dy)$$
$$+\frac{1}{2}\int_{t_n}^{t_{n+1}} ds \int_{\mathbb{R}} \overline{\mu}^f_{t_1,\dots,t_n,s}(dy)g''(y)\sigma^2(s,y),$$

or, differentiating with respect to t_{n+1}:

$$\frac{\partial}{\partial t}\overline{\mu}^f_{t_1,\dots,t_n,t}(dy) = \frac{1}{2}\frac{\partial^2}{\partial y^2}\left(\sigma^2(t,y)\overline{\mu}^f_{t_1,\dots,t_n,t}(dy)\right)$$

in the sense of distributions on $]t_n,+\infty[\times\mathbb{R}$. This is a Fokker-Planck equation on $[t_n,+\infty[$, with initial distribution $\overline{\mu}^f_{t_1,\dots,t_n,t_n}(dy)$. Since we assumed that $\mu_{t_1,\dots,t_n} = \nu_{t_1,\dots,t_n}$, then:

$$\overline{\mu}^f_{t_1,\dots,t_n,t_n} = \overline{\nu}^f_{t_1,\dots,t_n,t_n}.$$

Besides, we may assume without loss of generality that:

$$\mathbb{E}[f(X_{t_1},\dots,X_{t_n})] = \mathbb{E}[f(Y_{t_1},\dots,Y_{t_n})] = 1.$$

Then, $\overline{\mu}^f_{t_1,\dots,t_n,t}$ and $\overline{\nu}^f_{t_1,\dots,t_n,t}$ are actually probability measures for $t \geq t_n$. Moreover,

$$\int |y|\overline{\mu}^f_{t_1,\dots,t_n,t_n}(dy) = \mathbb{E}[f(X_{t_1},\dots,X_{t_n})|X_{t_n}|] < \infty.$$

Thus, we may apply M. Pierre's uniqueness theorem of Section 6.1 replacing $\mathbb{R}_+$ by $[t_n,+\infty[$, and infer that:

$$\overline{\mu}^f_{t_1,\dots,t_n,t} = \overline{\nu}^f_{t_1,\dots,t_n,t} \qquad \text{for } t \geq t_n.$$

Since f is arbitrary, we deduce from the monotone class theorem that

$$\mu_{t_1,\dots,t_n,t_{n+1}} = \nu_{t_1,\dots,t_n,t_{n+1}}.$$

Then, the uniqueness in law of a weak solution to (6.15) is proven, and the strong Markov property follows from classical arguments. □

Remark 6.2. 1) The proof that the equality $\mu_t = \nu_t$ for every $t \geq 0$ implies the uniqueness in law of the solution of equation (6.15) is classical. See for instance Stroock and Varadhan [164, Theorem 6.2.3, p.147] or Karatzas and Shreve [105, Proposition 4.27, p.326].

2) The uniqueness in law of the one-dimensional marginals of solutions to equation (6.15) is generally obtained under the hypothesis of uniform ellipticity, and assuming that the coefficients are bounded, hypotheses that are stronger that those of Theorem 6.2.

3) We may also refer to Stroock and Varadhan [164, Exercise 7.3.3, p.192] for another proof of the uniqueness in law of a weak solution to equation (6.15).

4) One may easily construct examples where the function σ vanishes, and where there exists an infinity of solutions to the equation (6.15), with different laws. This is for example the case when σ does not depend on s and satisfies:

- $\sigma(a) = 0$,
- there exists $\varepsilon > 0$ such that:

$$\int_{a-\varepsilon}^{a+\varepsilon} \frac{dy}{\sigma(y)} = +\infty,$$

- $\sigma(x) \geq 1$ for $|x-a| > \varepsilon$.

This example is due to I.V. Girsanov [70], see also Ikeda and Watanabe [86, p.184].

We now state an interesting application of Theorem 6.2 to peacocks.

Theorem 6.3. *Let $(X_t, t \geq 0)$ be a continuous process such that:*

(P_1) *For every function f of $\mathscr{C}^2$ class and with compact support,*

$$\mathbb{E}[f(X_t)] = f(a) + \frac{1}{2}\int_0^t \mathbb{E}[f''(X_s)\sigma^2(s, X_s)]ds.$$

(P_2) *For every $t > 0$,*

$$\int_0^t \mathbb{E}[\sigma^2(s, X_s)]ds < \infty.$$

Then, $(X_t, t \geq 0)$ is a peacock and the (unique in law) weak solution of the SDE (6.15):

$$Z_t = a + \int_0^t \sigma(s, Z_s)dB_s$$

is an associated martingale.

Proof. Let $p(t,dx)$ be the law of X_t. By (P_1), $(p(t,dx), t \geq 0)$ solves the Fokker-Planck equation (6.17) with initial condition δ_a. By Theorem 6.2, if $(Y_t, t \geq 0)$ is a weak solution of (6.15), one has

$$X_t \overset{(1.d)}{=} Y_t.$$

Thus, it simply remains to show that $(Y_t, t \geq 0)$ is a martingale. Now,

$$\mathbb{E}[\langle Y, Y \rangle_t] = \int_0^t \mathbb{E}[\sigma^2(s, Y_s)]ds = \int_0^t \mathbb{E}[\sigma^2(s, X_s)]ds < \infty$$

by (P_2). This proves that Y is a martingale (in H^2_{loc}). □

Remark 6.3. 1) We keep here the hypotheses of Theorem 6.3, but we no longer assume (P_2). What can be said about the local martingale $(Y_t, t \geq 0)$?

First, it is proven in Section 6.1 (cf. Remark 6.1) that for every $t \geq 0$,

$$\sup_{s \leq t} \int_{\mathbb{R}} |x| p(s, dx) < +\infty. \tag{6.19}$$

Hence, for every $t \geq 0$, there exists a constant $C(t) < \infty$ such that:

$$\sup_{s \leq t} \mathbb{E}[|Y_s|] \leq C(t)$$

i.e. $(Y_s, s \leq t)$ is bounded in L^1. Then, using Krickeberg's decomposition (cf. for instance [150, Ex. (1.49), p.136]), there exist two positive supermartingales $(Y_t^+, t \geq 0)$ and $(Y_t^-, t \geq 0)$ such that $Y = Y^+ - Y^-$. Note that $(Y_t, t \geq 0)$ might not be a true martingale.

Example: let $(R_t, t \geq 0)$ and $(R'_t, t \geq 0)$ be two independent 3-dimensional Bessel processes started at 0, and define, for $\varepsilon > 0$,

$$Y_t := \frac{1}{R_t} - \frac{1}{R'_t}, \quad t \geq \varepsilon$$

$(Y_t, t \geq \varepsilon)$ is a local martingale but not a true martingale; indeed, by scaling

$$Y_t \overset{(1.d)}{=} \frac{1}{\sqrt{t}} \left(\frac{1}{R_1} - \frac{1}{R'_1} \right) \overset{(1.d)}{=} \frac{1}{\sqrt{t}} X,$$

with X a centered r.v., hence, the marginals of $(Y_t, t \geq 0)$ are decreasing for the convex order, see Proposition 1.4. Such processes have been considered in [52].

2) In a close context, if $(X_t, t \geq 0)$ and $(M_t, t \geq 0)$ are two processes such that:

i) $(X_t, t \geq 0)$ is integrable and $\mathbb{E}[X_t]$ does not depend on t,

ii) $X_t \overset{(1.d)}{=} M_t$,

iii) $(M_t, t \geq 0)$ is a $\mathbb{R}_+$-valued continuous local martingale,

then, $(M_t, t \geq 0)$ is a true martingale (thus $(X_t, t \geq 0)$ is a peacock). Indeed, it is a positive supermartingale with constant expectation. This idea is developed in Section 6.5.

In the following two sections, we present applications of Theorem 6.3.

6.3 Martingales Associated to $(\varphi(B_t), t \geq 0)$

Let $(B_t, t \geq 0)$ be a Brownian motion started from 0. We know (see for example Exercise 1.23) that if $\varphi : \mathbb{R} \longmapsto \mathbb{R}$ is an odd and increasing function such that, for every $t \geq 0$, $\mathbb{E}[|\varphi(B_t)|] < \infty$, then $(\varphi(B_t), t \geq 0)$ is a peacock. We shall exhibit, in Chapter 7, Section 7.5, a martingale associated to this peacock, using Bass' method for the Skorokhod embedding. In the present section, at the cost of making some further hypotheses on φ, we shall associate a martingale to $(\varphi(B_t), t \geq 0)$ by means of Theorem 6.3.

A function $\varphi : \mathbb{R} \longrightarrow \mathbb{R}$ will be said to satisfy *Hypothesis* (Z) if:

a) φ is an odd $\mathscr{C}^2$ function,
b) for every $x \in \mathbb{R}$, $\varphi'(x) > 0$, and $\lim_{x \to +\infty} \varphi(x) = +\infty$,
c) for every $t > 0$: $\mathbb{E}[|\varphi''(B_t)|] < \infty$.

Lemma 6.4. *Let $\varphi : \mathbb{R} \longrightarrow \mathbb{R}$ be a function satisfying Hypothesis (Z). Then, for every $c > 0$:*

$$\lim_{|x| \longrightarrow +\infty} \varphi'(x) e^{-cx^2} = 0 \quad \textit{and} \quad \lim_{|x| \longrightarrow +\infty} \varphi(x) e^{-cx^2} = 0.$$

Proof. We have, for $A < B$,

$$\int_A^B \varphi'(x) e^{-cx^2} dx = \varphi'(A) \int_A^B e^{-cx^2} dx + \int_A^B dx\, \varphi''(x) \int_x^B e^{-cy^2} dy.$$

Now, $\displaystyle\int_x^{+\infty} e^{-cy^2} dy \underset{x \to +\infty}{\sim} \frac{1}{2cx} e^{-cx^2}$. Therefore, by Property *c*), we obtain, letting B tend to $+\infty$:

$$\int_A^{+\infty} \varphi'(x) e^{-cx^2} dx = \varphi'(A) \int_A^{+\infty} e^{-cx^2} dx + \int_A^{+\infty} dx\, \varphi''(x) \int_x^{+\infty} e^{-cy^2} dy < \infty,$$

and then: $\displaystyle\lim_{A \to +\infty} \varphi'(A) \int_A^{+\infty} e^{-cx^2} dx = 0$, or equivalently,

$$\lim_{x \to +\infty} \varphi'(x) \frac{1}{x} e^{-cx^2} = 0$$

which implies $\lim_{x \to +\infty} \varphi'(x) e^{-c'x^2} = 0$ for $c' > c$. Since c is arbitrary, $\lim_{x \to +\infty} \varphi'(x) e^{-cx^2} = 0$ and symmetrically, $\lim_{x \to -\infty} \varphi'(x) e^{-cx^2} = 0$.

Now, since from the previous point, $\mathbb{E}[|\varphi'(B_t)|] < \infty$ for every $t \geq 0$, the same proof implies $\lim_{|x| \to +\infty} \varphi(x) e^{-cx^2} = 0$. □

Let $\varphi : \mathbb{R} \longmapsto \mathbb{R}$ be a function satisfying Hypothesis (Z). By Property *b*), the inverse φ^{-1} of φ in the sense of composition of functions is well-defined as a $\mathscr{C}^2$ function on $\mathbb{R}$. We set, for $t > 0$ and $x \in \mathbb{R}$,

$$\sigma^2(t,x) := \frac{1}{t} (\varphi' \circ \varphi^{-1})(x) \exp\left(\frac{(\varphi^{-1}(x))^2}{2t} \right) \int_{\varphi^{-1}(|x|)}^{+\infty} \varphi'(y) y e^{-\frac{y^2}{2t}} dy \qquad (6.20)$$

and

$$\sigma^2(0,x) = \left(\varphi' \circ \varphi^{-1}(x)\right)^2. \tag{6.21}$$

Lemma 6.5. *Under Hypothesis* (Z)*, with* $\sigma \geq 0$ *defined by (6.20) and (6.21),* σ *is continuous and strictly positive on* $\mathbb{R}_+ \times \mathbb{R}$.

Proof. The strict positivity is clear by definition, in view of Property b). On the other hand, σ is obviously continuous on $]0,+\infty[\times\mathbb{R}$. Therefore, we only need to prove that $\lim_{t\to 0} \sigma^2(t,\varphi(x)) = (\varphi'(x))^2$, uniformly on compact sets in x. Now by integration by parts:

$$\sigma^2(t,\varphi(x)) = \left(\varphi'(x)\right)^2 + \varphi'(x)\int_{|x|}^{+\infty} \varphi''(y)\exp\left(-\frac{y^2-x^2}{2t}\right)dy. \tag{6.22}$$

Therefore,

$$|\sigma^2(t,\varphi(x)) - \left(\varphi'(x)\right)^2| \leq \varphi'(x)\int_{|x|}^{+\infty} |\varphi''(y)|\exp\left(-\frac{y^2-x^2}{2t}\right)dy,$$

and the desired result follows from the monotone convergence theorem (letting t tend to 0) and from Dini's lemma. □

We may now state the main result of this section.

Theorem 6.4. *Let* φ *satisfying Hypothesis* (Z) *and* $\sigma \geq 0$ *defined by (6.20) and (6.21). Then, the SDE*

$$Z_t = \int_0^t \sigma(s,Z_s)dB_s \tag{6.23}$$

admits a weak solution $(M_t, t \geq 0)$ *which is a martingale in* H^2_{loc}*, and is associated to the peacock* $(\varphi(B_t), t \geq 0)$.

Proof. We shall apply Theorem 6.3 to $X_t = \varphi(B_t)$, proving the properties (P_1) and (P_2).
i) Let f be a $\mathscr{C}^2$-function with compact support. From Itô's formula applied to $f \circ \varphi(B_t)$, we get:

$$\mathbb{E}[f(X_t)] = f(0) + \frac{1}{2}\int_0^t \mathbb{E}\left[f'(X_s)\varphi''(B_s) + f''(X_s)\left(\varphi'(B_s)\right)^2\right]ds. \tag{6.24}$$

We set:

$$\Lambda(t,x) := \int_x^{+\infty} \varphi''(y)e^{-\frac{y^2}{2t}}dy. \tag{6.25}$$

Then, we obtain after integrating by parts:

$$\mathbb{E}[f'(X_s)\varphi''(B_s)] = \mathbb{E}\left[f''(X_s)\varphi'(B_s)\Lambda(s,B_s)\exp\left(\frac{B_s^2}{2s}\right)\right]. \tag{6.26}$$

Remark that Λ is an even function. Then, plugging (6.26) into (6.24), we obtain by (6.22):

$$\mathbb{E}[f(X_t)] = f(0) + \frac{1}{2}\int_0^t \mathbb{E}[f''(X_s)\sigma^2(s,\varphi(B_s))]ds,$$

which is (P_1) since $\varphi(B_s) = X_s$.

ii) We now prove (P_2). We have, by (6.20),

$$\sigma^2(s,\varphi(x)) = \frac{1}{s}e^{\frac{x^2}{2s}}\varphi'(x)\int_{|x|}^{+\infty}\varphi'(y)ye^{-\frac{y^2}{2s}}dy.$$

Therefore,

$$\begin{aligned}\int_0^t \mathbb{E}[\sigma^2(s,X_s)]ds &= \frac{2}{\sqrt{2\pi}}\int_0^t\int_0^{+\infty} ds\,dx\,s^{-\frac{3}{2}}\varphi'(x)\int_x^{+\infty}\varphi'(y)ye^{-\frac{y^2}{2s}}dy \\ &= \frac{2}{\sqrt{2\pi}}\int_0^t\int_0^{+\infty} ds\,dy\,s^{-\frac{3}{2}}\varphi(y)\varphi'(y)ye^{-\frac{y^2}{2s}} \quad \text{(by Fubini)}.\end{aligned}$$

Now, by integration by parts,

$$\int_0^t s^{-\frac{3}{2}}e^{-\frac{y^2}{2s}}ds \le 2\sqrt{t}y^{-2}e^{-\frac{y^2}{2t}}$$

and it suffices to prove:

$$\int_0^{+\infty}\varphi'(y)\frac{\varphi(y)}{y}e^{-\frac{y^2}{2t}dy} < +\infty,$$

which follows from Lemma 6.4. □

We now show that, under an additional hypothesis, the SDE (6.23) admits a unique strong solution.

Proposition 6.1. *Suppose that φ satisfies Hypothesis (Z) as well as the additional assumption: for every $x \ge 0$, $\varphi''(x) \ge 0$. Then σ is locally Lipschitz continuous in x, uniformly on compact sets in t. Consequently, the SDE (6.23) admits a unique strong solution.*

Proof. Suppose that $\varphi''(x) \ge 0$ for every $x \ge 0$. Then, $\Lambda(t,x)$ defined in (6.25) is non negative. By (6.22),

$$\sigma(t,\varphi(x)) = \left(\varphi'(x)\left(\varphi'(x) + e^{\frac{x^2}{2t}}\Lambda(t,x)\right)\right)^{1/2}$$

and therefore, it suffices, by differentiation, to prove that the function:

$$D(t,x) := \frac{x}{t}e^{\frac{x^2}{2t}}\Lambda(t,x)$$

is bounded on every set of the form $]0,a]\times]0,b]$ with $a,b>0$. But, for $0<t\le a$ and $0<x\le b$, we have:

$$D(t,x)=\frac{x}{t}\int_0^{+\infty}\varphi''(x+z)\exp\left(-\frac{z^2}{2t}-\frac{xz}{t}\right)dz.$$

Decomposing this integral as $\int_0^b\ldots+\int_b^{+\infty}\ldots$, we write, with obvious notation: $D(t,x)=D_b^{(-)}(t,x)+D_b^{(+)}(t,x)$. Then, since $\frac{x}{t}\exp\left(-\frac{xz}{t}\right)\le\frac{1}{ez}$, we have, concerning $D_b^{(+)}$:

$$\begin{aligned}D_b^{(+)}(t,x)&\le\frac{1}{eb}\int_{x+b}^{+\infty}\varphi''(v)\exp\left(-\frac{(v-x)^2}{2t}\right)dv\\&\le\frac{1}{eb}\int_b^{+\infty}\varphi''(v)e^{-\frac{v^2}{8a}}dv<\infty\end{aligned}$$

since $v>x+b\Longrightarrow v\ge 2x\Longrightarrow v-x\ge\frac{v}{2}$, while, concerning $D_b^{(-)}(t,x)$:

$$D_b^{(-)}(t,x)\le\left(\sup_{r\in[0,2b]}\varphi''(r)\right)\frac{x}{t}\int_0^b\exp\left(-\frac{xz}{t}\right)dz\le\sup_{r\in[0,2b]}\varphi''(r).$$

Finally:

$$D(t,x)\le\sup_{r\in[0,2b]}\varphi''(r)+\frac{1}{eb}\int_b^{+\infty}\varphi''(v)e^{-\frac{v^2}{8a}}dv.$$

□

Remark 6.4. In the proof of Theorem 6.4, we look for σ such that the law of M_t, for fixed t, is identical to that of $\varphi(B_t)$; this is somewhat similar to Dupire (see [60]) who considers the SDE:

$$dM_t=\sigma(t,M_t)dB_t$$

and finds $\sigma^2(t,K)$ as the solution to the equation:

$$\frac{1}{2}\sigma^2(t,K)\frac{\partial^2 C}{\partial K^2}(t,K)=\frac{\partial C}{\partial t}(t,K)$$

where $C(t,K):=\mathbb{E}[(M_t-K)^+]$ is the price of an European call with maturity t and strike K.

6.4 Martingales Associated to Solutions of Some SDE's

Let $b,\tau:\mathbb{R}\longrightarrow\mathbb{R}$ denote two functions of $\mathscr{C}^\infty$ class such that:

i) τ is even and positive,
ii) b is odd and $\operatorname{sgn} b(x)=\operatorname{sgn}(x)$,

iii) all derivatives of b and τ are bounded and there exist c_1, c_2 and c_3 such that:

$$0 < c_1 \leq \inf_{x\in\mathbb{R}} \tau(x) \leq \sup_{x\in\mathbb{R}} \tau(x) \leq c_2 < \infty$$
$$\sup_{x\in\mathbb{R}} |b(x)| \leq c_3 < \infty.$$

Let $(X_t, t \geq 0)$ be the unique strong solution of the SDE:

$$Z_t = \int_0^t \tau(Z_s)dB_s + \int_0^t b(Z_s)ds. \tag{6.27}$$

We know from Exercise 1.27 that $(X_t, t \geq 0)$ is a peacock.

Under these hypotheses, the law of X_t, for every $t > 0$, admits a strictly positive density $p(t,x)$ of class $\mathscr{C}^{1,2}$ on $]0,+\infty[\times\mathbb{R}$. Let $\sigma : \mathbb{R}_+ \times \mathbb{R} \longrightarrow \mathbb{R}_+$ be defined by:

$$\begin{cases} \sigma(t,x) := \left(\tau^2(x) - \dfrac{2}{p(t,x)}\displaystyle\int_{-\infty}^x b(y)p(t,y)dy\right)^{1/2} & (t>0). \\ \sigma(0,x) := \tau(x) \end{cases} \tag{6.28}$$

It is easy to show that $y \longmapsto p(t,y)$ is an even function. Then, from Point *ii*):

$$-\frac{2}{p(t,x)}\int_{-\infty}^x b(y)p(t,y)dy \geq 0$$

and, a fortiori:

$$\sigma^2(t,x) = \tau^2(x) - \frac{2}{p(t,x)}\int_{-\infty}^x b(y)p(t,y)dy \geq 0.$$

Lemma 6.6. *Let us define* $h(t,x) := \dfrac{1}{p(t,x)}\displaystyle\int_{-\infty}^x b(y)p(t,y)dy$. *Then:*

1) *for every compact* K, $\sup_{x\in K} |h(t,x)| \xrightarrow[t\to 0]{} 0$,
2) *for every compact* K *and every* $t_0 > 0$, *there exists* $C(K,t_0)$ *such that:*

$$\sup_{t\leq t_0;\, x,y\in K} |h(t,x) - h(t,y)| \leq C(K,t_0)|x-y|$$

where we set $h(0,x) := 0$.

Proof. We equip $\mathbb{R}$ with the Riemannian metric δ defined by $d\delta^2 = \dfrac{1}{\tau^2(x)}dx^2$ and we denote by $\delta(x) = \delta(0,x)$ the distance from 0 to x. Observe that, from Hypothesis *iii*), this metric is equivalent to the Euclidean metric, since:

$$\frac{1}{c_2}|x| \leq \delta(x) := \left|\int_0^x \frac{dy}{\tau(y)}\right| \leq \frac{1}{c_1}|x|. \tag{6.29}$$

We recall the following facts:

1) There is the asymptotic result:

$$p(t,x) \underset{t\to 0}{=} \frac{1}{\sqrt{t}} e^{-\frac{\delta^2(x)}{2t}} \left(\alpha_0(x) + \mathscr{O}(t)\right) \tag{6.30}$$

where α_0 is a strictly positive function of $\mathscr{C}^1$ class. Furthermore, this equivalent is uniform on compact sets (in x).

2) The asymptotic result (6.30) may be differentiated:

$$\frac{\partial}{\partial x} p(t,x) \underset{t\to 0}{=} \frac{1}{t^{\frac{3}{2}}} e^{-\frac{\delta^2(x)}{2t}} \left(-\frac{\alpha_0(x)}{\tau(x)} \delta(x) + \mathscr{O}(t) \right). \tag{6.31}$$

This equivalent is also uniform on compact sets (in x).

3) Let $B(0,\rho)$ be the ball of center 0 and radius ρ, for the metric δ. Then:

$$\mathbb{P}\left(X_t \notin B(0,\rho)\right) \underset{t\to 0}{\sim} \frac{c}{\sqrt{t}} e^{-\frac{\rho^2}{2t}}. \tag{6.32}$$

We refer the reader to [18], [17, Theorem 4.1, p.173] and [10].

Now, observe that since p is of $\mathscr{C}^{1,2}$ class on $]0,+\infty[\times\mathbb{R}$, it is sufficient to study $h(t,x)$ in a neighborhood of $t=0$. Moreover, h being an even function of x, we may consider only the case $x<0$.

In the following computations, C denotes a positive constant which may change from a line to another.

Proof of Point 1) of Lemma 6.6. Let $a<0$ and $x\in[a,0]$. We write:

$$|h(t,x)| \leq \frac{1}{p(t,x)} \int_{-\infty}^{2a} |b(y)|p(t,y)dy + \frac{1}{p(t,x)} \int_{2a}^{x} |b(y)|p(t,y)dy = (1) + (2).$$

We study each term separately:

$$\begin{aligned}
(1) &\leq \frac{\|b\|_\infty}{p(t,x)} \left(\int_{-\infty}^{2a} p(t,y)dy + \int_{-2a}^{+\infty} p(t,y)dy \right) \\
&= \frac{\|b\|_\infty}{p(t,x)} \mathbb{P}(X_t \notin B(0,\delta(2a))) \quad \text{(from (6.32))} \\
&\underset{t\to 0}{\sim} \frac{\|b\|_\infty \sqrt{t}}{\alpha_0(x)} e^{\frac{\delta^2(x)}{2t}} \frac{c}{\sqrt{t}} e^{-\frac{\delta^2(2a)}{2t}} \\
&= \frac{c\|b\|_\infty}{\alpha_0(x)} \exp\left(\frac{-\delta^2(2a)+\delta^2(x)}{2t} \right) \underset{t\to 0}{\longrightarrow} 0
\end{aligned}$$

uniformly with respect to $x\in[a,0]$.

$$(2) \leq C \int_{2a}^{x} e^{\frac{1}{2t}\left(-\delta^2(y)+\delta^2(x)\right)} dx \underset{t\to 0}{\longrightarrow} 0$$

uniformly with respect to $x\in[a,0]$, from Dini's lemma, since for $y<x\leq 0$, $t \longrightarrow \exp\left(\frac{1}{2t}\left(-\delta^2(y)+\delta^2(x)\right)\right)$ is an increasing function of t. □

Proof of Point 2) of Lemma 6.6. Since $\frac{\partial}{\partial x}h(t,x) = -\frac{1}{p^2(t,x)}\left(\frac{\partial}{\partial x}p(t,x)\right)\int_{-\infty}^{x} b(y)$ $p(t,y)dy + b(x)$, it suffices, b being a bounded function, to see that, for every compact K and every $t_0 > 0$, there exists $\theta(K,t_0)$ such that:

$$\sup_{0<t\leq t_0;\, x\in K}\left|\frac{1}{p^2(t,x)}\left(\frac{\partial}{\partial x}p(t,x)\right)\int_{-\infty}^{x} b(y)p(t,y)dy\right| \leq \theta(K,t_0).$$

For $a < 0$ fixed and $x \in [a,0]$:

$$\begin{aligned}
&\left|\frac{1}{p^2(t,x)}\left(\frac{\partial}{\partial x}p(t,x)\right)\int_{-\infty}^{x} b(y)p(t,y)dy\right| \\
&\qquad\leq \left|\frac{1}{p^2(t,x)}\left(\frac{\partial}{\partial x}p(t,x)\right)\right|\left(\int_{-\infty}^{2a}|b(y)|p(t,y)dy + \int_{2a}^{x}|b(y)|p(t,y)dy\right) \\
&\qquad= (3)\ +\ (4).
\end{aligned}$$

Once again, we study each term separately:

$$\begin{aligned}
(3) &\leq \frac{\frac{1}{t^{3/2}}\exp\left(-\frac{\delta^2(x)}{2t}\right)\left(\frac{\alpha_0(x)}{\tau(x)}\delta(x) + \mathcal{O}(t)\right)}{\frac{1}{t}\exp\left(-\frac{\delta^2(x)}{t}\right)(\alpha_0(x) + \mathcal{O}(t))^2}\|b\|_\infty \frac{c}{\sqrt{t}}e^{-\frac{\delta^2(2a)}{2t}} \\
&\leq \frac{C}{t}\exp\left(\frac{-\delta^2(2a) + \delta^2(x)}{2t}\right)
\end{aligned}$$

since, as a continuous and strictly positive function on the segment $[a,0]$, α_0 is bounded from above and below by strictly positive constants. Thus, the function $t \longmapsto \frac{1}{t}\exp\left(\frac{-\delta^2(2a)+\delta^2(x)}{2t}\right)$ being bounded uniformly in $x \in [a,0]$, we deduce that there exists $\widetilde{C}_1$ such that $(3) \leq \widetilde{C}_1$.

$$\begin{aligned}
(4) &\leq \frac{\frac{1}{t^{3/2}}\exp\left(-\frac{\delta^2(x)}{2t}\right)\left(\frac{\alpha_0(x)}{\tau(x)}\delta(x) + \mathcal{O}(t)\right)}{\frac{1}{t}\exp\left(-\frac{\delta^2(x)}{t}\right)(\alpha_0(x) + \mathcal{O}(t))^2}\|b\|_\infty \int_{2a}^{x}\frac{1}{\sqrt{t}}e^{-\frac{\delta^2(y)}{2t}}(\alpha_0(y) + \mathcal{O}(t))\,dy \\
&\leq \frac{C}{t}\int_{2a}^{x}\exp\left(-\frac{\delta^2(y) - \delta^2(x)}{2t}\right)(\delta(x) + \mathcal{O}(t))\,dy \\
&\leq C(x-2a) + \frac{C}{t}\int_{2a}^{x}\exp\left(-\frac{1}{tc_2^2}(x-y)|x|\right)|x|dy \\
&\text{(from (6.29), since } \delta^2(y) - \delta^2(x) = (\delta(y) - \delta(x))(\delta(y) + \delta(x)) \\
&\leq C(x-2a) + C\int_{0}^{\frac{x-2a}{t}}\exp\left(-\frac{|x|v}{c_2^2}\right)|x|dv \\
&\leq C(x-2a) + C\frac{c_2^2}{|x|}|x| \leq \widetilde{C}_2.
\end{aligned}$$

□

Theorem 6.5. *Under the preceding notation and assumptions, the equation:*

$$Z_t = \int_0^t \sigma(s, Z_s) dB_s \tag{6.33}$$

admits a unique solution, which is strong, i.e. adapted to the filtration of B. We denote this solution by $(M_t, t \geq 0)$. Then, $(M_t, t \geq 0)$ is a martingale which is associated to the peacock $(X_t, t \geq 0)$ which is the solution of (6.27).

Proof. Here again, we shall apply Theorem 6.3 to $(X_t, t \geq 0)$ proving properties (P_1) and (P_2).

i) Let f be a $\mathscr{C}^2$ function with compact support. Itô's formula implies:

$$\mathbb{E}[f(X_t)] = f(0) + \frac{1}{2}\int_0^t \mathbb{E}[f''(X_s)\tau^2(X_s) + 2f'(X_s)b(X_s)]ds$$

and

$$\begin{aligned}\mathbb{E}[f'(X_s)b(X_s)] &= \int f'(x)b(x)p(s,x)dx \\ &= -\int f''(x)\left(\int_{-\infty}^x b(y)p(s,y)dy\right)dx \\ &= -\mathbb{E}[f''(X_s)h(s,X_s)]\end{aligned}$$

with $h(s,x) = \dfrac{1}{p(s,x)}\displaystyle\int_{-\infty}^x b(y)p(s,y)dy$. Therefore,

$$\begin{aligned}\mathbb{E}[f(X_t)] &= f(0) + \frac{1}{2}\int_0^t \mathbb{E}[f''(X_s)\left(\tau^2(X_s) - 2h(s,X_s)\right)]ds \\ &= f(0) + \frac{1}{2}\int_0^t \mathbb{E}[f''(X_s)\sigma^2(s,X_s)]ds,\end{aligned}$$

which is Property (P_1).

ii) We have:

$$\int_0^t \mathbb{E}[\sigma^2(s,X_s)]ds = \int_0^t ds \int_{-\infty}^{+\infty} dx \left(\tau^2(x)p(s,x) - 2\int_{-\infty}^x b(y)p(s,y)dy\right).$$

Then:

$$\int_0^t ds \int_{-\infty}^{+\infty} \tau^2(x)p(s,x)dx = \int_0^t \mathbb{E}[\tau^2(X_s)]ds \leq t\|\tau^2\|_\infty,$$

and by parity:

$$\begin{aligned}-\int_0^t ds \int_{-\infty}^{+\infty} dx \int_{-\infty}^x b(y)p(s,y)dy &= 2\int_0^t ds \int_0^{+\infty} dx \int_x^{+\infty} b(y)p(s,y)dy \\ &= 2\int_0^t ds \int_0^{+\infty} yb(y)p(s,y)dy \\ &\leq 2\|b\|_\infty \int_0^t \mathbb{E}[X_s^+]ds < \infty,\end{aligned}$$

since $\mathbb{E}[X_s^+] \leq \mathbb{E}[|X_s|] \leq \sqrt{s}\|\tau\|_\infty + s\|b\|_\infty < \infty$. □

Exercise 6.1 (Change of probability). Let $(\Omega = \mathscr{C}([0,+\infty[,\mathbb{R}),(X_t,t\geq 0),(\mathscr{F}_t, t\geq 0),W)$ be the canonical realization of Brownian motion started from 0. Let $\varphi : \mathbb{R} \longrightarrow \mathbb{R}_+$ be a function of $\mathscr{C}^2$ class which is strictly positive, convex, even, and such that $\dfrac{\varphi'}{\varphi}$ is bounded.

1) We define, for every $t \geq 0$:

$$M_t := \frac{\varphi(X_t)}{\varphi(0)} \exp\left(-\frac{1}{2}\int_0^t \frac{\varphi''}{\varphi}(X_s)ds\right).$$

Prove that $(M_t, t\geq 0)$ is a positive $((\mathscr{F}_t, t\geq 0), W)$-martingale such that $M_0 = 1$.

2) Let $\theta : \mathbb{R} \longrightarrow \mathbb{R}$ be a function of $\mathscr{C}^1$ class which is increasing and bounded. Prove that, for every $s \geq 0$:

$$\mathbb{E}\left[\theta(X_s)\varphi'(X_s)\exp\left(-\frac{1}{2}\int_0^s \frac{\varphi''}{\varphi}(X_u)du\right)\right] \geq 0$$

(*Hint*: $\mathbb{E}\left[\varphi'(X_s)\exp\left(-\frac{1}{2}\int_0^s \frac{\varphi''}{\varphi}(X_u)du\right)\right] = 0$, since φ' is odd.)

3) Let $\mathbb{Q}$ denote the probability defined on $(\Omega, \mathscr{F}_\infty = \bigvee_{t\geq 0} \mathscr{F}_t)$ and characterized by

$$\mathbb{Q}_{|\mathscr{F}_t} = M_t \centerdot W_{|\mathscr{F}_t}$$

i) Prove that, for every function $\psi \in \boldsymbol{C}$:

$$\mathbb{E}_{\mathbb{Q}}[\psi(X_t)] = \psi(0) + \mathbb{E}\left[\int_0^t M_s\left(\psi''(X_s) + \psi'(X_s)\frac{\varphi'}{\varphi}(X_s)\right)ds\right]$$

ii) Deduce that, under $\mathbb{Q}$, $(X_t, t\geq 0)$ is a peacock.

4) Prove that there exists a $\mathbb{Q}$-Brownian motion $(\beta_t, t\geq 0)$ started from 0 such that, under $\mathbb{Q}$:

$$X_t = \beta_t + \int_0^t \frac{\varphi'}{\varphi}(X_s)ds. \tag{6.34}$$

(*Hint*: apply Girsanov's theorem.)

We assume in the following that $\dfrac{\varphi''}{\varphi}$ is bounded. Recover the fact that $(X_t, t\geq 0)$ is a peacock under $\mathbb{Q}$. (*Hint*: apply the results of Exercise 1.27.)

5) Suppose that φ is of $\mathscr{C}^\infty$ class and let $(X_t, t\geq 0)$ be the solution of (6.34). Prove that Theorem 6.5 applies and exhibit a martingale associated to the peacock $(X_t, t\geq 0)$.

6) In this question we propose, in the previous situation and assuming furthermore that φ is of $\mathscr{C}^3$ class and $\dfrac{\varphi'''}{\varphi}$ is bounded, a "direct" proof of Lemma 6.6.

i) Prove that the density $p(t,\centerdot)$ of X_t is given by:

$$p(t,x) = \pi(t,x)\varphi(x)D(t,x)$$

with

$$\pi(t,x) := \frac{1}{\sqrt{2\pi t}} e^{-\frac{x^2}{2t}},$$

and

$$D(t,x) = \frac{1}{\varphi(0)} \mathbb{E}\left[\exp\left(-\frac{t}{2}\int_0^1 \frac{\varphi''}{\varphi}(xv + \sqrt{t}\widetilde{\beta}_v)dv\right)\right]$$

where $(\widetilde{\beta}_v, 0 \le v \le 1)$ denotes the standard Brownian bridge satisfying $\widetilde{\beta}_0 = \widetilde{\beta}_1 = 0$.

ii) Deduce then that:

$$h(t,x) := \frac{1}{p(t,x)} \int_{-\infty}^{x} \frac{\varphi'}{\varphi}(y) p(t,y) dy$$

satisfies Properties *i*) and *ii*) in Lemma 6.6.

Exercise 6.2 (Change of probability: an example with the local time at 0). Let $(\Omega = \mathscr{C}([0,+\infty[,\mathbb{R}), (X_t, t \ge 0), (\mathscr{F}_t, t \ge 0), W)$ be the canonical realization of a Brownian motion started from 0. We denote by $(L_t, t \ge 0)$ its local time process at 0. Let $\lambda > 0$.

1) *i*) Prove that the process $\left(M_t^{(\lambda)} := (1 + \lambda |X_t|) \exp(-\lambda L_t), t \ge 0\right)$ is a positive $((\mathscr{F}_t, t \ge 0), W)$-martingale which may be written:

$$M_t^{(\lambda)} = 1 + \lambda \int_0^t e^{-\lambda L_u} \mathrm{sgn}(X_u) dX_u.$$

ii) Prove that for every function $\theta : \mathbb{R} \longrightarrow \mathbb{R}$ of $\mathscr{C}^1$ class which is increasing and bounded:

$$\mathbb{E}\left[\theta(X_u) e^{-\lambda L_u} \mathrm{sgn}(X_u)\right] \ge 0.$$

(*Hint*: $\mathbb{E}\left[e^{-\lambda L_u} \mathrm{sgn}(X_u)\right] = 0$.)

2) Let $\mathbb{Q}^{(\lambda)}$ be the probability defined on $(\Omega, \mathscr{F}_\infty = \bigvee_{t \ge 0} \mathscr{F}_t)$ and characterized by:

$$\mathbb{Q}^{(\lambda)}_{|\mathscr{F}_t} = M_t^{(\lambda)} \cdot W_{|\mathscr{F}_t}$$

Prove that, for every function $\psi \in \mathbf{C}$:

$$\mathbb{E}_{\mathbb{Q}^{(\lambda)}}[\psi(X_t)] = \psi(0) + \mathbb{E}\left[\int_0^t M_u^{(\lambda)} \left(\frac{1}{2}\psi''(X_u)du + \lambda e^{-\lambda L_u} \mathrm{sgn}(X_u) \psi'(X_u)\right) du\right]$$

Deduce that, under $\mathbb{Q}^{(\lambda)}$, $(X_t, t \ge 0)$ is a peacock.

3) Prove that there exists a $\mathbb{Q}^{(\lambda)}$-Brownian motion $(\beta_t, t \ge 0)$ started from 0 such that, under $\mathbb{Q}^{(\lambda)}$:

$$X_t = \beta_t + \int_0^t \frac{\mathrm{sgn}(X_s)}{\frac{1}{\lambda} + |X_s|} ds. \tag{6.35}$$

(*Hint*: apply Girsanov's theorem.)

4) We now assume that $\lambda = 1$. Let $p(t,\centerdot)$ be the density of X_t under $\mathbb{Q}^{(1)}$.

i) Prove that:

$$p(t,x) = \frac{1}{\sqrt{2\pi t}}(1+|x|)e^{|x|}\int_{|x|/\sqrt{t}}^{+\infty} y\exp\left(-\frac{y^2}{2} - y\sqrt{t}\right)dy$$

(*Hint*: we recall that, as a consequence of the symmetry principle, the density f of the couple $(|X_t|, L_t)$ under W equals $f(a,b) = \sqrt{\frac{2}{\pi t^3}}(a+b)\exp\left(-\frac{(a+b)^2}{2t}\right)$ $1_{\{a>0\}}1_{\{b>0\}}$.)

ii) Deduce that:

$$p(t,x) = \frac{1}{\sqrt{2\pi t}}(1+|x|)e^{-\frac{x^2}{2t}}\left(1-\sqrt{t}\,l\left(\sqrt{t}+\frac{|x|}{\sqrt{t}}\right)\right)$$

with:

$$l(u) = \int_0^{+\infty} \exp\left(-\frac{y^2}{2} - yu\right)dy.$$

iii) Set

$$h(t,x) := \frac{1}{p(t,x)}\int_{|x|}^{+\infty}\frac{1}{1+y}p(t,y)dy.$$

Prove that h satisfies Properties *i*) and *ii*) of Lemma 6.6.

iv) Exhibit a martingale associated to the peacock $(X_t, t \geq 0)$ under $\mathbb{Q}^{(1)}$, as a solution of a SDE. (*Hint*: proceed as for Theorem 6.5.)

Remark 6.5. We mention here two results close to our topic, i.e. which consist, given a process, in finding another one having the same unidimensional marginals.

A result of Krylov and Gyöngy. We only state this result, whose proof can be found in [112] and [72].

Let $(B_t, t \geq 0)$ be a d-dimensional Brownian motion and $(\sigma(s), s \geq 0)$ and $(b(s), s \geq 0)$ two predictable processes taking values respectively in $\mathcal{M}_d$, the space of $d \times d$-matrices, and $\mathbb{R}^d$. We define the process:

$$\xi_t := \int_0^t \sigma(s)dB_s + \int_0^t b(s)ds.$$

Of course, we assume that σ and b are such that $(\xi_t, t \geq 0)$ is well-defined. We assume that there exists a constant $c > 0$ such that:

$$\forall t \geq 0, \forall x \in \mathbb{R}^d, \quad \sum_{i,j=1}^{d} \alpha_{i,j}(t,\omega)x_i x_j \geq c\sum_{i=1}^{d} x_i^2 \text{ a.s.}$$

with $\alpha_{ij} = (\sigma\sigma^*)_{i,j}$, $i,j = 1,\dots,d$. Under these hypotheses, we have:

Theorem 6.6 (N.V. Krylov, I. Gyöngy). *Let us consider the SDE:*

$$X_t = \int_0^t \tau(s,X_s)dB_s + \int_0^t a(s,X_s)ds$$

where

$$\tau(s,x) := (\mathbb{E}[\sigma\sigma^*(s)|\xi_s = x])^{1/2} \quad \text{and} \quad a(s,x) := \mathbb{E}[b(s)|\xi_s = x],$$

the square root being taken in the sense of positive matrices. Then, this SDE admits a weak solution $(X_t, t \geq 0)$ such that:

$$\xi_t \overset{(1.d)}{=} X_t.$$

Some generalizations of the previous theorem have been obtained, in the set-up of non continuous processes, by A. Bentata and R. Cont [22].

A non-linear process having the same one-dimensional marginals as Brownian motion. Let $(B_t = (B_t^{(1)}, B_t^{(2)}), t \geq 0)$ be a Brownian motion of dimension 2 started from 0 and let $b = (b^{(1)}, b^{(2)})$ be a function from $\mathbb{R}^2$ into $\mathbb{R}^2$ Lipschitz continuous (or locally Lipschitz continuous and such that, for every $x = (x_1, x_2) \in \mathbb{R}^2$, $x_1 b^{(1)}(x_1,x_2) + x_2 b^{(2)}(x_1,x_2) \leq 0$).

We consider the non-linear (in the sense of McKean-Vlasov) differential system of order 2:

$$\begin{cases} X_t^{(2)} = \displaystyle\int_0^t b^{(2)}(X_s)ds + B_t^{(2)} \\ X_t^{(1)} = \displaystyle\int_0^t b^{(1)}(X_s)ds - \int_0^t a(s, X_s^{(1)})ds + B_t^{(1)} \\ \text{with } a(s,x) := \mathbb{E}[b^{(1)}(X_s)|X_s^{(1)} = x]. \end{cases} \tag{S}$$

Theorem 6.7 ([119]). *Under the previous hypotheses:*
1) *The system (S) admits a unique solution in law.*
2) *The process $(X_t^{(1)}, t \geq 0)$ has the same unidimensional marginals as a Brownian motion started from 0.*

The first point of this theorem can be obtained either by a fixed point theorem, or by the introduction of a (linear) equation of n particles (i.e. a ordinary SDE of dimension $2n$) and letting n tend to $+\infty$.

Point 2) follows from Itô's formula. Indeed, if $f : \mathbb{R} \longrightarrow \mathbb{R}$ is of $\mathscr{C}^2$-class with compact support, and if $((X_t^{(1)}, X_t^{(2)}), t \geq 0)$ is solution of (S), we have:

$$\begin{aligned} \mathbb{E}\left[f(X_t^{(1)})\right] =& f(0) + \int_0^t \mathbb{E}\left[f'(X_s^{(1)})\left(b^{(1)}(X_s^{(1)}, X_s^{(2)}) - a(s, X_s^{(1)}\right) ds\right] \\ &+ \frac{1}{2}\int_0^t \mathbb{E}[f''(X_s^{(1)})]ds \end{aligned}$$

and

$$\mathbb{E}\left[f'(X_s^{(1)})\left(b^{(1)}(X_s^{(1)},X_s^{(2)})-a(s,X_s^{(1)})\right)\right]$$
$$=\mathbb{E}\left[f'(X_s^{(1)})\left(\mathbb{E}[b^{(1)}(X_s^{(1)},X_s^{(2)})|X_s^{(1)}]-a(s,X_s^{(1)})\right)\right]=0.$$

Hence,

$$\mathbb{E}\left[f(X_t^{(1)})\right]=f(0)+\frac{1}{2}\int_0^t \mathbb{E}[f''(X_s^{(1)})]ds$$

which easily implies that, if $p(t,dx)$ denotes the law of $X_t^{(1)}$,

$$\frac{\partial}{\partial t}p(t,dx)=\frac{1}{2}\frac{\partial^2}{\partial x^2}p(t,dx) \quad \text{(in the sense of distributions)}$$

and, since $X_0^{(1)}=0$ (and therefore $p(0,dx)=\delta_0(dx)$),

$$p(t,dx)=\frac{1}{\sqrt{2\pi t}}e^{-\frac{x^2}{2t}}dx.$$

We also refer the interested reader to [138] for a close thema.

6.5 A Variant of the Fokker-Planck Equation Method

In this section, we present a variant of the method introduced in Section 6.2, which is adapted to the study of $\mathbb{R}_+^*$-valued processes, with $\mathbb{R}_+^*=]0,+\infty[$. We first state the analogue of M. Pierre's uniqueness theorem given in Section 6.1, replacing $\mathbb{R}$ by $\mathbb{R}_+^*$.

Theorem 6.8. *Let $a:\mathbb{R}_+^*\times\mathbb{R}_+^*\longrightarrow\mathbb{R}_+$ be a Borel function satisfying the following hypotheses:*

(H_1) *For any $0<t<T$ and $0<r<R$, there exist real numbers $\varepsilon(t,T,r,R)>0$ and $m(T,r,R)<+\infty$ such that:*

$$\forall(s,x)\in[t,T]\times[r,R], \qquad a(s,x)\geq\varepsilon(t,T,r,R),$$

and

$$\forall(s,x)\in]0,T]\times[r,R], \qquad a(s,x)\leq m(T,r,R).$$

(H_2) *For all $T>0$, there exists $A(T)<+\infty$ such that:*

$$\forall(s,x)\in]0,T]\times[1,+\infty[, \qquad a(s,x)\leq A(T)x^2(\log(x)+1).$$

Let μ be a probability measure on $\mathbb{R}_+^*$ such that $\int x\mu(dx)<+\infty$. Then, there exists at most one family of probability measures $(p(t,dx),t\geq0)$ on $\mathbb{R}_+^*$ such that:

(FP_1) *$t \geq 0 \longmapsto p(t,dx)$ is weakly continuous.*

(FP_2) *The following Fokker-Planck equation is satisfied:*

$$\begin{cases} p(0,dx) = \mu(dx) \\ \dfrac{\partial p}{\partial t} - \dfrac{\partial^2}{\partial x^2}(ap) = 0 \\ \text{in the sense of distributions on } \mathbb{R}_+^* \times \mathbb{R}_+^*. \end{cases}$$

(FP_3) *For all $T > 0$, there exists $B(T) < +\infty$ such that:*

$$\forall t \in]0,T], \qquad \int_1^{+\infty} (\log x) p(t,dx) \leq B(T).$$

Proof. We follow the proof of Theorem 6.1. Let $(p(t,dx), t \geq 0)$ be a family of probability measures satisfying (FP_1), (FP_2) and (FP_3). We set:

$$P(t,dx) := \int_0^t a(s,x)p(s,dx)ds.$$

Then, as in the proof of Lemma 6.2, there exists a continuous function P on $\mathbb{R}_+ \times \mathbb{R}_+^*$ such that

$$P(t,dx) = P(t,x)dx,$$

$P(t,\centerdot)$ admits a right-continuous right-derivative $\frac{\partial P}{\partial x}(t,x)$ and there exists $C(t)$ such that:

$$\frac{\partial P}{\partial x}(t,x) = \int_{]x,+\infty[} \mu(du) - \int_{]x,+\infty[} p(t,du) + C(t). \tag{6.36}$$

In particular:

$$\lim_{x \to 0} \frac{\partial P}{\partial x}(t,x) = \lim_{x \to +\infty} \frac{\partial P}{\partial x}(t,x) = C(t),$$

and, as a consequence, the limit

$$P(t,0^+) := \lim_{x \to 0} P(t,x)$$

exists. Integrating (6.36), we obtain:

$$P(t,x) = P(t,0^+) + \int_0^{+\infty} (u \wedge x)\mu(du) - \int_0^{+\infty} (u \wedge x)p(t,du) + C(t)x. \tag{6.37}$$

By (6.37),

$$\lim_{x \to +\infty} \frac{P(t,x)}{x} = C(t) = \lim_{n \to +\infty} \frac{1}{n} \int_n^{2n} \frac{1}{x} P(t,dx).$$

Now, by Hypothesis (H_2),

$$\frac{1}{n} \int_n^{2n} \frac{1}{x} P(t,dx) \leq 4A(t)(\log(2n)+1) \int_0^t \int_n^{2n} p(s,dx)ds.$$

Since $\int_n^{2n} p(s,dx) \leq \frac{1}{\log n} \int_n^{2n} (\log x) p(s,dx)$, we obtain,

$$0 \leq C(t) \leq 4A(t) \lim_{n \to +\infty} \int_0^t \int_n^{2n} (\log x) p(s,dx) ds.$$

Using (FP_3), we obtain by dominated convergence $C(t) = 0$. From this point, we may follow the proof of Theorem 6.1 with obvious modifications. □

Exercise 6.3 (A variant of the uniqueness theorem for the Fokker-Planck equation on $\mathbb{R}_+^*$). Show that if one replaces, in the statement of Theorem 6.8, (H_2) by the stronger hypothesis:

For all $T > 0$, there exists $A(T) < +\infty$ such that:

$$\forall (s,x) \in]0,T] \times [1,+\infty[, \qquad a(s,x) \leq A(T)x^2,$$

then Theorem 6.8 holds without Condition (FP_3).

We state below the analogue of Theorem 6.3 for $\mathbb{R}_+^*$-valued peacocks.

Theorem 6.9. *Let $\sigma : \mathbb{R}_+ \times \mathbb{R}_+^* \longrightarrow \mathbb{R}_+$ satisfying the following conditions:*

(S_1) *σ is continuous on $\mathbb{R}_+ \times \mathbb{R}_+^*$.*
(S_2) *For all $t > 0$ and $x > 0$, $\sigma(t,x) > 0$.*
(S_3) *For all $T > 0$, there exists $A(T) < +\infty$ such that:*

$$\forall t \in [0,T],\ \forall x > 0, \qquad \sigma(t,x) \leq A(T)x(|\log x| + 1)^{\frac{1}{2}}.$$

Let $(X_t, t \geq 0)$ be a continuous $\mathbb{R}_+^*$-valued process such that:

(P_1) *For every function φ of $\mathscr{C}^2$ class and with compact support in $\mathbb{R}_+^*$:*

$$\mathbb{E}[\varphi(X_t)] = \varphi(1) + \frac{1}{2} \int_0^t \mathbb{E}\left[\varphi''(X_s)\sigma^2(s,X_s)\right] ds$$

(P_2) *For every $t \geq 0$, $\mathbb{E}[X_t] = 1$.*

Then, $(X_t, t \geq 0)$ is a peacock and an associated martingale is given as a $\mathbb{R}_+^*$-valued solution of

$$Z_t = 1 + \int_0^t \sigma(s,Z_s) dB_s. \tag{6.38}$$

Proof. Consider the SDE:

$$Y_t = \int_0^t e^{-Y_s}\sigma(s,e^{Y_s}) dB_s - \frac{1}{2} \int_0^t e^{-2Y_s}\sigma^2(s,e^{Y_s}) ds. \tag{6.39}$$

The coefficients of this SDE are continuous (by (S_1)) and, by (S_3), one has, for $0 \leq t \leq T$ and $x \in \mathbb{R}$,

$$e^{-x}\sigma(t,e^x) \leq A(T)(|x|+1)^{1/2} \leq A(T)(|x|+1)$$

and

$$e^{-2x}\sigma^2(t,e^x) \leq (A(T))^2(|x|+1).$$

Therefore, the SDE (6.39) admits a weak solution $(Y_t, t \geq 0)$ which is locally bounded in L^2. We set:

$$M_t := \exp(Y_t).$$

Then, $(M_t, t \geq 0)$ is a $\mathbb{R}_+^*$-valued solution to (6.38). We denote by $p(t,dx)$ the law of X_t, by $q(t,dx)$ the law of M_t, and we set, for $(t,x) \in \mathbb{R}_+^* \times \mathbb{R}_+^*$,

$$a(t,x) := \frac{1}{2}\sigma^2(t,x).$$

The conditions (S_1), (S_2) and (S_3) imply that a satisfies (H_1) and (H_2) in Theorem 6.8. Moreover, p and q satisfy obviously properties (FP_1) and (FP_2) of this theorem. Besides,

$$\int_1^{+\infty} (\log x)p(t,dx) \leq \int_1^{+\infty} xp(t,dx) \leq 1 \quad \text{by } (P_2),$$

and

$$\int_1^{+\infty} (\log x)q(t,dx) \leq \mathbb{E}[|Y_t|] \leq \left(\mathbb{E}[(Y_t)^2]\right)^{1/2},$$

and therefore $(p(t,dx), t \geq 0)$ and $(q(t,dx), t \geq 0)$ also satisfy (FP_3). Hence, by Theorem 6.8, $p = q$, which means:

$$M_t \overset{(1.d)}{=} X_t. \tag{6.40}$$

Finally, $(M_t, t \geq 0)$ is a positive local martingale, therefore a supermartingale. By (6.40) and (P_2),

$$\forall t \geq 0, \qquad \mathbb{E}[M_t] = 1$$

and consequently, $(M_t, t \geq 0)$ is a martingale which is, by (6.40), associated to $(X_t, t \geq 0)$. □

We now give an application of Theorem 6.9.

Theorem 6.10. *Let X be a random variable admitting a strictly positive log-concave density f satisfying:*

$$\lim_{|x| \to +\infty} \frac{\log f(x)}{|x|} = -\infty. \tag{6.41}$$

We set, for $t \geq 0$, $h(t) := \log\left(\mathbb{E}[\exp(tX)]\right)$ and

$$X_t := \exp\left(tX - h(t)\right).$$

We also set, for $t > 0$ *and* $x > 0$, $u(t,x) := \dfrac{\log x + h(t)}{t}$,

$$\sigma^2(t,x) := 2tx^2 \int_0^{+\infty} \frac{f(u(t,x)+z)}{f(u(t,x))}(z+u(t,x)-h'(t))e^{tz}dz \tag{6.42}$$

and, for $t = 0$ *and* $x > 0$

$$\sigma(0,x) = 0. \tag{6.43}$$

Then, $\sigma \geq 0$ *is well-defined by (6.42) and (6.43), and satisfies Conditions* (S_1), (S_2) *and* (S_3). *If* $(M_t, t \geq 0)$ *is a* $\mathbb{R}_+^*$*-valued solution of the SDE (6.38) (the existence of such a solution is ensured by Theorem 6.9), then* $(M_t, t \geq 0)$ *is a martingale associated to the peacock* $(X_t, t \geq 0)$.

Proof. 1) By (6.41), for every $c \geq 0$,

$$\lim_{|x|\to+\infty} e^{c|x|} f(x) = 0$$

In particular, h is well-defined as a $\mathscr{C}^\infty$-function on $[0,+\infty[$.

2) Let us show that, for $t > 0$ and $x > 0$, the RHS of (6.42) is strictly positive. We have:

$$x \int_0^{+\infty} f(u(t,x)+z)(z+u(t,x)-h'(t))e^{tz}dz$$
$$= \int_{u(t,x)}^{+\infty} (v-h'(t)) \exp(tv-h(t)) f(v)dv \tag{6.44}$$

after the change of variable $v = u(t,x)+z$. Let π_t be the probability measure on $\mathbb{R}$ with density:

$$\exp(tv-h(t))f(v).$$

Then, the integral (6.44) may be rewritten as

$$\mathbb{E}_{\pi_t}\left[(v-h'(t))1_{\{v-h'(t)\geq u(t,x)-h'(t)\}}\right],$$

and

$$\mathbb{E}_{\pi_t}\left[v-h'(t)\right] = \mathbb{E}[(X-h'(t))\exp(tX-h(t))]$$
$$= \frac{d}{dt}\mathbb{E}[X_t] = 0.$$

We remark that Lemma 1.1 may be specified in the following way.

Lemma 6.7. *Let* U *be a real-valued integrable r.v. such that* ess sup $U = +\infty$, ess inf $U = -\infty$ *and* $\mathbb{E}[U] = 0$. *Then, for every* $c \in \mathbb{R}$, $\mathbb{E}[U 1_{\{U \geq c\}}] > 0$.

The proof is easy and will be omitted. We now apply this lemma with $\mathbb{E} = \mathbb{E}_{\pi_t}$ and $U = v - h'(t)$. This proves that σ is well-defined by (6.42) and (6.43) and satisfies (S_2).

3) We now prove (S_3). We first suppose $x \geq 1$ and we set $c := \inf_{t \in]0,T]} \frac{h(t)}{t}$, which belongs to $\mathbb{R}$ since $\lim_{t \to 0} \frac{h(t)}{t} = h'(0) = \mathbb{E}[X]$. Let $\phi = -\log f$. Then ϕ is a convex function and therefore, for every $z \geq 0$, since for $x \geq 1$, $u(t,x) \geq c$:

$$\phi(u(t,x)+z) - \phi(u(t,x)) \geq \phi(c+z) - \phi(c).$$

Therefore, by (6.42),

$$\sigma^2(t,x) \leq 2tx^2 \int_0^{+\infty} \frac{f(c+z)}{f(c)} \left(z + \left| \frac{h(t)}{t} - h'(t) \right| + \frac{\log x}{t} \right) e^{tz} dz \tag{6.45}$$

and (6.45) implies (S_3) for $x \geq 1$. Likewise, we have by (6.42), and by the fact that $\mathbb{E}[(X - h'(t))X_t] = 0$:

$$\sigma^2(t,x) = 2tx^2 \int_0^{+\infty} \frac{f(u(t,x)-z)}{f(u(t,x))} \left(z - u(t,x) + h'(t) \right) e^{-tz} dz$$

and we may prove (S_3) for $x \leq 1$ as previously.

4) Obviously, σ is continuous on $\mathbb{R}_+^* \times \mathbb{R}_+^*$. We now prove the continuity at $t = 0$. We consider, for instance, the case $x \geq 1$. By (6.45) and (6.43), we see that it suffices to prove that, for all $\varepsilon > 0$,

$$\lim_{t \to 0} \int_0^{+\infty} \frac{f(u(t,x)+z)}{f(u(t,x))} e^{tz} dz = 0$$

uniformly with respect to $x \in [1+\varepsilon, +\infty[$. By (6.41), $\lim_{x \to +\infty} \phi'(x) = +\infty$, where ϕ' denotes the right derivative of $\phi = -\log f$. We have

$$\phi(u(t,x)+z) - \phi(u(t,x)) \geq \phi'(u(t,x))z$$

and therefore,

$$\int_0^{+\infty} \frac{f(u(t,x)+z)}{f(u(t,x))} e^{tz} dz \leq \frac{1}{\phi'(u(t,x)) - t} \leq \frac{1}{\phi'\left(\frac{\log(1+\varepsilon)+h(t)}{t}\right) - t}$$

and the desired result follows. Thus σ also satisfies (S_1).

5) To apply Theorem 6.9, it remains to prove that $(X_t, t \geq 0)$ satisfies (P_1). Let φ be a $\mathscr{C}^2$-function with compact support in $\mathbb{R}_+^*$. We have:

$$\begin{aligned}
\frac{d}{dt}\mathbb{E}[\varphi(X_t)] &= \mathbb{E}[(X - h'(t))X_t\varphi'(X_t)] \\
&= \int_{-\infty}^{+\infty} \varphi'\left(e^{tx-h(t)}\right)(x - h'(t))e^{tx-h(t)} f(x)dx \\
&= \frac{1}{t}\int_0^{+\infty} \varphi'(y)(u(t,y) - h'(t))f(u(t,y))dy \quad (\text{with } y = e^{tx-h(t)}) \\
&= \frac{1}{t}\int_0^{+\infty} dy\varphi''(y)\int_y^{+\infty}(u(t,z) - h'(t))f(u(t,z))dz \quad (\text{integration by parts}) \\
&= t\int_{-\infty}^{+\infty} dx\, e^{tx-h(t)} f(x)\varphi''\left(e^{tx-h(t)}\right)\frac{1}{f(x)}\int_x^{+\infty}(v - h'(t))e^{tv-h(t)} f(v)dv
\end{aligned}$$

(after the changes of variables $y = \exp(tx - h(t))$ and $z = \exp(tv - h(t))$).

Finally, we obtain:

$$\frac{d}{dt}\mathbb{E}[\varphi(X_t)] = t\mathbb{E}\left[\varphi''(X_t)\frac{X_t}{f(u(t,X_t))}\int_{u(t,X_t)}^{+\infty}(v - h'(t))e^{tv-h(t)} f(v)dv\right]$$

which is (P_1) since:

$$\begin{aligned}
&\frac{tx}{f(u(t,x))}\int_{u(t,x)}^{+\infty}(v - h'(t))e^{tv-h(t)} f(v)dv \\
&= tx^2\int_0^{+\infty}\frac{f(u(t,x)+z)}{f(u(t,x))}(z + u(t,x) - h'(t))e^{tz}dz \\
&= \frac{1}{2}\sigma^2(t,x).
\end{aligned}$$

Exercise 6.4. Determine explicitly σ and $(M_t, t \geq 0)$ when $f(x) = \frac{1}{\sqrt{2\pi}}e^{-\frac{x^2}{2}}$.

(*Answer*: $\sigma(t,x) = \sqrt{2t}x$, $M_t = \exp\left(\int_0^t \sqrt{2s}dB_s - \frac{t^2}{2}\right)$.)

Exercise 6.5 (A continuous martingale associated to $(\varphi(tX), t \geq 0)$). We assume, in this exercise, that X satisfies the same hypotheses as in Theorem 6.10. We still denote by f the log-concave density of X.

Let φ be a $\mathscr{C}^1$-function. We assume that there exist $M \geq 0$ and $c \geq 0$ such that, for every $x \in \mathbb{R}$,

$$0 < \varphi'(x) \leq Me^{c|x|},$$

and

$$\lim_{x\to+\infty}\varphi(x) = +\infty, \qquad \lim_{x\to-\infty}\varphi(x) = -\infty.$$

We assume moreover that, for all $t \geq 0$, $\mathbb{E}[\varphi(tX)] = \varphi(0)$. We set:

$$\sigma^2(t, \varphi(x)) = \frac{2t\varphi'(x)}{f\left(\frac{x}{t}\right)} \int_{\frac{x}{t}}^{+\infty} \varphi'(tu)uf(u)du, \qquad \text{for } t > 0$$

and

$$\sigma^2(0, \varphi(x)) = 0.$$

1) Prove that $\sigma \geq 0$ is well-defined by the above formula, σ is continuous on $\mathbb{R}_+ \times \mathbb{R}$ and σ is strictly positive on $\mathbb{R}_+^* \times \mathbb{R}$.
(*Hint*: the proof of the continuity at $t = 0$ is similar to the one of Point 4) in the proof of Theorem 6.10.)

2) Verify that:

$$\int_0^t \mathbb{E}[\sigma^2(s, \varphi(sX))]ds = \mathbb{E}[\varphi^2(tX)] - \varphi^2(0)$$

3) Prove that Theorem 6.3 applies with $X_t = \varphi(tX)$ and yields a continuous martingale associated to the peacock $(\varphi(tX), t \geq 0)$.

4) Compare with Section 6.3 and with Exercise 7.13.

6.6 Martingales Associated to the Processes $(\sqrt{t}X, t \geq 0)$

In this subsection, we adopt essentially the same attitude as in Section 6.3. The main difference is that, instead of considering the peacock $(\varphi(B_t), t \geq 0)$, we consider the peacock $(\sqrt{t}X, t \geq 0)$ for a generic centered r.v. X (see Proposition 1.4). In other terms, we replace the "Gaussian framework" underlying the peacock $(\varphi(B_t), t \geq 0)$ by the scaling property of the peacock $(\sqrt{t}X, t \geq 0)$.

6.6.1 Hypotheses and Notation

Let X denote a r.v. with density h such that:

$$\int_{-\infty}^{+\infty} |x|h(x)dx < \infty \quad \text{and} \quad \int_{-\infty}^{+\infty} xh(x)dx = 0 \tag{6.46}$$

i.e. $\mathbb{E}[|X|] < \infty$ and $\mathbb{E}[X] = 0$. We assume that there exist α and β with:

$$-\infty \leq \alpha < 0 < \beta \leq +\infty$$

such that:

i) h continuous and strictly positive on $]\alpha, \beta[$, (6.47)

ii) $h(x) = 0 \quad$ for $x \notin]\alpha, \beta[$. (6.48)

We now define $a_h : \mathbb{R} \longrightarrow \mathbb{R}_+$ via:

$$\begin{cases} a_h(y) := \dfrac{1}{h(y)} \displaystyle\int_y^{+\infty} zh(z)dz & \text{if } y \in]\alpha,\beta[\\ a_h(y) := 0 & \text{if } y \notin]\alpha,\beta[. \end{cases} \tag{6.49}$$

Thus, for every $y \in \mathbb{R}$:

$$h(y)a_h(y) = \int_y^{+\infty} zh(z)dz. \tag{6.50}$$

We assume furthermore that

$$a_h \text{ is Lipschitz continuous.} \tag{6.51}$$

Remark 6.6. 1) From (6.46), it is clear that a_h defined by (6.49) is positive on $\mathbb{R}$.
2) Under hypothesis (6.51), for every integer p:

$$\mathbb{E}\left[|X|^p\right] = \int_{-\infty}^{+\infty} |x|^p h(x)dx < \infty.$$

Indeed, from (6.50) and (6.51), there exist $C \geq 0$ and $D \geq 0$ such that, for every $y \in [0,\beta]$:

$$\int_y^{\beta} zh(z)dz \leq Cyh(y) + Dh(y). \tag{6.52}$$

Hence:

$$\int_0^{\beta} z^2 h(z)dz = \int_0^{\beta} dy \int_y^{\beta} zh(z)dz \leq \int_0^{\beta} (Cyh(y) + Dh(y))dy < \infty.$$

Symmetrically, $\int_{\alpha}^{0} z^2 h(z)dz < \infty$.

Multiplying (6.52) by y, we may iterate this reasoning and thus obtain the desired result.

Here is now a description of densities of probabilities h which satisfy (6.46), (6.47) and (6.48) and such that their associated function a_h is Lipschitz continuous.

Proposition 6.2. *Let $-\infty \leq \alpha < 0 < \beta \leq +\infty$ and h be a density of probability which is continuous and strictly positive on $]\alpha,\beta[$ and null outside $]\alpha,\beta[$. Then, h satisfies (6.46) and is such that a_h is Lipschitz continuous if and only if h may be written, for every $x \in]\alpha,\beta[$:*

$$h(x) = \frac{C}{l(x)} \exp\left(-\int_0^x \frac{u}{l(u)} du\right) \tag{6.53}$$

where C is a strictly positive constant and $l : \mathbb{R} \longrightarrow \mathbb{R}_+$ satisfies the following properties:

a) l is Lipschitz continuous,
b) l is strictly positive on $]\alpha,\beta[$, and null outside $]\alpha,\beta[$,

c) $\displaystyle\int_0^{\beta} \frac{u}{l(u)} du = \int_0^{-\alpha} \frac{u}{l(-u)} du = +\infty.$

Moreover, C and l are uniquely determined by h. Actually, $l = a_h$ *and* $C = \displaystyle\int_0^{\beta} xh(x)dx = -\int_{\alpha}^{0} xh(x)dx.$

Proof. 1) Let h be defined by:

$$h(x) = \frac{C}{l(x)} \exp\left(-\int_0^x \frac{u}{l(u)} du\right)$$

with l satisfying *a*), *b*) and *c*). For $\alpha < y < \beta$, we have:

$$\begin{aligned}\int_y^{\beta} xh(x)dx &= C\int_y^{\beta} \frac{x}{l(x)} \exp\left(-\int_0^x \frac{u}{l(u)} du\right) dx \\ &= C\left[-\exp\left(-\int_0^{\cdot} \frac{u}{l(u)} du\right)\right]_y^{\beta} \\ &= C\exp\left(-\int_0^y \frac{u}{l(u)} du\right) \quad \text{(from } c)). \end{aligned} \tag{6.54}$$

Letting y tend to α, we obtain, again from *c*):

$$\int_{\alpha}^{\beta} xh(x)dx = C\exp\left(-\int_0^{\alpha} \frac{u}{l(u)} du\right) = 0.$$

This proves that h satisfies (6.46). On the other hand, for $\alpha < x < \beta$:

$$\begin{aligned}a_h(x) &= \frac{l(x)}{C} \exp\left(\int_0^x \frac{u}{l(u)} du\right) \int_x^{\beta} zh(z)dz \\ &= \frac{l(x)}{C} \exp\left(\int_0^x \frac{u}{l(u)} du\right) C\exp\left(-\int_0^x \frac{u}{l(u)} du\right) \quad \text{(from (6.54))} \\ &= l(x).\end{aligned}$$

2) Conversely, suppose h satisfies (6.46) and a_h is Lipschitz continuous.
We have, for every $x \in]\alpha, \beta[$:

$$\begin{aligned}\int_0^x \frac{u}{a_h(u)} du = \int_0^x du \frac{uh(u)}{\int_u^{\beta} zh(z)dz} &= \left[-\log \int_{\cdot}^{\beta} zh(z)dz\right]_0^x \\ &= -\log\left(\int_x^{\beta} zh(z)dz\right) + \log\left(\int_0^{\beta} zh(z)dz\right). \end{aligned} \tag{6.55}$$

Thus:

$$\frac{1}{a_h(x)} \exp\left(-\int_0^x \frac{u}{a_h(u)} du\right) = \frac{h(x)}{\int_x^{\beta} zh(z)dz} \frac{\int_x^{\beta} zh(z)dz}{\int_0^{\beta} zh(z)dz}$$

and

$$h(x) = \frac{C}{a_h(x)} \exp\left(- \int_0^x \frac{u}{a_h(u)} du \right), \quad \text{with } C = \int_0^\beta zh(z)dz.$$

Hence, (6.53) holds with $l(x) = a_h(x)$ and $C = \int_0^\beta zh(z)dz$. In particular, $l = a_h$ satisfies $a)$ and $b)$ and we deduce $c)$ passing to the limit as x tends to β or α in (6.55). □

6.6.2 Examples

1) $l(x) = 1, x \in \mathbb{R}$; then $h(x) = \frac{1}{\sqrt{2\pi}} e^{-\frac{x^2}{2}}$.

2) $l(x) = 1 + |x|, x \in \mathbb{R}$; then $h(x) = \frac{1}{2} e^{-|x|}$.

3) $l(x) = (1+|x|)^\alpha, x \in \mathbb{R}, \alpha < 1$; then

$$h(x) = \frac{C(\alpha)}{(1+|x|)^\alpha} \exp\left(- \frac{(1+|x|)^{2-\alpha}}{2-\alpha} + \frac{(1+|x|)^{1-\alpha}}{1-\alpha} \right).$$

4) $l(x) = (1-|x|)^+$; then, $h(x) = \frac{1}{2(e-1)} e^{|x|} 1_{]-1,1[}(x)$.

5) Let $h : \mathbb{R} \longrightarrow \mathbb{R}_+$ even and defined on $\mathbb{R}_+$ by:

$$h(x) = C_1 \exp\left(- \int_0^x k(z)dz \right) \quad (x \geq 0)$$

with $C_1 > 0$ and where $k : \mathbb{R}_+ \longrightarrow \mathbb{R}_+$ is continuous and satisfies

$$k(z) \underset{z \to +\infty}{=} C_2 z^\alpha \left(1 + \mathcal{O}\left(\frac{1}{z} \right) \right) \quad (C_2 > 0, \alpha \geq 0).$$

We prove that $a_h(y) = \frac{1}{h(y)} \int_y^{+\infty} zh(z)dz$ *is Lipschitz continuous.* Since h is even, it suffices to prove that: $\sup_{y \geq 0} |a_h'(y)| < \infty$. We have:

$$\begin{aligned}
a_h'(y) &= -y - \frac{h'(y)}{h^2(y)} \int_y^{+\infty} zh(z)dz \\
&= -y + y^2 k(y) \int_1^{+\infty} \frac{xh(xy)}{h(y)} dx \\
&\quad \text{(after the change of variable } z = xy\text{)} \\
&= -y + y^2 k(y) \int_1^{+\infty} x dx \exp\left(- \int_y^{xy} k(u)du \right) \\
&= -y + y^2 k(y) \int_1^{+\infty} x dx \exp\left(- \frac{C_2}{1+\alpha} y^{1+\alpha} (x^{\alpha+1} - 1) \left(1 + \mathcal{O}\left(\frac{1}{y} \right) \right) \right) \\
&= -y + \frac{y^2 k(y)}{\alpha+1} \int_0^{+\infty} dz (1+z)^{\frac{1-\alpha}{1+\alpha}} \exp\left(- \frac{C_2 z}{1+\alpha} y^{\alpha+1} \left(1 + \mathcal{O}\left(\frac{1}{y} \right) \right) \right) \qquad (6.56) \\
&\quad \text{(after the change of variable } x^{\alpha+1} - 1 = z\text{).}
\end{aligned}$$

Now, we have to estimate, as $\lambda \to +\infty$:

$$I_\lambda := \int_0^{+\infty} (1+z)^{\frac{1-\alpha}{1+\alpha}} \exp(-\lambda z) dz$$

with $\lambda = \frac{C_2}{1+\alpha} y^{\alpha+1} \left(1 + \mathcal{O}\left(\frac{1}{y}\right)\right)$. But,

$$\begin{aligned}
I_\lambda &= \frac{1}{\lambda} \int_0^{+\infty} \left(1 + \frac{v}{\lambda}\right)^{\frac{1-\alpha}{1+\alpha}} e^{-v} dv \quad \text{(after the change of variable } v = \lambda z\text{)} \\
&= \frac{1}{\lambda} \left(\int_0^{+\infty} e^{-v} dv \left(1 + \int_0^{v/\lambda} \frac{1-\alpha}{1+\alpha} (1+u)^{-\frac{2\alpha}{1+\alpha}} du \right) \right) && (6.57) \\
&= \frac{1}{\lambda} \left(1 + \frac{1-\alpha}{\lambda(1+\alpha)} \int_0^{+\infty} \left(1 + \frac{u}{\lambda}\right)^{-\frac{2\alpha}{1+\alpha}} e^{-u} du \right) \\
&\quad \text{(after integration by parts)} \\
&= \frac{1}{\lambda} \left(1 + \mathcal{O}\left(\frac{1}{\lambda}\right)\right). && (6.58)
\end{aligned}$$

Hence, from (6.58) and (6.56),

$$\begin{aligned}
a_h'(y) = -y + \frac{C_2 y^2}{1+\alpha} y^\alpha \left(1 + \mathcal{O}\left(\frac{1}{y}\right)\right) \\
\times \left(\frac{1+\alpha}{C_2 y^{\alpha+1} \left(1 + \mathcal{O}\left(\frac{1}{y}\right)\right)} \left(1 + \mathcal{O}\left(\frac{1+\alpha}{C_2 y^{\alpha+1} \left(1 + \mathcal{O}\left(\frac{1}{y}\right)\right)} \right)\right)\right)
\end{aligned}$$

i.e., since $\alpha \geq 0$:

$$a_h'(y) = -y + y\left(1 + \mathcal{O}\left(\frac{1}{y}\right)\right)\left(1 + \mathcal{O}\left(\frac{1}{y^{\alpha+1}}\right)\right) = y\mathcal{O}\left(\frac{1}{y}\right)$$

which prove that $\sup_{y \geq 0} |a'(y)| < \infty$.

6) Let h be a symmetric function which vanishes outside $]-1,1[$, and such that:

$$\forall x \in [0,1[, \quad h(x) = C \exp\left(- \int_0^x k(t) dt \right) \quad (C > 0),$$

where $k : [0,1[\longrightarrow \mathbb{R}_+$ is a continuous function.

We assume that there exist two functions ε and η:

$$\varepsilon, \eta : [0,1[\longrightarrow \mathbb{R}_+$$

such that:

i) for every $t \in [0,1[$, $k(t) = \varepsilon(t)\eta(t)$,

ii) ε is increasing and:

$$0 < \varliminf_{t \to 1^-} \eta(t) \leq \varlimsup_{t \to 1^-} \eta(t) < \infty. \qquad (6.59)$$

Then a_h is Lipschitz continuous (and (6.46) is fulfilled). Indeed, since h is decreasing on $[0,1]$, for every $y \in [0,1[$, $\int_y^1 zh(z)dz \le h(y)(1-y)$ and therefore $\lim_{y\to 1^-} a_h(y) = 0$. We must then prove that $a'_h(y)$ is bounded on $[0,1[$. We have, for $y \in [0,1[$:

$$a'_h(y) = -y - \frac{h'(y)}{h^2(y)} \int_y^1 zh(z)dz.$$

Hence:

$$\sup_{y\in[0,1[} |a'_h(y)| \le 1 + \sup_{y\in[0,1[} k(y) \int_y^1 \frac{h(z)}{h(y)} dz.$$

Thus, we have to prove that:

$$\overline{\lim_{y\to 1^-}}\, k(y) \int_y^1 \exp\left(-\int_y^z k(u)du\right) dz < \infty.$$

Let $0 < c < \underline{\lim}_{t\to 1^-} \eta(t)$. Then, for every y in a neighborhood of 1, since ε is increasing:

$$\begin{aligned} k(y) \int_y^1 \exp\left(-\int_y^z k(u)du\right) dz &\le k(y) \int_y^1 \exp\left(-c\varepsilon(y)(z-y)\right) dz \\ &= \frac{\eta(y)}{c} \int_0^{c\varepsilon(y)(1-y)} e^{-t} dt \end{aligned}$$

after the change of variable $t = c\varepsilon(y)(z-y)$. Therefore, from (6.59):

$$\overline{\lim_{y\to 1^-}}\, k(y) \int_y^1 \exp\left(-\int_y^z k(u)du\right) dz \le \frac{1}{c} \overline{\lim_{t\to 1^-}}\, \eta(t) < \infty.$$

7) Let $h : \mathbb{R} \to \mathbb{R}_+$ be a symmetric function, strictly positive and of $\mathscr{C}^1$ class on $]-1,1[$, satisfying

$$h(x) = 0 \quad \text{for } x \notin]-1,1[,$$

and

$$h'(y) \underset{y\to 1^-}{\sim} \frac{C}{(1-y)^{\alpha+1}}$$

for some $C > 0$ and $\alpha \in]0,1[$. Then:

$$h(y) \underset{y\to 1^-}{\sim} \frac{C}{\alpha(1-y)^{\alpha}}$$

and it is easy to show that $a_h(y) \xrightarrow[y\to 1^-]{} 0$ and $\lim_{y\to 1^-} a'_h(y) = -1 - \frac{\alpha}{1-\alpha} = -\frac{1}{1-\alpha}$. In particular, a_h is Lipschitz continuous.

8) Let $-\infty < \alpha < 0 < \beta < +\infty$ and h be a function satisfying (6.46), (6.47) and (6.48). We assume moreover that h is $\mathscr{C}^1$ on $[\alpha,\beta]$, $h(\alpha) = h(\beta) = 0$, $h'(\alpha) > 0$ and $h'(\beta) < 0$. Then, a_h is Lipschitz continuous.

6.6.3 Main Result

Here is now the main result of this Section 6.6.

Theorem 6.11 (Madan-Yor ([125], p. 523–525)). *Under the hypotheses defined in Subsection 6.6.1:*
1) The SDE:

$$Z_t = \int_0^t \sqrt{a_h}\left(\frac{Z_s}{\sqrt{s}}\right) dB_s \tag{6.60}$$

enjoys the uniqueness in law property.

2) Let $(M_t, t \geq 0)$ denote a solution in law of (6.60). Then:

$$\textit{for every } c > 0, \quad (M_{ct}, t \geq 0) \overset{(law)}{=} (\sqrt{c}M_t, t \geq 0).$$

3) $(M_t, t \geq 0)$ is a continuous martingale associated to $(\sqrt{t}X, t \geq 0)$.

Proof. 1) We start with the following lemma.

Lemma 6.8. *There exist a continuous process $(Y_t, -\infty < t < +\infty)$ and a Brownian motion $(\beta_t, -\infty < t < +\infty)$ indexed by $\mathbb{R}$, such that:*

1) *for every $s, t \in \mathbb{R}$, $s < t$:*

$$Y_t = Y_s + \int_s^t \sqrt{a_h(Y_u)}d\beta_u - \frac{1}{2}\int_s^t Y_u du, \tag{6.61}$$

2) *for every t real, Y_t admits h as a density.*

Proof. Classically, for fixed s, (6.61) admits a unique strong solution (see [150, p.390]). Let us denote by L the infinitesimal generator of $(Y_t, t \geq s)$ and by L^* its formal adjoint:

$$L^*f(y) = \frac{1}{2}(a_h f)''(y) + \frac{1}{2}(\bullet f)'(y).$$

Then, we have:

$$\begin{aligned} L^*h(y) &= \frac{1}{2}\left(\frac{1}{h}\left(\int_{\bullet}^{+\infty} zh(z)dz\right) h(\bullet)\right)''(y) + \frac{1}{2}(\bullet h)'(y) \\ &= -\frac{1}{2}(\bullet h)'(y) + \frac{1}{2}(\bullet h)'(y) = 0. \end{aligned}$$

Thus the probability $h(y)dy$ is stationary for the process $(Y_t, t \geq s)$ for every s and we can define this process with index $t \in \mathbb{R}$ (see K. Itô and M. Nisio [89] or K. Itô [88]). □

2) We now define the process $(M_t, t \geq 0)$ by:

$$\begin{cases} M_t = \sqrt{t} Y_{\log t} & \text{if } t > 0, \\ M_0 = 0. \end{cases} \tag{6.62}$$

Then, $(M_t, t \geq 0)$ is a continuous martingale satisfying:

$$M_t = \int_0^t \sqrt{a_h}\left(\frac{M_s}{\sqrt{s}}\right) dB_s \tag{6.63}$$

for some suitable Brownian motion $(B_s, s \geq 0)$.

We prove (6.63). *i*) First, for $0 < s < t$, Itô's formula yields:

$$\begin{aligned} M_t &= M_s + \int_s^t \frac{1}{2\sqrt{u}} Y_{\log u} du + \int_s^t \sqrt{u} d(Y_{\log u}) \\ &= M_s + \int_s^t \frac{M_u}{2u} du + \int_s^t \sqrt{u}\left(-\frac{1}{2} Y_{\log u} \frac{du}{u}\right) + \int_s^t \sqrt{u}\, d\left(\int_0^{\log u} \sqrt{a_h(Y_v)} d\beta_v\right) \\ &= M_s + \int_s^t \frac{M_u}{2u} du - \int_s^t \frac{M_u}{2u} du + \int_s^t \sqrt{u}\sqrt{a_h(Y_{\log u})} d_u \beta_{\log u} \\ &= M_s + \int_s^t \sqrt{a_h}\left(\frac{M_u}{\sqrt{u}}\right) dB_u \end{aligned}$$

where $(B_u, u \geq 0)$, defined by $dB_u = \sqrt{u} d_u \beta_{\log u}$, is a Brownian motion. Indeed, it is a local martingale such that $d\langle B \rangle_t = dt$ since:

$$\langle \int_0^\cdot \sqrt{u} d_u \beta_{\log u} \rangle_t = \int_0^t u\, d\log u = \int_0^t du = t.$$

We now prove that $M_t \xrightarrow[t \to 0]{} 0$ *a.s.* From the inequality $0 \leq a_h(x) \leq C(1 + |x|)$ (a_h being Lipschitz continuous), we have, for $0 < s < t$:

$$\begin{aligned} \mathbb{E}\left[\left(\int_s^t \sqrt{a_h}\left(\frac{M_u}{\sqrt{u}}\right) dB_u\right)^2\right] &= \int_s^t \mathbb{E}\left[a_h\left(\frac{M_u}{\sqrt{u}}\right)\right] du \\ &\leq C \int_s^t \mathbb{E}\left[1 + \frac{|M_u|}{\sqrt{u}}\right] du \\ &= C \int_s^t (1 + \mathbb{E}[|X|]) du \\ &= C(t - s)(1 + \mathbb{E}[|X|]) < \infty \end{aligned}$$

from (6.62), and since, for every $u \in \mathbb{R}$, Y_u is a r.v. with density h. Hence, for every $s > 0$, $(M_u, u \geq s)$ is a martingale.

Let $D(a,b;[s,t])$ denote the number of downcrossings of $(M_u, s \leq u \leq t)$ on the spatial interval $[a,b]$. From Dubins' inequality (see [150, Chapter II, p. 66]):

$$\mathbb{E}\left[D(a,b;[s,t])\right] \leq \frac{|b| + \sup\limits_{u\in[s,t]} \mathbb{E}[|M_u|]}{b-a} = \frac{|b| + \sqrt{t}\int_{\mathbb{R}} |x|h(x)dx}{b-a} \quad \text{(from (6.62))}.$$

Thus, letting $s \downarrow 0$:

$$\mathbb{E}[D(a,b;]0,t])] \leq \frac{|b| + \sqrt{t}\mathbb{E}[|X|]}{b-a}.$$

This proves that, for every $a < b$, the number of downcrossings of $(M_u, 0 < u \leq t)$ on the spatial interval $[a,b]$ is a.s. finite. Thus, $(M_u, u > 0)$ admits an a.s. limit as $u \to 0$:

$$M_u \xrightarrow[u\to 0]{} M_0 \quad \text{a.s.}$$

But, for every $\varepsilon > 0$:

$$\mathbb{P}(|M_u| > \varepsilon) = \mathbb{P}(|Y_u| > \frac{\varepsilon}{\sqrt{u}}) \leq \frac{\sqrt{u}}{\varepsilon}\int_{\mathbb{R}} |x|h(x)dx.$$

Thus, $M_u \xrightarrow[u\to 0]{} 0$ in probability; hence $M_0 = 0$ (see, for a similar approach, J.Y. Calais and M. Génin [40]). We have proven the existence of a continuous solution to the equation (6.60). The unicity in law of the solution of (6.60) comes from the following fact: if $(Z_t, t \geq 0)$ is a solution of (6.60), then, $(Y_t := \exp(-\frac{t}{2})Z_{\exp t}, t \geq 0)$ is a solution of (6.61).

3) *We now prove Point 2) of Theorem 6.11.* First, it is clear that, for every t, $M_t \overset{(law)}{=} \sqrt{t}M_1$. Indeed, since $M_t = \sqrt{t}Y_{\log t}$, we have $M_t \overset{(law)}{=} \sqrt{t}X \overset{(law)}{=} \sqrt{t}M_1$.

Now, we prove the scaling property of $(M_t, t \geq 0)$. For every $c > 0$, we have:

$$\begin{aligned}\sqrt{c}M_{\frac{t}{c}} &= \sqrt{c}\int_0^{t/c} \sqrt{a_h}\left(\frac{M_u}{\sqrt{u}}\right) dB_u \\ &= \int_0^t \sqrt{a_h}\left(\sqrt{c}\frac{M_{v/c}}{\sqrt{v}}\right) d_v\sqrt{c}B_{v/c}\end{aligned}$$

after the change of variable $u = \frac{v}{c}$. This proves the desired scaling property, since $(\sqrt{c}B_{v/c}, v \geq 0)$ is a Brownian motion. □

6.7 Peacocks and the Porous Media Equation

Long Exercise 6.6. This problem is taken from the contents of [19]. Let $m > 0$. We are interested in the non linear system (in the sense of McKean-Vlasov):

$$\begin{cases} Y_t = \int_0^t u^m(s, Y_s) dB_s \\ Y_t \text{ admits } u(t, \centerdot) \text{ as a density.} \end{cases} \tag{S}$$

The following notation will be useful:

$$\beta := \frac{1}{2m+2}, \quad a_m := \left(\frac{m^{1/2}}{((2m+1)(m+1))^{1/2} B\left(\frac{2m+1}{2m}, \frac{1}{2}\right)} \right)^{\frac{m}{m+1}} \tag{6.64}$$

where B is the first Euler function:

$$B(x,y) = \frac{\Gamma(x)\Gamma(y)}{\Gamma(x+y)}, \quad (x,y > 0).$$

We also define:

$$\gamma_m := a_m \sqrt{\frac{(2m+1)(m+1)}{m}}$$

$$\varphi(x) := a_m^{1/m} \left(\left(1 - \left(\frac{x}{\gamma_m} \right)^2 \right)^+ \right)^{\frac{1}{2m}}.$$

1) Prove that φ is a probability density, with support in $[-\gamma_m, \gamma_m]$.

2) Prove that, if the pair $\{(Y_t, t \geq 0), (u(t, \centerdot), t \geq 0)\}$ solves (S), then u satisfies:

$$\begin{cases} \frac{\partial u}{\partial t} = \frac{1}{2} \frac{\partial^2 (u^{2m+1})}{\partial x^2}(t,x), \\ u(0, \centerdot) = \delta_0. \end{cases} \tag{PM}$$

This equation is the porous media equation.

3) *i*) Prove that the unique solution of equation (PM) is given by:

$$u(t,x) = \frac{1}{t^\beta} \varphi\left(\frac{x}{t^\beta}\right). \tag{6.65}$$

(see G. I. Barenblatt [13] for the explicit expression of this solution, and P. Bénilan [21] for uniqueness).

ii) Let, for every $t > 0$, X_t denote a r.v. with density $u(t, \centerdot)$. Prove that $(t^\beta X_1, t \geq 0)$ is a peacock and that:

$$X_t \overset{(1.d)}{=} t^\beta X_1.$$

Our aim is now to find, via the SDE method, a martingale associated with this peacock. We consider the non homogeneous SDE:

$$M_t = \int_0^t \frac{1}{s^{m\beta}} \varphi^m \left(\frac{M_s}{s^\beta} \right) dB_s. \tag{6.66}$$

We wish to show that (6.66) admits a unique solution in law.

4) Consider, for $s \in \mathbb{R}$, the SDE:

$$Z_t = Z_s + \int_s^t \varphi^m(Z_u) dB_u - \beta \int_s^t Z_u du \quad (t \geq s). \tag{6.67}$$

i) Prove that, for every real s, this SDE admits a unique solution, which is strong. (*Hint*: consider $\psi(Z_t)$ where $\psi(x) := \int_0^x \frac{dy}{\varphi^m(y)}$ and reduce the discussion to that of Legendre processes (c.f. [150, Chapter VIII, p.357, Case 3]).)

ii) Prove that the probability φ is invariant for the process defined in (6.67). Deduce that there exist a process $(Z_t, -\infty < t < +\infty)$ indexed by $\mathbb{R}$ and a Brownian motion $(\gamma_u, -\infty < u < +\infty)$ such that, for $s < t$:

$$Z_t = Z_s + \int_s^t \varphi^m(Z_u) d\gamma_u - \beta \int_s^t Z_u du$$

and such that, for every t, the law of Z_t admits φ as its density.

5) We now define the process $(M_t, t \geq 0)$ via:

$$\begin{cases} M_t := t^\beta Z_{\log t} & \text{if } t > 0, \\ M_0 := 0. \end{cases} \tag{6.68}$$

i) Prove that $(M_t, t \geq 0)$ has continuous paths. (*Hint*: use Dubins' inequality)

ii) Prove that there exists a Brownian motion $(B_u, u \geq 0)$ such that:

$$M_t = \int_0^t \frac{1}{s^{m\beta}} \varphi^m \left(\frac{M_s}{s^\beta} \right) dB_s.$$

iii) Prove that (6.66) has a unique solution in law. (*Hint*: if $(V_t, t \geq 0)$ is a solution of (6.66), study the process $(U_t := e^{-t\beta} V_{e^t}, t \in \mathbb{R})$.)

iv) Prove that $(M_t, t \geq 0)$ defined by (6.68) is a martingale associated to the peacock $(t^\beta X_1, t \geq 0)$, i.e. satisfies $M_t \overset{(1.d)}{=} t^\beta X_1$, and that, for every $c > 0$:

$$(c^\beta M_t, t \geq 0) \overset{(law)}{=} (M_{tc}, t \geq 0).$$

v) Prove that:

$$\overline{\lim_{t \to +\infty}}\, Z_t = \gamma_m \text{ a.s.}, \qquad \underline{\lim}_{t \to +\infty}\, Z_t = -\gamma_m$$

$$\overline{\lim_{t \to +\infty}}\, M_t = +\infty \text{ a.s.}, \qquad \underline{\lim}_{t \to +\infty}\, M_t = -\infty.$$

Let us define:

$$\begin{aligned} T :&= \inf\{u \geq 0; Z_u = \gamma_m \text{ or } Z_t = -\gamma_m\} \\ &= +\infty \text{ if this set is empty.} \end{aligned}$$

Prove that $T = +\infty$ a.s. Deduce that $(M_t, t \geq 0)$ is recurrent.

vi) Let us define:

$$a(y) = \frac{1}{\varphi(y)} \int_y^{\gamma_m} z\varphi(z)dz \quad \text{if } y \in]-\gamma_m, \gamma_m[,$$
$$= 0 \quad \text{if } y \notin]-\gamma_m, \gamma_m[.$$

Prove that $a(y) = (m+1)\varphi^{2m}(y)$ and that a is Lipschitz continuous.

Remark 6.7. When $m \to 0$ (i.e. $\beta \to \frac{1}{2}$), the results of this problem appear as a particular case of Theorem 6.11. Observe that, for $m = 0$, the porous media equation (PM) reduces to the heat equation. Similar results to those presented in this problem can be obtained for $-1 < m < 0$. Equation (PM) is then called the rapid diffusion equation, and its fundamental solution (given by (6.65)) no longer has compact support, see [103, p. 177].

Notes and Comments

The uniqueness Theorem of Section 6.1 is due to M. Pierre. Theorem 6.11 is due to D. Madan and M. Yor. The problem of Section 6.7 (peacocks and porous media equation) is taken from S. Benachour and al. [19]. All other results of this chapter are new.

7

The Skorokhod Embedding (SE) Method

Abstract Several Skorokhod embeddings – which are presented in a table in the introduction of this chapter – allow to associate martingales to certain peacocks.

7.1 Introduction

7.1.1 Skorokhod Embeddings[1]

For ease of the reader, we recall briefly the following facts:

Consider X a real valued, integrable and centered random variable. Realizing a *Skorokhod embedding* (SE) of X into Brownian motion B, consists in constructing a stopping time τ with respect to some filtration $(\mathscr{F}_t)$, for which B is a Brownian motion, such that:

(Sk1) $B_\tau \overset{(law)}{=} X$.

(Sk2) $(B_{u\wedge\tau}, u \geq 0)$ is a uniformly integrable $(\mathscr{F}_t)$ martingale.

There are many ways to realize such a Skorokhod embedding. J. Obłój [136] numbered twenty one methods scattered in the literature. These methods separate (at least) in two kinds:

- the time τ is a stopping time relative to the natural filtration $(\mathscr{B}_t)$ of the Brownian motion B,
- the time τ is a stopping time relative to an enlargement $(\mathscr{F}_t)$ of the natural filtration of the Brownian motion B, such that B is a $(\mathscr{F}_t)$ martingale.

In the second case, the stopping time τ is called a *randomized* stopping time. We call the corresponding embedding a *randomized Skorokhod embedding*.

[1] This problem was first addressed by Skorokhod, in a slightly different manner, in [161].

Hirsch F., Profeta C., Roynette B., Yor M.:
Peacocks and Associated Martingales, with Explicit Constructions

7.1.2 Skorokhod Embedding and 1-Martingales

Consider $(X_t, t \geq 0)$ a peacock such that, for every $t \geq 0$, $\mathbb{E}[X_t] = 0$. We denote by ν_t the law of X_t. Assume that, for each t, we know how to realize a Skorokhod embedding of ν_t (randomized or not) with the help of a stopping time τ_t, such that:

- $\nu_t \sim B_{\tau_t}$, i.e. B_{τ_t} has law ν_t,
- for every $t \geq 0$, $(B_{u \wedge \tau_t}, u \geq 0)$ is a uniformly integrable martingale,
- the family $(\tau_t, t \geq 0)$ is a.s. increasing, i.e. for every $s \leq t$, $\tau_s \leq \tau_t$ a.s.

Then, $(M_t := B_{\tau_t}, t \geq 0)$ is a martingale such that:

$$M_t \overset{(1.d)}{=} X_t.$$

In other terms, $(M_t, t \geq 0)$ is a martingale associated to the peacock $(X_t, t \geq 0)$. We shall now illustrate this ideal program for which the main difficulty is to find an *increasing* family of stopping times $(\tau_t, t \geq 0)$. Note that this kind of method has also been used by D. Hobson in a more financial setup [85].

7.1.3 Organization of this Chapter

• In Section 7.2, we present a Skorokhod embedding which seems to be new. This randomized SE makes it possible to associate a martingale to any peacock of the form $(\sqrt{t}X, t \geq 0)$ where X is an integrable centered r.v. Moreover, this martingale $(M_t, t \geq 0)$ enjoys the Brownian scaling property:

$$\forall c > 0, \quad (M_{c^2 t}, t \geq 0) \overset{(law)}{=} (cM_t, t \geq 0). \tag{7.1}$$

• In Section 7.3, we present a Hall-Breiman method which also allows to associate to any peacock of the form $(\sqrt{t}X, t \geq 0)$, with X an integrable centered r.v, a martingale $(M_t, t \geq 0)$ which enjoys the Brownian scaling property (7.1).

• In Section 7.4, we use the Azéma-Yor method for SE. It is a non-randomized SE, which under some extra-hypotheses on the law of X, makes it possible to associate to the peacock $(\sqrt{t}X, t \geq 0)$, with X an integrable centered r.v, a martingale $(M_t, t \geq 0)$ which enjoys the Brownian scaling property (7.1).

• To begin Section 7.5, we use Bass' embedding method in order to associate a martingale to the peacock $(\varphi(B_t), t \geq 0)$, where φ is an odd increasing function such that, for every $t \geq 0$, $\mathbb{E}[|\varphi(B_t)|] < \infty$.

This chapter ends with three more parts written in a different spirit:

• At the end of Section 7.5, a relationship between Bougerol's identity and Bass' embedding method is established and studied.

• Section 7.6 uses the notion of Ocone martingales to associate a martingale to certain peacocks.

Table 7.1. Some SE's which allow construction of associated martingales

Peacock	*Hypotheses*	*SE*	*References*
$(tX, t \geq 0)$ with X integrable and centered	No further hypothesis	Our new SE	Theorem 7.6
	No further hypothesis	Hall-Breiman SE	Theorem 7.8
	$(M{\cdot}Y)$	Azéma-Yor SE	Theorem 7.10 Lemma 7.3 Theorem 7.13 Theorem 7.14
	X with density h such that $x \mapsto \log(\|x\|h(x))$ is concave on $]-\infty, 0[$ and on $]0, +\infty[$	Bertoin-Le Jan SE	Exercise 7.14
	$X = \varphi(G)$ with G a reduced and centered gaussian r.v. and φ odd, increasing, and log-concave on $]0, +\infty[$	Bass SE	Theorem 7.19
$(\varphi(B_t), t \geq 0)$ with φ odd, increasing and $\mathbb{E}[\|\varphi(B_t)\|] < \infty$	φ of $\mathscr{C}^2$ class and $\lim\limits_{x \to +\infty} \varphi(x) = +\infty$	Hall-Breiman SE	Exercise 7.5
	No further hypothesis	Bass SE	Theorem 7.17
$\left(\dfrac{e^{tX}}{\mathbb{E}[e^{tX}]}, t \geq 0\right)$ with $\forall t \geq 0, \mathbb{E}[\|X\|e^{tX}] < \infty$	$x \mapsto \mathbb{P}(X \geq x)$ log-concave	Azéma-Yor SE	Exercise 7.12
$(\varphi(tX), t \geq 0)$ with $\sup\limits_{0 \leq s \leq t} \mathbb{E}[\|X\|\varphi'(sX)] < \infty$ and $\mathbb{E}[X\varphi'(tX)] = 0$	$x \longmapsto \mathbb{P}(X \geq x)$ log-concave	Azéma-Yor SE	Exercise 7.13

• Finally, "Student's peacocks" are studied, in the framework of a problem, in Section 7.7.

We sum-up the main Skorokhod embeddings in Table 7.1.

7.2 Construction of Randomized SE's

Let X be an integrable and centered r.v. We recall that, from Proposition 1.4, $(\sqrt{t}X, t \geq 0)$ is a peacock. Our aim is now to construct a martingale $(M_t, t \geq 0)$ which satisfies:

i) $(M_t, t \geq 0)$ is associated to $(X_t, t \geq 0)$, i.e:

$$M_t \overset{(1.d)}{=} \sqrt{t}X, \tag{7.2}$$

ii) $(M_t, t \geq 0)$ enjoys the scaling Brownian property:

$$\forall c > 0, \quad (M_{c^2t}, t \geq 0) \overset{(law)}{=} (cM_t, t \geq 0). \tag{7.3}$$

We now present a (seemingly) new SE, in the spirit of the original Skorokhod construction and of the Hall (or Hall-Breiman) method (see Section 7.3 below). Our method of randomized SE will ensure directly that the family of stopping times $(\tau_t, t \geq 0)$ (where τ_t is associated to μ_t, the law of $\sqrt{t}X$) is increasing.

Here is the content of this Section 7.2

- In Subsection 7.2.1, we consider a real valued, integrable and centered random variable X . We prove that there exist an $\mathbb{R}_+$-valued random variable V and an $\mathbb{R}^*_-$-valued random variable W, with V and W independent and independent of $(B_u, u \geq 0)$, such that, denoting:

$$\tau = \inf\{u \geq 0 \,;\, B_u = V \text{ or } B_u = W\},$$

property (Sk1) is satisfied by this randomized stopping time τ, i.e: $B_\tau \overset{(law)}{=} X$. To prove this result we use, as an essential tool, the Schauder-Tychonoff fixed point theorem (see [59]). A remarkable feature of our SE is that, despite the use of Schauder-Tychonoff fixed point theorem, the laws of V and W are *uniquely determined* by the law of X.
- In Subsection 7.2.2, we prove that the stopping time τ defined in Subsection 7.2.1 satisfies (Sk2), i.e: the martingale $B^\tau := (B_{u\wedge\tau}, u \geq 0)$ is uniformly integrable. Moreover, for every $p \geq 1$, we state conditions ensuring that B^τ is a martingale belonging to the space H^p which consists of all martingales $(M_t, t \geq 0)$ such that $\sup_{t\geq 0}|M_t| \in L^p$. Recall that, for $p > 1$, H^p consists simply of uniformly integrable martingales $(M_t, t \geq 0)$ such that $M_\infty \in L^p$, since then, from Doob's L^p inequality:

$$\mathbb{E}\left[\left(\sup_{t\geq 0}|M_t|\right)^p\right] \leq \left(\frac{p}{p-1}\right)^p \mathbb{E}[|M_\infty|^p]. \tag{7.4}$$

- In Subsection 7.2.3, we follow the method presented in Subsection 7.1.2, and construct an increasing family of randomized stopping times $(\tau_t, t \geq 0)$, such that $(B_{\tau_t}, t \geq 0)$ is a martingale satisfying properties (7.2) and (7.3).

7.2.1 Randomized SE's

7.2.1.1 Notation

We denote by $\mathbb{R}_+$ (resp. $\mathbb{R}^*_-$) the interval $[0, +\infty[$ (resp. $]-\infty, 0[$), and by $\mathscr{M}_+$ (resp. $\mathscr{M}_-$) the set of positive finite measures on $\mathbb{R}_+$ (resp. $\mathbb{R}^*_-$), equipped with the weak topology:

$$\sigma(\mathscr{M}_+\,,\,\mathscr{C}^0(\mathbb{R}_+)) \quad (\text{resp. } \sigma(\mathscr{M}_-\,,\,\mathscr{C}^0(\mathbb{R}^*_-))\,)$$

where $\mathscr{C}^0(\mathbb{R}_+)$ (resp. $\mathscr{C}^0(\mathbb{R}^*_-)$) denotes the space of continuous functions on $\mathbb{R}_+$ (resp. $\mathbb{R}^*_-$) tending to 0 at $+\infty$ (resp. at 0 and at $-\infty$).

$B = (B_u, u \geq 0)$ denotes a standard Brownian motion started from 0.

In the sequel we consider a real valued, integrable, centered random variable X, the law of which we denote by μ . The restrictions of μ to $\mathbb{R}_+$ and $\mathbb{R}^*_-$ are denoted respectively by μ_+ and μ_-.

7.2.1.2 Existence of a Randomized Stopping Time

This subsection is devoted to the proof of the following SE.

Theorem 7.1. *i) There exist an $\mathbb{R}_+$-valued random variable V and an $\mathbb{R}^*_-$-valued random variable W, V and W being independent and independent of $(B_u, u \geq 0)$, such that, setting*

$$\tau = \inf\{u \geq 0\,;\, B_u = V \text{ or } B_u = W\}\,,$$

one has: $B_\tau \overset{(law)}{=} X$.

ii) Denoting by γ_+ (resp. γ_-) the law of V (resp. W), then:

$$\mu_+ \leq \gamma_+ \ll \mu_+ \qquad \text{and} \qquad \mu_- \leq \gamma_- \ll \mu_-\,.$$

Moreover,

$$\mathbb{E}[V \wedge (-W)] \leq \mathbb{E}[|X|] \leq 2\,\mathbb{E}[V \wedge (-W)] \tag{7.5}$$

and, for every $p > 1$,

$$\frac{1}{2}\mathbb{E}\left[(V \wedge (-W))\left(V^{p-1} + (-W)^{p-1}\right)\right] \\ \leq \mathbb{E}[|X|^p] \leq \mathbb{E}\left[(V \wedge (-W))\left(V^{p-1} + (-W)^{p-1}\right)\right]. \tag{7.6}$$

Proof. In the following, we exclude the case $\mu = \delta_0$, the Dirac measure at 0. Otherwise, it suffices to set: $V = 0$. Then, *i)* is satisfied since $\tau = 0$, and *ii)* is also satisfied except the property $\gamma_- \ll \mu_-$ (since $\mu_- = 0$).

1) We first recall the following classical result: Let $b < 0 \leq a$ and

$$T_{b,a} = \inf\{u \geq 0\,;\, B_u = a \text{ or } B_u = b\}. \tag{7.7}$$

Then,

$$\mathbb{P}(B_{T_{b,a}} = a) = \frac{-b}{a-b} \qquad \text{and} \qquad \mathbb{P}(B_{T_{b,a}} = b) = \frac{a}{a-b}\,.$$

2) Let V and W be respectively an $\mathbb{R}_+$-valued random variable and an $\mathbb{R}^*_-$-valued random variable, V and W being independent and independent of B, and let τ, γ_+, γ_- be defined as in the statement of the theorem. As a direct consequence of Point 1, we obtain that $B_\tau \overset{(law)}{=} X$ if and only if:

$$\mu_+(dv) = \left(\int_{\mathbb{R}^*_-} \frac{-w}{v-w}\,\gamma_-(dw)\right)\gamma_+(dv) \quad \text{on } \mathbb{R}_+ \tag{7.8}$$

$$\mu_-(dw) = \left(\int_{\mathbb{R}_+} \frac{v}{v-w}\,\gamma_+(dv)\right)\gamma_-(dw) \quad \text{on } \mathbb{R}^*_-. \tag{7.9}$$

As γ_+ and γ_- are probabilities, the above equations entail:

$$\gamma_+(\mathrm{d}v) = \mu_+(\mathrm{d}v) + \left(\int_{\mathbb{R}_-^*} \frac{v}{v-w}\, \gamma_-(\mathrm{d}w)\right) \gamma_+(\mathrm{d}v) \quad \text{on } \mathbb{R}_+ \tag{7.10}$$

$$\gamma_-(\mathrm{d}w) = \mu_-(\mathrm{d}w) + \left(\int_{\mathbb{R}_+} \frac{-w}{v-w}\, \gamma_+(\mathrm{d}v)\right) \gamma_-(\mathrm{d}w) \quad \text{on } \mathbb{R}_-^*. \tag{7.11}$$

To prove Point *i*) of the theorem, we shall now solve this system of equations (7.10) and (7.11) using a fixed point method, and then we shall verify that the solution thus obtained is a pair of probabilities, which will entail (7.8) and (7.9).

3) We now introduce some further notation. If $(a,b) \in \mathscr{M}_+ \times \mathscr{M}_-$ and $\varepsilon > 0$, we set

$$a(\varepsilon) = \int 1_{]0,\varepsilon[}(v)\, a(\mathrm{d}v) \quad \text{and} \quad b(\varepsilon) = \int 1_{]-\varepsilon,0[}(w)\, b(\mathrm{d}w)\,.$$

We also set: $m_+ = \int \mu_+(\mathrm{d}v)$, $m_- = \int \mu_-(\mathrm{d}w)$. We note that, since μ is centered and is not the Dirac measure at 0, then $m_+ > 0$ and $m_- > 0$. We then define:

$$\rho(\varepsilon) := 4 \sup\left(\mu_+(\varepsilon)\, m_+^{-1}, \mu_-(\varepsilon)\, m_-^{-1}\right)$$

and

$$\begin{aligned}\Theta \;:=\; &\Big\{(a,b) \in \mathscr{M}_+ \times \mathscr{M}_- \,;\, a \geq \mu_+, b \geq \mu_-, \int a(\mathrm{d}v) + \int b(\mathrm{d}w) \leq 2 \\ &\text{and for every } \varepsilon \leq \varepsilon_0,\;\; a(\varepsilon) \leq \rho(\varepsilon) \;\text{ and }\; b(\varepsilon) \leq \rho(\varepsilon)\Big\}\end{aligned}$$

where ε_0 will be defined in the sequel.

Finally, we define $\Gamma = (\Gamma_+, \Gamma_-) : \mathscr{M}_+ \times \mathscr{M}_- \longrightarrow \mathscr{M}_+ \times \mathscr{M}_-$ by:

$$\begin{aligned}\Gamma_+(a,b)(\mathrm{d}v) &= \mu_+(\mathrm{d}v) + \left(\int_{\mathbb{R}_-^*} \frac{v}{v-w}\, b(\mathrm{d}w)\right) a(\mathrm{d}v), \\ \Gamma_-(a,b)(\mathrm{d}w) &= \mu_-(\mathrm{d}w) + \left(\int_{\mathbb{R}_+} \frac{-w}{v-w}\, a(\mathrm{d}v)\right) b(\mathrm{d}w).\end{aligned}$$

□

Lemma 7.1. *Θ is a convex compact subset of $\mathscr{M}_+ \times \mathscr{M}_-$ (equipped with the product of the weak topologies), and $\Gamma(\Theta) \subset \Theta$.*

Proof. The first part is clear. Suppose that $(a,b) \in \Theta$. By definition of Γ, we have:

$$\Gamma_+(a,b) \geq \mu_+ \;,\; \Gamma_-(a,b) \geq \mu_-$$

and

$$\int \Gamma_+(a,b)(\mathrm{d}v) + \int \Gamma_-(a,b)(\mathrm{d}w) = 1 + \left(\int a(\mathrm{d}v)\right)\left(\int b(\mathrm{d}w)\right). \tag{7.12}$$

Consequently,

$$\int \Gamma_+(a,b)(\mathrm{d}v) + \int \Gamma_-(a,b)(\mathrm{d}w) \leq 2.$$

Besides, since $(a,b) \in \Theta$,

$$\int a(\mathrm{d}v) \le 2 - m_- \,, \int b(\mathrm{d}w) \le 2 - m_+ . \tag{7.13}$$

On the other hand,

$$\begin{aligned}\Gamma_+(a,b)(\varepsilon) &= \mu_+(\varepsilon) + \int 1_{]0,\varepsilon[}(v)\,a(\mathrm{d}v) \int 1_{]-v,0]}(w)\,\frac{v}{v-w}\,b(\mathrm{d}w) \\ &\quad + \int 1_{]0,\varepsilon[}(v)\,a(\mathrm{d}v) \int 1_{]-\infty,-v]}(w)\,\frac{v}{v-w}\,b(\mathrm{d}w)\,.\end{aligned}$$

Since $\dfrac{v}{v-w} \le 1$, and $\dfrac{v}{v-w} \le 1/2$ if $w \le -v$, taking into account (7.13) we obtain:

$$\Gamma_+(a,b)(\varepsilon) \le \mu_+(\varepsilon) + a(\varepsilon)b(\varepsilon) + a(\varepsilon)\left(1 - \frac{m_+}{2}\right).$$

Hence,

$$\Gamma_+(a,b)(\varepsilon) \le \rho^2(\varepsilon) + \rho(\varepsilon)\left(1 - \frac{m_+}{2}\right) + \mu_+(\varepsilon)\,.$$

In order to deduce from the preceding that: $\Gamma_+(a,b)(\varepsilon) \le \rho(\varepsilon)$, it suffices to prove:

$$\rho^2(\varepsilon) - \frac{m_+}{2}\rho(\varepsilon) + \mu_+(\varepsilon) \le 0$$

or

$$\rho(\varepsilon) \in \left[\frac{1}{4}(m_+ - \sqrt{m_+^2 - 16\,\mu_+(\varepsilon)}), \frac{1}{4}(m_+ + \sqrt{m_+^2 - 16\,\mu_+(\varepsilon)})\right],$$

which is satisfied for $\varepsilon \le \varepsilon_0$ for some choice of ε_0, by definition of ρ. The proof that $\Gamma_-(a,b)(\varepsilon) \le \rho(\varepsilon)$ is similar. □

Lemma 7.2. *The restriction of the map Γ to Θ is continuous.*

Proof. We first prove the continuity of Γ_+. For $\varepsilon > 0$, we denote by h_ε a continuous function on $\mathbb{R}_-^*$ satisfying:

$$h_\varepsilon(w) = 0 \;\text{ for }\; -\varepsilon < w < 0 \;,\; h_\varepsilon(w) = 1 \text{ for } w < -2\,\varepsilon$$

and, for every $w < 0$, $0 \le h_\varepsilon(w) \le 1$. We set: $\Gamma_+^\varepsilon(a,b) = \Gamma_+(a, h_\varepsilon\, b)$. Then, $\Gamma_+^\varepsilon(a,b) \le \Gamma_+(a,b)$ and

$$0 \le \int \Gamma_+(a,b)(\mathrm{d}v) - \int \Gamma_+^\varepsilon(a,b)(\mathrm{d}v) \le 2\rho(2\,\varepsilon)\,,$$

which tends to 0 as ε tends to 0. Therefore, by uniform approximation, it suffices to prove the continuity of the map Γ_+^ε.

Let (a_n, b_n) be a sequence in Θ, weakly converging to (a,b), and let $\varphi \in \mathscr{C}^0(\mathbb{R}_+)$. It is easy to see that the set:

$$\left\{\frac{v\,\varphi(v)}{v - \centerdot}\,h_\varepsilon(\centerdot)\;;\; v \ge 0\right\}$$

is relatively compact in the Banach space $\mathscr{C}^0(\mathbb{R}_-^*)$. Consequently,

$$\lim_{n\to\infty}\int \frac{v\,\varphi(v)}{v-w}\,h_\varepsilon(w)\,b_n(\mathrm{d}w)=\int \frac{v\,\varphi(v)}{v-w}\,h_\varepsilon(w)\,b(\mathrm{d}w) \tag{7.14}$$

uniformly with respect to v. Since

$$\left|\int \frac{v\,\varphi(v)}{v-w}\,h_\varepsilon(w)\,b_n(\mathrm{d}w)\right| \le 2\,|\varphi(v)|\,, \tag{7.15}$$

then

$$\left\{\int \frac{v\,\varphi(v)}{v-w}\,h_\varepsilon(w)\,b_n(\mathrm{d}w)\,;\,n\ge 0\right\}$$

is relatively compact in the Banach space $\mathscr{C}^0(\mathbb{R}_+)$. Therefore,

$$\lim_{n\to\infty}\int \varphi(v)\,\Gamma_+^\varepsilon(a_n,b_p)(\mathrm{d}v)=\int \varphi(v)\,\Gamma_+^\varepsilon(a,b_p)(\mathrm{d}v)$$

uniformly with respect to p, and, by (7.14) and (7.15):

$$\lim_{n\to\infty}\int \varphi(v)\,\Gamma_+^\varepsilon(a,b_n)(\mathrm{d}v)=\int \varphi(v)\,\Gamma_+^\varepsilon(a,b)(\mathrm{d}v)\,.$$

Finally,

$$\lim_{n\to\infty}\int \varphi(v)\,\Gamma_+^\varepsilon(a_n,b_n)(\mathrm{d}v)=\int \varphi(v)\,\Gamma_+^\varepsilon(a,b)(\mathrm{d}v)\,,$$

which proves the desired result.

The proof of the continuity of Γ_- is similar, but simpler since no approximation procedure is needed. □

As a consequence of Lemma 7.1 and Lemma 7.2, we may apply the Schauder-Tychonoff fixed point theorem (see, for instance, [59, Theorem V.10.5]), which yields the existence of a pair $(\gamma_+,\gamma_-)\in\Theta$ satisfying (7.10) and (7.11). We set

$$\alpha_+=\int \gamma_+(\mathrm{d}v)\,,\ \alpha_-=\int \gamma_-(\mathrm{d}w)$$

and we shall now prove that $\alpha_+=\alpha_-=1$.

4) By (7.12) applied to $(a,b)=(\gamma_+,\gamma_-)$, we obtain:

$$\alpha_++\alpha_-=1+\alpha_+\,\alpha_-$$

and therefore, $\alpha_+=1$ or $\alpha_-=1$. Suppose, for instance, $\alpha_+=1$. Since $\alpha_++\alpha_-\le 2$, then $\alpha_-\le 1$. We now suppose $\alpha_-<1$. By (7.10), $\gamma_+\le\mu_++\alpha_-\,\gamma_+$, and hence, $\gamma_+\le(1-\alpha_-)^{-1}\,\mu_+$. Consequently,

$$\int v\,\gamma_+(\mathrm{d}v)\le(1-\alpha_-)^{-1}\int v\,\mu_+(\mathrm{d}v)<\infty\,.$$

We deduce from (7.10) and (7.11) that, for every $r > 0$,

$$\int_0^\infty v\,\gamma_+(\mathrm{d}v) + \int_{-r}^0 w\,\gamma_-(\mathrm{d}w)$$
$$= \varepsilon_1(r) + \varepsilon_2(r) + \int_0^\infty \gamma_+(\mathrm{d}v) \int_{-r}^0 \gamma_-(\mathrm{d}w)(v+w) \quad (7.16)$$

with

$$\varepsilon_1(r) = \int_{-r}^{+\infty} x\,\mu(\mathrm{d}x) \qquad \text{and}$$
$$\varepsilon_2(r) = \int_0^\infty \gamma_+(\mathrm{d}v) \int_{-\infty}^{-r} \gamma_-(\mathrm{d}w) \frac{v^2}{v-w}\,.$$

Since X is centered, $\lim_{r\to+\infty} \varepsilon_1(r) = 0$. On the other hand,

$$\varepsilon_2(r) \le \left(\int v\,\gamma_+(\mathrm{d}v)\right)\left(\int_{-\infty}^{-r} \gamma_-(\mathrm{d}w)\right)$$

and therefore, $\lim_{r\to\infty} \varepsilon_2(r) = 0$. Since $\alpha_+ = 1$, we deduce from (7.16):

$$\left(\int v\,\gamma_+(\mathrm{d}v)\right)\left(1 - \int_{-r}^0 \gamma_-(\mathrm{d}w)\right) = \varepsilon_1(r) + \varepsilon_2(r)\,.$$

Since μ is not the Dirac measure at 0, then $\gamma_+(]0,+\infty[) > 0$. Therefore, letting r tend to ∞, we obtain $\alpha_- = 1$, which contradicts the assumption $\alpha_- < 1$. Thus, $\alpha_- = 1$ and $\alpha_+ = 1$.

5) We now prove point *ii*). We have already seen: $\gamma_+ \ge \mu_+$ and $\gamma_- \ge \mu_-$. The property: $\gamma_+ \ll \mu_+$ follows directly from (7.8). More precisely, the Radon-Nikodym density of γ_+ with respect to μ_+ is given by:

$$\left(\int_{\mathbb{R}_-^*} \frac{-w}{v-w}\,\gamma_-(\mathrm{d}w)\right)^{-1},$$

which is well defined since γ_- is a probability and $\dfrac{-w}{v-w}$ is > 0 for $w < 0$ and $v \ge 0$. Now, since μ is not the Dirac measure at 0, $\gamma_+(]0,+\infty[) > 0$. By (7.9), this easily entails the property: $\gamma_- \ll \mu_-$, the Radon-Nikodym density of γ_- with respect to μ_- being given by:

$$\left(\int_{\mathbb{R}_+} \frac{v}{v-w}\,\gamma_+(\mathrm{d}v)\right)^{-1}.$$

On the other hand, we have for $v \ge 0$ and $w < 0$,

$$\frac{1}{2}(v \wedge (-w)) \le \frac{-vw}{v-w} \le v \wedge (-w). \quad (7.17)$$

Moreover, we deduce from (7.8) and (7.9)

$$\mathbb{E}\left[|X|^p\right] = \iint \frac{-vw}{v-w}\left(v^{p-1} + (-w)^{p-1}\right)\gamma_+(dv)\gamma_-(dw) \tag{7.18}$$

for every $p \geq 1$. Then, (7.5) and (7.6) in Theorem 7.1 follow directly from (7.17) and (7.18). □

We have obtained an existence theorem, thanks to the application of the Schauder-Tychonoff fixed point theorem, which, of course, says nothing about the uniqueness of the pair (γ_+, γ_-) of probabilities satisfying the conditions (7.8) and (7.9). However, the following theorem states that this uniqueness holds.

Theorem 7.2. *Assume $\mu \neq \delta_0$. Then the laws of the r.v.'s V and W satisfying Point i) in Theorem 7.1 are uniquely determined by μ.*

Proof. Consider $\left(\gamma_+^{(j)}, \gamma_-^{(j)}\right)$, $j = 1, 2$, two pairs of probabilities in $\mathscr{M}_+ \times \mathscr{M}_-$ satisfying (7.8) and (7.9). We set, for $j = 1, 2$, $v \geq 0$ and $w < 0$,

$$a^{(j)}(v) = \int_{\mathbb{R}_-^*} \frac{-w}{v-w}\gamma_-^{(j)}(dw), \tag{7.19}$$

$$b^{(j)}(w) = \int_{\mathbb{R}_+} \frac{v}{v-w}\gamma_+^{(j)}(dv). \tag{7.20}$$

By (7.8) and (7.9), we have:

$$\gamma_+^{(j)} = \frac{1}{a^{(j)}}\mu_+ \quad \text{and} \quad \gamma_-^{(j)} = \frac{1}{b^{(j)}}\mu_- . \tag{7.21}$$

On the other hand, the following obvious equality holds:

$$\iint_{\mathbb{R}_+\times\mathbb{R}_-^*} \frac{v-w}{v-w}\left(\gamma_+^{(1)}(dv) + \gamma_+^{(2)}(dv)\right)\left(\gamma_-^{(1)}(dw) + \gamma_-^{(2)}(dw)\right) = 4. \tag{7.22}$$

Therefore, developing (7.22) and using (7.19), (7.20) and (7.21), we obtain:

$$\begin{aligned}\int_{\mathbb{R}+} &\left(a^{(1)}(v) + a^{(2)}(v)\right)\left(\frac{1}{a^{(1)}(v)} + \frac{1}{a^{(2)}(v)}\right)\mu_+(dv)\\ &+ \int_{\mathbb{R}_-^*}\left(b^{(1)}(w) + b^{(2)}(w)\right)\left(\frac{1}{b^{(1)}(w)} + \frac{1}{b^{(2)}(w)}\right)\mu_-(dw) = 4.\end{aligned} \tag{7.23}$$

Now, for $x > 0$, $x + \frac{1}{x} \geq 2$, and $x + \frac{1}{x} = 2$ if and only if $x = 1$. Therefore,

$$\left(a^{(1)}(v) + a^{(2)}(v)\right)\left(\frac{1}{a^{(1)}(v)} + \frac{1}{a^{(2)}(v)}\right) \geq 4$$

and $\left(a^{(1)}(v)+a^{(2)}(v)\right)\left(\frac{1}{a^{(1)}(v)}+\frac{1}{a^{(2)}(v)}\right)=4$ if and only if $a^{(1)}(v)=a^{(2)}(v)$, and similarly with $b^{(1)}(w)$ and $b^{(2)}(w)$. Since μ is a probability, we deduce from (7.23) and the preceding that:

$$a^{(1)}(v)=a^{(2)}(v)\ \ \mu_+\text{-a.s.} \quad \text{and} \quad b^{(1)}(w)=b^{(2)}(w)\ \ \mu_-\text{-a.s.}$$

We then deduce from (7.21):

$$\gamma_+^{(1)}=\gamma_+^{(2)} \quad \text{and} \quad \gamma_-^{(1)}=\gamma_-^{(2)},$$

which is the desired result. □

Remark 7.1. We have:

$$\forall v\geq 0,\ \forall w<0, \qquad \frac{-w}{v-w}\geq \frac{1}{(v\vee 1)}\frac{-w}{1-w}\,.$$

Therefore, by (7.8), for $p>1$:

$$\mathbb{E}[V^{p-1}]\leq \left(\int \frac{-w}{1-w}\,\gamma_-(dw)\right)^{-1}\int (v\vee 1)\,v^{p-1}\,\mu_+(dv)\,,$$

and similarly for $\mathbb{E}[(-W)^{p-1}]$. Consequently,

$$\mathbb{E}[|X|^p]<\infty \Longrightarrow \mathbb{E}[V^{p-1}]<\infty \quad \text{and} \quad \mathbb{E}[(-W)^{p-1}]<\infty\,.$$

However, in general the converse does not hold (see Example 7.3 below), but it holds if $p\geq 2$ (see Remark 7.3).

7.2.1.3 Some Examples

In this subsection, we develop some explicit examples. We keep the previous notation. For $x\in\mathbb{R}$, δ_x denotes the Dirac measure at x.

Example 7.1. Let $0<\alpha<1$ and $x>0$. We define $\mu_+=\alpha\,\delta_x$ and we take for μ_- any measure in $\mathscr{M}_-$ such that:

$$\int \mu_-(dw)=1-\alpha \qquad \text{and} \qquad \int w\,\mu_-(dw)=-\alpha\,x\,.$$

Then, the unique pair of probabilities (γ_+,γ_-) satisfying (7.8) and (7.9) is given by:

$$\gamma_+=\delta_x \qquad \text{and} \qquad \gamma_-(dw)=\left(1-\frac{w}{x}\right)\mu_-(dw)\,.$$

Example 7.2. Let $0<\alpha<1$ and $0<x<y$. We consider a symmetric measure μ such that:

$$\mu_+=\frac{1}{2}(\alpha\,\delta_x+(1-\alpha)\,\delta_y)\,.$$

By an easy computation, the unique pair of probabilities (γ_+, γ_-) satisfying (7.8) and (7.9) is shown to be:

$$\gamma_+ = \frac{y - \sqrt{(1-\alpha)y^2 + \alpha x^2}}{y-x}\,\delta_x + \frac{-x + \sqrt{(1-\alpha)y^2 + \alpha x^2}}{y-x}\,\delta_y$$

and $\gamma_-(\mathrm{d}w) = \gamma_+(-\mathrm{d}w)$.

Example 7.3. Let $0 < \alpha < 1$ and $0 < \beta < 1$ such that $\alpha + \beta > 1$. We define μ by:

$$\mu_+(\mathrm{d}v) = \frac{\sin\alpha\pi}{\pi}\,\frac{v^{\alpha-1}}{(1+v^\beta)(1+2v^\alpha\cos\alpha\pi + v^{2\alpha})}\,\mathrm{d}v$$

and

$$\mu_-(\mathrm{d}w) = \frac{\sin\beta\pi}{\pi}\,\frac{(-w)^{\beta-1}}{(1+(-w)^\alpha)(1+2(-w)^\beta\cos\beta\pi + (-w)^{2\beta})}\,\mathrm{d}w\,.$$

Then, the unique pair of probabilities (γ_+, γ_-) satisfying (7.8) and (7.9) is given by:

$$\gamma_+(\mathrm{d}v) = \frac{\sin\alpha\pi}{\pi}\,\frac{v^{\alpha-1}}{1+2v^\alpha\cos\alpha\pi + v^{2\alpha}}\,\mathrm{d}v = (1+v^\beta)\mu_+(dv)$$

and

$$\gamma_-(\mathrm{d}w) = \frac{\sin\beta\pi}{\pi}\,\frac{(-w)^{\beta-1}}{1+2(-w)^\beta\cos\beta\pi + (-w)^{2\beta}}\,\mathrm{d}w = (1+(-w)^\alpha)\mu_-(dw)\,.$$

This follows from the classical formula for the Laplace transform of the resolvent R_1 of parameter 1 of a stable subordinator of index α:

$$\frac{1}{1+v^\alpha} = \frac{\sin\alpha\pi}{\pi}\int_0^{+\infty}\frac{w^\alpha}{(v+w)(1+2w^\alpha\cos\alpha\pi + w^{2\alpha})}\,\mathrm{d}w\,,$$

(see Chaumont-Yor [46, Exercise 4.21: Mellin and Stieltjes transforms of stable variables], where it is proven that $\dfrac{1}{1+v^\alpha} = \mathbb{E}\left[\dfrac{1}{1+vX^{1/\alpha}}\right]$ with X a r.v. such that $\mathbb{P}(X \in dy) = \dfrac{\sin(\pi\alpha)}{\pi\alpha}\,\dfrac{1}{y^2+2y\cos(\pi\alpha)+1}1_{\{y>0\}}dy$).

We note that, in this example, if $p > 1$, the condition: $\mathbb{E}[|X|^p] < \infty$ is satisfied if and only if $p < \alpha + \beta$, whereas the conditions: $\mathbb{E}[V^{p-1}] < \infty$ and $\mathbb{E}[(-W)^{p-1}] < \infty$ are satisfied if and only if $p < 1 + \alpha\wedge\beta$. Now, $\alpha + \beta < 1 + \alpha\wedge\beta$ since $\alpha\vee\beta < 1$. This illustrates Remark 7.1.

Exercise 7.1. Let μ be the symmetric measure such that:

$$\mu_+(\mathrm{d}v) = \frac{2}{\pi}(1+v^2)^{-2}(1+\frac{2}{\pi}v\log v)\,\mathrm{d}v\,.$$

Prove that the unique pair of probabilities (γ_+, γ_-) satisfying (7.8) and (7.9) is given by:

$$\gamma_+(dv) = \frac{2}{\pi}(1+v^2)^{-1}\,\mathrm{d}v$$

and $\gamma_-(\mathrm{d}w) = \gamma_+(-\mathrm{d}w)$.

Exercise 7.2. Let μ be the symmetric measure such that:

$$\mu_+(\mathrm{d}v) = \frac{1}{\pi}\left(\frac{1}{\sqrt{v(1-v)}} - \frac{1}{\sqrt{1-v^2}}\right) 1_{]0,1[}(v)\,\mathrm{d}v\,.$$

Prove that the unique pair of probabilities (γ_+,γ_-) satisfying (7.8) and (7.9) is given by:

$$\gamma_+(dv) = \frac{1}{\pi}\frac{1}{\sqrt{v(1-v)}} 1_{]0,1[}(v)\,\mathrm{d}v$$

and $\gamma_-(\mathrm{d}w) = \gamma_+(-\mathrm{d}w)$. Observe that γ_+ is the Arcsine law.
(*Hint*: one may use the following formula:

$$\frac{1}{\pi}\int_0^1 \frac{w}{v+w}\frac{1}{\sqrt{w(1-w)}}\,\mathrm{d}w = 1-\sqrt{\frac{v}{1+v}}\,,$$

which can be found in [25, (1.18) and (1.23)].)

7.2.2 Uniform Integrability

In this subsection, we consider again X, an integrable, centered, real-valued r.v. and we keep the notation of Theorem 7.1. We now study the uniform integrability property of the martingale: $B^\tau := (B_{u\wedge\tau}, u \geq 0)$.

Theorem 7.3. *The martingale B^τ is uniformly integrable. Moreover, if $\mathbb{E}[\phi(X)] < \infty$ where $\phi : \mathbb{R} \to \mathbb{R}_+$ is defined by $\phi(x) = |x|\log^+(|x|)$, then, the martingale B^τ belongs to H^1, i.e. $\mathbb{E}\left[\sup_{u\geq 0}|B_u^\tau|\right] < \infty$.*

Proof. 1) We denote by $\mathbb{E}_{W,V}$ the expectation with respect to the law of (W,V), and by $\mathbb{E}_B$ the expectation with respect to the law of the Brownian motion B. We have, for every $\lambda \geq 0$:

$$\begin{aligned}
&\sup_{u\geq 0}\mathbb{E}\left[|B_{u\wedge\tau}|1_{\{|B_{u\wedge\tau}|>\lambda\}}\right] \\
&\quad= \sup_{u\geq 0}\mathbb{E}_{W,V}\left[\mathbb{E}_B\left[|B_{u\wedge T_{W,V}}|1_{\{|B_{u\wedge T_{W,V}}|>\lambda\}}\right]\right] \quad \text{(with } T_{b,a} \text{ defined by (7.7))} \\
&\quad\leq \mathbb{E}_{W,V}\left[\sup_{u\geq 0}\mathbb{E}_B\left[|B_{u\wedge T_{W,V}}|1_{\{|B_{u\wedge T_{W,V}}|>\lambda\}}\right]\right] \\
&\quad\leq \mathbb{E}_{W,V}\left[1_{\{V\vee(-W)>\lambda\}}\sup_{u\geq 0}\mathbb{E}_B\left[|B_{u\wedge T_{W,V}}|\right]\right] \quad \text{(since } |B_{u\wedge T_{W,V}}| \leq V\vee(-W)) \\
&\quad= \mathbb{E}_{W,V}\left[1_{\{V\vee(-W)>\lambda\}}\mathbb{E}_B\left[|B_{T_{W,V}}|\right]\right] \\
&\qquad \text{(since, for fixed } W \text{ and } V, (B_{u\wedge T_{W,V}}, u\geq 0) \text{ is a bounded martingale)} \\
&\quad= \mathbb{E}[1_{\{V\vee(-W)>\lambda\}}|B_\tau|].
\end{aligned}$$

Since $\mathbb{E}[|B_\tau|] = \mathbb{E}[|X|] < \infty$, we finally obtain by dominated convergence:

$$\lim_{\lambda \to +\infty} \sup_{u \geq 0} \mathbb{E}\left[|B_{u\wedge\tau}| 1_{\{|B_{u\wedge\tau}|>\lambda\}}\right] = 0$$

which yields the desired uniform integrability.

2) We now suppose that $\mathbb{E}[\phi(X)] < \infty$. Applying the previous computation of Point 1) to the submartingale $(\phi(B_u^\tau), u \geq 0)$ (ϕ is convex), we obtain

$$\begin{aligned}\sup_{u\geq 0} \mathbb{E}\left[\phi(B_u^\tau)\right] &= \lim_{u\to+\infty} \uparrow \mathbb{E}\left[\phi(B_u^\tau)\right] = \mathbb{E}\left[\phi(B_\tau)\right] \\ &= \mathbb{E}\left[\phi(X)\right] < \infty. \end{aligned} \tag{7.24}$$

Note that, under the hypothesis $\mathbb{E}\left[\phi(X)\right] < \infty$, (7.24) yields another proof that B^τ is a uniformly integrable martingale ([127, Chapitre 2, Théorème T22]).

On the other hand, from Doob's $L\log L$ inequality [150, p.55],

$$\mathbb{E}\left[\sup_{u\geq 0} |B_u^\tau|\right] \leq \frac{e}{e-1}\left(1 + \sup_{u\geq 0} \mathbb{E}\left[\phi(B_u^\tau)\right]\right) = \frac{e}{e-1}\left(1 + \mathbb{E}\left[\phi(X)\right]\right) < \infty$$

from (7.24). Therefore, B^τ belongs to H^1. Actually, the martingale B^τ belongs to the $L\log L$ class (cf. [150, Exercise 1.16]). □

Remark 7.2. The Azéma-Gundy-Yor [6, Théorème 1] criterion for uniform integrability of an L^1 bounded martingale also yields:

$$\lim_{\lambda\to+\infty} \lambda \mathbb{P}\left(\sqrt{\tau} \geq \lambda\right) = 0.$$

We now complete Theorem 7.3 when the r.v. X admits moments of order $p > 1$. We start with $p = 2$.

Theorem 7.4. *The following properties are equivalent:*

i) $\mathbb{E}[V] < \infty$ *and* $\mathbb{E}[-W] < \infty$,
ii) $\mathbb{E}[X^2] < \infty$,
iii) $\mathbb{E}[\tau] < \infty$.
iv) The martingale B^τ *is in* H^2.

Moreover, when these properties are satisfied, then

$$\mathbb{E}[X^2] = \mathbb{E}[V]\,\mathbb{E}[-W] = \mathbb{E}[\tau].$$

Proof. • We deduce from (7.8) and (7.9) by addition:

$$\mathbb{E}[X^2] = \mathbb{E}[V]\,\mathbb{E}[-W].$$

This entails the equivalence of properties *i)* and *ii)*.

• On the other hand, if $b \geq 0$ and $a < 0$, the martingale $\left(B^2_{u\wedge T_{a,b}} - (u \wedge T_{a,b}), u \geq 0\right)$ is uniformly integrable and hence, $\mathbb{E}[T_{a,b}] = \mathbb{E}\left[B^2_{T_{a,b}}\right] = -ab$. Consequently,

$$\mathbb{E}[\tau] = \mathbb{E}[T_{W,V}] = -\mathbb{E}[WV] = \mathbb{E}[V]\mathbb{E}[-W].$$

This shows that properties *i*) and *iii*) are equivalent.

• By Doob's L^2 inequality,

$$\mathbb{E}\left[\left(\sup_{u\geq 0} |B^\tau_u|\right)^2\right] \leq 4 \sup_{u\geq 0} \mathbb{E}\left[(B^\tau_u)^2\right] = 4\mathbb{E}[\tau].$$

Hence, *iii*) $\Longrightarrow$ *iv*).

• The converse follows from:

$$\mathbb{E}[t \wedge \tau] = \mathbb{E}\left[(B^\tau_t)^2\right] \leq \mathbb{E}\left[\left(\sup_{u\geq 0} |B^\tau_u|\right)^2\right],$$

upon letting t tend to $+\infty$. Therefore:

$$\mathbb{E}[\tau] \leq \mathbb{E}\left[\left(\sup_{u\geq 0} |B^\tau_u|\right)^2\right] \leq 4\mathbb{E}[\tau].$$

□

We now replace the L^2 space by L^p for $p > 1$.

Theorem 7.5. *Let $p > 1$. The following properties are equivalent:*

i) $\mathbb{E}[(V \wedge (-W))(V^{p-1} + (-W)^{p-1})] < \infty$,
ii) $\mathbb{E}[|X|^p] < \infty$,
iii) $\mathbb{E}[\tau^{p/2}] < \infty$,
iv) *the martingale B^τ is in H^p.*

Proof. 1) By (7.6), properties *i*) and *ii*) are equivalent.
2) Since we know from Theorem 7.3 that B^τ is uniformly integrable under the condition that X is integrable and centered, then properties *ii*), *iii*) and *iv*) are well-known to be equivalent, see (7.4). □

Remark 7.3. If $p \geq 2$, the property $\mathbb{E}[|X|^p] < \infty$ is equivalent to: $\mathbb{E}[V^{p-1}] < \infty$ and $\mathbb{E}[(-W)^{p-1}] < \infty$. This is proven in Theorem 7.4 for $p = 2$.

Now, suppose $p > 2$. We saw in Remark 7.1 that:

$$\mathbb{E}[|X|^p] < \infty \Longrightarrow \mathbb{E}[V^{p-1}] < \infty \text{ and } \mathbb{E}[(-W)^{p-1}] < \infty.$$

Conversely, suppose $\mathbb{E}[V^{p-1}] < \infty$ and $\mathbb{E}[(-W)^{p-1}] < \infty$. In particular, $\mathbb{E}[V] < \infty$ and $\mathbb{E}[(-W)] < \infty$. We deduce from (7.8) and (7.9):

$$\mathbb{E}[|X|^p] \leq \mathbb{E}[-W]\,\mathbb{E}[V^{p-1}] + \mathbb{E}[V]\,\mathbb{E}[(-W)^{p-1}]$$

which entails $\mathbb{E}[|X|^p] < \infty$.

7.2.3 Construction of a Self-Similar Martingale Associated to the Process $(\sqrt{t}X, t \geq 0)$

In this section, we consider a real valued, centered, random variable X. Let V, W be as in Theorem 7.1. We set:

$$\tau_t = \inf\{u \geq 0\,;\, B_u = \sqrt{t}\,V \;\text{ or }\; B_u = \sqrt{t}\,W\}\,.$$

Theorem 7.6. *i) The process $(B_{\tau_t}, t \geq 0)$ is a left-continuous martingale such that, for every fixed t, $B_{\tau_t} \overset{(law)}{=} \sqrt{t}X$.*
ii) For any $c > 0$,

$$(B_{\tau_{c^2 t}}, t \geq 0) \overset{(law)}{=} (c B_{\tau_t}, t \geq 0).$$

iii) The process $(B_{\tau_t}, t \geq 0)$ is an inhomogeneous Markov process.
In particular, $(B_{\tau_t}, t \geq 0)$ is a martingale associated to the peacock $(\sqrt{t}X, t \geq 0)$.

Proof. 1) The definition of times τ_t and the continuity of B easily entails that the process $(\tau_t, t \geq 0)$ is a left-continuous increasing process. As a consequence, $(B_{\tau_t}, t \geq 0)$ is a left-continuous process.

2) Since, for a given $t \geq 0$, $(M_u := B_{u \wedge \tau_t}, u \geq 0)$ is a uniformly integrable martingale, and for $s < t$, $\tau_s \leq \tau_t$, then $(B_{\tau_t}, t \geq 0)$ is a martingale.

Let, for $c > 0$, $(B^{(c)}_t := c B_{c^{-2} t}, t \geq 0)$, and denote by $(\tau^{(c)}_t, t \geq 0)$ the family of stopping times associated with the Brownian motion $B^{(c)}$. In other words,

$$\tau^{(c)}_t = \inf\{u \geq 0\,;\, B^{(c)}_u = \sqrt{t}\,V \;\text{ or }\; B^{(c)}_u = \sqrt{t}\,W\}\,.$$

We easily obtain, for every $t \geq 0$, $\tau^{(c)}_t = c^2\,\tau_{c^{-2} t}$ and then, $B^{(c)}_{\tau^{(c)}_{c^2 t}} = c\,B_{\tau_t}$, which proves point *ii)* since $(B^{(c)}_t, t \geq 0) \overset{(law)}{=} (B_t, t \geq 0)$. Moreover, since $B_{\tau_1} \overset{(law)}{=} X$, it also holds that, for every $t \geq 0$, $B_{\tau_t} \overset{(law)}{=} \sqrt{t}X$.

3) We shall use the strong Markov property of Brownian motion B in $\mathbb{R}$. We may define $\widetilde{\tau}_t$ by:

$$\widetilde{\tau}_t = \inf\{u \geq 0\,;\, B_u \notin]\sqrt{t}\,W, \sqrt{t}\,V[\}\,.$$

(Note that $\widetilde{\tau}_t = \tau_t$ under $\mathbb{P}_0$, whereas, if $x \neq 0$, then $\widetilde{\tau}_t \neq \tau_t$ under $\mathbb{P}_x$.) For $s < t$, making use of the standard notation for time translation operators (θ_u), one has:

$$\widetilde{\tau}_t = \widetilde{\tau}_s + \widetilde{\tau}_t \circ \theta_{\widetilde{\tau}_s}$$

and therefore $B_{\widetilde{\tau}_t} \circ \theta_{\widetilde{\tau}_s} = B_{\widetilde{\tau}_t}$. Then, by the strong Markov property, we obtain that, for f a bounded Borel function,

$$\mathbb{E}[f(B_{\widetilde{\tau}_t}) \mid \mathscr{F}_{\widetilde{\tau}_s}] = \mathbb{E}_{B_{\widetilde{\tau}_s}}[f(B_{\widetilde{\tau}_t})]\,.$$

This proves point *iii*). More precisely, the transition semi-group: $(P_{s,t}, 0 \leq s < t)$ is given by:

$$P_{s,t}f(x) = \mathbb{E}\left[\left(f(\sqrt{t}V)\frac{x-\sqrt{t}W}{\sqrt{t}(V-W)} + f(\sqrt{t}W)\frac{-x+\sqrt{t}V}{\sqrt{t}(V-W)}\right) 1_{]\sqrt{t}W,\sqrt{t}V[}(x)\right]$$
$$+f(x)\mathbb{P}\left(x \notin]\sqrt{t}W, \sqrt{t}V[\right) .$$

Thus, $(P_{s,t}, 0 \leq s < t)$ is a transition semi-group of a very special kind since, actually, $P_{s,t}$ does not depend on $s \in [0,t[$. □

7.3 Hall's Method for SE

For this method, we refer the reader to W. Hall [73, 74], L. Breiman [38] and J. Oblój [136, p.354-355]. Let X be a r.v. with law μ. We assume that:

$$\mathbb{E}[|X|] = \int_{-\infty}^{+\infty} |x|\mu(dx) < \infty \quad \text{and} \quad \mathbb{E}[X] = \int_{-\infty}^{+\infty} x\mu(dx) = 0. \tag{7.25}$$

Observe that, from (7.25):

$$\int_{0}^{+\infty} x\mu(dx) = -\int_{-\infty}^{0} x\mu(dx). \tag{7.26}$$

Let (V,W) be a couple of r.v.'s with values in $\mathbb{R}_+ \times \mathbb{R}_-^*$ whose law is given by:

$$\gamma(dv, dw) = \frac{v-w}{\int_0^{+\infty} x\mu(dx)} 1_{\{w<0\leq v\}}\mu(dv)\mu(dw). \tag{7.27}$$

(Observe that (7.26) implies that γ is a probability on $\mathbb{R}_+ \times \mathbb{R}_-^*$.) We assume that the couple (V,W) is independent from the Brownian motion $(B_u, u \geq 0)$ started from 0, and we define:

$$\tau := \inf\{u \geq 0; B_u = V \ \text{ or } B_u = W\}. \tag{7.28}$$

Theorem 7.7. 1) *The law of* B_τ *is* μ.
2) $(B_{u\wedge\tau}, u \geq 0)$ *is a uniformly integrable martingale.*

Remark 7.4. Hall's method, as the one described in Section 7.2, is a randomized method. Although they look similar, they present two main differences:

i) In Section 7.2, the r.v.'s V and W are independent, while they are not in Hall's method.
ii) In Hall's method, the law of the couple (V,W) is explicit, whereas it is not in the method described in Section 7.2.

Proof. 1) *We first prove Point 1).* We recall that, for $b < 0 \le a$, with:

$$T_{b,a} := \inf\{u \ge 0; B_u = a \text{ or } B_u = b\}, \tag{7.29}$$

we have:

$$\mathbb{P}(B_{T_{b,a}} = a) = \frac{-b}{a-b} \qquad \text{and} \qquad \mathbb{P}(B_{T_{b,a}} = b) = \frac{a}{a-b}. \tag{7.30}$$

Now, let $f : \mathbb{R} \longrightarrow \mathbb{R}$ be a Borel bounded function. From (7.30), by conditioning with respect to V and W, we obtain:

$$\begin{aligned}
\mathbb{E}[f(B_\tau)] &= \int\int d\gamma(dv,dw)\left(f(v)\frac{-w}{v-w} + f(w)\frac{v}{v-w}\right) \\
&= \int\int 1_{\{w<0\le v\}}\left(f(v)\frac{-w}{v-w} + f(w)\frac{v}{v-w}\right)(v-w)\frac{\mu(dv)\mu(dw)}{\int_0^{+\infty} x\mu(dx)} \\
&= \frac{1}{\int_0^{+\infty} x\mu(dx)}\int\int 1_{\{w<0\le v\}}(-wf(v)+vf(w))\,\mu(dv)\mu(dw) \\
&= \frac{1}{\int_0^{+\infty} x\mu(dx)}\left(-\int 1_{\{w<0\}} w\mu(dw)\int 1_{\{v\ge 0\}} f(v)\mu(dv)\right. \\
&\qquad \left. + \int 1_{\{v\ge 0\}} v\mu(dv)\int 1_{\{w<0\}} f(w)\mu(dw)\right) \\
&= \int_{-\infty}^{+\infty} f(x)\mu(dx) \quad \text{(from (7.26)).}
\end{aligned}$$

2) *We now prove that $(B_{u\wedge\tau}, u \ge 0)$ is uniformly integrable.* We denote by $\mathbb{E}_{V,W}$ the expectation with respect to the law of (V,W), and by $\mathbb{E}_B$ the expectation with respect to the law of the Brownian motion B. We have, for every $\lambda \ge 0$:

$$\begin{aligned}
&\sup_{u\ge 0}\mathbb{E}\left[|B_{u\wedge\tau}|1_{\{|B_{u\wedge\tau}|>\lambda\}}\right] \\
&\quad = \sup_{u\ge 0}\mathbb{E}_{V,W}\left[\mathbb{E}_B\left[|B_{u\wedge T_{W,V}}|1_{\{|B_{u\wedge T_{W,V}}|>\lambda\}}\right]\right] \quad \text{(with } T_{W,V} \text{ defined by (7.29))} \\
&\quad \le \mathbb{E}_{V,W}\left[\sup_{u\ge 0}\mathbb{E}_B\left[|B_{u\wedge T_{W,V}}|1_{\{|B_{u\wedge T_{W,V}}|>\lambda\}}\right]\right] \\
&\quad \le \mathbb{E}_{V,W}\left[1_{\{V\vee(-W)>\lambda\}}\sup_{u\ge 0}\mathbb{E}_B\left[|B_{u\wedge T_{W,V}}|\right]\right] \quad \text{(since } |B_{u\wedge T_{W,V}}| \le V\vee(-W)\text{)} \\
&\quad = \mathbb{E}_{V,W}\left[1_{\{V\vee(-W)>\lambda\}}\mathbb{E}_B\left[|B_{T_{W,V}}|\right]\right] \\
&\qquad \text{(since, for fixed } V \text{ and } W, (B_{u\wedge T_{W,V}}, u\ge 0) \text{ is a bounded martingale)} \\
&\quad = \frac{2}{\int_0^{+\infty} x\mu(dx)}\int\int 1_{\{v\vee(-w)>\lambda\}}\frac{-vw}{v-w}\gamma(dv,dw) \\
&\quad = \frac{2}{\int_0^{+\infty} x\mu(dx)}\int\int 1_{\{v\vee(-w)>\lambda\}}1_{\{w<0\le v\}}(-vw)\mu(dv)\mu(dw) \\
&\quad \xrightarrow[\lambda\to+\infty]{} 0
\end{aligned}$$

by dominated convergence, since

$$\int\int 1_{\{w\leq 0\leq v\}}(-vw)\mu(dv)\mu(dw) = \left(\int_0^{+\infty} x\mu(dx)\right)^2 < \infty.$$

□

Now, let X satisfying (7.25). As indicated previously, we shall associate to the peacock $(\sqrt{t}X, t\geq 0)$ a martingale by using Hall's embedding method. More precisely, let (V,W) be defined by (7.27) and let, for every $t\geq 0$:

$$\tau_t := \inf\{u\geq 0; B_u = \sqrt{t}V \text{ or } B_u = \sqrt{t}W\}.$$

Then, the family $(\tau_t, t\geq 0)$ is clearly increasing, and we have:

Theorem 7.8. *i) The process $(B_{\tau_t}, t\geq 0)$ is a left-continuous martingale associated to the peacock $(\sqrt{t}X, t\geq 0)$.*
ii) For any $c\geq 0$:

$$(B_{\tau_{c^2t}}, t\geq 0) \overset{(law)}{=} (cB_{\tau_t}, t\geq 0).$$

iii) The process $(B_{\tau_t}, t\geq 0)$ is an inhomogeneous Markov process.

The proof of Theorem 7.8 is analogous to that of Theorem 7.6, hence is omitted.

Exercise 7.3 (Necessary and sufficient conditions for the stopped Brownian motion $(B_{u\wedge\tau}, u\geq 0)$ to belong to H^p). With the notation of Theorem 7.7:

1) Prove that, for every $p\geq 0$:

$$\mathbb{E}[V^p] < \infty \Longleftrightarrow \mathbb{E}\left[(X^+)^{p+1}\right] < \infty,$$

and

$$\mathbb{E}[W^p] < \infty \Longleftrightarrow \mathbb{E}\left[(X^-)^{p+1}\right] < \infty.$$

2) Prove that, if $p>1$, the following properties are equivalent:

i) the martingale $(B_{u\wedge\tau}, u\geq 0)$ is in H^p,
ii) $\mathbb{E}\left[\tau^{p/2}\right] < +\infty$,
iii) $\mathbb{E}\left[|X|^p\right] < \infty$,
iv) $\mathbb{E}\left[V^{p-1}\right] < \infty$ and $\mathbb{E}\left[(-W)^{p-1}\right] < \infty$.

Exercise 7.4 (A martingale associated to $(tX, t\geq 0)$, with X symmetric). Let X be an integrable r.v. such that $X \overset{(law)}{=} -X$.

1) Let $(B_u, u\geq 0)$ be a standard Brownian motion started from 0, independent of X. Let:

$$T_t := \inf\{u\geq 0;\ |B_u| = t\}.$$

Prove that $(M_t := B_{T_t}X, t\geq 0)$ is a martingale associated to $(tX, t\geq 0)$.

2) Prove that $(M_t, t \geq 0)$ is Markovian and admits, as transition semi-group, $(P_{s,t}, 0 < s \leq t)$ defined by:

$$P_{s,t}f(x) = \frac{1}{2}\left[f\left(\frac{t}{s}x\right)\left(1+\frac{s}{t}\right) + f\left(-\frac{t}{s}x\right)\left(1-\frac{s}{t}\right)\right].$$

3) Prove that $(M_t, t \geq 0)$ is 1-self similar, i.e.: for every $c > 0$, $(M_{ct}, t \geq 0) \overset{(law)}{=} (cM_t, t \geq 0)$.

4) (M. Émery) Let $(N_t, t \geq 0)$ be a standard Poisson process, independent of X. We set

$$\widetilde{M}_t := t(-1)^{N_{\frac{\log t}{2}}} X \qquad \text{for } t \geq 1.$$

Prove that $(\widetilde{M}_t, t \geq 1)$ is a martingale associated to $(tX, t \geq 1)$.

5) Prove that $(\widetilde{M}_t, t \geq 1)$ is Markovian and admits, as transition semi-group, $(P_{s,t}, 1 \leq s \leq t)$ where $P_{s,t}$ is defined in Question 2). Conclude that $(\widetilde{M}_t, t \geq 1) \overset{(law)}{=} (M_t, t \geq 1)$.

Exercise 7.5 (A martingale associated to $(\varphi(B_t), t \geq 0)$). We take back the notation of Exercise 5.7. Let $(B_u, u \geq 0)$ a Brownian motion started from 0, $(X_t, t \geq 0)$ a positive diffusion satisfying the hypotheses of Exercise 5.7 and Y a symmetric and integrable r.v. We assume that $(B_u, u \geq 0)$, $(X_t, t \geq 0)$ and Y are independent.

Let us define, for every $t \geq 0$:

$$(V^{(t)}, W^{(t)}) := \left(|Y| \inf_{s \geq t} \overline{X}_s, \ -|Y| \inf_{s \geq t} \overline{X}_s\right)$$

and

$$T^{(t)} := \inf\{u \geq 0; B_u = V^{(t)} \text{ or } B_u = W^{(t)}\}$$
$$(= \inf\{u \geq 0; |B_u| = |Y| \inf_{s \geq t} \overline{X}_s\}).$$

1) Prove that $B_{T^{(t)}} \overset{(law)}{=} Y \inf_{s \geq t} \overline{X}_s \ (\overset{(law)}{=} YX_t)$.
2) Prove that $(B_{u \wedge T^{(t)}}, u \geq 0)$ is a uniformly integrable martingale.
3) Deduce that $(B_{T^{(t)}}, t \geq 0)$ is a martingale associated to $(YX_t, t \geq 0)$.
4) *Application*: let φ be a function of $\mathscr{C}^2$ class which is strictly increasing, odd, and such that, for every $t \geq 0$, $\mathbb{E}[|\varphi(B_t)|] < +\infty$ and $\lim_{x \to +\infty} \varphi(x) = +\infty$.

i) Exhibit a martingale associated to $(\varphi(B_t), t \geq 0)$.

(*Hint*: $\varphi(B_t) \overset{(1.d)}{=} \varepsilon|\varphi(B_t)|$ where ε is a symmetric Bernoulli r.v. taking the values ± 1, independent from $(B_t, t \geq 0)$; prove then that $(X_t := |\varphi(B_t)|, t \geq 0)$ satisfies the hypotheses of Exercise 5.7, in particular, the assumption $\lim_{x \to +\infty} \varphi(x) = +\infty$ implies that $m([0, +\infty[) = +\infty$.)

ii) Compute the infinitesimal generator of $(\overline{X}_t := \overline{\varphi(B_t)}, t \geq 0)$.
(*Hint*: compute first the speed measure and scale function of $\overline{X}$, see [150, Chapter VII, Exercise (3.20)], $\overline{L} = \frac{1}{2}\left(\varphi' \circ \varphi^{-1}\right)^2 \frac{\partial^2}{\partial x^2} + \left(\frac{\varphi' \circ \varphi^{-1}}{\varphi^{-1}} + \frac{\varphi'' \circ \varphi^{-1}}{2}\right)\frac{\partial}{\partial x}$.)

iii) Compute explicitly $\overline{L}$ in the two following cases:
$(X_t^{(1)} = |\sinh(B_t)|, t \geq 0)$ and $(X_t^{(2)} := |B_t^{2n+1}|, t \geq 0)$, $(n \in \mathbb{N})$.

(Answer:

$$\overline{L}^{(1)} = \frac{1}{2}(x^2+1)\frac{\partial^2}{\partial x^2} + \left(\frac{\sqrt{x^2+1}}{\operatorname{Argsh}(x)} + \frac{x}{2}\right)\frac{\partial}{\partial x}$$

and

$$\overline{L}^{(2)} = (2n+1)\left(\frac{1}{2}x^{\frac{4n}{2n+1}}\frac{\partial^2}{\partial x^2} + (n+1)x^{\frac{2n-1}{2n+1}}\frac{\partial}{\partial x}\right).\Bigg)$$

Exercise 7.6 (On a two-parameter semi-group $P_{s,t}$ which does not depend on s for $s<t$). We keep the hypotheses and notation of this Section 7.3, (except that we replace systematically $\sqrt{t}$ by t to simplify).

1) Prove that $(X_t = B_{\tau_t}, t \geq 0)$ is a (inhomogeneous) Markov process whose transition semi-group $P_{s,t}$ is given by:

$$\begin{aligned} &P_{s,t}f(x) \\ &= \mathbb{E}\left[\left(f(tV)\frac{x-tW}{t(V-W)} + f(tW)\frac{-x+tV}{t(V-W)}\right)1_{]tW,tV[}(x)\right] \\ &+ f(x)\,\mathbb{P}(x \notin]tW,tV[)\ . \end{aligned}$$

for every $x \in \mathbb{R}$, $s<t$, and f Borel and bounded. (*Hint*: see the proof of Theorem 7.6.) Observe that, in particular, for $s \leq t$, $P_{s,t}$ does not depend on s. Therefore, there exists, for every $t \geq 0$, a Markovian kernel Q_t such that $P_{s,t} = Q_t$.

2) Prove that there exists a family of r.v. $(Z_x, x \in \mathbb{R})$ such that, for every $t>0$ and $x \in \mathbb{R}$,

$$Q_t f(x) = \mathbb{E}\left[f(tZ_{\frac{x}{t}})\right].$$

3) Prove that for every $x \in \mathbb{R}$ and $0 < c \leq 1$:

$$Z_x \overset{(law)}{=} Z^{(1)}_{cZ^{(2)}_{x/c}} \tag{7.31}$$

where, in the right-hand side, $Z^{(1)}$ and $Z^{(2)}$ are two independent families of r.v.'s both having the same marginals as the family Z. (*Hint*: use the relation $P_{s,u}P_{u,t} = P_{s,t}$ for $s<u<t$.)

4) Let us define, for λ and y real:

$$\varphi(\lambda,y) := \mathbb{E}\left[e^{i\lambda Z_y}\right].$$

i) Show that, for every $c \in]0,1]$, and every $\lambda, y \in \mathbb{R}$:

$$\mathbb{E}[\varphi(\lambda, cZ_{y/c})] = \varphi(\lambda,y).$$

(*Hint*: use the Markov property of the process $(B_{\tau_t}, t \geq 0)$.)

ii) Recover then formula (7.31).

5) We recall that the law γ of the couple (V,W) is given by:

$$\gamma(dv,dw) = \frac{v-w}{\int_0^{+\infty} x\mu(dx)} 1_{\{w<0\leq v\}}\mu(dv)\mu(dw).$$

i) Prove that the law of Z_0 is μ.

ii) Prove that the law ν_y of Z_y is, for $y>0$ given by:

$$\nu_y(dz) = \left(\frac{\mu^-}{C}\int_{[0,y]} a\mu(da) + \mu([0,y])\right)\delta_y(dz) + \left(\frac{y\mu^-}{C}+1\right)1_{\{z>y\}}\mu(dz)$$
$$+\frac{1}{C}\left(-y\mu(]y,+\infty[) + \int_{]y,+\infty[} a\mu(da)\right)1_{\{z<0\}}\mu(dz)$$

where $C := \int_0^{+\infty} a\mu(da) = -\int_{-\infty}^0 a\mu(da)$ and $\mu^- = \mu(]-\infty,0[)$.

Verify that $\nu_y(\mathbb{R}) = 1$.

iii) Prove that, for $y<0$, the law ν_y of Z_y is given by:

$$\nu_y(dz) = \left(\mu([y,0[) - \frac{\mu^+}{C}\int_{[y,0[} a\mu(da)\right)\delta_y(dz) + \left(\frac{-y\mu^+}{C}+1\right)1_{\{z<y\}}\mu(dz)$$
$$+\frac{1}{C}\left(y\mu(]-\infty,y[) - \int_{]-\infty,y[} a\mu(da)\right)1_{\{z\geq 0\}}\mu(dz)$$

where $\mu^+ := 1-\mu^- = \mu([0,+\infty[)$.

6) We now assume that $\mu(dx) = \alpha\delta_{-a}(dx) + \beta\delta_b(dx)$, with $a,b>0$, $\alpha+\beta=1$ and $a\alpha = b\beta$.

i) Compute explicitly the law of Z_y.

(*Hint*: for $y \in [-a,b]$, $\nu_y(dz) = \frac{\beta}{a}\left((y+a)\delta_b(dz) + (b-y)\delta_{-a}(dz)\right)$ and for $y \notin [-a,b]$, $\nu_y(dz) = \delta_y(dz)$.)

ii) Verify directly by inspection formula (7.31) in this particular case.

7.4 The Azéma-Yor Algorithm for SE

Our purpose here is the same as in the two previous Sections 7.2 and 7.3: to associate a martingale to the peacock $(\sqrt{t}X, t\geq 0)$, with X an integrable and centered r.v. We shall use the Azéma-Yor method for SE. This method presents two main differences with both previous methods:

i) It is a non-randomized method.

ii) Let τ_t be the stopping time associated by this method to μ_t, the law of $\sqrt{t}X$. In general, the family of stopping times $(\tau_t, t \geq 0)$ is not increasing. We shall see that the increase of $(\tau_t, t \geq 0)$ is equivalent to a condition on the law μ of X, the $(M{\cdot}Y)$ condition that we describe below.

7.4.1 Description of this Algorithm

Let μ be a probability on $\mathbb{R}$ such that:

$$\int_{-\infty}^{+\infty} |x|\mu(dx) < \infty \quad \text{and} \quad \int_{-\infty}^{+\infty} x\mu(dx) = 0. \tag{7.32}$$

We define its Hardy-Littlewood function by:

$$\psi_\mu(x) = \frac{1}{\mu([x,+\infty[)} \int_{[x,+\infty[} y\mu(dy). \tag{7.33}$$

In case there exists $x \geq 0$ such that $\mu([x,+\infty[) = 0$, we set $\alpha = \inf\{x \geq 0; \mu([x,+\infty[) = 0\}$ and $\psi_\mu(x) = \alpha$ for $x \geq \alpha$. Note that ψ_μ is positive, left-continuous, increasing, and, for every $x \geq 0$, $\psi_\mu(x) \geq x \wedge \alpha$. Let $(B_t, t \geq 0)$ be a standard Brownian motion started at 0. Azéma-Yor [7, 8] introduced the stopping time:

$$T_\mu := \inf\{t \geq 0; S_t \geq \psi_\mu(B_t)\}$$

where $S_t := \sup\limits_{s \leq t} B_s$ and showed:

Theorem 7.9 (Azéma-Yor [7]). 1) *$(B_{t\wedge T_\mu}, t \geq 0)$ is a uniformly integrable martingale.*
2) *The law of B_{T_μ} is μ, i.e.: $B_{T_\mu} \sim \mu$.*

To prove Theorem 7.9, Azéma-Yor make use of the martingales:

$$\left(\varphi(S_t)(S_t - B_t) + \int_{S_t}^{+\infty} \varphi(x)dx,\ t \geq 0\right)$$

for any $\varphi \in L^1(\mathbb{R}_+, dx)$. Rogers [151] shows how to derive Theorem 7.9 from excursion theory, while Jeulin-Yor [98] develop a number of results about the laws of $\int_0^{T_\mu} h(B_s)ds$ for a generic function h. See also [41] for recent developments about Azéma-Yor martingales.

7.4.2 Use of this Algorithm

Our aim is to construct a martingale $(M_t, t \geq 0)$ such that:

i) $(M_t, t \geq 0)$ is associated to the peacock $(\sqrt{t}X, t \geq 0)$, i.e.

$$M_t \overset{(1.d)}{=} \sqrt{t}X,$$

ii) $(M_t, t \geq 0)$ enjoys the Brownian scaling property, i.e,

$$\forall c > 0, \quad (M_{c^2 t}, t \geq 0) \overset{(law)}{=} (cM_t, t \geq 0).$$

Here are the analogues of Theorems 7.6 and 7.8 in this new set-up.

Theorem 7.10. *Let X be an integrable and centered r.v. with law μ. Let, for every $t \geq 0$, $\widetilde{X}_t := \sqrt{t}X$. Denote by μ_t the law of $\widetilde{X}_t$ and by ψ_t $(= \psi_{\mu_t})$ the Hardy-Littlewood function attached to μ_t:*

$$\psi_t(x) = \frac{1}{\mu_t([x, +\infty[)} \int_{[x,+\infty[} y\mu_t(dy) = \sqrt{t}\psi_1\left(\frac{x}{\sqrt{t}}\right)$$

and by $T_t^{(\mu)}$ the Azéma-Yor stopping time (which we shall also denote T_{ψ_t}):

$$T_t^{(\mu)} := \inf\{u \geq 0; S_u \geq \psi_t(B_u)\} \tag{7.34}$$

(Theorem 7.9 asserts that $B_{T_t^{(\mu)}} \sim \mu_t$). We assume furthermore:

$$t \longmapsto T_t^{(\mu)} \text{ is a.s. increasing.} \tag{I}$$

Then:
1) The process $\left(M_t^{(\mu)} := B_{T_t^{(\mu)}}, t \geq 0\right)$ is a martingale, and an (inhomogeneous) Markov process, associated to the peacock $(\sqrt{t}X, t \geq 0)$.
2) The process $\left(M_t^{(\mu)}, t \geq 0\right)$ enjoys the Brownian scaling property, i.e for every $c > 0$:

$$\left(M_{c^2 t}^{(\mu)}, t \geq 0\right) \overset{(law)}{=} \left(cM_t^{(\mu)}, t \geq 0\right).$$

Proof. Point 1) is clear: the proof is identical to that of Theorems 7.6 and 7.8. We also refer the reader to [125] for the computation of the infinitesimal generator of $(M_t^{(\mu)}, t \geq 0)$. It is therefore sufficient to prove Point 2). Let $c > 0$ be fixed.
i) From the scaling property of Brownian motion:

$$(S_{c^2 t}, B_{c^2 t}, t \geq 0) \overset{(law)}{=} (cS_t, cB_t, t \geq 0),$$

and the definition (7.34) of T_{ψ_t}, we deduce that:

$$\left(B_{T_{\psi_t}}, t \geq 0\right) \overset{(law)}{=} \left(cB_{T_{\psi_t^{(c)}}}, t \geq 0\right) \tag{7.35}$$

with $\psi_t^{(c)}(x) := \frac{1}{c}\psi_t(cx)$.

ii) An elementary computation yields:

$$\psi_t(x) = \sqrt{t}\,\psi\left(\frac{x}{\sqrt{t}}\right) \tag{7.36}$$

with $\psi := \psi_1 = \psi_\mu$. We obtain from (7.36) that:

$$\psi_{c^2t}^{(c)}(x) = \frac{1}{c}\psi_{c^2t}(cx) = \sqrt{t}\,\psi\left(\frac{x}{\sqrt{t}}\right) = \psi_t(x). \tag{7.37}$$

Finally, gathering *i*) and *ii*), it holds:

$$\begin{aligned}\left(M_{c^2t}^{(\mu)} := B_{T_{\psi_{c^2t}}}, t \geq 0\right) &\overset{(law)}{=} \left(cB_{T_{\psi_{c^2t}^{(c)}}}, t \geq 0\right) \quad \text{(from (7.35))}\\ &\overset{(law)}{=} \left(cB_{T_{\psi_t}}, t \geq 0\right) \quad \text{(from (7.37))}\\ &= \left(cM_t^{(\mu)}, t \geq 0\right).\end{aligned}$$

□

Let us give a few examples for which Theorem 7.10 applies:

Example 7.4 (see [130] where the following examples are studied in details).

i) The dam-drawdown example:

$$\mu_t(dx) = \frac{1}{\sqrt{t}}\exp\left(-\frac{1}{\sqrt{t}}(x+\sqrt{t})\right)1_{[-\sqrt{t},+\infty[}(x)dx$$

which yields the stopping time $T_t := \inf\{u \geq 0; S_u - B_u = \sqrt{t}\}$. Recall that from Lévy's theorem, $(S_u - B_u, u \geq 0)$ is a reflected Brownian motion.

ii) The "BES(3)-Pitman" example:

$$\mu_t(dx) = \frac{1}{2\sqrt{t}}1_{[-\sqrt{t},\sqrt{t}]}(x)dx$$

which corresponds to the stopping time $T_t := \inf\{u \geq 0; 2S_u - B_u = \sqrt{t}\}$. Recall that from Pitman's Theorem [142], $(2S_u - B_u, u \geq 0)$ is distributed as a Bessel process of dimension 3 started from 0.

iii) The Azéma-Yor "fan", which is a generalization of the two previous examples:

$$\mu_t^{(\alpha)}(dx) = \frac{\alpha}{\sqrt{t}}\left(\alpha - \frac{(1-\alpha)x}{\sqrt{t}}\right)^{\frac{2\alpha-1}{1-\alpha}} 1_{\left[-\sqrt{t},\frac{\alpha\sqrt{t}}{1-\alpha}\right]}(x)dx, \qquad (0 < \alpha < 1)$$

which yields to the stopping time $T_t^{(\alpha)} := \inf\{u \geq 0; S_u = \alpha(B_u + \sqrt{t})\}$. Example *i*) is obtained by letting $\alpha \to 1^-$.

7.4.3 The $(M\cdot Y)$ Condition

Theorem 7.10 highlights the importance of condition (I) for our search of martingales satisfying conditions (7.2) and (7.3). We now wish to be able to read "directly" from the measure μ whether (I) is satisfied or not. The answer to this question is presented in the following lemma.

Lemma 7.3 (Madan-Yor [125], Lemma 3). *Let $X \sim \mu$ satisfy (7.32). We define the Madan-Yor function attached to μ:*

$$D_\mu(x) := \frac{x\overline{\mu}(x)}{\int_{[x,+\infty[} y\mu(dy)} = \frac{x}{\psi_\mu(x)} \quad \text{with } \overline{\mu}(x) := \mathbb{P}(X \geq x) = \int_{[x,+\infty[} \mu(dy).$$

Then (I) is satisfied if and only if:

$$x \longmapsto D_\mu(x) \text{ is increasing on } \mathbb{R}_+. \qquad (M\cdot Y)$$

Proof. Condition (I) is equivalent to the increase, for any given $x \in \mathbb{R}$, of the function $t \longmapsto \psi_t(x)$. From (7.36), $\psi_t(x) = \sqrt{t}\psi\left(\frac{x}{\sqrt{t}}\right)$, hence, if $x \leq 0$, since ψ is a positive and increasing function, $t \longmapsto \psi_t(x)$ is increasing. For $x > 0$, we set $a_t = \frac{x}{\sqrt{t}}$; thus:

$$\psi_t(x) = \sqrt{t}\psi\left(\frac{x}{\sqrt{t}}\right) = x\frac{\psi(a_t)}{a_t},$$

and, $t \longmapsto a_t$ being a decreasing function of t, condition (I) is equivalent to the increase of the function $a \in \mathbb{R}_+ \longmapsto \frac{a}{\psi(a)} = D_\mu(a)$. □

A remarkable feature of this result is that (I) only depends on the restriction of μ to $\mathbb{R}_+$. This is inherited from the asymmetric character of the Azéma-Yor construction in which $\mathbb{R}_+$, via $(S_u, u \geq 0)$, plays a special role.

Remark 7.5 (Due to P. Vallois). The process $(S_{T_t^{(\mu)}}, t \geq 0)$ is always stochastically increasing (see Exercise 1.25), even without the $(M\cdot Y)$ hypothesis.

It is easy to prove (see [7, Proposition 3.6]) that, for every $x \geq 0$:

$$\mathbb{P}\left(S_{T_\mu} \geq x\right) = \exp\left(-\int_0^x \frac{dy}{y - \psi_\mu^{-1}(y)}\right)$$

where ψ_μ^{-1} is the right-continuous inverse of ψ_μ. Replacing μ by μ_t in this formula, and using $\psi_{\mu_t}^{-1}(y) = \sqrt{t}\psi_\mu^{-1}\left(\frac{y}{\sqrt{t}}\right)$, we obtain:

$$\mathbb{P}\left(S_{T_{\mu_t}} \geq x\right) = \exp\left(-\int_0^x \frac{dy}{y - \psi_{\mu_t}^{-1}(y)}\right)$$

$$= \exp\left(-\int_0^x \frac{dy}{y - \sqrt{t}\,\psi_{\widetilde{\mu}}^{-1}\left(\frac{y}{\sqrt{t}}\right)}\right)$$

$$= \exp\left(-\int_0^{\frac{x}{\sqrt{t}}} \frac{du}{u - \psi_{\widetilde{\mu}}^{-1}(u)}\right).$$

Thus, for every $x \geq 0$, the function $t \longmapsto \mathbb{P}\left(S_{T_{\mu_t}} \geq x\right)$ is increasing, which proves the announced property. Finally, note that, under $(M{\cdot}Y)$, this process is a.s. increasing, which of course implies that it is stochastically increasing.

To ease up reading, we now provide an outline of the contents of the remainder of Section 7.4. Our aim is now to:

- exhibit conditions equivalent to $(M{\cdot}Y)$,
- give some easy conditions which imply $(M{\cdot}Y)$, i.e. such that the family of stopping-times $(\tau_t, t \geq 0)$ is increasing, in order to apply Theorem 7.10.

7.4.4 The $(M \cdot Y)$ Condition under a Support Hypothesis

Let $\widetilde{\mu}$ be a probability on $\mathbb{R}$ satisfying (7.32). Since our aim is to obtain conditions equivalent to $(M{\cdot}Y)$, i.e.:

$$x \longmapsto D_{\widetilde{\mu}}(x) := \frac{x\int_{[x,+\infty[}\widetilde{\mu}(dy)}{\int_{[x,+\infty[} y\widetilde{\mu}(dy)} \quad \text{increases on } \mathbb{R}_+, \tag{7.38}$$

it suffices to study $D_{\widetilde{\mu}}$ on $\mathbb{R}_+$. Clearly, this function (on $\mathbb{R}_+$) depends only on the restriction of $\widetilde{\mu}$ to $\mathbb{R}_+$, which we denote by μ. Observe that $(M{\cdot}Y)$ remains unchanged if we replace μ by $\lambda\mu$ for λ a positive constant.

Besides, we shall restrict our study to the case where $\widetilde{\mu}$ is carried by $]-\infty, k]$, i.e. where $\mu = \widetilde{\mu}_{|\mathbb{R}_+}$ is a subprobability on $[0,k]$. To simplify further, but without loss of generality, we shall take $k = 1$ and assume that 1 belongs to the support of μ. In Subsection 7.4.6, we shall study briefly the case where μ is a measure whose support is $\mathbb{R}_+$.

7.4.4.1 Notation

We now present some notation which shall be in force throughout the remainder of this subsection. Let μ be a positive measure on $[0,1]$, with finite total mass, and whose support contains 1. We denote by and , respectively its tail and its double tail functions:

$$\overline{\mu}(x) = \int_{[x,1]} \mu(dy) = \mu([x,1])$$

$$\text{and} \quad \overline{\overline{\mu}}(x) = \int_x^1 \overline{\mu}(y)dy = \int_x^1 dy \int_y^1 \mu(dz) = \int_x^1 \mu(dz)(z-x).$$

Note that $\overline{\mu}$ is left-continuous, $\overline{\overline{\mu}}$ is continuous, and $\overline{\mu}$ and $\overline{\overline{\mu}}$ are both decreasing functions. Furthermore, it is not difficult to see that a function $\Lambda : [0,1] \longrightarrow \mathbb{R}_+$ is the double tail function of a positive finite measure on $[0,1]$ if and only if Λ is a convex function on $[0,1]$, left-differentiable at 1, right-differentiable at 0, and satisfying $\Lambda(1)=0$.

We also define the tails ratio u_μ attached to μ:

$$u_\mu(x) = \overline{\mu}(x)/\overline{\overline{\mu}}(x), \qquad x \in [0,1[.$$

Here is now a lemma of general interest which bears upon positive measures:

Lemma 7.4 (Pierre [139] or Revuz-Yor [150], Chapter VI, Lemma 5.1).
1) *For every* $x \in [0,1[$*:*

$$\overline{\mu}(x) = \overline{\overline{\mu}}(0) u_\mu(x) \exp\left(-\int_0^x u_\mu(y)dy\right) \tag{7.39}$$

and u_μ *is left-continuous.*

2) *Let* $v : [0,1[\longrightarrow \mathbb{R}_+$ *be a left-continuous function such that, for all* $x \in [0,1[$*:*

$$\overline{\mu}(x) = \overline{\overline{\mu}}(0) v(x) \exp\left(-\int_0^x v(y)dy\right).$$

Then, $v = u_\mu$.

Proof. 1) We first prove Point 1). For $x \in [0,1[$, we have:

$$-\int_0^x \frac{\overline{\mu}(y)}{\overline{\overline{\mu}}(y)}dy = \int_0^x \frac{d\overline{\overline{\mu}}(y)}{\overline{\overline{\mu}}(y)} = \left[\log \overline{\overline{\mu}}(y)\right]_0^x = \log \overline{\overline{\mu}}(x) - \log \overline{\overline{\mu}}(0),$$

hence,

$$\overline{\overline{\mu}}(0) u_\mu(x) \exp\left(-\int_0^x u_\mu(y)dy\right) = \overline{\overline{\mu}}(0) \frac{\overline{\mu}(x)}{\overline{\overline{\mu}}(x)} \exp\left(\log \frac{\overline{\overline{\mu}}(x)}{\overline{\overline{\mu}}(0)}\right) = \overline{\mu}(x).$$

2) We now prove Point 2). Let $U_\mu(x) := \int_0^x u_\mu(y)dy$ and $V(x) := \int_0^x v(y)dy$. Relation (7.39) implies:

$$\begin{aligned} u_\mu(x) \exp(-U_\mu(x)) &= v(x) \exp(-V(x)), \quad \text{i.e.} \\ (\exp(-U_\mu(x)))' &= (\exp(-V(x)))', \quad \text{hence} \\ \exp(-U_\mu(x)) &= \exp(-V(x)) + c. \end{aligned}$$

(The above derivatives actually denote left-derivatives.) Now, since $U_\mu(0) = V(0) = 0$, we obtain $c = 0$ and $U_\mu = V$. Then, differentiating, and using the fact that u_μ and v are left-continuous, we obtain: $u_\mu = v$. □

Table 7.2. Some functions attached to μ

$\mu(dx)$	*A finite positive measure on* $[0,1]$ *whose support contains 1*
$\overline{\mu}(a)=\mu([a,1])$	Tail function attached to μ
$\overline{\overline{\mu}}(a)=\int_a^1 \overline{\mu}(x)dx$	Double tail function attached to μ
$u_\mu(a)=\overline{\mu}(a)/\overline{\overline{\mu}}(a)=\frac{1}{\psi_\mu(a)-a}$	Tails ratio function attached to μ
$\psi_\mu(a)=\frac{1}{\overline{\mu}(a)}\int_{[a,1]} x\mu(dx)$	Hardy-Littlewood function attached to μ
$D_\mu(a)=\frac{a}{\psi_\mu(a)}=\frac{au_\mu(a)}{au_\mu(a)+1}$	Madan-Yor function attached to μ

Remark 7.6. 1) $u_\mu^2-u'_\mu$ is a positive measure on $]0,1]$. Note that if $\mu(dx)=h(x)dx$, then $\overline{\mu}$ is differentiable a.e. and, from (7.39), so is u_μ:

$$u_\mu^2-u'_\mu=(\overline{\mu}/\overline{\overline{\mu}})^2-\left(\frac{\overline{\mu}^2-h\overline{\overline{\mu}}}{\overline{\overline{\mu}}^2}\right)=h/\overline{\overline{\mu}}\geq 0.$$

2) For every $x\in[0,1[$,

$$\int_0^x u_\mu(y)dy<\infty \quad \text{and} \quad \int_0^{1^-} u_\mu(y)dy=\infty. \tag{7.40}$$

Indeed, from (7.39), we deduce that: for every $x\in[0,1[$,

$$\overline{\overline{\mu}}(x)=\overline{\overline{\mu}}(0)\exp\left(-\int_0^x u_\mu(y)dy\right),$$

and (7.40) follows from: $\overline{\overline{\mu}}(0)=\int_{[0,1]} y\mu(dy)>0$ and $\lim_{x\to 1^-}\overline{\overline{\mu}}(x)=0$.

We just attached to μ several functions: $\overline{\mu}$, $\overline{\overline{\mu}}$, D_μ, ψ_μ and u_μ. Here is, under the form of an exercise, some relations between these functions, which we sum up in Table 7.2.

Exercise 7.7 (Elementary properties of D_μ and ψ_μ).
1) *i)* Prove that ψ_μ and D_μ are left-continuous on $]0,1]$ and that, for every $x\in[0,1]$:

$$x\leq\psi_\mu(x)\leq 1 \quad \text{and} \quad x\leq D_\mu(x)\leq 1. \tag{7.41}$$

ii) Prove that ψ_μ is increasing.
iii) Prove that $D_\mu(1)=1$ and $D_\mu(0)=0$.

2) Prove that, for every $x\in[0,1[$:

i) $\overline{\overline{\mu}}(x)=\overline{\mu}(x)\left(\psi_\mu(x)-x\right)$, *ii)* $u_\mu(x) = \dfrac{1}{\psi_\mu(x) - x}$,

iii) $D_\mu(x) = \dfrac{x u_\mu(x)}{x u_\mu(x) + 1}$.

3) We now assume that μ admits a density h. Prove that:

i) if h is continuous at 0, then: $D'_\mu(0^+) = \overline{\mu}(0)/\overline{\overline{\mu}}(0)$,

ii) if h is continuous at 1, and $h(1) > 0$, then: $D'_\mu(1^-) = \dfrac{1}{2}$,

iii) if, in a neighborhood of 1, h admits the development:

$h(1-x) \underset{x\to 0}{=} Cx^\alpha + o(x^\alpha)$, with $C, \alpha > 0$ then: $D'_\mu(1^-) = \dfrac{1}{2+\alpha}$.

(*Hint*: these three properties are consequences of the following formula, which holds at every continuity point of h:

$$\frac{D'_\mu(x)}{D_\mu(x)} = \frac{1}{x} - h(x)\frac{1 - D_\mu(x)}{\overline{\mu}(x)}.\Bigg)$$

Now, we assume that the support of $\widetilde{\mu}$, the law of X, is included in $]-\infty, 1]$. Let μ be the restriction of $\widetilde{\mu}$ to $[0,1]$. Here is another formulation of the $(M{\cdot}Y)$ condition.

Theorem 7.11. *Let μ be a finite positive measure on $[0,1]$ whose support contains 1, and u_μ its tails ratio. The following assertions are equivalent:*

i) D_μ is increasing on $[0,1]$, i.e. $(M{\cdot}Y)$ holds.
ii) There exists a probability measure ν_μ on $]0,1[$ such that:

$$\forall a \in [0,1], \quad D_\mu(a) = \nu_\mu(]0,a[). \tag{7.42}$$

iii) $a \longrightarrow a u_\mu(a)$ is an increasing function on $[0,1[$.

Proof. Of course, the equivalence between *i)* and *ii)* holds, since $D_\mu(0) = 0$ and $D_\mu(1) = 1$. As for the equivalence between *i)* and *iii)*, it follows from the relation:

$$D_\mu(a) = \frac{a u_\mu(a)}{a u_\mu(a) + 1}.$$

□

Remark 7.7. 1) The probability measure ν_μ defined via (7.42) enjoys some particular properties. Indeed, from (7.42), it satisfies

$$\frac{\nu_\mu(]0,a[)}{a} = \frac{D_\mu(a)}{a} = \frac{1}{\psi_\mu(a)}.$$

Thus, since the function ψ_μ is increasing on $[0,1]$, the function $a \longmapsto \dfrac{\nu_\mu(]0,a[)}{a}$ is decreasing on $[0,1]$, and $\lim\limits_{a\to 0+} \dfrac{\nu_\mu(]0,a[)}{a} = \dfrac{1}{\psi_\mu(0)}$.

2) From Exercise 7.7, we have $\nu_\mu(]0,a[) = D_\mu(a) = \dfrac{au_\mu(a)}{au_\mu(a)+1}$, hence, for every $a \in]0,1[$:

$$u_\mu(a) = \frac{\nu_\mu(]0,a[)}{a\nu_\mu([a,1[)},$$

and, in particular, $\nu_\mu([a,1[) > 0$. Thus, with the help of (7.40), ν_μ necessarily satisfies the relation:

$$\int^{1^-} \frac{da}{\nu_\mu([a,1[)} = +\infty.$$

3) The function D_μ is characterized by its values on $]0,1[$ (since $D_\mu(0) = 0$ and $D_\mu(1) = 1$). Hence, D_μ only depends on the values of ψ_μ on $]0,1[$, and therefore, D_μ only depends on the restriction of μ to $]0,1]$. The value of $\mu(\{0\})$ is irrelevant for the $(M{\cdot}Y)$ condition.

4) Theorem 7.11 invites to ask for the following question: given a probability measure ν on $]0,1[$, under which conditions on ν does there exist a positive measure μ on $[0,1]$ with finite total mass such that μ satisfies $(M{\cdot}Y)$? In particular, are the conditions given in Points 1) and 2) of this Remark 7.7 sufficient? The answer to this question is positive. We refer the interested reader to [78, Theorem 3] for a precise formulation of this answer.

Example 7.5. Let $\beta \geq 2\alpha > 0$ and $\nu(dx) = g(x)dx$ with

$$g(x) = \frac{\beta - 2\alpha x}{\beta - \alpha} 1_{]0,1[}(x).$$

Then, if we define μ by

$$\overline{\mu}(x) := \frac{\beta - \alpha x}{\beta}(1-x)^{\frac{\alpha}{\beta - 2\alpha}} \left(1 - \frac{\alpha x}{\beta - \alpha}\right)^{-\frac{\beta - \alpha}{\beta - 2\alpha}},$$

we have: $D_\mu(x) = \nu(]0,x[) = \dfrac{\beta x - \alpha x^2}{\beta - \alpha}$. In particular, letting α tend to 0, we obtain that $\mu(dx) = \delta_1(dx)$, the Dirac measure at 1, and $\nu_\mu(dx) = 1_{]0,1[}(x)dx$.

In Theorem 7.11, we have parameterized the measure μ by its tail function $\overline{\mu}$ and its tail ratio u_μ (see Lemma 7.4). Here is another parametrization of μ which provides an equivalent statement to that of Theorem 7.11.

Exercise 7.8 (Another version of $(M \cdot Y)$ condition). Prove the following Theorem, which is equivalent to Theorem 7.11.

Theorem 7.12. *Let μ be a finite positive measure on $[0,1]$ whose support contains 1. Then, μ satisfies $(M{\cdot}Y)$ (i.e. D_μ is increasing on $[0,1[$) if and only if there exists a fonction $\alpha_\mu :]0,1[\longrightarrow \mathbb{R}_+$ such that:*

i) α_μ *is an increasing left-continuous function on* $]0,1[$,
ii) $\left(\alpha_\mu^2(x)+\alpha_\mu(x)\right)dx - x\,d\alpha_\mu(x)$ *is a positive measure on* $]0,1[$,
iii) $\lim\limits_{x\to 0^+} \dfrac{\alpha_\mu(x)}{x} < \infty$, *and* $\displaystyle\int^{1^-} \alpha_\mu(x)dx = +\infty$

and such that:

$$\overline{\overline{\mu}}(x) = \overline{\overline{\mu}}(0)\exp\left(-\int_0^x \frac{\alpha_\mu(y)}{y}dy\right). \tag{7.43}$$

Exercise 7.9 (A generic example of probabilities which satisfy ($M\cdot Y$)). Let ρ be a positive measure on $]0,+\infty[$ such that $\int_0^{+\infty} y\rho(dy) < \infty$. To ρ, we attach the measure ν defined on $]0,1[$ by:

$$\nu(]0,a[) = \frac{1}{Z}\int_0^{+\infty}(1-e^{-ay})\rho(dy) \quad (0<a<1)$$

where $Z := \displaystyle\int_0^{+\infty}(1-e^{-y})\rho(dy)$ is such that $\nu(]0,1[) = 1$. Note that the function $a \longmapsto \nu(]0,a[)$ is of $\mathscr{C}^1$ class on $]0,1[$.

1) *i)* Prove that

$$\frac{\nu(]0,a[)}{a} = \frac{1}{Z}\int_0^{+\infty} e^{-au}\overline{\rho}(u)du$$

with $\overline{\rho}(u) = \rho([u,+\infty[)$.

ii) Deduce that $a \longmapsto \dfrac{\nu(]0,a[)}{a}$ is decreasing on $]0,1[$ and that:

$$\lim_{a\to 0^+}\frac{\nu(]0,a[)}{a} = \frac{1}{Z}\int_0^{+\infty} y\rho(dy) =: c_\nu.$$

iii) Prove that $\lim\limits_{a\to 1^-}\dfrac{\nu([a,1[)}{1-a} = \dfrac{1}{Z}\displaystyle\int_0^{+\infty} ye^{-y}\rho(dy) > 0$, hence $\displaystyle\int^1 \frac{da}{\nu([a,1[)} = +\infty$.

2) Let us define:

$$u^{(\nu)}(a) := \frac{\nu(]0,a[)}{a(1-\nu(]0,a[))}.$$

i) Prove that: $\lim\limits_{a\to 0^+} u^{(\nu)}(a) = c_\nu$.

ii) Prove that $\left(u^{(\nu)}\right)^2 - \left(u^{(\nu)}\right)'$ is a positive measure.

iii) Show that:

$$\forall a<1, \qquad \int_0^a u^{(\nu)}(y)dy < +\infty \quad \text{and} \quad \int_0^{1^-} u^{(\nu)}(y)dy = +\infty.$$

3) Let us define:

$$m(a) := \frac{1}{c_\nu}u^{(\nu)}(a)\exp\left(-\int_0^a u^{(\nu)}(y)dy\right).$$

Prove that m is decreasing, $\lim_{a\to 0^+} m(a) = 1$, and $\lim_{a\to 1^-} m(a) = 0$. Hence, there exists a probability measure μ on $[0,1]$ such that:

$$\overline{\mu}(a) := \mu([a,1]) = m(a) \quad (0 \le a \le 1)$$

and such that $m(\{0\}) = 0$.

4) $i)$ Prove that $\overline{\overline{\mu}}(0) = \frac{1}{c_V}$.

$ii)$ Prove that $D_\mu(a) = \nu(]0,a[)$ $(0 < a < 1)$ and that μ satisfies the $(M{\cdot}Y)$ condition.

Example 7.6 (Applications of Exercise 7.9). 1) Let $\rho(dx) = e^{-\lambda x}dx$ $(\lambda > 0)$. Then $\nu(]0,a[) = \frac{(\lambda+1)a}{\lambda+a}$ $(a \in [0,1])$ and

$$\overline{\mu}(a) = \frac{1}{1-a}\exp\left(-\int_0^a \frac{\lambda+1}{\lambda}\frac{dx}{1-x}\right) = (1-a)^{1/\lambda}.$$

2) Let $\rho(dx) = \mathbb{P}(\Gamma \in dx)$ where Γ is a positive r.v. with finite expectation. Then, we have

$$\nu(]0,a[) = \mathbb{P}\left(\frac{\mathfrak{e}}{\Gamma} \le a \middle| \frac{\mathfrak{e}}{\Gamma} \le 1\right)$$

where $\mathfrak{e}$ is a standard exponential r.v. independent from Γ. In this case, we also have:

$$\begin{aligned}\frac{1}{\psi_\mu(a)} = \frac{\nu(]0,a[)}{a} &= K\int_0^{+\infty} e^{-ax}\mathbb{P}(\Gamma > x)dx \\ &= K\mathbb{E}\left[\int_0^{\Gamma} e^{-ax}dx\right] = \frac{K}{a}\mathbb{E}\left[1-e^{-a\Gamma}\right]\end{aligned}$$

where $K = 1/\mathbb{E}\left[1-e^{-\Gamma}\right]$. Consequently, the Madan-Yor function

$$D_\mu(a) = \frac{a}{\psi_\mu(a)} = K\mathbb{E}\left[1-e^{-a\Gamma}\right]$$

is the Lévy exponent of a compound Poisson process.

7.4.5 Some Sufficient Conditions for $(M\cdot Y)$

Throughout this subsection, we consider a positive finite measure μ on $\mathbb{R}_+$ which admits a density, denoted by h. Our aim is to give some sufficient conditions on h which ensure that $(M{\cdot}Y)$ holds. We start with a general lemma which takes up Madan-Yor condition as given in [125, Theorem 4] (this is Condition $iii)$ below):

Proposition 7.1. *Let h be a strictly positive function of $\mathscr{C}^1$ class on $]0,l[$ $(0 < l \le +\infty)$. The three following conditions are equivalent:*

$i)$ for every $c \in]0,1[$, $a \longmapsto \frac{h(a)}{h(ac)}$ is a decreasing function,

ii) the function $\varepsilon(y) := -\dfrac{yh'(y)}{h(y)}$ *is increasing,*

iii) $h(a) = e^{-V(a)}$ *where* $a \longmapsto aV'(a)$ *is an increasing function.*

We denote this condition by (S_0).

Moreover, V *and* ε *are related, for any* $a, b \in]0, l[$, *by:*

$$V(a) - V(b) = \int_b^a dy \frac{\varepsilon(y)}{y},$$

so that:

$$h(a) = h(b) \exp\left(-\int_b^a \frac{\varepsilon(y)}{y} dy\right).$$

Remark 7.8. Here are some general observations about condition (S_0):

- if both h_1 and h_2 satisfy condition (S_0), then so does $h_1 h_2$,
- if h satisfies condition (S_0), then, for every $\alpha \in \mathbb{R}$ and $\beta \geq 0$, so does $a \longmapsto a^\alpha h(a^\beta)$,
- as an example, we note that the Laplace transform $h(a) = \mathbb{E}\left[e^{-aX}\right]$ of a positive self-decomposable r.v. X satisfies condition $i)$. Indeed, by definition, for every $c \in [0, 1]$, there exists a positive r.v. $X^{(c)}$ independent from X such that:

$$X \overset{(law)}{=} cX + X^{(c)}.$$

  Taking Laplace transforms for both sides, we obtain:

$$h(a) := \mathbb{E}\left[e^{-aX}\right] = \mathbb{E}\left[e^{-acX}\right] \mathbb{E}\left[e^{-aX^{(c)}}\right],$$

  which can be rewritten:

$$\frac{h(a)}{h(ac)} = \mathbb{E}\left[e^{-aX^{(c)}}\right],$$

- we note that in Theorem 5 of Madan-Yor [125], the second and third observations above are used jointly, as the authors remark that the function: $k(a) := \mathbb{E}\left[e^{-a^2 X}\right] = h(a^2)$ for X positive and self-decomposable satisfies (S_0).

Proof. 1) *We prove that i)* $\Longleftrightarrow$ *ii).*

- The implication $ii) \Longrightarrow i)$ is clear. Indeed, for $c \in]0, 1[$, we write:

$$\frac{h(a)}{h(ac)} = \exp\left(-\int_{ac}^a \frac{\varepsilon(y)}{y} dy\right) = \exp\left(-\int_c^1 \frac{\varepsilon(ax)}{x} dx\right) \tag{7.44}$$

  which is a decreasing function of a since ε is increasing and $0 < c < 1$.
- We now prove that $i) \Longrightarrow ii)$. From (7.44), we know that for every $c \in]0, 1[$, $a \longmapsto \int_{ac}^a \frac{\varepsilon(x)}{x} dx$ is an increasing function. Therefore, by differentiation,

$$\forall a \in]0, l[, \ \forall c \in]0, 1[, \quad \varepsilon(a) - \varepsilon(ac) \geq 0$$

  which proves that ε is an increasing function.

2) *We prove that ii)* $\Longleftrightarrow$ *iii)*. From the two representations of h, we deduce that

$$V(a) = \int_b^a \frac{\varepsilon(y)}{y} dy - \ln h(b),$$

which gives, by differentiation:

$$aV'(a) = \varepsilon(a). \tag{7.45}$$

This ends the proof of Proposition 7.1. □

In the following, we shall once again restrict our attention to probabilities μ on $[0,1]$, and shall assume that they admit a density h which is strictly positive in a neighborhood of 1 (so that 1 belongs to the support of μ). We now give a first set of sufficient conditions (including (S_0)) which encompass most of the examples we shall deal with in the next section.

Theorem 7.13. *We assume that the density h of μ is continuous on $]0,1[$. Then, the following conditions imply $(M{\cdot}Y)$:*

(S_0) *h is strictly positive on $]0,1[$ and satisfies condition i) of Proposition 7.1 (with $l=1$).*

(S_1) *For every $a \in]0,1[$*

$$\overline{\mu}(a) := \int_a^1 h(x)dx \geq a(1-a)h(a).$$

(S_1') *The function $a \longmapsto a^2 h(a)$ is increasing on $]0,1[$.*

(S_2) *The function $a \longmapsto \log(a\overline{\mu}(a))$ is concave on $]0,1[$ and* $\lim_{a\to 1^-} (1-a)h(a) = 0$.

Proof. 1) *We first prove:* $(S_0) \Longrightarrow (M{\cdot}Y)$. We write for $a \in]0,1[$:

$$\frac{1}{D_\mu(a)} = \frac{\int_a^1 yh(y)dy}{a\overline{\mu}(a)} = \frac{[-y\overline{\mu}(y)]_a^1 + \int_a^1 \overline{\mu}(y)dy}{a\overline{\mu}(a)} = 1 + \int_1^{1/a} \frac{\overline{\mu}(ax)}{\overline{\mu}(a)} dx.$$

Clearly, $(M{\cdot}Y)$ is implied by the property: for all $x > 1$, $a \longmapsto \dfrac{\overline{\mu}(ax)}{\overline{\mu}(a)}$ is a decreasing function on $\left]0, \frac{1}{x}\right[$. Differentiating with respect to a, we obtain:

$$\frac{\partial}{\partial a}\left(\frac{\overline{\mu}(ax)}{\overline{\mu}(a)}\right) = \frac{-xh(ax)\overline{\mu}(a) + h(a)\overline{\mu}(ax)}{(\overline{\mu}(a))^2}.$$

We then rewrite the numerator as:

$$h(a)\int_{ax}^1 h(y)dy - xh(ax)\int_a^1 h(u)du$$

$$= xh(a)\int_a^{1/x} h(ux)du - xh(ax)\int_a^1 h(u)du$$

$$= xh(a)\int_a^{1/x} h(u)\left(\frac{h(ux)}{h(u)} - \frac{h(ax)}{h(a)}\right)du - xh(ax)\int_{1/x}^1 h(u)du \le 0$$

from assertion $i)$ of Proposition 7.1, since for $x > 1$, the function $u \longmapsto \dfrac{h(ux)}{h(u)} = \dfrac{h(ux)}{h\left(ux\frac{1}{x}\right)}$ is decreasing.

2) *We now prove:* $(S_1) \Longrightarrow (M{\cdot}Y)$. We must prove that under (S_1), the function $D_\mu(a) := \dfrac{a\overline{\mu}(a)}{\int_a^1 xh(x)dx}$ is increasing. Elementary computations lead, for $a \in]0,1[$, to:

$$\frac{D'_\mu(a)}{D_\mu(a)} = \frac{1}{a} - h(a)\frac{1-D_\mu(a)}{\overline{\mu}(a)}. \tag{7.46}$$

From (S_1) and (7.41):

$$0 \le \frac{h(a)}{\overline{\mu}(a)}(1-D_\mu(a)) \le \frac{1}{a(1-a)}(1-a) = \frac{1}{a}.$$

Hence, from (7.46):

$$\frac{D'_\mu(a)}{D_\mu(a)} \ge \frac{1}{a} - \frac{1}{a} = 0.$$

3) *We then prove:* $(S'_1) \Longrightarrow (S_1)$, *hence* $(M{\cdot}Y)$ *holds*. We have, for $a \in]0,1[$:

$$\begin{aligned}\overline{\mu}(a) := \int_a^1 h(x)dx &= \int_a^1 \frac{x^2h(x)}{x^2}dx \\ &\ge a^2h(a)\int_a^1 \frac{1}{x^2}dx \quad \text{(since } x \longmapsto x^2h(x) \text{ is increasing.)} \\ &= a^2h(a)\left(\frac{1}{a} - 1\right) = ah(a)(1-a).\end{aligned}$$

4) *We finally prove:* $(S_2) \Longrightarrow (M{\cdot}Y)$. We set $\theta(a) := \log(a\overline{\mu}(a))$. Since

$$\int_a^1 th(t)dt = a\overline{\mu}(a) + \int_a^1 \overline{\mu}(t)dt$$

by integration by parts, we have, for $a \in]0,1[$,

$$D_\mu(a) = \frac{e^{\theta(a)}}{e^{\theta(a)} + \int_a^1 \frac{1}{t}e^{\theta(t)}dt}.$$

Therefore, we must prove that the function $a \longmapsto e^{-\theta(a)} \int_a^1 \frac{1}{t} e^{\theta(t)} dt$ is decreasing. Differentiating this function, we need to prove:

$$k(a) := \theta'(a) \int_a^1 \frac{1}{t} e^{\theta(t)} dt + \frac{1}{a} e^{\theta(a)} \geq 0.$$

Now, since $\lim_{a \to 1^-} \theta(a) = -\infty$, an integration by parts gives:

$$k(a) = \int_a^1 \frac{dt}{t} e^{\theta(t)} (\theta'(a) - \theta'(t)) + \int_a^1 \frac{1}{t^2} e^{\theta(t)} dt,$$

and, θ' being a decreasing function, this expression of k shows that it is also a decreasing function. Therefore, it remains to prove that $\lim_{a \to 1^-} k(a) \geq 0$, which is implied by:

$$\lim_{a \to 1^-} \frac{\overline{\mu}(a) - ah(a)}{\overline{\mu}(a)} \int_a^1 \overline{\mu}(t) dt \geq 0$$

or

$$\lim_{a \to 1^-} \frac{h(a)}{\overline{\mu}(a)} \int_a^1 \overline{\mu}(t) dt = 0.$$

Since $\int_a^1 \overline{\mu}(t) dt \leq (1-a)\overline{\mu}(a)$, the result follows from the assumption $\lim_{a \to 1^-} (1 - a)h(a) = 0$. □

Here are now some alternative conditions which ensure that $(M{\cdot}Y)$ is satisfied.

Exercise 7.10 (Sufficient conditions for $(M{\cdot}Y)$). We assume that μ admits a density h of $\mathscr{C}^1$ class on $]0,1[$ which is strictly positive in a neighborhood of 1.

1) Prove that each of the following conditions implies $(M{\cdot}Y)$:

(S_3) $\quad a \longmapsto a^3 h'(a)$ is increasing on $]0,1[$.
(*Hint*: let $m := \lim_{a \to 0^+} a^3 h'(a) \geq -\infty$; if $m < 0$, then $h(x) \geq A_1 + \frac{A_2}{x^2}$ which is absurd; hence $m \geq 0$ and h is increasing; prove then that $a \longmapsto a^2 h(a)$ is increasing and apply (S_1').)

(S_4) $\quad a \longmapsto a^3 h'(a)$ is decreasing on $]0,1[$.
(*Hint*: let $\widehat{\mu}(a) := \int_{[a,1]} x\mu(dx)$; thus $D_\mu(a) := \frac{a\overline{\mu}(a)}{\widehat{\mu}(a)}$ and $D_\mu'(a) \geq 0$ is equivalent to $\gamma(a) := \overline{\mu}(a)\widehat{\mu}(a) + a^2 h(a)\overline{\mu}(a) - ah(a)\widehat{\mu}(a) \geq 0$; then prove that $\gamma(1^-) = 0$, $\gamma'(1^-) = 0$ and that γ is convex.)

(S_4') $\quad h$ is decreasing and concave.
(*Hint*: (S_4') implies (S_4).)

(S_5) $\quad h$ is a decreasing function and $a \longmapsto \frac{ah(a)}{1-a}$ is increasing on $]0,1[$.
(*Hint*: D_μ is increasing if $\rho(a) := \frac{\overline{\mu}(a)\widehat{\mu}(a)}{ah(a)} + a\overline{\mu}(a) - \widehat{\mu}(a) \geq 0$; now prove that $\lim_{a \to 1} \rho(a) = 0$ and that $\rho' \leq 0$.)

(S_5') $\quad 0 \geq h'(x) \geq -4h(x)$.
(*Hint*: (S_5') implies (S_5).)

2) Let us suppose that h' is bounded and negative. Then prove that, for large enough c, $\mu^{(c)}(dx) = (h(x)+c)dx$ satisfies condition (S_5'), hence $(M{\cdot}Y)$.

3) Prove that (S_1') implies (S_2).

4) Prove that (S_4') implies both (S_0) and (S_5).

Here is a diagram summing up all the relations discussed in Theorem 7.13 and Exercise 7.10.

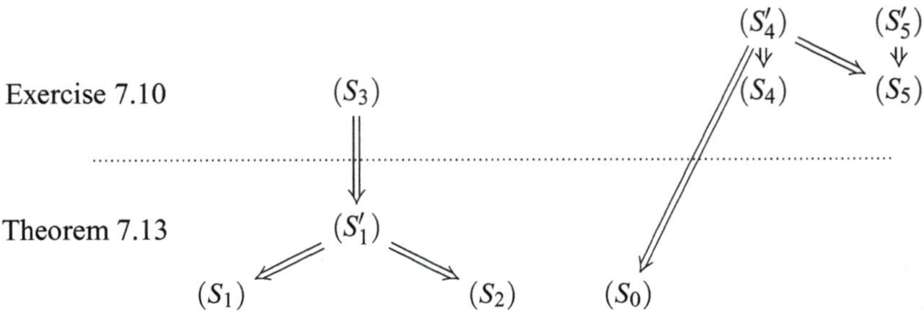

Exercise 7.11 ($(M{\cdot}Y)$ condition: examples and counterexample). In this exercise, we assume that μ admits a density h: $\mu(dx) = h(x)dx$. Prove that the following examples of measures μ satisfy the $(M{\cdot}Y)$ condition:

1) Beta laws: $h(x) = x^{\alpha}(1-x)^{\beta} 1_{]0,1[}(x)$ $(\alpha, \beta > -1)$.

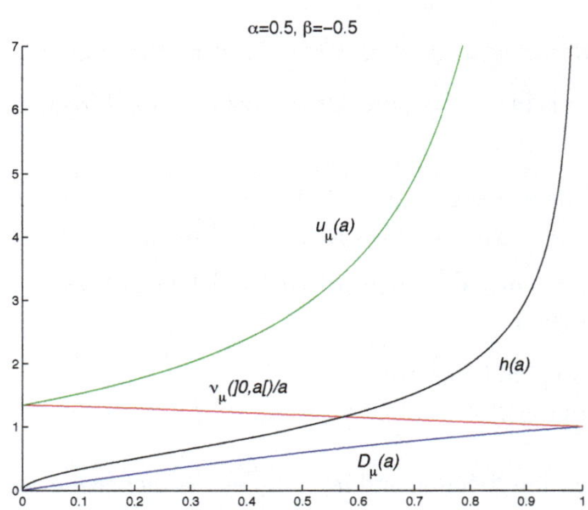

Fig. 7.1. Graphs for $h(x) = \sqrt{\dfrac{x}{(1-x)}}\, 1_{[0,1[}(x)$ $(\alpha = \frac{1}{2},\ \beta = -\frac{1}{2})$

2) *i)* $h(x) = \dfrac{x^\alpha}{(1+x^2)^\beta} 1_{[0,1]}(x)$ $(\alpha > -1, \beta \in \mathbb{R})$.

ii) $h(x) = \dfrac{x^\alpha}{(1-x^2)^\beta} 1_{[0,1]}(x)$ $(\alpha > -1, \beta < 1)$.

3) $h(x) = |\cos(\pi x)|^m 1_{[0,1]}(x)$ $(m \in \mathbb{R}_+)$.
(*Hint*: prove that h satisfies condition (S_1). More generally, every function which is symmetric with respect to the axis $x = \frac{1}{2}$, and is decreasing for $x \leq \frac{1}{2}$ and increasing for $x \geq \frac{1}{2}$, satisfies condition (S_1).)

4) $h(x) = x^\alpha e^{-x^\lambda} 1_{]0,1]}(x)$ $(\alpha > -1, \lambda \in \mathbb{R})$.

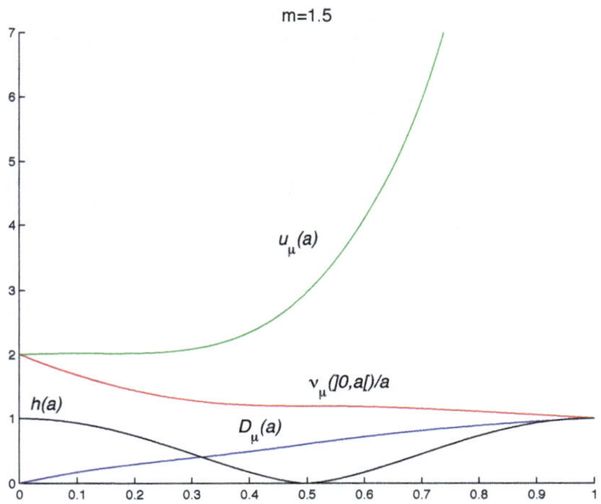

Fig. 7.2. Graphs for the Beta law $h(x) = |\cos(\pi x)|^{3/2} 1_{[0,1]}(x)$ $(m = \frac{3}{2})$

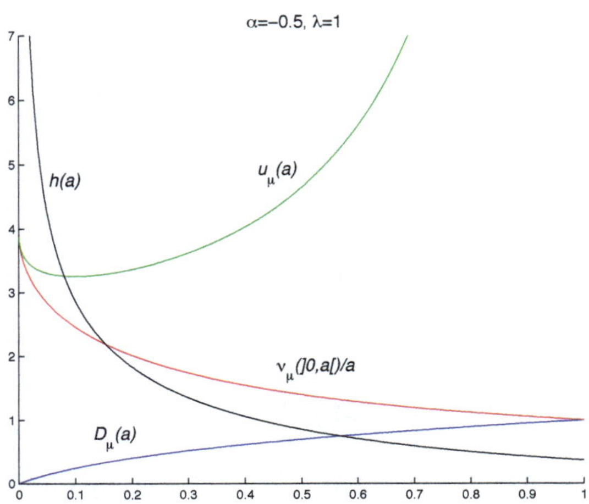

Fig. 7.3. Graphs for $h(x) = \dfrac{e^{-x}}{\sqrt{x}} 1_{]0,1]}(x)$ $(\alpha = -\frac{1}{2},\ \lambda = 1)$

5) *An example where $(M\cdot Y)$ is not satisfied.* Let $h(x) = c1_{[0,p[}(x) + e1_{[p,1]}(x)$, $(c,e \geq 0, p \in]0,1[)$.

i) Prove that, for $a < p$:

$$D_\mu(a) := \frac{2a\,(c(p-a)+e(1-p))}{c(p^2-a^2)+e(1-p^2)}.$$

ii) Prove that D_μ is $\mathscr{C}^\infty$ on $[0,p[$ and that, for $a < p$:

$$D'_\mu(a) = 2\frac{c^2p(p-a)^2+e^2(1-p)^2(1+p)+ec(1-p)\left((p-a)^2+p^2+p-2a\right)}{\left(c(p^2-a^2)+e(1-p^2)\right)^2}$$

and

$$D'_\mu(p^-) = 2\frac{e^2(1-p)^2\left(1+p\left(1-\frac{c}{e}\right)\right)}{e^2(1-p^2)^2} = 2\frac{1+p\left(1-\frac{c}{e}\right)}{(1+p)^2}.$$

iii) Deduce that, for $\frac{c}{e}$ large enough, $D'_\mu(p^-) < 0$, hence D_μ is not increasing on $[0,1]$. Note that, if $e \geq c$ (h is increasing), then $D'_\mu \geq 0$ (see condition (S'_1)), and that D_μ is increasing if and only if $D'_\mu(p^-) \geq 0$, i.e. $\frac{c}{e} \leq 1 + \frac{1}{p}$.

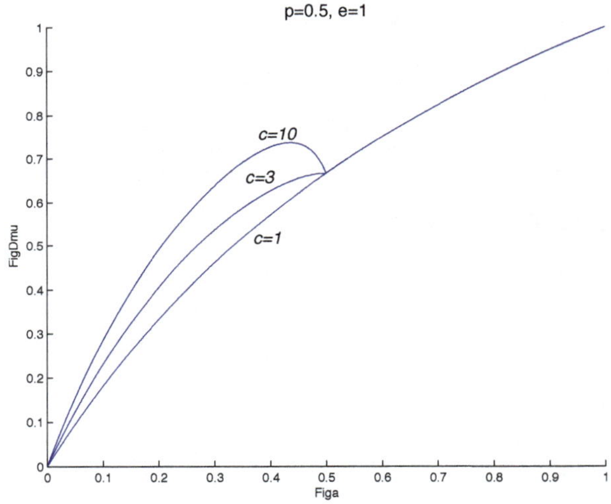

Fig. 7.4. Graph of D_μ for $h(x) = c1_{[0,1/2[}(x) + 1_{[1/2,1]}(x)$

7.4.6 Study when the Support of μ is $\mathbb{R}_+$

In this subsection, we assume that $\mu(dx) = h(x)dx$ is a positive measure whose density h is strictly positive a.e. on $\mathbb{R}_+$. The following theorem gives sufficient conditions on h for the function D_μ to be increasing (i.e. for μ to satisfy $(M\cdot Y)$) and converging to 1 when a tends to $+\infty$.

Theorem 7.14. *We assume that μ admits a density h on $\mathbb{R}_+$ which satisfies (S_0) (see Proposition 7.1, with $l = +\infty$), i.e. h is a strictly positive function of $\mathscr{C}^1$ class on $\mathbb{R}_+$ such that:*

$$\forall c \in]0,1], \quad a \longmapsto \frac{h(a)}{h(ac)} \text{ is a decreasing function on } \mathbb{R}_+.$$

1) *Then, there exists $\rho > 2$ (possibly $+\infty$) such that:*

$$\forall\, c \in]0,1[, \quad \lim_{a\to+\infty} \frac{h(a)}{h(ac)} = c^{\rho}. \tag{7.47}$$

Furthermore:

$$\rho = \lim_{a\to+\infty} \varepsilon(a) = \lim_{a\to+\infty} aV'(a).$$

2) *D_μ is an increasing function which converges towards L with:*

- *if $\rho < +\infty$, then $L = \dfrac{\rho-2}{\rho-1}$,*
- *if $\rho = +\infty$, then $L = 1$.*

In particular, if $\rho = +\infty$, then, there exists a probability measure ν_μ such that:

$$D_\mu(a) = \nu_\mu(]0,a[), \qquad a \geq 0.$$

Remark 7.9. 1) Point 1) of Theorem 7.14 casts a new light on Proposition 7.1. Indeed, from (7.47), we see that h is a regularly varying function in the sense of Karamata, and Proposition 7.1 looks like a version of Karamata's representation Theorem (see [30, Chapter 1, Theorems 1.3.1 and 1.4.1]).

2) The property that the function $a \longmapsto \dfrac{h(a)}{h(ac)}$ is decreasing is not necessary to obtain the existence of a limit for $D_\mu(a)$ when $a \to +\infty$, see [30, Theorem 8.1.4].

Proof. 1) *We first prove Point 1).* We assume that h satisfies (S_0) on $\mathbb{R}_+$. Therefore the decreasing limit $\gamma_c := \lim\limits_{a\to+\infty} \dfrac{h(a)}{h(ac)}$ exists and belongs to $[0,1]$. Then, for all $c,d \in]0,1[$:

$$\gamma_{cd} = \lim_{a\to+\infty} \frac{h(a)}{h(acd)} = \lim_{a\to+\infty} \frac{h(a)}{h(ac)} \frac{h(ac)}{h(acd)} = \gamma_c \gamma_d.$$

This implies that $\gamma_c = c^{\rho}$ with $\rho \in \mathbb{R}_+$. Now, let $\eta(a) = \displaystyle\int_a^{+\infty} yh(y)dy$. For $A > 1$ and $a > 0$, we have

$$\eta(a) = a^2 \int_1^{+\infty} zh(az)dz \geq a^2 \int_1^{A} z \frac{h(az)}{h(z)} h(z)dz$$

$$\geq a^2 \frac{h(aA)}{h(A)} \int_1^A zh(z)dz \tag{7.48}$$

$$\xrightarrow[A\to+\infty]{} a^{2-\rho} \int_1^{+\infty} zh(z)dz.$$

Letting a tend to $+\infty$, we obtain, since $\eta(a) \xrightarrow[a\to+\infty]{} 0$, that necessarily $\rho > 2$. Then, passing to the limit in (7.44), we obtain:

$$c^\rho = \exp\left(-\int_c^1 \frac{\varepsilon(+\infty)}{y} dy\right), \quad \text{i.e. } \varepsilon(+\infty) = \rho.$$

The last equality is a direct consequence of (7.45).

2) *We now prove that D_μ is increasing and converges towards L as $a \to +\infty$.* Let us denote $\overline{\mu}(a) = \displaystyle\int_a^{+\infty} h(y)dy$. Then:

$$\frac{1}{D_\mu(a)} = \frac{[-y\overline{\mu}(y)]_a^{+\infty} + \int_a^{+\infty} \overline{\mu}(y)dy}{a\overline{\mu}(a)} = 1 + \int_1^{+\infty} \frac{\overline{\mu}(ax)}{\overline{\mu}(a)} dx. \tag{7.49}$$

Now, the proof of the increase of D_μ is exactly the same as that of the implication $(S_0) \Longrightarrow (M\cdot Y)$ (see Theorem 7.13). Then, D_μ being bounded by 1, it converges towards a limit L, and it remains to identify L. We write, for $x > 1$:

$$\frac{\overline{\mu}(a)}{\overline{\mu}(ax)} = \frac{\displaystyle\int_a^{+\infty} h(y)dy}{\displaystyle\int_{ax}^{+\infty} h(y)dy} = \frac{\displaystyle\int_1^{+\infty} h(au)du}{\displaystyle\int_x^{+\infty} h(au)du}$$

$$= \frac{\displaystyle\int_1^{x} h(au)du}{\displaystyle\int_x^{+\infty} h(au)du} + 1$$

$$= \frac{\displaystyle\int_1^{x} \frac{h(ax\frac{u}{x})}{h(ax)} du}{\displaystyle\int_x^{+\infty} \frac{h(au)}{h\left(au\frac{x}{u}\right)} du} + 1 \xrightarrow[a\to+\infty]{} \frac{\displaystyle\int_1^{x} \left(\frac{x}{u}\right)^\rho du}{\displaystyle\int_x^{+\infty} \left(\frac{x}{u}\right)^\rho du} + 1$$

from (7.47). Now, we must discuss different cases:

- if $\rho = +\infty$, then $\displaystyle\lim_{a\to+\infty} \frac{\overline{\mu}(a)}{\overline{\mu}(ax)} = +\infty$, and plugging this limit into (7.49), we obtain $L = 1$,
- if $\rho < +\infty$, we obtain:

$$\lim_{a\to+\infty} \frac{\overline{\mu}(a)}{\overline{\mu}(ax)} = \frac{1}{x^{1-\rho}}.$$

Plugging this into (7.49), we obtain:

$$\frac{1}{L} = 1 + \int_1^{+\infty} \frac{dx}{x^{\rho-1}} = 1 - \frac{1}{2-\rho} = \frac{1-\rho}{2-\rho}.$$

□

Remark 7.10. More generally, for $p \geq 1$, there is the equivalence:

$$\int^{+\infty} y^p h(y)dy < \infty \Longleftrightarrow \rho > p+1.$$

The implication $\Longrightarrow$ can be proven by computations similar to (7.48). Conversely, since $\varepsilon(y)$ tends to ρ when y tends to $+\infty$, there exists $A > 0$ and $\theta > 0$ such that: $\forall y \geq A$, $\varepsilon(y) \geq p+1+\theta$. Then applying Proposition 7.1, we obtain:

$$\begin{aligned} h(a) = h(A)\exp\left(-\int_A^a \frac{\varepsilon(y)}{y}dy\right) &\leq h(A)\exp\left(-(p+1+\theta)\int_A^a \frac{dy}{y}\right) \\ &= h(A)\left(\frac{A}{a}\right)^{p+1+\theta}. \end{aligned}$$

We note in particular that μ admits moments of all orders if and only if $\rho = +\infty$.

Exercise 7.12 (A martingale associated to $\left(\frac{e^{tX}}{\mathbb{E}[e^{tX}]}, t \geq 0\right)$). Let X be a r.v. with density f, which we assume to be strictly positive on $\mathbb{R}$, and such that, for every $t \geq 0$, $\mathbb{E}[|X|e^{tX}] < \infty$. Let $h : \mathbb{R}_+ \longmapsto \mathbb{R}$ be the $\mathscr{C}^1$-class function defined by $h(t) := \log(\mathbb{E}[e^{tX}])$. We know from Theorem 1.7 that $(A_t := e^{tX-h(t)}, t \geq 0)$ is a peacock. We shall now associate a martingale to this peacock thanks to Azéma-Yor SE.

1) Let ψ_t be the Hardy-Littlewood function associated to the r.v. $A_t - 1$:

$$\psi_t(x) = x + \frac{\mathbb{E}[(A_t - 1 - x)^+]}{\mathbb{P}(A_t - 1 \geq x)}.$$

i) Prove that, with $z = 1 + x$:

$$\begin{aligned} &(\mathbb{P}(A_t \geq z))^2 \frac{\partial \psi_t}{\partial t}(z) \\ &= \left(\int_{\frac{\log z + h(t)}{t}}^{+\infty} (y - h'(t))e^{ty-h(t)} f(y)dy\right)\left(\int_{\frac{\log z + h(t)}{t}}^{+\infty} f(y)dy\right) \\ &+ \left(\int_{\frac{\log z + h(t)}{t}}^{+\infty} (e^{ty-h(t)} - z)f(y)dy\right)\left(-\frac{\log z + h(t)}{t^2} + \frac{h'(t)}{t}\right) f\left(\frac{\log z + h(t)}{t}\right), \end{aligned}$$

or, with $a := \frac{\log z + h(t)}{t}$:

$$\begin{aligned} (\mathbb{P}(A_t \geq z))^2 \frac{\partial \psi_t}{\partial t}(z) = &\left(\int_a^{+\infty} (y - h'(t))e^{ty-h(t)} f(y)dy\right)\left(\int_a^{+\infty} f(y)dy\right) \\ &+ \left(\int_a^{+\infty} (e^{ty-h(t)} - e^{ta-h(t)})f(y)dy\right)\left(-\frac{a}{t} + \frac{h'(t)}{t}\right) f(a). \end{aligned}$$

ii) Prove that

$$\frac{1}{t}\int_a^{+\infty}(e^{ty-h(t)}-e^{ta-h(t)})f(y)dy=\int_a^{+\infty}e^{tu-h(t)}du\int_u^{+\infty}f(y)dy$$

and deduce that:

$$\begin{aligned}
&(\mathbb{P}(A_t\geq z))^2\frac{\partial\psi_t}{\partial t}(z)\\
&=f(a)\int_a^{+\infty}(y-h'(t))e^{ty-h(t)}f(y)\left(\frac{\int_a^{+\infty}f(u)du}{f(a)}-\frac{\int_y^{+\infty}f(u)du}{f(y)}\right)dy\\
&\quad+\int_a^{+\infty}dye^{ty-h(t)}f(a)(y-a)\int_y^{+\infty}f(u)du\\
&\geq f(a)\int_a^{+\infty}(y-h'(t))e^{ty-h(t)}f(y)\left(\frac{\int_a^{+\infty}f(u)du}{f(a)}-\frac{\int_y^{+\infty}f(u)du}{f(y)}\right)dy.
\end{aligned}$$

2) We now assume that $x\longmapsto\int_x^{+\infty}f(u)du$ is log-concave (cf. Definition 1.14)

i) Deduce that $y\longmapsto-\dfrac{\int_y^{+\infty}f(u)du}{f(y)}$ is increasing and that $\dfrac{\partial\psi_t}{\partial t}(x)\geq 0$.

(*Hint*: observe that $\mathbb{E}[(X-h'(t))e^{tX-h(t)}]=0$; let $\theta(y):=\left(\frac{\int_a^{+\infty}f(u)du}{f(a)}-\frac{\int_y^{+\infty}f(u)du}{f(y)}\right)$ $1_{\{y\geq a\}}$, verify that θ is positive and increasing, and apply Lemma 1.1.)

ii) With the notation of Subsection 7.4.1, let

$$T_t:=\inf\{u\geq 0;\ S_u\geq\psi_t(B_u)\}.$$

Prove that $(B_{T_t},t\geq 0)$ is a martingale associated to $(A_t-1,t\geq 0)$.

3) *i*) Prove that the hypothesis f is log-concave implies that $x\longmapsto\int_x^{+\infty}f(u)du$ is log-concave.

(*Hint*: observe that the two-variables function $(x,y)\longmapsto f(y)1_{\{y\geq x\}}$ is log-concave as soon as f is log-concave, and apply Prékopa's Theorem, which is recalled just before Lemma 7.6 below.)

ii) Prove that one can replace the hypothesis, f strictly positive on $\mathbb{R}$ by, f strictly positive on an interval $]\alpha,\beta[$ and $f\equiv 0$ on $]\alpha,\beta[^c$.

iii) Compare the results obtained in this exercise with Theorem 6.10.

Exercise 7.13 (A martingale associated to $(\varphi(tX),t\geq 0)$). 1) Let $\varphi:\mathbb{R}\longmapsto\mathbb{R}$ be an increasing function of $\mathscr{C}^1$ class, and let X be a r.v. We assume that, for every $t\geq 0$,

$$\sup_{0\leq s\leq t}\mathbb{E}[|X|\varphi'(sX)]<\infty\quad\text{and}\quad\mathbb{E}[X\varphi'(tX)]=0.$$

i) Prove that $(\varphi(tX),t\geq 0)$ is a peacock.

ii) Assume that $x \longmapsto \mathbb{P}(X \geq x)$ is a log-concave function on $\mathbb{R}$ (cf. Definition 1.14). Prove that X admits a density f which is strictly positive on an interval $]\alpha, \beta[$, and vanishes on $]\alpha, \beta[^c$ (the law of X may possibly have an atom at β, but this has no consequence on the sequel of this exercise). Using then the same method as in Exercise 7.12, show that the Azéma-Yor SE allows to construct a martingale associated to the peacock $(\varphi(tX), t \geq 0)$. Compare with Exercise 6.5.

2) Let now X be a centered r.v.

i) Deduce from the preceding that the condition

$$x \longmapsto \mathbb{P}(X \geq x) \ \text{ is log-concave on } \mathbb{R}_+$$

implies the $(M{\cdot}Y)$ condition. Compare with Condition (S_2) in Theorem 7.13.

ii) Suppose that X admits a density which is a function of $\mathscr{C}^1$ class, strictly positive, decreasing and log-concave on $\mathbb{R}_+$. Use then Theorem 7.14 to give another proof that the $(M{\cdot}Y)$ condition is satisfied.

Exercise 7.14 (Bertoin-Le Jan method for SE). Here, we shall use the Bertoin-Le Jan SE to associate a martingale to $(\sqrt{t}X, t \geq 0)$. We start by recalling this method.

1) Let $(B_t, t \geq 0)$ be a Brownian motion started from 0. We denote by $(L_u^x; u \geq 0, x \in \mathbb{R})$ its bicontinuous family of local times. Let X be a r.v. with law μ such that:

$$\mathbb{E}[|X|] = \int_{-\infty}^{+\infty} |x|\mu(dx) < +\infty \quad \text{and} \quad \mathbb{E}[X] = \int_{-\infty}^{+\infty} x\mu(dx) = 0.$$

We assume moreover that the support of μ is $\mathbb{R}$. We define:

$$\gamma := 2\int_0^{+\infty} y\mu(dy) = -2\int_{-\infty}^{0} y\mu(dy) \tag{7.50}$$

and $\rho : \mathbb{R} \longrightarrow \mathbb{R}_+$ by:

$$\rho(x) := \begin{cases} 2\displaystyle\int_x^{+\infty} (a-x)\mu(da) & \text{if } x \geq 0 \\ 2\displaystyle\int_{-\infty}^{x} (x-a)\mu(da) & \text{if } x \leq 0. \end{cases} \tag{7.51}$$

Let $T^{(\mu)}$ be the stopping time defined by:

$$T^{(\mu)} := \inf\left\{u \geq 0; \gamma \int_{-\infty}^{+\infty} \frac{L_u^x}{\rho(x)}\mu(dx) > L_u^0\right\}. \tag{7.52}$$

Theorem 7.15 (Bertoin-Le Jan [26]). *Under the preceding hypotheses:*

1) *The law of $B_{T^{(\mu)}}$ is μ, i.e.:*

$$B_{T^{(\mu)}} \overset{(law)}{=} X.$$

2) *$(B_{u \wedge T^{(\mu)}}, u \geq 0)$ is a uniformly integrable martingale.*

2) Let μ_t be the law of $\sqrt{t}X$. To μ_t, we attach $\gamma(t)$, $\rho_t(x)$ and $T^{(\mu_t)}$ respectively given by (7.50), (7.51) and (7.52), upon replacing μ by μ_t.

i) Prove that $\gamma(t) = \sqrt{t}\gamma$ and $\rho_t(x) = \sqrt{t}\rho\left(\frac{x}{\sqrt{t}}\right)$.

ii) Let $A(t,u) := \gamma(t)\int_{-\infty}^{+\infty} \frac{L_u^x}{\rho_t(x)}\mu_t(dx)$. Prove that if for every u, the function $t \longmapsto A(t,u)$ is decreasing, then the family of stopping times $(T^{(\mu_t)}, t \geq 0)$ is a.s. increasing. Deduce that under this hypothesis, $(B_{T^{(\mu_t)}}, t \geq 0)$ is a martingale associated to $(\sqrt{t}X, t \geq 0)$.

iii) We now assume that μ admits a density h: $\mu(dx) = h(x)dx$. We may assume that $h(x) > 0$ for every $x \in \mathbb{R}$. Prove that

$$A(t,u) = \gamma \int_{-\infty}^{+\infty} \frac{|\frac{x}{\sqrt{t}}|h\left(\frac{x}{\sqrt{t}}\right)}{|x|\rho\left(\frac{x}{\sqrt{t}}\right)} L_u^x\, dx.$$

Let $F : \mathbb{R} \longrightarrow \mathbb{R}$ defined by: $F(z) = \frac{zh(z)}{\rho(z)}$. Prove that if F is increasing on $\mathbb{R}$, then, for every u, $t \longmapsto A(t,u)$ is decreasing.

iv) We now assume that h is such that:

$$\begin{gathered}\text{the function } y \longmapsto \log(|y|h(y)) \text{ is concave}\\ \text{on each interval }]0,+\infty[\text{ and }]-\infty,0[.\end{gathered} \tag{LC}$$

Prove that under hypothesis (LC), F is increasing.
(*Hint*: prove that, for $y > 0$:

$$\frac{1}{2}\rho^2(y)F'(y) = \int_y^{+\infty} da \int_a^{+\infty} yh(y)h(z)\left(\frac{1}{y} + \frac{h'(y)}{h(y)} - \frac{h'(z)}{h(z)}\right) dz,$$

where the derivatives denote right derivatives (which are well defined according to (LC)). Deduce then from (LC) that for $0 < y \leq z$:

$$\frac{1}{y} + \frac{h'(y)}{h(y)} \geq \frac{1}{z} + \frac{h'(z)}{h(z)} \geq \frac{h'(z)}{h(z)}.$$

Use a similar argument for $y < 0$.)

v) Prove that under (LC), $(M_t := B_{T^{(\mu_t)}}, t \geq 0)$ is a martingale which is associated to $(\sqrt{t}X, t \geq 0)$, and that this martingale enjoys the Brownian scaling property:

$$\forall c > 0, \quad (\sqrt{c}M_t, t \geq 0) \overset{(law)}{=} (M_{ct}, t \geq 0).$$

(*Hint*: use the scaling property:

$$\forall c > 0, \quad ((\sqrt{c}B_t, \sqrt{c}L_t^x), t \geq 0) \overset{(law)}{=} ((B_{ct}, L_{ct}^{x\sqrt{c}}), t \geq 0).)$$

3) We now assume that the law of X is symmetric, i.e. $h(x) = h(-x)$. Let $0 < \alpha \leq 1$ and $\psi^{(\alpha)}(x) := (\mathrm{sgn}(x))|x|^\alpha$.

i) Prove that $\left(X_t^{(\alpha)} := \sqrt{t}\psi^{(\alpha)}(X), t \geq 0\right)$ is a peacock.

ii) Let $\mu_t^{(\alpha)}$ denote the law of $\sqrt{t}\psi^{(\alpha)}(X)$ and define, for $z > 0$:

$$F^{(\alpha)}(z) := \frac{zh(z)}{\int_z^{+\infty} \left(\psi^{(\alpha)}\right)'(a)\overline{\mu}(a)da}$$

where $\overline{\mu}(a) := \displaystyle\int_a^{+\infty} h(x)dx$. Prove that if $F^{(\alpha)}$ is increasing on $]0, +\infty[$, then the family of stopping times $(T^{(\mu_t^{(\alpha)})}, t \geq 0)$ is a.s. increasing.

iii) Prove that if moreover μ satisfies (LC), then $F^{(\alpha)}$ is increasing on $]0, +\infty[$. (*Hint*: write $F^{(\alpha)}(z) = \dfrac{zh(z)}{\int_z^{+\infty}\overline{\mu}(a)da} \dfrac{\int_z^{+\infty}\overline{\mu}(a)da}{\int_z^{+\infty}\left(\psi^{(\alpha)}\right)'(a)\overline{\mu}(a)da}$).

iv) Deduce that if μ is symmetric and satisfies (LC), then $(M_t^{(\alpha)} := B_{T^{(\mu_t^{(\alpha)})}}, t \geq 0)$ is a martingale associated to $(X_t^{(\alpha)}, t \geq 0)$ which enjoys the Brownian scaling property.

Let us mention that Vallois [167] has proposed another SE method involving local times, but its application to the construction of martingales does not seem easy.

7.5 Bass' Method

We denote by $\mathbf{I}_0$ the set of odd increasing functions $\varphi : \mathbb{R} \longrightarrow \mathbb{R}$. In this section, we shall:

- describe briefly Bass' method for SE,
- use this method to associate a martingale to $(\varphi(B_t), t \geq 0)$, with $\varphi \in \mathbf{I}_0$,
- use this method, once again, to associate a martingale to the process $(\sqrt{t}X, t \geq 0)$, when X is an integrable and centered r.v. satisfying some extra hypotheses.

Finally, we shall use Bougerol's identity to associate a martingale to the peacock $(\sinh(B_t), t \geq 0)$.

7.5.1 Bass' Method for SE

Throughout this section, G shall always denote a reduced gaussian r.v.

7.5.1.1 Notation

Let μ be a probability measure on $\mathbb{R}$ such that:

$$\int_{-\infty}^{+\infty} |x|\mu(dx) < \infty \quad \text{and} \quad \int_{-\infty}^{+\infty} x\mu(dx) = 0. \tag{7.53}$$

Then, there exists an increasing function $\theta : \mathbb{R} \longrightarrow \mathbb{R}$ such that the r.v. $\theta(G)$ has law μ. Indeed, let:

$$F(x) := \mu(]-\infty, x[), \qquad F^{-1}(y) := \inf\{x; F(x) \geq y\}$$

and define

$$\mathscr{N}(x) := \frac{1}{\sqrt{2\pi}} \int_{-\infty}^{x} e^{-\frac{y^2}{2}} dy.$$

Now, if we define θ by:

$$\theta(x) := F^{-1} \circ \mathscr{N}(x), \qquad x \in \mathbb{R} \tag{7.54}$$

then, the law of $\theta(G)$ is μ. If μ differs from the Dirac measure at 0, then θ is not identically constant. Moreover, from (7.53):

$$\frac{1}{\sqrt{2\pi}} \int_{-\infty}^{+\infty} |\theta(x)| e^{-\frac{x^2}{2}} dx < \infty \quad \text{and} \quad \frac{1}{\sqrt{2\pi}} \int_{-\infty}^{+\infty} \theta(x) e^{-\frac{x^2}{2}} dx = 0. \tag{7.55}$$

Remark 7.11. Conversely, let θ be an increasing, left-continuous function satisfying (7.55). Let μ be the law of $\theta(G)$ and $F(x) := \mu(]-\infty, x[)$. It is not difficult to see that μ satisfies (7.53) and (7.54). Therefore, there exists a bijective correspondence between the probability measures μ satisfying (7.53) and the increasing, left-continuous functions satisfying (7.55). This correspondence is given by:

$$\begin{cases} \mu = \text{ law of } \theta(G) \\ \theta = F^{-1} \circ \mathscr{N}. \end{cases}$$

If θ is only increasing, then denoting by θ_l the left-continuous regularized function of θ, then $\theta(G) \overset{(law)}{=} \theta_l(G)$ since G has a density, and thus $\theta_l = F^{-1} \circ \mathscr{N}$ with F the cumulative distribution function of $\theta(G)$.

We assume that μ differs from the Dirac measure at 0, and we define, for $0 \leq s < 1$:

$$\begin{aligned} b(s,x) &= \frac{1}{\sqrt{2\pi}} \int_{-\infty}^{+\infty} \theta(x + y\sqrt{1-s}) e^{-\frac{y^2}{2}} dy \qquad (7.56) \\ &= Q_{1-s}\theta(x) \end{aligned}$$

where $(Q_u, u \geq 0)$ denotes the heat semi-group. Observe that under hypothesis (7.55), b is a function of $\mathscr{C}^\infty$ class on $]0,1[\times\mathbb{R}$. This easily follows from the identity $b(s,x) = \dfrac{1}{\sqrt{2\pi}} \displaystyle\int_{-\infty}^{+\infty} \theta(z) e^{-\frac{(z-x)^2}{2(1-s)}} \frac{dz}{\sqrt{1-s}}$ and the inequality, for z large enough, $\displaystyle\sup_{|x| \leq x_0} |z-x|^k e^{-\frac{(z-x)^2}{2(1-s)}} \leq e^{-\frac{z^2}{2}}$. We also define, for $0 \leq s < 1$:

$$a(s,x) := \frac{\partial b}{\partial x}(s,x) = \frac{1}{\sqrt{2\pi(1-s)}} \int_{-\infty}^{+\infty} \theta(x+y\sqrt{1-s}) y e^{-\frac{y^2}{2}} dy \tag{7.57}$$

$$= \frac{1}{\sqrt{2\pi}} \int_{-\infty}^{+\infty} \theta'(x+y\sqrt{1-s}) e^{-\frac{y^2}{2}} dy$$

when θ is differentiable.

In the following, we shall denote by a prime the derivative with respect to the spatial variable x, and we shall write (7.57) under the form:

$$a(s,x) = b'(s,x) = \frac{1}{\sqrt{2\pi}} \int_{-\infty}^{+\infty} \theta'(x+y\sqrt{1-s}) e^{-\frac{y^2}{2}} dy$$

$$= \frac{1}{\sqrt{2\pi(1-s)}} \int_{-\infty}^{+\infty} e^{-\frac{(z-x)^2}{2(1-s)}} d\theta(z) \tag{7.58}$$

even when θ is not differentiable. This is a natural idea, since, in that case, the Stieltjes integral is well-defined. We deduce from (7.58) that $b(s,\centerdot)$, for $s \in [0,1[$, is continuous and strictly increasing. Thus, it admits an inverse in the sense of composition of functions that we shall denote by c:

$$b(s,c(s,x)) = c(s,b(s,x)) = x. \tag{7.59}$$

Then, it is not difficult to see that c is uniformly Lipschitz continuous in s, for (s,x) varying in a compact of its domain of definition, and that a is of $\mathcal{C}^\infty$ class on $[0,1[\times\mathbb{R}$ and strictly positive.

7.5.1.2 The Bass Method

Now, let $(B_u, u \geq 0)$ be a Brownian motion started from 0, and consider the ordinary differential equation:

$$\begin{cases} \dfrac{dV_u}{du} & = \dfrac{1}{a^2(V_u, c(V_u, B_u))}, \qquad V_u < 1 \\ V_0 & = 0 \end{cases} . \tag{7.60}$$

Of course, since $a(s,x)$ is only defined for $0 \leq s < 1$, this equation is only defined for $V_u < 1$. It admits, from Subsection 7.5.1.1, a unique solution, which is increasing. Let us define $\tau^{(\mu)}$, the stopping time associated to μ, or equivalently to θ, by:

$$\tau^{(\mu)} := \sup\{u; V_u < 1\}.$$

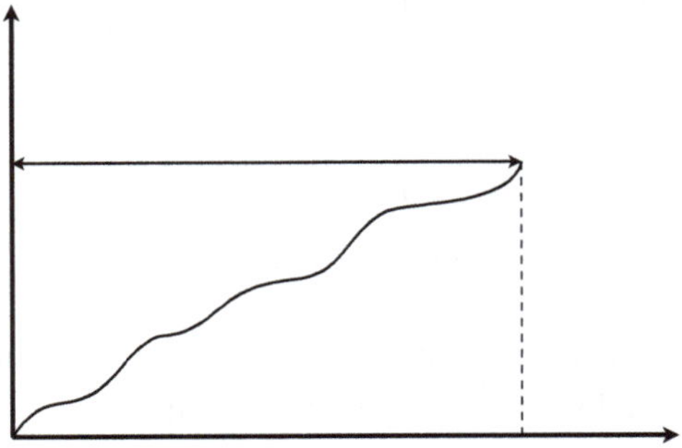

Fig. 7.5. Stopping time $\tau^{(\mu)}$ defined by Bass' method

Here is Bass' result:

Theorem 7.16 (Bass [16]). 1) *The law of* $B_{\tau^{(\mu)}}$ *is* μ.
2) $(B_{u\wedge\tau^{(\mu)}}, u \geq 0)$ *is a uniformly integrable martingale.*

Corollary 7.1. *Let* $(\theta(t,\centerdot), t \geq 0)$ *be a family of functions from* $\mathbb{R}$ *to* $\mathbb{R}$*, which are increasing in x, not identically constant, and such that, for every* $t \geq 0$*:*

$$\mathbb{E}[|\theta(t,G)|] < \infty \quad \textit{and} \quad \mathbb{E}[\theta(t,G)] = 0.$$

Let, for every $t \geq 0$*,* τ_t *be the stopping time associated to* $x \longmapsto \theta(t,x)$ *by Bass' method. We assume furthermore that:*

$$(\tau_t, t \geq 0) \textit{ is an increasing family of stopping times.}$$

Then, $(B_{\tau_t}, t \geq 0)$ *is a martingale such that:*

$$B_{\tau_t} \overset{(1.d)}{=} \theta(t,G).$$

7.5.2 A Martingale Associated to $(\varphi(B_t), t \geq 0)$ for φ Odd and Increasing

Let $\varphi : \mathbb{R} \longrightarrow \mathbb{R}$ be an increasing, odd function such that, for every $t \geq 0, \mathbb{E}[|\varphi(B_t)|] < \infty$. We know, from Theorem 1.3 (see also Exercise 1.23) that $(\varphi(B_t), t \geq 0)$ is a peacock. We shall now, φ being given, exhibit a martingale associated to the peacock $(\varphi(B_t), t \geq 0)$. Let $\theta(t,x) := \varphi(\sqrt{t}x)$. It is clear that:

$$\mathbb{E}[|\theta(t,G)|] < \infty \quad \text{and} \quad \mathbb{E}[\theta(t,G)] = 0.$$

Therefore, from Theorem 7.16, for every fixed $t \geq 0$, we can associate to the function $\theta(t,\centerdot)$ a stopping time τ_t from Bass' method.

Theorem 7.17. *Under the preceding hypotheses, the family of stopping times* $(\tau_t, t \geq 0)$ *is a.s. increasing. Therefore,* $(B_{\tau_t}, t \geq 0)$ *is a martingale associated to* $(\varphi(B_t), t \geq 0)$, *i.e.*

$$B_{\tau_t} \overset{(1.d)}{=} \varphi\left(\sqrt{t}G\right) \overset{(1.d)}{=} \varphi(B_t)$$

Proof. We adopt the following notation. For $u > 0$ and $x \in \mathbb{R}$, we set:

$$\begin{aligned}\beta(u,x) &:= \frac{1}{\sqrt{2\pi u}} \int_{-\infty}^{+\infty} \varphi(y) \exp\left(-\frac{(y-x)^2}{2u}\right) dy \\ &= \frac{1}{\sqrt{2\pi}} \int_{-\infty}^{+\infty} \varphi(x+\sqrt{u}z) e^{-\frac{z^2}{2}} dz,\end{aligned}$$

and

$$\alpha(u,x) := \beta'(u,x) = \frac{1}{\sqrt{2\pi u}} \int_{-\infty}^{+\infty} \exp\left(-\frac{(y-x)^2}{2u}\right) d\varphi(y).$$

Since φ is not identically equal to 0 (if $\varphi \equiv 0$, Theorem 7.17 is trivial), $\alpha(u,x) > 0$. We then denote by $\gamma(u,\centerdot)$ the inverse function of $\beta(u,\centerdot)$, i.e. $\beta(u, \gamma(u,x)) = x$.

Let us start with a preliminary lemma.

Lemma 7.5. *Let* φ *be an odd, increasing function such that, for every* $t \geq 0$, $\mathbb{E}[|\varphi(B_t)|] < \infty$. *Then, for every* $u > 0$ *and every* $x \in \mathbb{R}$:

$$\left(\beta'(u,x)\right)^2 \geq u\left(\left(\beta''(u,x)\right)^2 - \beta'(u,x)\beta'''(u,x)\right) \tag{7.61}$$

where the derivatives are taken with respect to x.

Proof. Assume first that φ is of $\mathscr{C}^2$ class. Relation (7.61) is then equivalent to:

$$\begin{aligned}\left(\int_{-\infty}^{+\infty} \varphi'(x+\sqrt{u}z) e^{-\frac{z^2}{2}} dz\right)^2 &\geq \left(\int_{-\infty}^{+\infty} \varphi'(x+\sqrt{u}z) z e^{-\frac{z^2}{2}} dz\right)^2 \\ &- \left(\int_{-\infty}^{+\infty} \varphi(x+\sqrt{u}z) z e^{-\frac{z^2}{2}} dz\right)\left(\int_{-\infty}^{+\infty} \varphi''(x+\sqrt{u}z) z e^{-\frac{z^2}{2}} dz\right).\end{aligned} \tag{7.62}$$

We have, since $\varphi' \geq 0$:

$$\begin{aligned}0 &\leq \int_{-\infty}^{+\infty} \varphi'(x+\sqrt{u}z) e^{-\frac{z^2}{2}} (\lambda z - 1)^2 dz \\ &= \lambda^2 \int_{-\infty}^{+\infty} \varphi'(x+\sqrt{u}z) z^2 e^{-\frac{z^2}{2}} dz - 2\lambda \int_{-\infty}^{+\infty} \varphi'(x+\sqrt{u}z) z e^{-\frac{z^2}{2}} dz \\ &\quad + \int_{-\infty}^{+\infty} \varphi'(x+\sqrt{u}z) e^{-\frac{z^2}{2}} dz.\end{aligned}$$

Since this last quantity is quadratic polynomial in λ and positive, we deduce that:

$$\begin{aligned}
&\left(\int_{-\infty}^{+\infty} \varphi'(x+\sqrt{u}z) z e^{-\frac{z^2}{2}} dz\right)^2 \\
&\quad \le \left(\int_{-\infty}^{+\infty} \varphi'(x+\sqrt{u}z) z^2 e^{-\frac{z^2}{2}} dz\right)\left(\int_{-\infty}^{+\infty} \varphi'(x+\sqrt{u}z) e^{-\frac{z^2}{2}} dz\right) \\
&\quad = \left(\int_{-\infty}^{+\infty} \varphi'(x+\sqrt{u}z) e^{-\frac{z^2}{2}} dz\right)\left(\int_{-\infty}^{+\infty} \left(\varphi'(x+\sqrt{u}z)+\sqrt{u}z\varphi''(x+\sqrt{u}z)\right) e^{-\frac{z^2}{2}} dz\right) \\
&\quad = \left(\int_{-\infty}^{+\infty} \varphi'(x+\sqrt{u}z) e^{-\frac{z^2}{2}} dz\right)^2 \\
&\qquad + \left(\int_{-\infty}^{+\infty} \varphi'(x+\sqrt{u}z) e^{-\frac{z^2}{2}} dz\right)\left(\int_{-\infty}^{+\infty} \sqrt{u}\varphi''(x+\sqrt{u}z) z e^{-\frac{z^2}{2}} dz\right).
\end{aligned}$$

It only remains to notice, integrating by part, that:

$$\int_{-\infty}^{+\infty} \varphi'(x+\sqrt{u}z) e^{-\frac{z^2}{2}} dz = \frac{1}{\sqrt{u}} \int_{-\infty}^{+\infty} \varphi(x+\sqrt{u}z) z e^{-\frac{z^2}{2}} dz$$

to obtain (7.62). It is then easy to remove the hypothesis of $\mathscr{C}^2$ class of φ by integrating by part (7.62), in order to remove the terms φ' and φ'', so that only φ appears in the inequality. □

End of the proof of Theorem 7.17. 1) Taking back the notation of Subsection 7.5.1, and replacing θ by $\varphi(\sqrt{t}\,\bullet)$ in (7.56) and (7.57), we have:

$$\begin{aligned}
b(t,s,x) &:= \beta(t(1-s), \sqrt{t}x), \\
a(t,s,x) &:= \sqrt{t}\alpha\left(t(1-s), \sqrt{t}x\right), \\
c(t,s,x) &:= \frac{1}{\sqrt{t}}\gamma(t(1-s), x).
\end{aligned} \tag{7.63}$$

2) Let $V^{(t)}$ be the solution of the ordinary differential equation:

$$\begin{cases} \dfrac{dV_u^{(t)}}{du} &= \dfrac{1}{a^2(t, V_u^{(t)}, c(t, V_u^{(t)}, B_u))}, \quad V_u^{(t)} < 1 \\ V_0^{(t)} &= 0 \end{cases}. \tag{7.64}$$

If we prove that, for every $(s,x) \in [0,1[\times\mathbb{R}$, $t \longmapsto \dfrac{1}{a^2(t,s,c(t,s,x))}$ is a decreasing function of t, then the comparison theorem for ODE will imply that, for $t_1 \le t_2$, $V_u^{(t_1)} \ge V_u^{(t_2)}$, hence $\tau_{t_1} \le \tau_{t_2}$, which is the desired result.

3) *Let us prove that* $t \longmapsto a^2(t,s,c(t,s,x))$ *is increasing.* We have, from (7.63), setting $v = t(s-1)$ and $z = \gamma(v,x)$:

$$2\sqrt{t}\frac{\partial}{\partial t}a(t,s,c(t,s,x)) = \alpha(v,z) + 2v\left(\frac{\partial}{\partial v}\alpha(v,z) + \alpha'(v,z)\frac{\partial}{\partial v}\gamma(v,x)\right).$$

But, from elementary properties of the heat semi-group:

$$\frac{\partial}{\partial v}\alpha(v,z) = \frac{1}{2}\beta'''(v,z) \quad \text{and} \quad \frac{\partial}{\partial v}\gamma(v,z) = -\frac{1}{\beta'(v,z)}\frac{\partial}{\partial v}\beta(v,z) = -\frac{1}{2}\frac{\beta''}{\beta'}(v,z).$$

Therefore,

$$2\sqrt{t}\frac{\partial}{\partial t}a(t,s,c(t,s,x)) = \frac{1}{\beta'}\left((\beta')^2 - v\left((\beta'')^2 - \beta'\beta'''\right)\right) \geq 0$$

from Lemma 7.5. □

Remark 7.12. Let $\varphi : \mathbb{R} \longrightarrow \mathbb{R}$ be an increasing, odd function such that, for every $t \geq 0$, $\mathbb{E}[|\varphi(B_t)|] < \infty$. Then, for every $t \geq 0$, $\mathbb{E}[\varphi(B_t)] = 0$ since φ is odd. Conversely, let $\varphi : \mathbb{R} \longrightarrow \mathbb{R}$ be an increasing function such that for every $t \geq 0$, $\mathbb{E}[|\varphi(B_t)|] < \infty$ and $\mathbb{E}[\varphi(B_t)] = 0$. Then, $\varphi(-x) = -\varphi(-x)$ a.e. If we assume that φ is left-continuous on $]0,+\infty[$ and right-continuous on $]-\infty,0[$ (this can always be assumed without changing the law of $\varphi(B_t)$, see Remark 7.11), then φ is odd.

Proof. Since $\mathbb{E}[\varphi(B_0)] = 0$, $\varphi(0) = 0$, and since φ is increasing, φ is positive for $x \geq 0$ and negative for $x \leq 0$. Furthermore, we have, for every $t \geq 0$:

$$\frac{1}{\sqrt{2\pi t}}\int_{-\infty}^{+\infty} \varphi(x)e^{-\frac{x^2}{2t}}\,dx = 0,$$

hence, for every $\lambda > 0$:

$$\begin{aligned}
0 &= \int_0^{+\infty} e^{-\lambda t}\frac{dt}{\sqrt{2\pi t}}\int_{-\infty}^{+\infty}\varphi(x)e^{-\frac{x^2}{2t}}\,dx \\
&= \int_{-\infty}^{+\infty}\varphi(x)dx\int_0^{+\infty} e^{-\lambda t-\frac{x^2}{2t}}\frac{dt}{\sqrt{2\pi t}} \\
&= \frac{1}{\sqrt{4\pi\lambda}}\int_{-\infty}^{+\infty}\varphi(x)e^{-|x|\sqrt{2\lambda}}dx \quad \text{(cf. Lebedev [117, p. 118]).}
\end{aligned}$$

Therefore:

$$\int_0^{+\infty} e^{-x\sqrt{2\lambda}}\varphi(x)dx = \int_0^{+\infty} e^{-x\sqrt{2\lambda}}\left(-\varphi(-x)\right)dx.$$

Observe now that both functions $x \longmapsto \varphi(x)$ and $x \longmapsto -\varphi(-x)$ are defined on $\mathbb{R}_+$, positive, and have the same Laplace transform. Consequently, they are a.e. equal: $\varphi(x) = -\varphi(-x)$ a.e. Hence, if we assume that φ is left-continuous on $]0,+\infty[$ and right-continuous on $]-\infty,0[$, φ is an odd function. □

Example 7.7. Let $\varphi(x) = \sinh(\lambda x)$ ($\lambda \geq 0$). φ is an increasing, odd function which satisfies $\mathbb{E}[|\sinh(\lambda B_t)|] < \infty$ for every $t \geq 0$. Therefore, Theorem 7.17 makes it possible to associate to $(\sinh(\lambda B_t), t \geq 0)$ a martingale $(B_{\tau_t}, t \geq 0)$. In this situation, all the formulae can be written explicitly. Indeed, with the notation of the proof of Theorem 7.17, we have:

$$\begin{aligned} b(t,s,x) &= \frac{1}{\sqrt{2\pi}} \int_{-\infty}^{+\infty} \sinh\left(\lambda x\sqrt{t} + \lambda y\sqrt{t(1-s)}\right) e^{-\frac{y^2}{2}} dy \\ &= \sinh(\lambda x\sqrt{t}) \exp\left(\frac{\lambda^2 t(1-s)}{2}\right), \\ c(t,s,x) &= \frac{1}{\lambda\sqrt{t}} \operatorname{Argsh}\left(x \exp\left(\frac{\lambda^2 t(s-1)}{2}\right)\right), \end{aligned}$$

hence,

$$a^2(t,s,c(t,s,x)) = \lambda^2 t \left(e^{\lambda^2 t(1-s)} + x^2\right).$$

It is then clear that for every $s \in [0,1[$ and every $x \in \mathbb{R}$, the mapping $t \longmapsto a^2(t,s,c(t,s,x))$ is increasing on $\mathbb{R}_+$.

Remark 7.13. Let φ be a strictly increasing odd function such that, for every $t \geq 0$, $\mathbb{E}[|\varphi(B_t)|] < \infty$, and G be a reduced centered Gaussian r.v. independent from the Brownian motion $(B_u, u \geq 0)$. We define

$$T^{(t)} := \inf\{u \geq 0; |B_u| = |\varphi(\sqrt{t}G)|\}.$$

Then, $(B_{T^{(t)}}, t \geq 0)$ is a martingale associated to $(\varphi(B_t), t \geq 0)$.

7.5.3 Martingale Associated to $(\sqrt{t}X, t \geq 0)$

In the preceding sections, we have associated to $(\sqrt{t}X, t \geq 0)$, for X an integrable and centered r.v., a martingale, thanks to a SE. We shall now show how Bass' method also makes it possible to realize such a construction.

Let μ be a probability measure on $\mathbb{R}$ satisfying (7.53) and φ be an increasing function such that $\varphi(G)$ has law μ (see Subsection 7.5.1). We define, for every $t \geq 0$, $\theta(t,x) := \sqrt{t}\varphi(x)$. Let also, for every $s \in [0,1]$:

$$b(s,x) := \frac{1}{\sqrt{2\pi}} \int_{-\infty}^{+\infty} \varphi(x + y\sqrt{1-s}) e^{-\frac{y^2}{2}} dy, \quad a(s,x) := b'(s,x)$$

and $c(s,\bullet)$ the inverse of $b(s,\bullet)$: $b(s,c(s,x)) = x$.

Likewise, we define:

$$b(t,s,x) := \frac{1}{\sqrt{2\pi}} \int_{-\infty}^{+\infty} \theta(t, x + y\sqrt{1-s}) e^{-\frac{y^2}{2}} dy = \sqrt{t}\, b(s,x),$$

$a(t,s,x) := \sqrt{t}b'(s,x)$, and $c(t,s,.)$ the inverse of $b(t,s,.)$, i.e. $b(t,s,c(t,s,x)) = x$. Clearly:

$$c(t,s,x) = c\left(s, \frac{x}{\sqrt{t}}\right). \tag{7.65}$$

Since $\mathbb{E}[|\theta(t,G)|] = \mathbb{E}[\sqrt{t}|\varphi(G)|] < \infty$ and $\mathbb{E}[\theta(t,G)] = \sqrt{t}\mathbb{E}[\varphi(G)] = 0$, we may, for every $t \geq 0$, define the stopping time τ_t associated to the function $\theta(t,.)$ by Bass' method.

Theorem 7.18. *Assume that, for every* $(s,x) \in]0,1[\times\mathbb{R}$*:*

$$b(s,x)b''(s,x) \leq (b'(s,x))^2. \tag{7.66}$$

Then, the family of stopping times $(\tau_t, t \geq 0)$ *is increasing, and therefore* $(B_{\tau_t}, t \geq 0)$ *is a martingale associated to* $(\sqrt{t}\varphi(G), t \geq 0)$*:*

$$B_{\tau_t} \overset{(1.d)}{=} \sqrt{t}\varphi(G) \sim \mu_t$$

where μ_t *is the law of* $\sqrt{t}\varphi(G)$*.*

Proof. Applying the comparison theorem for ODE, it is sufficient, to prove that, for every $(s,x) \in [0,1[\times\mathbb{R}$, $t \longmapsto \dfrac{1}{a^2(t,s,c(t,s,x))}$ is a decreasing function. Now, since from (7.65),

$$\sqrt{t}b\left(s, c\left(s, \frac{x}{\sqrt{t}}\right)\right) = x$$

we obtain, differentiating with respect to x:

$$a(t,s,c(t,s,x)) = \frac{\sqrt{t}}{c'\left(s, \frac{x}{\sqrt{t}}\right)}.$$

Therefore:

$$\frac{\partial}{\partial t}a(t,s,c(t,s,x)) = \frac{1}{2\sqrt{t}c'\left(s, \frac{x}{\sqrt{t}}\right)} + \frac{xc''\left(s, \frac{x}{\sqrt{t}}\right)}{2t\left(c'\left(s, \frac{x}{\sqrt{t}}\right)\right)^2},$$

and, setting $y = \dfrac{x}{\sqrt{t}}$, the inequality $\dfrac{\partial}{\partial t}a(t,s,c(t,s,x)) \geq 0$ is equivalent to:

$$\Delta := c'(s,y) + yc''(s,y) \geq 0 \tag{7.67}$$

since $c' \geq 0$. On the other hand, differentiating the relation $b(s,c(s,y)) = y$ with respect to y, we obtain successively:

$$b'(s,c(s,y))c'(s,y) = 1, \quad \text{and}$$
$$b''(s,c(s,y))\,(c'(s,y))^2 + b'(s,c(s,y))c''(s,y) = 0.$$

Plugging the values of c' and c'' back into (7.67), we obtain:

$$\Delta = \frac{1}{b'(s,c(s,y))} - \frac{b(s,c(s,y))b''(s,c(s,y))}{(b'(s,c(s,y)))^3},$$

and, since $b' \geq 0$, the condition $\Delta \geq 0$ is equivalent, with $z = c(s,y)$, to

$$b(s,z)b''(s,z) \leq \left(b'(s,z)\right)^2,$$

i.e. to (7.66). □

The drawback of Theorem 7.18 is that condition (7.66) relies on $b(s,x)$, for every $s \in [0,1]$. It would be desirable to obtain a condition relying only on φ, and implying (7.66). This is the purpose of the end of this subsection. Recall that a positive function f defined on a convex D in $\mathbb{R}^n$ is said to be *log-concave* if:

$$\forall x,y \in D,\ \forall \lambda \in]0,1[, \quad f(\lambda x + (1-\lambda)y) \geq (f(x))^{\lambda}\,(f(y))^{1-\lambda}.$$

This condition is equivalent to: the set $S_f := \{x \in D; f(x) > 0\}$ is convex and $\log f$ is a concave function on S_f (see also Chapter 1 Subsection 1.4.2.3 for an equivalent notion in the case $D = \mathbb{R}$, that of PF_2 functions). A classical theorem, due to Prékopa [146], admits, as a particular case, the following result.
Let $f :]0,+\infty[\times]0,+\infty[\longrightarrow \mathbb{R}_+$ be a log-concave function on $]0,+\infty[\times\,]0,+\infty[$ such that, for every $x > 0$:

$$F(x) := \int_0^{+\infty} f(x,y)dy < \infty.$$

Then, F is log-concave on $]0,+\infty[$. This result allows to obtain the following lemma:

Lemma 7.6. *Let $\varphi : \mathbb{R} \longrightarrow \mathbb{R}$ be an odd function which is positive on $]0,+\infty[$. We assume that:*

$$\varphi \text{ is log concave on }]0,+\infty[. \tag{7.68}$$

Then, condition (7.66) is satisfied.

Proof. Since φ is odd, we have, from the definition of b:

$$b(s,x) = \frac{2}{\sqrt{2\pi(1-s)}} \int_0^{+\infty} \exp\left(-\frac{x^2+y^2}{2(1-s)}\right) \sinh\left(\frac{yx}{1-s}\right) \varphi(y)dy,$$

and we must prove (7.66), i.e. that $x \longmapsto b(s,x)$ is log concave. From Prékopa's Theorem applied with the function $f(x,y) = \varphi(y)e^{-\frac{x^2+y^2}{2(1-s)}} \sinh\left(\frac{xy}{1-s}\right)$, it suffices to see that the function $\phi :]0,+\infty[\times]0,+\infty[\longrightarrow \mathbb{R}_+$ defined by:

$$\phi(x,y) = -\frac{x^2}{2} - \frac{y^2}{2} + \log(\sinh(xy))$$

is concave. But,

$$\frac{\partial^2}{\partial x^2}\phi(x,y) = -1 - \frac{y^2}{\sinh^2(yx)}, \quad \frac{\partial^2}{\partial y^2}\phi(x,y) = -1 - \frac{x^2}{\sinh^2(yx)}$$
$$\frac{\partial^2}{\partial x \partial y}\phi(x,y) = \frac{\partial^2}{\partial y \partial x}\phi(x,y) = \frac{\cosh(yx)}{\sinh(yx)} - \frac{xy}{\sinh^2(yx)}.$$

Since $\frac{\partial^2}{\partial x^2}\phi(x,y) < 0$ and $\frac{\partial^2}{\partial y^2}\phi(x,y) < 0$, it suffices to prove that

$$\frac{\partial^2}{\partial x^2}\phi(x,y)\frac{\partial^2}{\partial y^2}\phi(x,y) - \left(\frac{\partial^2}{\partial x \partial y}\phi(x,y)\right)^2 > 0,$$

or,

$$\frac{x^2+y^2}{\sinh^2(yx)} + 1 - \frac{\cosh^2(yx)}{\sinh^2(yx)} + \frac{2yx\cosh(yx)}{\sinh^3(yx)} > 0,$$

a relation which is implied by the inequality $\sinh(u) \leq 2u\cosh(u)$ $(u \geq 0)$. □

Gathering Theorem 7.18 and Lemma 7.6, we have obtained the following result.

Theorem 7.19. *Let μ be a symmetric measure on $\mathbb{R}$ satisfying (7.53), and let φ be an increasing function such that the law of $\varphi(G)$ is μ. We assume that*

$$\varphi \textit{ is log-concave on }]0,+\infty[.$$

Let $(\tau_t, t \geq 0)$ be the family of stopping times associated to $(\sqrt{t}\varphi(G), t \geq 0)$ by Bass' method. Then, $(\tau_t, t \geq 0)$ is a.s. increasing, and $(B_{\tau_t}, t \geq 0)$ is a martingale associated to $(\sqrt{t}\varphi(G), t \geq 0)$:

$$B_{\tau_t} \overset{(1.d)}{=} \sqrt{t}\varphi(G) \sim \mu_t$$

where μ_t is the law of $\sqrt{t}\varphi(G)$.

Proof. It is sufficient to observe that, as μ is a symmetric measure, φ is odd. But, since φ is increasing, it is positive on $]0,+\infty[$, and we can apply Lemma 7.6 and Theorem 7.18. □

Example 7.8. Here are some examples where Theorem 7.19 applies:

1) $\varphi(x) = \lambda x$; then $\tau_t = \lambda^2 t$ and $B_{\tau_t} = B_{\lambda^2 t} \overset{(1.d)}{=} \lambda\sqrt{t}B_1$.
2) $\varphi(x) = \mathrm{sgn}(x)|x|^\alpha$ $(\alpha > 0)$.
3) $\varphi(x) = \sinh(\lambda x)$; the property that φ is log-concave is equivalent to: $\cosh^2(\lambda x) \geq \sinh^2(\lambda x)$.

7.5.4 The Peacock $(\sinh(\lambda B_t), t \geq 0)$ and Bougerol's Identity

Since, for $\lambda \geq 0$, the function $x \longmapsto \sinh(\lambda x)$ is odd, increasing, and satisfies $\mathbb{E}[|\sinh(\lambda B_t)|] < +\infty$ for every $t \geq 0$, Theorem 7.17 (see also Example 7.7) makes it possible to associate, via Bass' method, a martingale to the peacock $(\sinh(\lambda B_t), t \geq 0)$. Bougerol's identity allows to obtain another associated martingale.

Theorem 7.20 (P. Bougerol [36]). *Let $(\beta_s, s \geq 0)$ and $(\gamma_s, s \geq 0)$ be two independent Brownian motions, started at 0. Then:*

$$\sinh(\lambda B_t) \overset{(1.d)}{=} \lambda \int_0^t e^{\lambda \beta_s} d\gamma_s. \tag{7.69}$$

(7.69) is known as Bougerol's identity, and $\left(\lambda \int_0^t e^{\lambda \beta_s} d\gamma_s, t \geq 0\right)$ is a martingale associated to $(\sinh(\lambda B_t), t \geq 0)$.

Proof. We now prove (7.69) by a time reversal argument as developed in Alili-Dufresne-Yor [3]. Indeed, since $(\beta_s, 0 \leq s \leq t) \overset{(law)}{=} (\beta_t - \beta_{t-s}, 0 \leq s \leq t)$, and similarly for γ, we have, for $t \geq 0$ fixed:

$$\begin{aligned}\lambda \int_0^t e^{\lambda \beta_s} d\gamma_s &\overset{(law)}{=} \lambda \int_0^t e^{\lambda(\beta_t - \beta_{t-s})} d_s(\gamma_t - \gamma_{t-s}) \\ &= \lambda e^{\lambda \beta_t} \int_0^t e^{-\lambda \beta_u}(-d\gamma_u)\end{aligned} \tag{7.70}$$

after the change of variable $t - s = u$. We write $\widetilde{\gamma}_u = -\gamma_u$. Then, from Itô's formula:

$$\begin{aligned}Y_t^{(\lambda)} &:= \lambda e^{\lambda \beta_t} \int_0^t e^{-\lambda \beta_u} d\widetilde{\gamma}_u \\ &= \lambda \int_0^t e^{\lambda \beta_u - \lambda \beta_u} d\widetilde{\gamma}_u + \int_0^t Y_s^{(\lambda)} e^{-\lambda \beta_s}(\lambda e^{\lambda \beta_s} d\beta_s + \frac{\lambda^2}{2} e^{\lambda \beta_s} ds) \\ &= \lambda \widetilde{\gamma}_t + \int_0^t Y_s^{(\lambda)} \left(\lambda d\beta_s + \frac{\lambda^2}{2} ds\right).\end{aligned}$$

Hence, the process $(Y_t^{(\lambda)}, t \geq 0)$ is a diffusion with infinitesimal generator $L^{(\lambda)}$ given by:

$$L^{(\lambda)} f(y) = \frac{\lambda^2}{2}(1 + y^2) f''(y) + \frac{\lambda^2}{2} y f'(y).$$

On the other hand, Itô's formula applied to $\left(X_t^{(\lambda)} := \sinh(\lambda B_t), t \geq 0\right)$ yields:

$$X_t^{(\lambda)} = \int_0^t \lambda \sqrt{1 + \left(X_s^{(\lambda)}\right)^2} dB_s + \frac{\lambda^2}{2} \int_0^t X_s^{(\lambda)} ds.$$

Hence, $(X_t^{(\lambda)}, t \geq 0)$ is a diffusion with infinitesimal generator:

$$\widetilde{L}^{(\lambda)} f(x) = \frac{\lambda^2}{2}(1 + x^2) f''(x) + \frac{\lambda^2}{2} x f'(x).$$

Thus, since $Y_0^{(\lambda)} = X_0^{(\lambda)} = 0$, the processes $(X_t^{(\lambda)}, t \geq 0)$ and $(Y_t^{(\lambda)}, t \geq 0)$ have the same law: this implies Bougerol's identity. □

Remark 7.14. Bougerol's identity may be written:

$$\sinh(\lambda B_t) \overset{(1.d)}{=} \lambda \widetilde{\beta}_{A_t^{(\lambda)}}$$

where, on the RHS, $(A_t^{(\lambda)} := \int_0^t e^{2\lambda \beta_s} ds, t \geq 0)$ is an increasing process independent from the Brownian motion $(\widetilde{\beta}_u, u \geq 0)$. Thus, the martingale $(\lambda \widetilde{\beta}_{A_t^{(\lambda)}}, t \geq 0)$ is an Ocone martingale, see Section 7.6 below.

Here is now a generalization of Bougerol's Theorem.

Theorem 7.21 (L. Alili, D. Dufresne, M. Yor [3]). 1) *Let $(\beta_s, s \geq 0)$ and $(\gamma_s, s \geq 0)$ be two independent Brownian motions, started at 0, and μ, ν two reals. We define:*

$$Y_t^{\mu,\nu} := e^{\beta_t + \mu t} \int_0^t e^{-\beta_s - \mu s} d(\gamma_s + \nu s). \tag{7.71}$$

Then $(Y_t^{\mu,\nu}, t \geq 0) \overset{(law)}{=} (\sinh(X_t^{\mu,\nu}), t \geq 0)$ where $(X_t^{\mu,\nu}, t \geq 0)$ is a diffusion started from 0 whose infinitesimal generator is given by:

$$\frac{1}{2} \frac{\partial^2}{\partial x^2} + \left(\mu \tanh(x) + \frac{\nu}{\cosh(x)} \right) \frac{\partial}{\partial x}. \tag{7.72}$$

2) *Let ε be a symmetric Bernoulli r.v. ($\mathbb{P}(\varepsilon = 1) = \mathbb{P}(\varepsilon = -1) = \frac{1}{2}$) independent from the Brownian motion $(B_t, t \geq 0)$. Then:*

i) $(\sinh(B_t + \varepsilon t), t \geq 0)$ *is a peacock.*
ii) There is the identity in law:

$$(\sinh(B_t + \varepsilon t), t \geq 0) \overset{(law)}{=} (\sinh(X_t^{1,0}), t \geq 0). \tag{7.73}$$

3) *The martingale $\left(M_t := \int_0^t e^{\beta_s + s} d\gamma_s, t \geq 0 \right)$ is associated to the peacock $(\sinh(B_t + t\varepsilon), t \geq 0)$:*

$$\sinh(B_t + t\varepsilon) \overset{(1.d)}{=} \int_0^t e^{\beta_s + s} d\gamma_s$$

or

$$\sinh(B_t + t\varepsilon) \overset{(1.d)}{=} \widetilde{\gamma}_{A_t}$$

with $A_t := \int_0^t e^{2(\beta_s + s)} ds$ independent from the Brownian motion $(\widetilde{\gamma}_u, u \geq 0)$.

Proof. 1) Point 1) relies on Itô's formula: the pattern is the same as in the proof of Theorem 7.20.

2) The fact that $(\sinh(B_t+\varepsilon t), t\geq 0)$ is a peacock comes from Corollary 1.1, since the function sinh belongs to **J** (see Theorem 1.3).

3) *We now prove (7.73).* Let $\mathbb{P}$ be the law of $(B_t+\varepsilon t, t\geq 0)$ and W be the Wiener measure. Cameron-Martin's formula yields:

$$\mathbb{P}_{|\mathscr{F}_t} = e^{-\frac{t}{2}}\cosh(B_t)\centerdot W_{|\mathscr{F}_t}.$$

On the other hand, since the law $\mathbb{Q}$ of $(X_t^{1,0}, t\geq 0)$ is that of the solution of:

$$Z_t = B_t + \int_0^t \tanh(Z_s)ds \qquad \text{(from (7.72), with } \mu=1 \text{ and } \nu=0),$$

Girsanov's formula implies:

$$\mathbb{Q}_{|\mathscr{F}_t} = M_t\centerdot W_{|\mathscr{F}_t}$$

with

$$M_t := \exp\left(\int_0^t \tanh(B_s)dB_s - \frac{1}{2}\int_0^t \tanh^2(B_s)ds\right). \tag{7.74}$$

From Itô's formula, we deduce:

$$\log\cosh(B_t) = \int_0^t \tanh(B_s)dB_s + \frac{1}{2}\int_0^t \left(1-\tanh^2(B_s)\right)ds, \tag{7.75}$$

and, plugging (7.75) in (7.74), we obtain:

$$\begin{aligned} M_t &= \exp\left(\log\cosh(B_t) - \frac{1}{2}\int_0^t \left(1-\tanh^2(B_s)+\tanh^2(B_s)\right)ds\right) \\ &= e^{-\frac{t}{2}}\cosh(B_t), \quad \text{hence, } \mathbb{P}=\mathbb{Q}. \end{aligned}$$

4) Point 3) of Theorem 7.21 can be obtained from (7.71) by time reversal, in the same way as in (7.70). □

For every λ real, $\left(e^{-\frac{\lambda^2}{2}t}\sinh(\lambda B_t), t\geq 0\right)$ is a peacock since, a fortiori, it is a martingale. Hence the following natural question arises: for which λ and μ, is $(e^{\mu t}\sinh(\lambda B_t), t\geq 0)$ a peacock? We answer this question in the next Proposition.

Proposition 7.2. *The process $(e^{\mu t}\sinh(\lambda B_t), t\geq 0)$ is a peacock if and only if $\mu\geq -\frac{\lambda^2}{2}$.*

Proof. 1) If $\mu\geq -\frac{\lambda^2}{2}$, we have, for $s<t$:

$$e^{\mu t}\sinh(\lambda B_t) = e^{(\mu+\frac{\lambda^2}{2})t}\left(\sinh(\lambda B_t)e^{-\frac{\lambda^2}{2}t}\right)$$

$$\overset{\text{(c)}}{\geq} e^{(\mu+\frac{\lambda^2}{2})s}\left(\sinh(\lambda B_t)e^{-\frac{\lambda^2}{2}t}\right)$$

(from Proposition 1.4 since $\mathbb{E}\left[\sinh(\lambda B_t)e^{-\frac{\lambda^2}{2}t}\right] = 0$)

$$\overset{\text{(c)}}{\geq} e^{(\mu+\frac{\lambda^2}{2})s}\left(\sinh(\lambda B_s)e^{-\frac{\lambda^2}{2}s}\right)$$

(since $(\sinh(\lambda B_u)e^{-\frac{\lambda^2}{2}u}, u \geq 0)$ is a martingale).

2) From the Itô-Tanaka formula, we have:

$$\mathbb{E}[|\sinh(\lambda B_t)|] = \mathbb{E}[\sinh(\lambda |B_t|)] = e^{\frac{\lambda^2}{2}t}\lambda \int_0^t \frac{ds}{\sqrt{2\pi s}} e^{-\frac{\lambda^2 s}{2}}.$$

Thus:

$$\mathbb{E}[|e^{\mu t}\sinh(\lambda B_t)|] \underset{t\to+\infty}{\sim} \lambda e^{(\frac{\lambda^2}{2}+\mu)t} \int_0^{+\infty} \frac{ds}{\sqrt{2\pi s}} e^{-\frac{\lambda^2 s}{2}} ds$$

and, if $\mu < -\frac{\lambda^2}{2}$,

$$\mathbb{E}[|e^{\mu t}\sinh(\lambda B_t)|] \xrightarrow[t\to+\infty]{} 0. \tag{7.76}$$

But, since $x \longmapsto |x|$ is convex, if $(e^{\mu t}\sinh(\lambda B_t), t \geq 0)$ were a peacock, the LHS of (7.76) would increase in t. Hence, $(e^{\mu t}\sinh(\lambda B_t), t \geq 0)$ is not a peacock for $\mu < -\frac{\lambda^2}{2}$.

□

7.6 The Ocone Martingale Method

7.6.1 Description of the Method

Definition 7.1. A continuous martingale $(M_t, t \geq 0)$ is an Ocone martingale if the Dubins-Schwarz representation of $(M_t, t \geq 0)$:

$$(M_t = B_{A_t}, t \geq 0)$$

writes as follows:

$(B_u, u \geq 0)$ is a Brownian motion started from 0, independent from the increasing process $(A_t \equiv \langle M \rangle_t, t \geq 0)$.

Observe that then:

$$\mathbb{E}[|M_t|] = \mathbb{E}[|B_{A_t}|] = \mathbb{E}\left[\sqrt{A_t}|B_1|\right] = \sqrt{\frac{2}{\pi}}\mathbb{E}\left[\sqrt{A_t}\right] < +\infty.$$

For a number of properties and characterizations of Ocone martingales, see Dubins-Emery-Yor [57] or Chaumont-Vostrikova [45].

Let $\varphi : \mathbb{R} \longrightarrow \mathbb{R}$ be an increasing, odd function such that, for every $t \geq 0$, $(\varphi(B_t), t \geq 0)$ is integrable. We know, from Theorem 1.3, that $(\varphi(B_t), t \geq 0)$ is a peacock and we managed (cf. Theorem 7.17) to associate to $(\varphi(B_t), t \geq 0)$ a martingale via Bass' embedding. Our aim is now to find an Ocone martingale $(M_t, t \geq 0)$ associated to $(\varphi(B_t), t \geq 0)$.

We proceed as follows: assume that we have found a continuous increasing process $(H_t^{(\varphi)}, t \geq 0)$ independent from $(B_u, u \geq 0)$ and such that:

$$\varphi(B_t) \overset{(1.d)}{=} H_t^{(\varphi)} B_1.$$

Assume also that, for every $t \geq 0$, $\mathbb{E}[H_t^{(\varphi)}] < +\infty$. Then: $\left(M_t^{(\varphi)} := B_{(H_t^{(\varphi)})^2}, t \geq 0\right)$ is an Ocone martingale and:

$$M_t^{(\varphi)} \overset{(1.d)}{=} H_t^{(\varphi)} B_1 \overset{(1.d)}{=} \varphi(B_t).$$

Thus, $(M_t^{(\varphi)}, t \geq 0)$ is a martingale associated to the peacock $(\varphi(B_t), t \geq 0)$.

7.6.2 Illustration

Unfortunately, we do not know how to implement this principle for every odd and increasing function φ, but only for $\varphi(x) = x^{[c]}$ with:

$$x^{[c]} := \operatorname{sgn}(x)|x|^c.$$

Theorem 7.22. *Let $c \geq 1$. 1) There exists a positive r.v. H_c such that:*

$$B_1^{[c]} \overset{(law)}{=} H_c \centerdot B_1 \tag{7.77}$$

where, on the RHS of (7.77), H_c and B_1 are independent.

2) Denote $(M_t^{[c]} := H_c \centerdot B_{t^c}, t \geq 0)$ and $(\widetilde{M}_t^{[c]} := B_{H_c^2 t^c}, t \geq 0)$ where H_c is independent from $(B_u, u \geq 0)$. Then, $(M_t^{[c]}, t \geq 0)$ and $(\widetilde{M}_t^{[c]}, t \geq 0)$ are two Ocone martingales associated to $(B_t^{[c]} := sgn(B_t)|B_t|^c, t \geq 0)$, i.e:

$$B_t^{[c]} \overset{(1.d)}{=} M_t^{[c]} \overset{(1.d)}{=} \widetilde{M}_t^{[c]}.$$

Proof. Point 2) is an easy consequence of Point 1). Indeed:

$$\begin{aligned} B_t^{[c]} &\overset{(1.d)}{=} t^{\frac{c}{2}} B_1^{[c]} \overset{(1.d)}{=} t^{\frac{c}{2}} H_c B_1 \\ &\overset{(1.d)}{=} B_{H_c^2 t^c} \overset{(1.d)}{=} H_c B_{t^c}. \end{aligned}$$

We now prove Point 1). *a*) It is easy to show that H_c satisfies (7.77) if and only if, for every $y \geq 0$:

$$\alpha y^{\frac{\alpha-1}{2}} e^{-\frac{y^\alpha}{2}} = \mathbb{E}\left[\frac{1}{H_c} e^{-\frac{y}{2H_c^2}}\right] \tag{7.78}$$

with $\alpha = \dfrac{1}{c}$. Thus, the existence of H_c shall be proven once we know that the LHS of (7.78) is a Laplace transform in the variable y. Note that the corresponding measure, for $c > 1$, has infinite mass since $\mathbb{E}\left[\frac{1}{H_c}\right] = \lim_{y \to 0} \alpha y^{\frac{\alpha-1}{2}} = +\infty$ because $\alpha < 1$.

b) Denote by $(S_u^{(\alpha)}, u \geq 0)$ the standard α-stable subordinator, i.e. the subordinator such that, for every $\lambda, u \geq 0$:

$$\mathbb{E}\left[e^{-\lambda S_u^{(\alpha)}}\right] = \exp(-u\lambda^\alpha). \tag{7.79}$$

The LHS of (7.78) writes:

$$\begin{aligned}
\alpha y^{\frac{\alpha-1}{2}} e^{-\frac{y^\alpha}{2}} &= \frac{\alpha}{(y^\alpha)^\beta} e^{-\frac{y^\alpha}{2}} \quad \text{with } \beta = \frac{c-1}{2} = \frac{1}{2}\left(\frac{1}{\alpha} - 1\right) \\
&= \frac{\alpha}{2^\beta} \frac{1}{\left(\frac{y^\alpha}{2}\right)^\beta} e^{-\frac{y^\alpha}{2}} \\
&= \frac{\alpha}{2^\beta \Gamma(\beta)} \int_0^{+\infty} t^{\beta-1} e^{-(t+1)\frac{y^\alpha}{2}} dt \\
&\quad \text{(from the classical "trick": } \frac{1}{r^\beta} = \frac{1}{\Gamma(\beta)} \int_0^{+\infty} t^{\beta-1} e^{-tr} dt) \\
&= \frac{\alpha}{2^\beta \Gamma(\beta)} \int_1^{+\infty} (v-1)^{\beta-1} e^{-v\frac{y^\alpha}{2}} dv \\
&\quad \text{(after the change of variable } t = v-1) \\
&= \frac{\alpha}{2^\beta \Gamma(\beta)} \int_1^{+\infty} (v-1)^{\beta-1} \mathbb{E}\left[e^{-yS_{\frac{v}{2}}^{(\alpha)}}\right] dv
\end{aligned} \tag{7.80}$$

which implies the existence of H_c, since the RHS of (7.80) is a Laplace transform in the variable y. □

Proposition 7.3 (Properties of H_c). *The law of the positive r.v. H_c is characterized by: for every $f : \mathbb{R}_+ \longrightarrow \mathbb{R}_+$ Borel and bounded:*

$$\mathbb{E}\left[f(2H_c^2)\right] = C_\alpha \mathbb{E}\left[\frac{1}{\sqrt{S_1^{(\alpha)}}} f\left(\frac{2(\Lambda_{1/2,\beta})^{1/\alpha}}{S_1^{(\alpha)}}\right)\right]$$

where $\Lambda_{1/2,\beta}$ is a beta r.v. with parameter $\left(\frac{1}{2}, \beta = \frac{c-1}{2}\right)$, $\alpha = \dfrac{1}{c}$, and $C_\alpha = \dfrac{1}{\mathbb{E}\left[\sqrt{S_1^{(\alpha)}}\right]}$.

Proof. From (7.78), (7.80), and the injectivity of the Laplace transform, we have, for every f Borel and positive:

$$\mathbb{E}\left[\frac{1}{H_c}f\left(\frac{1}{2H_c^2}\right)\right] = \frac{\alpha}{2^\beta\Gamma(\beta)}\int_1^{+\infty}(v-1)^{\beta-1}\mathbb{E}\left[f(S_{v/2}^{(\alpha)})\right]dv.$$

Replacing $f(x)$ by $f\left(\frac{1}{x}\right)\frac{1}{\sqrt{x}}$, we obtain:

$$\mathbb{E}\left[f(2H_c^2)\right] = \frac{\alpha}{2^{\beta+1/2}\Gamma(\beta)}\int_1^{+\infty}(v-1)^{\beta-1}\mathbb{E}\left[\frac{1}{\sqrt{S_{v/2}^{(\alpha)}}}f\left(\frac{1}{S_{v/2}^{(\alpha)}}\right)\right]dv.$$

Since by scaling (from (7.79)):

$$S_{v/2}^{(\alpha)} \overset{(law)}{=} \left(\frac{v}{2}\right)^{1/\alpha} S_1^{(\alpha)},$$

we obtain:

$$\begin{aligned}\mathbb{E}\left[f(2H_c^2)\right] &= \frac{\alpha}{2^{\frac{1}{2\alpha}}\Gamma(\beta)}\int_1^{+\infty}\frac{(v-1)^{\beta-1}}{(v/2)^{\frac{1}{2\alpha}}}\mathbb{E}\left[\frac{1}{\sqrt{S_1^{(\alpha)}}}f\left(\frac{1}{\left(\frac{v}{2}\right)^{1/\alpha}S_1^{(\alpha)}}\right)\right]dv \\ &= C_\alpha\int_0^1(1-w)^{\beta-1}w^{-1/2}\mathbb{E}\left[\frac{1}{\sqrt{S_1^{(\alpha)}}}f\left(\frac{(2w)^{1/\alpha}}{S_1^{(\alpha)}}\right)\right]dw\end{aligned}$$

after making the change of variable $v = 1/w$. This is Proposition 7.3. □

Remark 7.15. Let $\varphi : \mathbb{R} \longrightarrow \mathbb{R}$ be an odd, increasing function and let φ^{-1} its inverse in the sense of composition of functions, i.e.:

$$\varphi \circ \varphi^{-1} = \varphi^{-1} \circ \varphi = Id.$$

Assume that there exist $c > 0$ and a positive r.v. Z such that, for every $y \geq 0$:

$$(\varphi^{-1})'(\sqrt{y}) = c\mathbb{E}\left[e^{-yZ}\right]. \tag{7.81}$$

Under this hypothesis (7.81), we may prove that there exists, for every $t \geq 0$, a positive r.v. H_t such that:

$$\varphi(B_t) \overset{(1.d)}{=} \varphi\left(\sqrt{t}B_1\right) \overset{(1.d)}{=} H_t \cdot B_1$$

where, on the RHS, H_t and B_1 are independent. Unfortunately, we do not know how to prove, under some reasonable hypothesis, that $(H_t, t \geq 0)$ is increasing (at least stochastically), except for $\varphi(x) = x^{[c]}$, with $c \geq 1$. When $\varphi(x) = \sinh(x)$, i.e.

$(\varphi^{-1})'(\sqrt{y}) = \dfrac{1}{\sqrt{1+y}}$, hypothesis (7.81) is then satisfied with a gamma r.v. of parameter $\dfrac{1}{2}$. In this easy situation, we do not know how to prove that $(H_t, t \geq 0)$ is increasing. But going back to the notation of Remark 7.14, we find:

$$\sinh(B_t) \overset{(1.d)}{=} \sinh\left(\sqrt{t}B_1\right) \overset{(1.d)}{=} \widetilde{\beta}_{A_t^{(1)}} \overset{(1.d)}{=} \sqrt{A_t^{(1)}}\,\widetilde{\beta}_1$$

and we might choose $H_t = \sqrt{A_t^{(1)}}$ which is an increasing process.

Exercise 7.15 (Ocone martingales and diffusions). We take back the notation of Exercises 5.7 and 7.5. Let N be a reduced Gaussian r.v. independent from $(X_t, t \geq 0)$.

i) Exhibit an Ocone martingale associated to $(Z_t = NX_t, t \geq 0)$. (*Hint*: use (5.2).)

ii) Let $(B_t, t \geq 0)$ be a Brownian motion started from 0 and independent from the process $(X_t, t \geq 0)$. Prove that $(U_t := B_t X_t, t \geq 0)$ is a peacock and exhibit an Ocone martingale associated to $(U_t, t \geq 0)$.

Exercise 7.16 (Integrals of exponentials of Ocone martingales are peacocks).
• So far, we have presented numerous examples of processes $(N_u, u \geq 0)$ (cf. Section 2.1) such that, for every finite signed measure σ on $\mathbb{R}_+$:

$$\left(X_\lambda^{(\sigma)} := \int_0^{+\infty} \frac{e^{\lambda N_u}}{\mathbb{E}[e^{\lambda N_u}]}\sigma(du),\ \lambda \geq 0\right) \quad \text{is a peacock.}$$

In this situation, the factor $\dfrac{1}{\mathbb{E}[e^{\lambda N_u}]}$ ensures that $X^{(\sigma)}$ has a constant expectation. Now, when $(N_u, u \geq 0)$ is a continuous martingale started from 0, we may replace the factor $\dfrac{1}{\mathbb{E}[e^{\lambda N_u}]}$ by $\exp\left(-\frac{\lambda^2}{2}\langle N\rangle_u\right)$ which shall play the same role since (under some suitable integrability conditions) $(\exp(\lambda N_u - \frac{\lambda^2}{2}\langle N\rangle_u), u \geq 0)$ is a martingale. This leads to the following question:
• Let $(M_u, u \geq 0)$ be a continuous martingale started from 0 such that $(\exp(\lambda M_u - \frac{\lambda^2}{2}\langle M\rangle_u), u \geq 0)$ is a martingale. Under which conditions on M and σ is

$$\left(X_\lambda^{(\sigma)} := \int_0^{+\infty} e^{\lambda M_u - \frac{\lambda^2}{2}\langle M\rangle_u}\sigma(du),\ \lambda \geq 0\right) \quad \text{a peacock?}$$

The present exercise 7.16 gives a partial answer to this question.

Let $(M_u := B_{A_u}, u \geq 0)$ be an Ocone martingale, i.e.:

- $(B_t, t \geq 0)$ is a Brownian motion started at 0,
- $(A_u, u \geq 0)$ is an increasing process independent from $(B_t, t \geq 0)$.

1) Prove that for every $\lambda \in \mathbb{R}$, $(N_u^{(\lambda)} := e^{\lambda B_{A_u} - \frac{\lambda^2}{2} A_u}, u \geq 0)$ is a martingale.
2) Prove that for every finite signed measure σ on $\mathbb{R}_+$:

$$\left(X_\lambda^{(\sigma)} := \int_0^{+\infty} e^{\lambda B_{A_u} - \frac{\lambda^2}{2} A_u} \sigma(du),\ \lambda \geq 0 \right) \quad \text{is a peacock.}$$

(*Hint*: write, with obvious notation:

$$\mathbb{E}[\psi(X_\lambda^{(\sigma)})] = \mathbb{E}_A \left[\mathbb{E}_B[\psi(X_\lambda^{(\sigma)})] \right]$$

and apply Theorem 2.1.)

3) Prove that the result of Question 2) still holds without the hypothesis: $(A_u, u \geq 0)$ is increasing (note that, in this case, $(M_u = B_{A_u}, u \geq 0)$ is no longer a martingale).

4) Prove that

$$M_\lambda^{(\sigma)} := \int_{\mathbb{R}_+} \exp\left(W_{\lambda^2, A_u} - \frac{\lambda^2}{2} A_u \right) \sigma(du), \quad \lambda \geq 0$$

is a martingale associated to $(X_\lambda^{(\sigma)}, \lambda \geq 0)$.
(*Hint*: $M^{(\sigma)}$ is a $(\mathcal{G}_\lambda, \lambda \geq 0)$-martingale, with $\mathcal{G}_\lambda := \sigma(A_u, u \geq 0; W_{\mu,t}, t \geq 0, \mu \leq \lambda^2)$.)

5) Let $L = (L_t, t \geq 0)$ be a self-decomposable Lévy process such that, for every $\lambda, t \geq 0$, $\mathbb{E}[e^{\lambda L_t}] := \exp(t\phi_L(\lambda)) < \infty$. Let $(A_u, u \geq 0)$ be a process independent from L and σ a finite signed measure on $\mathbb{R}_+$.

Show that

$$Y_\lambda^{(\sigma)} := \int_{\mathbb{R}_+} \exp(\lambda L_{A_u} - A_u \phi_L(\lambda))\, \sigma(du), \quad \lambda \geq 0$$

is a peacock, and that

$$M_\lambda^{(\sigma)} := \int_{\mathbb{R}_+} \exp\left(\lambda \widehat{S}^{(L)}_{A_u,\lambda} - A_u \phi_L(\lambda) \right) \sigma(du), \quad \lambda \geq 0$$

is an associated martingale, where $(A_u, u \geq 0)$ is independent from $\widehat{S}^{(L)}$, the Sato sheet extending L.

One can compare this exercise with Exercises 2.1 and 2.22.

7.7 Student's Peacocks

We end up this chapter with a problem about Student's variables and related peacocks. This problem involves several methods developed so far to prove that certain processes are peacocks.

Long Exercise 7.17 (Student's peacocks). The aim of this problem is to describe several martingales associated to $(\sqrt{t}X, t \geq 0)$ for X a Student r.v. This problem originates essentially from D. Madan and M. Yor [125].

1) *Preliminaries.* *a)* A r.v. $X^{(\nu)}$ $(\nu > 0)$ is called a Student r.v. of parameter ν if its density $h_{X^{(\nu)}}$ is given by:

$$h_{X^{(\nu)}}(x) = \frac{C_\nu}{(1+x^2)^{\nu+1/2}}.$$

Prove that $C_\nu = \dfrac{\Gamma(\nu+1/2)}{\Gamma(\nu)\Gamma(1/2)}$.

Prove that $\mathbb{E}[|X^{(\nu)}|] < +\infty$ as soon as $\nu > 1/2$ and then $\mathbb{E}[X^{(\nu)}] = 0$. Deduce that, for $\nu > \frac{1}{2}$, $(\sqrt{t}X^{(\nu)}, t \geq 0)$ is a peacock.

b) Let $G_0, G_1, \dots, G_d$ be $(d+1)$ independent reduced Gaussian r.v.'s. Prove that:

$$X^{(d/2)} := \frac{G_0}{\sqrt{\sum_{i=1}^d G_i^2}}$$

follows a Student distribution of parameter $\frac{d}{2}$. (In the literature, one finds rather the terminology: Student's distribution with parameter d).

c) Let γ_ν be a gamma r.v. of parameter ν, i.e. whose density function is given by:

$$h_{\gamma_\nu}(x) = \frac{1}{\Gamma(\nu)} e^{-x} x^{\nu-1} \quad (x > 0).$$

Prove that γ_ν is self-decomposable. (*Hint*: $\mathbb{E}\left[e^{-\lambda\gamma_\nu}\right] = \dfrac{1}{(1+\lambda)^\nu} = e^{-\nu\log(1+\lambda)} = \exp\left(-\nu\int_0^{+\infty}(1-e^{-\lambda x})\frac{dx}{x}e^{-x}\right)$.)

d) Let $(B_t, t \geq 0)$ be a Brownian motion started from 0 and independent from γ_ν. Show that $B_{\frac{1}{2\gamma_\nu}}$ is a Student r.v. of parameter ν.

e) We recall the following result, due to Kent [107]. Let us denote $(R_t^{(a,\mu)}, t \geq 0)$ a Bessel process of index μ (i.e. of dimension $d = 2(1+\mu)$) started at a. Then, for $\nu > 0$:

$$T_0(R^{(1,-\nu)}) \overset{(law)}{=} \frac{1}{2\gamma_\nu}$$

with $T_0(R^{(1,-\nu)}) := \inf\{s \geq 0; R_s^{(1,-\nu)} = 0\}$. Use Williams' time reversal result:

$$\left(R_u^{(0,\nu)}, 0 \leq u \leq L_1(R^{(0,\nu)})\right) \overset{(law)}{=} \left(R^{(1,-\nu)}_{T_0(R^{(1,-\nu)})-u}, 0 \leq u \leq T_0(R^{(1,-\nu)})\right)$$

with $L_1(R^{(0,\nu)}) := \sup\{s \geq 0; R_s^{(0,\nu)} = 1\}$ to prove Getoor's result [68]:

$$L_1(R^{(0,\nu)}) \overset{(law)}{=} \frac{1}{2\gamma_\nu}$$

(see also Exercise 5.7 for another proof of the results of this question).

f) Let $(\beta_s, s \geq 0)$ be a Brownian motion started from 0 and $\nu > 0$. We define $A_t^{(-\nu)} := \int_0^t e^{2(\beta_s - \nu s)} ds$. Use Lamperti's representation (see [150], Exercise 1.28, p.452)

$$\exp(B_t - \nu t) = \rho_{A_t^{(-\nu)}}$$

where ρ_u, defined for $u < A_\infty^{(-\nu)}$, is a Bessel process of index $-\nu$ started from 1, to prove that:

$$A_\infty^{(-\nu)} := \int_0^{+\infty} e^{2(\beta_s - \nu s)} ds \overset{(law)}{=} T_0(R^{(1,-\nu)}) \overset{(law)}{=} \frac{1}{2\gamma_\nu}. \tag{7.82}$$

Formula (7.82) is due to Dufresne [58], the previous proof being a sketch of a second proof due to Yor [180].

2) *A first martingale associated to* $(\sqrt{t}X^{(\nu)}, t \geq 0)$ $(\nu > 1/2)$. We recall the following result (see Exercise 5.7): let $(R_u^{(0,\nu)}, u \geq 0)$ be a Bessel process of index $\nu > 0$ started from 0. Then, the process

$$\left(L_a^{(0,\nu)} := \sup\{u \geq 0;\ R_u^{(0,\nu)} = a\}, a \geq 0\right) \tag{7.83}$$

is a process with independent increments.

a) Prove that:

$$\left(L_{ca}^{(0,\nu)}, a \geq 0\right) \overset{(law)}{=} \left(c^2 L_a^{(0,\nu)}, a \geq 0\right). \tag{7.84}$$

Deduce that $\left(L_a^{(0,\nu)}, a \geq 0\right)$ is a 2-Sato process (see Chapter 5, Definition 5.2).

b) Let $(B_s, s \geq 0)$ be a Brownian motion independent from $(R_u^{(0,\nu)}, u \geq 0)$. Prove that:

$$B_{L_{\sqrt{t}}^{(0,\nu)}} \overset{(1.d)}{=} B_{tL_1^{(0,\nu)}} \overset{(1.d)}{=} \sqrt{t} B_{L_1^{(0,\nu)}} \overset{(1.d)}{=} \sqrt{t} X^{(\nu)}. \tag{7.85}$$

Deduce that the two processes $\left(B_{L_{\sqrt{t}}^{(0,\nu)}}, t \geq 0\right)$ and $(\sqrt{t}X^{(\nu)}, t \geq 0)$ are associated.

(*Remark*: for $\nu > \frac{1}{2}$, $\left(B_{L_{\sqrt{t}}^{(0,\nu)}}, t \geq 0\right)$ is a martingale associated to the peacock $(\sqrt{t}X^{(\nu)}, t \geq 0)$. *Note that this martingale is purely discontinuous.*)

c) Deduce from (7.83) and (7.85) that $X^{(\nu)}$ is self decomposable.

3) *A second martingale associated to* $(\sqrt{t}X^{(\nu)}, t \geq 0)$ $(\nu > \frac{1}{2})$. *a*) Using (7.82), prove that:

$$A_{\infty}^{(-\nu,t)} := \int_0^{+\infty} \exp\left(2\left(\beta_{\frac{s}{t}} - \nu\frac{s}{t}\right)\right) ds \overset{(law)}{=} \frac{t}{2\gamma_\nu}.$$

b) Let $(B_u, u \geq 0)$ be a Brownian motion independent from $(\beta_v, v \geq 0)$. Prove that:

$$\begin{aligned}\sqrt{t}X^{(\nu)} &\overset{(1.d)}{=} B_{\frac{t}{2\gamma_\nu}} \overset{(1.d)}{=} B_{\int_0^{+\infty} \exp\left(2(\beta_{\frac{s}{t}} - \nu\frac{s}{t})\right)ds} \\ &\overset{(1.d)}{=} \int_0^{+\infty} e^{\beta_{\frac{s}{t}} - \nu\frac{s}{t}} dB_s \\ &\overset{(1.d)}{=} \int_0^{+\infty} e^{\beta_u - \nu u} dB_{tu} \\ &\overset{(1.d)}{=} \int_0^{+\infty} e^{\beta_u - \nu u} d_u W_{u,t}\end{aligned}$$

where $(W_{u,s};\ u,s \geq 0)$ is the standard Brownian sheet, see Subsection 2.2.2.

c) Prove that, for $\nu > \frac{1}{2}$, $\left(\int_0^{+\infty} e^{\beta_u - \nu u} d_u W_{u,t}, t \geq 0\right)$ is a $(\mathcal{G}_t := \sigma(W_{u,s}, u \geq 0, s \leq t);\ t \geq 0)$ continuous martingale associated to $(\sqrt{t}X^{(\nu)}, t \geq 0)$.

4) *A martingale associated to* $(\sqrt{t}X^{(\nu)}, t \geq 0)$ $(\nu > \frac{1}{2})$ *from the Azéma-Yor SE.* Let μ be a probability whose support is $\mathbb{R}$, and such that:

$$\int_{\mathbb{R}} |x| \mu(dx) < +\infty \quad \text{and} \int_{\mathbb{R}} x\mu(dx) = 0. \tag{7.86}$$

Let ψ_μ be the Hardy-Littlewood function associated to μ (see (7.33)):

$$\psi_\mu(x) = \frac{\int_x^{+\infty} y\mu(dy)}{\int_x^{+\infty} \mu(dy)} = \frac{\int_x^{+\infty} y\mu(dy)}{\overline{\mu}(x)}$$

with $\overline{\mu}(x) = \mu([x, +\infty[)$. We recall that ψ_μ is positive, increasing and that, for all x real, $\psi_\mu(x) \geq x$.

a) Let $(B_s, s \geq 0)$ be a Brownian motion started from 0 and denote $(S_t := \sup_{s \leq t} B_s, t \geq 0)$ its one-sided supremum process. Let, for every $t \geq 0$:

$$T_t := \sup\{u \geq 0;\ S_u = B_u + \sqrt{B_u^2 + 2t}\}. \tag{7.87}$$

Prove that $(T_t, t \geq 0)$ is an increasing family of a.s. finite stopping times such that $T_t \overset{(1.d)}{=} tT_1$.

Let $\left(N_u := \frac{1}{2}S_u^2 - S_u(S_u - B_u), u \geq 0\right)$. Prove that $(N_u, u \geq 0)$ is a martingale such that $T_t = \inf\{u \geq 0;\ N_u = -t\}$.

b) We shall now show the existence of a family of probabilities $(\nu_t, t \geq 0)$ on $\mathbb{R}$ satisfying (7.86) such that:

$$T_t := \inf\{u \geq 0;\ S_u \geq \psi_{\nu_t}(B_s)\}$$

with $\psi_{\nu_t}(x) = \frac{1}{\overline{\nu_t}(x)} \int_x^{+\infty} y\nu_t(dy)$ the Hardy-Littlewood function attached to ν_t. In other terms, we shall show that the T_t's are the times obtained from the Azéma-Yor algorithm to realize the Skorokhod embedding of ν_t. To this end, from (7.87), we must find ν_t such that:

$$\psi_{\nu_t}(x) = x + \sqrt{x^2 + 2t}. \tag{7.88}$$

From Lemma 7.4, and the relation $u_{\nu_t}(x) = \frac{1}{\psi_{\nu_t}(x) - x} = \frac{1}{\sqrt{x^2+2t}}$, (7.88) is equivalent to:

$$\overline{\nu_t}(x) = \frac{C}{\sqrt{x^2+2t}} \exp\left(-\int_0^x \frac{dy}{\sqrt{y^2+2t}} \right).$$

c) Prove that:

$$\frac{C}{\sqrt{x^2+2t}} \exp\left(-\int_0^x \frac{dy}{\sqrt{y^2+2t}} \right) = \frac{\widetilde{C}}{\sqrt{x^2+2t}\,(x+\sqrt{x^2+2t})}. \tag{7.89}$$

Prove that, in (7.89), $\widetilde{C}$ is necessarily equal to t and that $C = \sqrt{\frac{t}{2}}$.
(*Hint*: one may let x tend to $-\infty$ in (7.89).)

Deduce that the measure ν_t is given by:

$$\nu_t(dx) = \frac{t}{(x^2+2t)^{3/2}} dx.$$

Prove that the r.v. $\sqrt{2t}X^{(1)}$ follows the law ν_t.

Deduce that the martingale $(B_{T_{\frac{t}{2}}}, t \geq 0)$ is associated to the peacock $(\sqrt{t}X^{(1)}, t \geq 0)$, i.e.:

$$B_{T_{\frac{t}{2}}} \overset{(1.d)}{=} \sqrt{t}X^{(1)}.$$

Note that we have also:

$$B_{T_{\frac{t}{2}}} \overset{(1.d)}{=} \sqrt{t}X^{(1)} \overset{(1.d)}{=} \sqrt{t}\frac{G_0}{\sqrt{G_1^2+G_2^2}} \overset{(1.d)}{=} \sqrt{\frac{t}{2\mathfrak{e}}}G_0 \overset{(1.d)}{=} \beta_{\frac{t}{2\mathfrak{e}}} \tag{7.90}$$

where, in (7.90), $\mathfrak{e}$ is a standard exponential r.v., independent from G_0, and independent from the Brownian motion $(\beta_u, u \geq 0)$. (Use $G_1^2 + G_2^2 \overset{(law)}{=} 2\mathfrak{e}$).

d) We shall now generalize the results obtained in questions 4)*b*) and 4)*c*) by replacing $X^{(1)}$ by $X^{(\nu)}$, for $\nu > 1/2$.

Let $\mu_t^{(\nu)}$ be the law of $\sqrt{2t}X^{(\nu)}$. Prove that

$$\mu_t^{(\nu)}(dx) = \frac{C_\nu (2t)^\nu}{(x^2+2t)^{\nu+1/2}} dx \quad \text{with } C_\nu = \frac{1}{B(\nu, 1/2)},$$

where B is the first Euler function. Prove that the family $(\mu_t^{(\nu)}, t \geq 0)$ enjoy the $(M{\cdot}Y)$ property.

(*Hint*: use Remark 7.8 upon noticing that $h_{X^{(\nu)}}(y) = C_\nu \mathbb{E}\left[e^{-y^2 \gamma_{\nu+1/2}}\right]$ and that, for any $a > 0$, γ_a is self-decomposable.)

Deduce that:

$$T_t^{(\nu)} := \inf\{u \geq 0;\ S_u \geq \psi_{\mu_t^{(\nu)}}(B_u)\} \tag{7.91}$$

is an increasing family of stopping times, and that $(B_{T^{(\nu)}_{\frac{t}{2}}}, t \geq 0)$ is a martingale associated to the peacock $(\sqrt{t}X^{(\nu)}, t \geq 0)$, $\nu > 1/2$.

To give an explicit expression for (7.91), we need to compute:

$$\psi_{\mu_t^{(\nu)}}(x) = \frac{\displaystyle\int_x^{+\infty} \frac{y\,dy}{(2t+y^2)^{\nu+1/2}}}{\displaystyle\int_x^{+\infty} \frac{dy}{(2t+y^2)^{\nu+1/2}}} =: \frac{N_t^{(\nu)}(x)}{D_t^{(\nu)}(x)}.$$

Prove that: $N_t^{(\nu)}(x) = \dfrac{1}{2\nu-1} \dfrac{1}{(2t+x^2)^{\nu-\frac{1}{2}}}$ and

$$2t(2\nu+1)D_t^{(\nu+1)}(x) = 2\nu D_t^{(\nu)}(x) - \frac{x}{(2t+x^2)^{\nu+\frac{1}{2}}}.$$

Deduce from these formulae, since $\psi_{\mu_t^{(1)}}(x) = x + \sqrt{x^2+2t}$ (from (7.88)), the value of $\psi_{\mu_t^{(\nu)}}(x)$. For non integer values of ν, one may use, for $x \geq 0$:

$$\begin{aligned}\int_x^{+\infty} \frac{dy}{(1+y^2)^{\nu+\frac{1}{2}}} &= \frac{1}{2}\int_{x^2}^{+\infty} \frac{dz}{\sqrt{z}(1+z)^{\nu+\frac{1}{2}}} \\ &= \frac{2}{\nu x^\nu} F_{2,1}\left(\frac{1}{2}, \nu, \nu+1; -\frac{1}{x^2}\right)\end{aligned}$$

(see [71], p.285, 3.195), and for $x < 0$:

$$\int_x^{+\infty} \frac{C_\nu dy}{(1+y^2)^{\nu+\frac{1}{2}}} = 1 - \int_{-x}^{+\infty} \frac{C_\nu dy}{(1+y^2)^{\nu+\frac{1}{2}}}.$$

5) *A generalization to the function* $x \longrightarrow x + (|x|^\alpha + 2)^{1/\alpha}$. We now want to extend the results of 4) by replacing the function $\psi_{\mu_1}(x) = x + \sqrt{x^2+2}$ by $x \longrightarrow x + (|x|^\alpha + 2)^{1/\alpha}$.

a) Prove that there exists, for all $\alpha > 0$, an integrable and centered r.v. $Y^{(\alpha)}$ whose law $\rho^{(\alpha)}$ is characterized by:

$$\psi_{\rho^{(\alpha)}}(x) := \frac{\int_x^{+\infty} y\rho^{(\alpha)}(dy)}{\int_x^{+\infty} \rho^{(\alpha)}(dy)} = x + (|x|^\alpha + 2)^{1/\alpha}.$$

(*Hint*: prove that there exists a constant C_α such that:

$$x \longrightarrow \frac{C_\alpha}{(|x|^\alpha + 2)^{1/\alpha}} \exp\left(- \int_0^x \frac{dy}{(|y|^\alpha + 2)^{1/\alpha}} \right)$$

is a positive and decreasing function converging to 1 at $-\infty$ and to 0 at $+\infty$.)

b) Prove that, if $\rho_t^{(\alpha)}$ denotes the law of $\sqrt{2t}Y^{(\alpha)}$, then:

$$\psi_{\rho_t^{(\alpha)}}(x) = \sqrt{2t}\,\psi_{\rho_1^{(\alpha)}}\left(\frac{x}{\sqrt{2t}}\right).$$

c) Prove that

$$T_t^{(\alpha)} := \inf\{u \geq 0;\, S_u \geq \psi_{\rho_t^{(\alpha)}}(B_u)\}$$

is an increasing family of stopping times, and that the martingale $(B_{T_{t/2}^{(\alpha)}}, t \geq 0)$ is associated to the peacock $(\sqrt{t}Y^{(\alpha)}, t \geq 0)$

d) Compute $\rho_t^{(\alpha)}$ the density of the r.v. $\sqrt{2t}Y^{(\alpha)}$. (*Hint*: use the relation (cf [71], p.284, 3.194):

$$\int_0^u \frac{x^{b-1}}{(1+\lambda x)^a} dx = \frac{u^b}{b} F_{2,1}(a, b; 1+b, -\lambda u) \quad (u, b > 0).$$

Notes and Comments

The use of the Azéma-Yor algorithm to construct a martingale associated to a peacock of the form $(\sqrt{t}X, t \geq 0)$ appears for the first time, in a slightly different manner, in Madan and Yor [125]. The results of Sections 7.2 and 7.4 are taken from Hirsch *et al.* [78]. The embedding method of Section 7.3 is due to Hall (see [73, 74], and also [136, 38]). Remark 7.5 about Azéma-Yor's embedding is due to Vallois. Theorem 7.15 and Exercise 7.14 (due to Bogso) rely on Bertoin-Le Jan embedding method [26], while Exercise 7.12 is taken from Profeta's thesis [147]. The embedding method of Section 7.5 is due to Bass (Theorem 7.16), Theorem 7.20 is due to Bougerol [36], and its generalization (Theorem 7.21) is taken from [3]. The long exercise of Section 7.7 (Student's peacock) is issued from Madan and Yor [125]. All other results in this chapter are new.

8

Comparison of Multidimensional Marginals

Abstract We compare, in a Gaussian setting, the multidimensional marginals of some peacocks and some 1-martingales which are associated to them via several methods.

8.1 Introduction

8.1.1 Motivation

Let $(B_s, s \geq 0)$ be a Brownian motion started from 0 and $(W_{u,t},\ u,t \geq 0)$ a Brownian sheet. For every ν real, we define:

$$\left(X_t := \frac{1}{t} \int_0^t e^{\nu B_s - \frac{\nu^2}{2} s} ds = \int_0^1 e^{\nu B_{tu} - \frac{\nu^2}{2} tu} du, t \geq 0 \right) \tag{8.1}$$

and

$$\left(X_t^{(S)} := \int_0^1 e^{\nu W_{u,t} - \frac{\nu^2}{2} ut} du \right) \tag{8.2}$$

((S) for Brownian *S*heet). We have seen in Chapter 2 that $(X_t^{(S)}, t \geq 0)$ is a martingale, which is associated to the peacock $(X_t, t \geq 0)$, i.e.:

$$X_t \stackrel{(1.d)}{=} X_t^{(S)}.$$

In particular, for every k real and $t \geq 0$:

$$\mathbb{E}\left[(k - X_t)^+\right] = \mathbb{E}\left[(k - X_t^{(S)})^+\right] \text{ and } \mathbb{E}\left[(X_t - k)^+\right] = \mathbb{E}\left[(X_t^{(S)} - k)^+\right]. \tag{8.3}$$

But are we able, for every $n \in \mathbb{N}$ and every n-tuples $0 < t_1 < \ldots < t_n$ and $0 < k_1 < \ldots < k_n$, to compare:

$$\mathbb{E}\left[\prod_{i=1}^n (k_i - X_{t_i})^+\right] \quad \text{and} \quad \mathbb{E}\left[\prod_{i=1}^n (k_i - X_{t_i}^{(S)})^+\right] \tag{8.4}$$

Hirsch F., Profeta C., Roynette B., Yor M.:
Peacocks and Associated Martingales, with Explicit Constructions

and

$$\mathbb{E}\left[\prod_{i=1}^{n}(X_{t_i}-k_i)^+\right] \quad \text{and} \quad \mathbb{E}\left[\prod_{i=1}^{n}(X_{t_i}^{(S)}-k_i)^+\right] ? \tag{8.5}$$

Unlike (8.3) which only involves the one-dimensional marginals of $(X_t, t \geq 0)$ and $(X_t^{(S)}, t \geq 0)$, (8.4) and (8.5) involve the n-marginals of both processes. That is why we need to define a notion of order between processes – the upper orthant order and the lower orthant order – which involve all finite dimensional marginals. This is the purpose of the following Subsection 8.1.2 dedicated to the upper orthant order and then of Section 8.3 dedicated to the lower orthant order. Then, we will be able to answer the questions raised by (8.4) and (8.5), and others similar questions (see Theorem 8.1, 8.2 and 8.3 and Corollary 8.3.)

8.1.2 The Upper Orthant Order

Let $X = (X_1, X_2, \dots, X_p)$ and $Y = (Y_1, Y_2, \dots, Y_p)$ be two $\mathbb{R}^p$-valued random vectors.

Definition 8.1 (see M. Shaked and J. Shantikumar [157], p. 140). X is said to be smaller than Y in the upper orthant (notation: $X \underset{u.o}{\leq} Y$) if one of the three following equivalent conditions is satisfied:

1) For every p-tuple $\lambda_1, \lambda_2, \dots, \lambda_p$ of reals:

$$\mathbb{P}(X_1 > \lambda_1, X_2 > \lambda_2, \dots, X_p > \lambda_p) \leq \mathbb{P}(Y_1 > \lambda_1, Y_2 > \lambda_2, \dots, Y_p > \lambda_p). \tag{8.6}$$

2) For every p-tuple $l_1, l_2, \dots, l_p$ of nonnegative increasing functions:

$$\mathbb{E}\left[\prod_{i=1}^{p} l_i(X_i)\right] \leq \mathbb{E}\left[\prod_{i=1}^{p} l_i(Y_i)\right]. \tag{8.7}$$

3) For every distribution function $F : \mathbb{R}^p \to \mathbb{R}_+$:

$$\mathbb{E}\left[F(X_1, \dots, X_p)\right] \leq \mathbb{E}\left[F(Y_1, \dots, Y_p)\right]. \tag{8.8}$$

We recall that a function $F : \mathbb{R}^p \longrightarrow \mathbb{R}_+$ is a distribution function if there exists a probability measure μ on $\mathbb{R}^p$ such that:

$$F(\lambda_1, \lambda_2, \dots, \lambda_p) = \mu\left(\prod_{i=1}^{p}]-\infty, \lambda_i]\right) \tag{8.9}$$

and if such a F is of class $\mathscr{C}^2$, then:

$$\text{for every } i, j = 1, 2, \dots, p \text{ and } i \neq j\text{:} \quad \frac{\partial^2 F}{\partial \lambda_i \partial \lambda_j} \geq 0. \tag{8.10}$$

Definition 8.2. A *process* $(X_t, t \geq 0)$ is smaller than a process $(Y_t, t \geq 0)$ for the upper orthant order (notation: $(X_t, t \geq 0) \underset{u.o}{\leq} (Y_t, t \geq 0)$) if, for every integer p and every $0 \leq t_1 < t_2 < \ldots < t_p$:

$$(X_{t_1}, X_{t_2}, \ldots, X_{t_p}) \underset{u.o}{\leq} (Y_{t_1}, Y_{t_2}, \ldots, Y_{t_p}). \tag{8.11}$$

Of course, if X and Y are two càdlàg processes, (8.11) is equivalent to

for every $h : \mathbb{R} \longrightarrow \mathbb{R}$ càdlàg:

$$\mathbb{P}(\text{for every } t \geq 0, X_t \geq h(t)) \leq \mathbb{P}(\text{for every } t \geq 0, Y_t \geq h(t)) \tag{8.12}$$

(see M. Shaked and J. Shanthikumar, [157], p.140-141 for the equivalence asserted in Definition 8.2).

Let now $(B_s, s \geq 0)$ be a Brownian motion started at 0 and let $\left(X_t^{(\nu)} := \frac{1}{t}\int_0^t e^{\nu B_s - \frac{\nu^2}{2}s} ds, t \geq 0\right)$. We showed in Chapter 2 that $(X_t^{(\nu)}, t \geq 0)$ is a peacock and we associated to $(X_t^{(\nu)}, t \geq 0)$ via several methods – Brownian sheet method, time reversal method, scaling method – several martingales. The aim of this chapter is to compare – in the sense of the upper orthant order – the multidimensional marginals of the peacock $(X_t^{(\nu)}, t \geq 0)$ and of these associated martingales.

8.2 A General Comparison Theorem for the Upper Orthant Order

8.2.1 Notation

We first set some notation and hypothesis:
a) for every integer n and every $(n+1)$ reals $x_0, x_1, \ldots, x_n$, we denote by $\omega^{[n, x_0, \ldots, x_n]}$ the element of $\mathscr{C}([0,1], \mathbb{R})$ such that:

- $\omega^{[n, x_0, \ldots, x_n]}$ is affine on every interval $\left[\frac{k}{n}, \frac{k+1}{n}\right]$ $(k = 0, 1, \ldots, n-1)$,
- for every $k = 0, 1, \ldots, n$, $\omega^{[n, x_0, \ldots, x_n]}\left(\frac{k}{n}\right) = x_k$.

b) For every $t \geq 0$, we consider an operator $H_t : \mathscr{C}([0,1], \mathbb{R}) \longrightarrow \mathscr{C}([0,1], \mathbb{R})$ which we assume to be of class $\mathscr{C}^2$ for every fixed t (the Banach space $\mathscr{C}([0,1], \mathbb{R})$ being equipped with the uniform norm). In particular, H_t is continuous and, for every $u \in [0,1]$, the map:

$$(x_0, x_1, \ldots, x_n) \longmapsto H_t(\omega^{[n, x_0, x_1, \ldots, x_n]})(u) \quad \text{is } \mathscr{C}^2. \tag{8.13}$$

We also assume that, for every $t \geq 0$, H_t is increasing (or decreasing), i.e., if $\omega(u) \leq \widetilde{\omega}(u)$ for every $u \in [0,1]$, then

$$H_t(\omega)(u) \leq H_t(\widetilde{\omega})(u) \text{ for every } u \in [0,1] \tag{8.14}$$

$$(\text{or} \quad H_t(\omega)(u) \geq H_t(\widetilde{\omega})(u) \text{ for every } u \in [0,1]). \tag{8.15}$$

For example, the operator $H_t^{(\nu)}$ defined by

$$H_t^{(\nu)}(\omega)(u) := \exp\left(\nu\omega(u) - \frac{\nu^2}{2}tu\right) \tag{8.16}$$

satisfies (8.13) and (8.14) (or (8.15)) for every $\nu \in \mathbb{R}$.

c) For $i = 1,2$ and $s \geq 0$, $(G_u^{(s,i)}, 0 \leq u \leq 1)$ denotes a centered Gaussian process started at 0, with continuous paths.

Definition 8.3. We say that $\mathbb{G}^{(2)} := (G_u^{(s,2)};\, 0 \leq u \leq 1, s \geq 0)$ is dominated by $\mathbb{G}^{(1)} := (G_u^{(s,1)};\, 0 \leq u \leq 1, s \geq 0)$ and we write

$$\mathbb{G}^{(1)} > \mathbb{G}^{(2)} \tag{8.17}$$

if:

1) For any $s \geq 0$,

$$(G_u^{(s,1)}, 0 \leq u \leq 1) \overset{(law)}{=} (G_u^{(s,2)}, 0 \leq u \leq 1). \tag{8.18}$$

2) For every $s \neq t$ and $u, v \in [0,1]$:

$$\mathbb{E}\left[G_u^{(s,1)} G_v^{(t,1)}\right] \geq \mathbb{E}\left[G_u^{(s,2)} G_v^{(t,2)}\right]. \tag{8.19}$$

8.2.2 The General Comparison Theorem

Theorem 8.1. *With the previous notation and hypothesis, let us define, for* $i = 1,2$*:*

$$\left(X_t^{(i)} := \int_0^1 H_t(G_{\bullet}^{(t,i)})(u)du, t \geq 0\right). \tag{8.20}$$

Let us assume that $\mathbb{G}^{(1)} > \mathbb{G}^{(2)}$. *Then:*

1) $X_t^{(1)} \overset{(1.d)}{=} X_t^{(2)}$. (8.21)

2) $(X_t^{(1)}, t \geq 0) \underset{u.o}{\geq} (X_t^{(2)}, t \geq 0)$. (8.22)

Remark 8.1. 1) The notation $H_t(G_{\bullet}^{(t,i)})$ stands for $H_t(\omega)$ with ω defined by:

$$\text{for every } \; v \in [0,1], \quad \omega(v) = G_v^{(t,i)}.$$

2) *a*) In the particular case where

$$(G_u^{(t,1)}, 0 \leq u \leq 1) = (B_{ut}, 0 \leq u \leq 1)$$

and

$$H_t^{(\nu)}(\omega)(u) = \exp\left(\nu\omega(u) - \frac{\nu^2}{2}tu\right)$$

we have

$$\begin{aligned} X_t^{(1)} &= \int_0^1 \exp\left(\nu B_{ut} - \frac{\nu^2}{2}ut\right) du \\ &= \frac{1}{t}\int_0^t \exp\left(\nu B_s - \frac{\nu^2}{2}s\right) ds \end{aligned}$$

(after the change of variable $ut = s$).

b) In the particular case where:

$$(G_u^{(t,2)}, 0 \leq u \leq 1) = (W_{u,t}, 0 \leq u \leq 1)$$

(with $(W_{u,t}, u \geq 0, t \geq 0)$ the standard Brownian sheet, see Chapter 2) and H_t is still defined by (8.16), we have

$$X_t^{(2)} = \int_0^1 \exp\left(\nu W_{u,t} - \frac{\nu^2}{2}ut\right) du.$$

Proof. Point 1) is clear since, from (8.18),

$$(G_u^{(t,1)}, 0 \leq u \leq 1) \overset{(law)}{=} (G_u^{(t,2)}, 0 \leq u \leq 1).$$

We now prove Point 2) of Theorem 8.1. 1) We start by proving that, if $F : \mathbb{R}^2 \longrightarrow \mathbb{R}_+$ is the distribution function of a 2-dimensional random vector, then, for $s < t$:

$$\mathbb{E}\left[F(X_s^{(1)}, X_t^{(1)})\right] \geq \mathbb{E}\left[F(X_s^{(2)}, X_t^{(2)})\right]. \tag{8.23}$$

Without loss of generality, by approximation, we may assume that F is of class $\mathscr{C}^2$.

i) Since H_s and H_t are continuous (as functions of $\omega \in \mathscr{C}([0,1], \mathbb{R})$), we have:

$$\begin{aligned} \mathbb{E}\left[F(X_s^{(1)}, X_t^{(1)})\right] = \lim_{n \to +\infty} \mathbb{E}\Bigg[F\Bigg(&\int_0^1 H_s(\omega^{[n, G_{\frac{0}{n}}^{(s,1)}, \dots, G_{\frac{n}{n}}^{(s,1)}]})(u)du, \\ &\int_0^1 H_t(\omega^{[n, G_{\frac{0}{n}}^{(t,1)}, \dots, G_{\frac{n}{n}}^{(t,1)}]})(u)du\Bigg)\Bigg] \end{aligned} \tag{8.24}$$

and an identical formula when replacing the index 1 by 2.

ii) Thus, relation (8.23) will be proven if we establish, for every n:

$$\mathbb{E}\left[L\left(G_{\frac{0}{n}}^{(s,1)},\dots,G_{\frac{k}{n}}^{(s,1)},\dots,G_{\frac{n}{n}}^{(s,1)};\ G_{\frac{0}{n}}^{(t,1)},\dots,G_{\frac{k'}{n}}^{(t,1)},\dots,G_{\frac{n}{n}}^{(t,1)}\right)\right]$$
$$\geq \mathbb{E}\left[L\left(G_{\frac{0}{n}}^{(s,2)},\dots,G_{\frac{k}{n}}^{(s,2)},\dots,G_{\frac{n}{n}}^{(s,2)};\ G_{\frac{0}{n}}^{(t,2)},\dots,G_{\frac{k'}{n}}^{(t,2)},\dots,G_{\frac{n}{n}}^{(t,2)}\right)\right] \quad (8.25)$$

where $L:\mathbb{R}^{2(n+1)}\longrightarrow\mathbb{R}$ is defined by:

$$L(x_0,\dots,x_n;\ y_0,\dots,y_n)$$
$$=\mathbb{E}\left[F\left(\int_0^1 H_s\left(\omega^{[n,x_0,\dots,x_n]}\right)(u)du,\int_0^1 H_t\left(\omega^{[n,y_0,\dots,y_n]}\right)(u)du\right)\right]. \quad (8.26)$$

iii) We now observe that, for $i=0,\dots,n$ and $j=0,\dots,n$, we have:

$$\frac{\partial^2 L}{\partial x_i\partial y_i}\geq 0. \quad (8.27)$$

Indeed,

$$\frac{\partial^2 L}{\partial x_i\partial y_i}=$$
$$\mathbb{E}\left[\frac{\partial^2 F}{\partial a\partial b}\times\left(\frac{\partial}{\partial x_i}\int_0^1 H_s(\omega^{[n,x_0,\dots,x_n]})(u)du\right)\left(\frac{\partial}{\partial y_i}\int_0^1 H_t(\omega^{[n,y_0,\dots,y_n]})(u)du\right)\right]. \quad (8.28)$$

But, F being a distribution function, $\dfrac{\partial^2 F}{\partial a\partial b}\geq 0$ and, from hypothesis (8.14) or (8.15), for every $i=0,1,\dots,n$ and $j=0,1,\dots,n$:

$$\operatorname{sgn}\left(\frac{\partial}{\partial x_i}\int_0^1 H_s(\omega^{[n,x_0,\dots,x_n]})(u)du\right)=\operatorname{sgn}\left(\frac{\partial}{\partial y_i}\int_0^1 H_t(\omega^{[n,y_0,\dots,y_n]})(u)du\right) \quad (8.29)$$

(under (8.14) this sign is positive, and negative under (8.15)).

iv) We now note that, from (8.18) and (8.19), we have, for every s,t and every k and k':

$$\mathbb{E}\left[G_{\frac{k}{n}}^{(s,1)}G_{\frac{k'}{n}}^{(s,1)}\right]=\mathbb{E}\left[G_{\frac{k}{n}}^{(s,2)}G_{\frac{k'}{n}}^{(s,2)}\right] \quad (8.30)$$

$$\mathbb{E}\left[G_{\frac{k}{n}}^{(t,1)}G_{\frac{k'}{n}}^{(t,1)}\right]=\mathbb{E}\left[G_{\frac{k}{n}}^{(t,2)}G_{\frac{k'}{n}}^{(t,2)}\right] \quad (8.31)$$

$$\mathbb{E}\left[G_{\frac{k}{n}}^{(s,1)}G_{\frac{k'}{n}}^{(t,1)}\right]\geq\mathbb{E}\left[G_{\frac{k}{n}}^{(s,2)}G_{\frac{k'}{n}}^{(t,2)}\right]. \quad (8.32)$$

Hence, from (8.26), (8.27), (8.30), (8.31) and (8.32) we deduce (8.25) (and (8.23)) by applying Slepian's Lemma, which we now recall:

Lemma 8.1 (Slepian's lemma; see Proposition 1.5). *Let* $U=(U_0,\dots,U_n,U_{n+1},\dots,U_{2n+1})$ *and* $V=(V_0,\dots,V_n,V_{n+1},\dots,V_{2n+1})$ *be two centered Gaussian vectors such that:*

1) *For either* $0 \leq i,j \leq n$*, or for* $n+1 \leq i,j \leq 2n+1$*:*

$$\mathbb{E}[U_iU_j] = \mathbb{E}[V_iV_j]. \tag{8.33}$$

2) *For either* $0 \leq i \leq n$ *and* $n+1 \leq j \leq 2n+1$*, or for* $n+1 \leq i \leq 2n+1$ *and* $0 \leq j \leq n$*:*

$$\mathbb{E}[U_iU_j] \geq \mathbb{E}[V_iV_j]. \tag{8.34}$$

On the other hand, let L *be a* $\mathscr{C}^2$*-function from* $\mathbb{R}^{2n+2}$ *to* $\mathbb{R}$ *such that, either for every* $0 \leq i \leq n$ *and* $n+1 \leq j \leq 2n+1$*, or for* $n+1 \leq i \leq 2n+1$ *and* $0 \leq j \leq n$*:*

$$\frac{\partial^2 L}{\partial u_i \partial u_j} \geq 0. \tag{8.35}$$

Then:

$$\mathbb{E}[L(U_0,\dots,U_n,U_{n+1},\dots,U_{2n+1})] \geq \mathbb{E}[L(V_0,\dots,V_n,V_{n+1},\dots,V_{2n+1})].$$

We obtain (8.25) by applying Slepian's lemma to

$$U = \left(G^{(s,1)}_{\frac{0}{n}},\dots,G^{(s,1)}_{\frac{k}{n}},\dots,G^{(s,1)}_{\frac{n}{n}};\, G^{(t,1)}_{\frac{0}{n}},\dots,G^{(t,1)}_{\frac{k'}{n}},\dots,G^{(t,1)}_{\frac{n}{n}}\right)$$

and

$$V = \left(G^{(s,2)}_{\frac{0}{n}},\dots,G^{(s,2)}_{\frac{k}{n}},\dots,G^{(s,2)}_{\frac{n}{n}};\, G^{(t,2)}_{\frac{0}{n}},\dots,G^{(t,2)}_{\frac{k'}{n}},\dots,G^{(t,2)}_{\frac{n}{n}}\right).$$

2) We now prove that, for every p, every p-tuple $t_1 < t_2 < \dots < t_p$, and $F : \mathbb{R}^p \longrightarrow \mathbb{R}_+$ any distribution function of a p-dimensional random vector, we have:

$$\mathbb{E}\left[F\left(X^{(1)}_{t_1},X^{(1)}_{t_2},\dots,X^{(1)}_{t_p}\right)\right] \geq \mathbb{E}\left[F\left(X^{(2)}_{t_1},X^{(2)}_{t_2},\dots,X^{(2)}_{t_p}\right)\right]. \tag{8.36}$$

We may assume, once again without loss of generality, that F is of class $\mathscr{C}^2$. The proof of (8.36) is then very similar to the one of (8.23), upon noticing that:

- For every $1 \leq l,r \leq p$ and $l \neq r$:

$$\frac{\partial^2 F}{\partial a_l \partial a_r} \geq 0$$

  since F is a distribution function.
- Slepian's Lemma only involves the covariances, i.e. the expectations of products taken two by two of Gaussian processes $(G^{(t_i,1)}_u, 0 \leq u \leq 1)$ and $(G^{(t_i,2)}_u, 0 \leq u \leq 1)$, $i = 1,\dots,p$. □

Corollary 8.1. *Under the notation and hypotheses of Theorem 8.1:*

1) *For every n, every n-tuple $k_1,\dots,k_n$ of reals and every n-tuple $0<t_1<t_2<\dots<t_n$, we have:*

$$\mathbb{E}\left[\prod_{i=1}^{n}(X_{t_i}^{(1)}-k_i)^+\right]\geq\mathbb{E}\left[\prod_{i=1}^{n}(X_{t_i}^{(2)}-k_i)^+\right]. \tag{8.37}$$

2) *For every $T>0$, for every continuous function h from $[0,T]$ to $\mathbb{R}$ we have:*

$$\mathbb{P}\left(\forall t\leq T,\, X_t^{(1)}\geq h(t)\right)\geq\mathbb{P}\left(\forall t\leq T,\, X_t^{(2)}\geq h(t)\right) \tag{8.38}$$

Corollary 8.1 is an immediate consequence of Theorem 8.1, and of (8.7) and (8.12).

8.3 A General Comparison Theorem for the Lower Orthant Order

The analogue of Definition 8.1 is, for the lower orthant order:

Definition 8.4 (see M. Shaked and J. Shantikumar [157], p. 140). Let $X=(X_1,X_2,\dots,X_p)$ and $Y=(Y_1,Y_2,\dots,Y_p)$ be two $\mathbb{R}^p$-valued random vectors. X is said to be smaller than Y for the lower orthant (notation $X\underset{l.o}{\leq}Y$) if one of the three following equivalent conditions is satisfied:

1) For every p-tuple $\lambda_1,\lambda_2,\dots,\lambda_p$ of reals:

$$\mathbb{P}(X_1\leq\lambda_1,X_2\leq\lambda_2,\dots,X_p\leq\lambda_p)\leq\mathbb{P}(Y_1\leq\lambda_1,Y_2\leq\lambda_2,\dots,Y_p\leq\lambda_p). \tag{8.39}$$

2) For every p-tuple $h_1,h_2,\dots,h_p$ of nonnegative decreasing functions:

$$\mathbb{E}\left[\prod_{i=1}^{p}h_i(X_i)\right]\geq\mathbb{E}\left[\prod_{i=1}^{p}h_i(Y_i)\right]. \tag{8.40}$$

3) For every survival function $\overline{F}:\mathbb{R}^p\to\mathbb{R}_+$:

$$\mathbb{E}\left[\overline{F}(X_1,\dots,X_p)\right]\geq\mathbb{E}\left[\overline{F}(Y_1,\dots,Y_p)\right]. \tag{8.41}$$

We recall (see [157]) that a function $\overline{F}:\mathbb{R}^p\longrightarrow\mathbb{R}_+$ is a survival function if there exists a probability measure μ on $\mathbb{R}^p$ such that:

$$\overline{F}(\lambda_1,\lambda_2,\dots,\lambda_p)=\mu\left(\prod_{i=1}^{p}[\lambda_i,+\infty[\right) \tag{8.42}$$

and if such a $\overline{F}$ is of class $\mathscr{C}^2$, then:

$$\text{for every } i,j=1,2,\dots,p \text{ and } i\neq j:\quad \frac{\partial^2\overline{F}}{\partial\lambda_i\partial\lambda_j}\geq 0. \tag{8.43}$$

Definition 8.5. A process $(X_t, t \geq 0)$ is smaller than a process $(Y_t, t \geq 0)$ for the lower orthant order (notation: $(X_t, t \geq 0) \underset{l.o}{\leq} (Y_t, t \geq 0)$) if for every integer p and $0 < t_1 < \ldots < t_p$:

$$(X_{t_1}, X_{t_2}, \ldots, X_{t_p}) \underset{l.o}{\leq} (Y_{t_1}, Y_{t_2}, \ldots, Y_{t_p}). \tag{8.44}$$

Of course, if X and Y are two càdlàg processes, (8.44) is equivalent to

for every $h : \mathbb{R} \longrightarrow \mathbb{R}$ càdlàg:

$$\mathbb{P}(\forall t \geq 0,\ X_t \leq h(t)) \leq \mathbb{P}(\forall t \geq 0,\ Y_t \leq h(t)). \tag{8.45}$$

Observe that, if $X \underset{l.o}{\leq} Y \underset{u.o}{\leq} X$, then

$$X_t \overset{(1.d)}{=} Y_t.$$

Indeed, for $l : \mathbb{R} \longmapsto [0,1]$, increasing, we have, from (8.40) and (8.7), for every $t \geq 0$

$$\mathbb{E}\left[(1-l)(X_t)\right] \geq \mathbb{E}\left[(1-l)(Y_t)\right] \geq \mathbb{E}\left[(1-l)(X_t)\right] \quad \text{i.e.} \quad \mathbb{E}\left[l(X_t)\right] = \mathbb{E}\left[l(Y_t)\right].$$

We now state the analogous of Theorem 8.1 for the lower orthant order.

Theorem 8.2. *Under the notation and hypotheses of Theorem 8.1, we have:*

1) $X_t^{(1)} \overset{(1.d)}{=} X_t^{(2)}$. (8.46)
2) $(X_t^{(1)}, t \geq 0) \underset{l.o}{\leq} (X_t^{(2)}, t \geq 0)$. (8.47)

We omit the proof of this theorem, which is very close to the proof of Theorem 8.1 (we use, instead of $\dfrac{\partial^2 F}{\partial a \partial b} \geq 0$ the fact that, if $\overline{F}$ is a survival function of two variables, $\dfrac{\partial^2 \overline{F}}{\partial a \partial b} \geq 0$). Observe that the order in (8.47) is the opposite of the order in (8.22).

Corollary 8.2. *Under the notation and hypotheses of Theorem 8.1:*

1) *For every n, every n-tuple $k_1, \ldots, k_n$ of reals and every n-tuple $0 < t_1 < t_2 < \ldots < t_n$, we have:*
$$\mathbb{E}\left[\prod_{i=1}^{n} (k_i - X_{t_i}^{(1)})^+\right] \geq \mathbb{E}\left[\prod_{i=1}^{n} (k_i - X_{t_i}^{(2)})^+\right] \tag{8.48}$$
(observe that the order in (8.48) is the same as the order in (8.37)).
2) *For every $T > 0$, for every continuous function h from $[0,T]$ to $\mathbb{R}$ we have:*
$$\mathbb{P}\left(\forall t \leq T,\ X_t^{(1)} \leq h(t)\right) \geq \mathbb{P}\left(\forall t \leq T,\ X_t^{(2)} \leq h(t)\right). \tag{8.49}$$

8.4 Comparison of Multidimensional Marginals in a Brownian Setting

Let $(B_s, s \geq 0)$ be a Brownian motion started at 0 and $\nu \in \mathbb{R}$. Let us define

$$\left(X_t := \frac{1}{t}\int_0^t e^{\nu B_s - \frac{\nu^2}{2}s} ds = \int_0^1 e^{\nu B_{ut} - \frac{\nu^2}{2}ut} du, t \geq 0\right). \tag{8.50}$$

We know (see Theorem 1.4) that $(X_t, t \geq 0)$ is a peacock. Let us now define

$$a) \quad \left(X_t^{(S)} := \int_0^1 e^{\nu W_{u,t} - \frac{\nu^2 ut}{2}} du, t \geq 0\right) \tag{8.51}$$

where $(W_{u,t}, u, t \geq 0)$ is the Brownian sheet. (S) stands for Brownian *S*heet. We know that $(X_t^{(S)}, t \geq 0)$ is a martingale associated to the peacock $(X_t, t \geq 0)$ (see Chapter 2).

$$b) \quad \left(X_t^{(R)} := \int_0^1 e^{\nu(B_t - B_{(1-u)t}) - \frac{\nu^2 ut}{2}} du, t \geq 0\right) \tag{8.52}$$

(R) stands here for time Reversal; we know that $(X_t^{(R)}, t \geq 0)$ is a martingale associated to the peacock $(X_t, t \geq 0)$ (see Theorem 3.2).

$$c) \quad \left(X_t^{(Sc)} := \int_0^1 e^{\nu\sqrt{t}B_u - \frac{\nu^2 ut}{2}} du, t \geq 0\right) \tag{8.53}$$

(Sc) stands for *Sc*aling. $(X_t^{(Sc)}, t \geq 0)$ is a peacock.

8.4.1 Comparison of Multidimensional Marginals

We want to compare the multidimensional marginals of the four processes $(X_t, t \geq 0)$, $(X_t^{(S)}, t \geq 0)$, $(X_t^{(R)}, t \geq 0)$ and $(X_t^{(Sc)}, t \geq 0)$.

Theorem 8.3. *With the previous notation, we have:*

1)
$$X_t \overset{(1.d)}{=} X_t^{(S)} \overset{(1.d)}{=} X_t^{(R)} \overset{(1.d)}{=} X_t^{(Sc)}. \tag{8.54}$$

2)
$$(X_t, t \geq 0) \underset{u.o}{\geq} (X_t^{(S)}, t \geq 0) \underset{u.o}{\geq} (X_t^{(R)}, t \geq 0) \tag{8.55}$$

$$(X_t^{(Sc)}, t \geq 0) \underset{u.o}{\geq} (X_t^{(S)}, t \geq 0) \underset{u.o}{\geq} (X_t^{(R)}, t \geq 0). \tag{8.56}$$

Proof. • Point 1) of Theorem 8.3 is clear since, for every $t \geq 0$,

$$(B_{ut}, 0 \leq u \leq 1) \overset{(law)}{=} (W_{u,t}, 0 \leq u \leq 1) \overset{(law)}{=} (B_t - B_{(1-u)t}, 0 \leq u \leq 1) \tag{8.57}$$
$$\overset{(law)}{=} (\sqrt{t}B_u, 0 \leq u \leq 1).$$

• *We now prove Point 2) of Theorem 8.3.* We shall apply Theorem 8.1 with H_t defined by:

$$\text{for } \nu \in \mathbb{R}, \quad H_t^{(\nu)}(\omega)(u) := \exp\left(\nu\omega(u) - \frac{\nu^2}{2}ut\right). \tag{8.58}$$

It is clear that $H_t^{(\nu)}$ satisfies hypothesis *b)* of Section 8.2. In particular, $H_t^{(\nu)}$ is increasing if $\nu > 0$, and decreasing if $\nu < 0$. Hence, Point 2) of Theorem 8.3 is a consequence of the following lemma.

Lemma 8.2.
1) *For every* $s,t \geq 0$ *and* $0 \leq u,v \leq 1$*:*

$$\mathbb{E}\left[B_{us}B_{vt}\right] \geq \mathbb{E}\left[W_{u,s}W_{v,t}\right]. \tag{8.59}$$

2) *For every* $s,t \geq 0$ *and* $0 \leq u,v \leq 1$*:*

$$\mathbb{E}\left[W_{u,s}W_{v,t}\right] \geq \mathbb{E}\left[(B_s - B_{(1-u)s})(B_t - B_{(1-v)t})\right]. \tag{8.60}$$

3) *For every* $s,t \geq 0$ *and* $0 \leq u,v \leq 1$*:*

$$\mathbb{E}\left[\sqrt{s}B_u\sqrt{t}B_v\right] \geq \mathbb{E}\left[W_{u,s}W_{v,t}\right]. \tag{8.61}$$

In other terms, with

$$\mathbb{G} := (B_{us};\ 0 \leq u \leq 1, s \geq 0);\ \mathbb{G}^{(S)} := (W_{u,s};\ 0 \leq u \leq 1, s \geq 0)$$
$$\mathbb{G}^{(R)} := (B_s - B_{(1-u)s};\ 0 \leq u \leq 1, s \geq 0); \mathbb{G}^{(Sc)} := (\sqrt{s}B_u;\ 0 \leq u \leq 1, s \geq 0)$$

we have, from Definition 8.3:

$$\mathbb{G} > \mathbb{G}^{(S)} > \mathbb{G}^{(R)} \quad \textit{and} \quad \mathbb{G}^{(Sc)} > \mathbb{G}^{(S)} > \mathbb{G}^{(R)}.$$

4) *On the other hand, there is no domination relation between* $\mathbb{G}$ *and* $\mathbb{G}^{(Sc)}$.

Proof. We first prove (8.61). Since $\mathbb{E}[B_uB_v] = u \wedge v$ and $\mathbb{E}[W_{u,s}W_{v,t}] = (u \wedge v)(s \wedge t)$, (8.61) is equivalent to $\sqrt{st}(u \wedge v) \geq (s \wedge t)(u \wedge v)$, i.e. to $\sqrt{st} \geq s \wedge t$.

We then prove (8.59). We have $us \geq (u \wedge v)(s \wedge t)$ and $vt \geq (u \wedge v)(s \wedge t)$, thus $us \wedge vt \geq (u \wedge v)(s \wedge t)$, i.e. (8.59).

We now prove (8.60). We easily have:

$$\mathbb{E}\left[(B_s - B_{(1-u)s})(B_t - B_{(1-v)t})\right] = \lambda\left\{[(1-u)s, s] \cap [(1-v)t, t]\right\} \tag{8.62}$$

where $\lambda\left\{[(1-u)s, s] \cap [(1-v)t, t]\right\}$ denotes the Lebesgue measure of the interval $[(1-u)s, s] \cap [(1-v)t, t]$. Let us suppose that $s \geq t$.

a) If $s < (1-v)t$ or $t < (1-u)s$, then $[(1-u)s,s] \cap [(1-v)t,t] = \emptyset$ and (8.60) is obvious.

b) Let us suppose that $s \geq (1-v)t$ and $t \geq (1-u)s$. Then:

$$[(1-u)s,s] \cap [(1-v)t,t] = [(1-u)s \vee (1-v)t, s \wedge t]$$

and

$$(s \wedge t - (1-u)s \vee (1-v)t) \leq s \wedge t\,(1-(1-u) \vee (1-v)) = (s \wedge t)(u \wedge v),$$

which proves (8.60).

Proof of Point 4) of Lemma 8.2. The processes $(\sqrt{s}B_u, \sqrt{t}B_v;\ 0 \leq u,v \leq 1)$ and $(B_{us}, B_{vt};\ 0 \leq v,u \geq 1)$ cannot be compared, since for all $s,t \geq 0$ and $u,v \in [0,1]$, neither $\mathbb{E}\left[\sqrt{s}B_u\sqrt{t}B_v\right] \leq \mathbb{E}\left[B_{us}B_{vt}\right]$ nor $\mathbb{E}\left[\sqrt{s}B_u\sqrt{t}B_v\right] \geq \mathbb{E}\left[B_{us}B_{vt}\right]$ hold.

Indeed, for $u = v$, we have:

$$\mathbb{E}\left[\sqrt{s}B_u\sqrt{t}B_v\right] = \sqrt{st}u \geq (s \wedge t)u = \mathbb{E}\left[B_{us}B_{vt}\right]$$

and, for $v = 1$, $s = \frac{1}{2}$, $t = \frac{1}{8}$ and $u \longrightarrow 0$:

$$\mathbb{E}\left[\sqrt{s}B_u\sqrt{t}B_v\right] \underset{u \to 0}{\sim} \frac{u}{4}$$

whereas,

$$\mathbb{E}\left[B_{us}B_{vt}\right] = \frac{u}{2} \wedge \frac{1}{8} \underset{u \to 0}{\sim} \frac{u}{2},$$

and $\dfrac{u}{2} \geq \dfrac{u}{4}$. This ends the proof of Lemma 8.2 and therefore of Theorem 8.3. □

Of course, the analogue of Theorem 8.3 holds with the lower orthant order:

Theorem 8.4. *With the notation of (8.50) ... (8.53):*

1) $$X_t \overset{(1.d)}{=} X_t^{(S)} \overset{(1.d)}{=} X_t^{(R)} \overset{(1.d)}{=} X_t^{(Sc)}. \tag{8.63}$$

2) $$(X_t, t \geq 0) \underset{l.o}{\leq} (X_t^{(S)}, t \geq 0) \underset{l.o}{\leq} (X_t^{(R)}, t \geq 0) \tag{8.64}$$

$$(X_t^{(Sc)}, t \geq 0) \underset{l.o}{\leq} (X_t^{(S)}, t \geq 0) \underset{l.o}{\leq} (X_t^{(R)}, t \geq 0). \tag{8.65}$$

Corollary 8.3. *For every n, every n-tuple $k_1,\dots,k_n$ and every $0<t_1<\dots<t_n$:*

$$1)\quad \mathbb{E}\left[\prod_{i=1}^n (X_{t_i}-k_i)^+\right] \geq \mathbb{E}\left[\prod_{i=1}^n (X_{t_i}^{(S)}-k_i)^+\right] \geq \mathbb{E}\left[\prod_{i=1}^n (X_{t_i}^{(R)}-k_i)^+\right]. \tag{8.66}$$

$$2)\quad \mathbb{E}\left[\prod_{i=1}^n (X_{t_i}^{(Sc)}-k_i)^+\right] \geq \mathbb{E}\left[\prod_{i=1}^n (X_{t_i}^{(S)}-k_i)^+\right] \geq \mathbb{E}\left[\prod_{i=1}^n (X_{t_i}^{(R)}-k_i)^+\right]. \tag{8.67}$$

$$3)\quad \mathbb{E}\left[\prod_{i=1}^n (k_i-X_{t_i})^+\right] \geq \mathbb{E}\left[\prod_{i=1}^n (k_i-X_{t_i}^{(S)})^+\right] \geq \mathbb{E}\left[\prod_{i=1}^n (k_i-X_{t_i}^{(R)})^+\right]. \tag{8.68}$$

$$4)\quad \mathbb{E}\left[\prod_{i=1}^n (k_i-X_{t_i}^{(Sc)})^+\right] \geq \mathbb{E}\left[\prod_{i=1}^n (k_i-X_{t_i}^{(S)})^+\right] \geq \mathbb{E}\left[\prod_{i=1}^n (k_i-X_{t_i}^{(R)})^+\right]. \tag{8.69}$$

8.4.2 Comparison of Variations

We now want to compare, for $s \neq t$ and $q \geq 1$, $\mathbb{E}[|X_t - X_s|^q]$, $\mathbb{E}\left[|X_t^{(S)} - X_s^{(S)}|^q\right]$, $\mathbb{E}\left[|X_t^{(R)} - X_s^{(R)}|^q\right]$ and $\mathbb{E}\left[|X_t^{(Sc)} - X_s^{(Sc)}|^q\right]$.

Theorem 8.5. *For every $q \geq 1$ and $s,t \geq 0$, we have:*

$$1)\qquad \mathbb{E}[|X_t - X_s|^q] \leq \mathbb{E}\left[|X_t^{(S)} - X_s^{(S)}|^q\right] \leq \mathbb{E}\left[|X_t^{(R)} - X_s^{(R)}|^q\right]. \tag{8.70}$$

$$2)\qquad \mathbb{E}\left[|X_t^{(Sc)} - X_s^{(Sc)}|^q\right] \leq \mathbb{E}\left[|X_t^{(S)} - X_s^{(S)}|^q\right] \leq \mathbb{E}\left[|X_t^{(R)} - X_s^{(R)}|^q\right]. \tag{8.71}$$

Remark 8.2. 1) The order in (8.70) and (8.71) is the inverse of the one in (8.48).
2) Since $(X_u^{(S)}, u \geq 0)$ is a martingale, we have, for every $0 \leq t_0 \leq t_1 \leq \dots \leq t_n = T$:

$$\sum_{j=0}^{n-1} \mathbb{E}\left[(X_{t_{j+1}}^{(S)} - X_{t_j}^{(S)})^2\right] = \sum_{j=0}^{n-1} \mathbb{E}\left[(X_{t_{j+1}}^{(S)})^2\right] - \mathbb{E}\left[(X_{t_j}^{(S)})^2\right] = \mathbb{E}\left[(X_T^{(S)})^2\right] - 1,$$

whereas

$$\sum_{j=0}^{n-1} \mathbb{E}\left[(X_{t_{j+1}} - X_{t_j})^2\right] \leq C_1(T)\mathbb{E}\left[\sup_{j,u\leq 1} |B_{ut_{j+1}} - B_{ut_j}| \exp\left(|\nu| C_2(T) \sup_{v \leq T} |B_v|\right)\right]$$
$$\xrightarrow[\sup_j |t_{j+1}-t_j| \to 0]{} 0.$$

Proof. We first observe that $X_t = \lim_{n\to+\infty} X_t^{(n)}$ with $X_t^{(n)} := \frac{1}{n}\sum_{k=0}^{n-1} e^{\nu B_{\frac{kt}{n}} - \frac{\nu^2 kt}{2n}}$. Since, for every $k > 0$, we have

$$\mathbb{E}\left[\sup_{n,t\leq T} |X_t^{(n)}|^k\right] < \infty$$

(and the same relations for $X^{(S)}$, $X^{(R)}$ and $X^{(Sc)}$), it suffices to prove, with obvious notation, that:

$$\mathbb{E}\left[|X_t^{(n)} - X_s^{(n)}|^q\right] \leq \mathbb{E}\left[|X_t^{(S,n)} - X_s^{(S,n)}|^q\right] \leq \mathbb{E}\left[|X_t^{(R,n)} - X_s^{(R,n)}|^q\right] \tag{8.72}$$

and

$$\mathbb{E}\left[|X_t^{(Sc,n)} - X_s^{(Sc,n)}|^q\right] \leq \mathbb{E}\left[|X_t^{(S,n)} - X_s^{(S,n)}|^q\right] \leq \mathbb{E}\left[|X_t^{(R,n)} - X_s^{(R,n)}|^q\right]. \tag{8.73}$$

But:

$$|X_t^{(n)} - X_s^{(n)}|^q = L\left(B_{\frac{0}{n}s}, B_{\frac{1}{n}s}, \dots, B_{\frac{n-1}{n}s}; B_{\frac{0}{n}t}, B_{\frac{1}{n}t}, \dots, B_{\frac{n-1}{n}t}\right) \tag{8.74}$$

with $L : \mathbb{R}^{2n} \to \mathbb{R}$ defined by:

$$L(x_0, x_1, \dots, x_{n-1}; y_0, y_1, \dots, y_{n-1}) = \left|\frac{1}{n}\sum_{k=0}^{n-1} e^{\nu x_k - \frac{\nu^2 ks}{2n}} - e^{\nu y_k - \frac{\nu^2 kt}{2n}}\right|^q \tag{8.75}$$

and analogous relations for $X^{(S,n)}$, $X^{(R,n)}$ and $X^{(Sc,n)}$ obtained by replacing in (8.74) $B_{\frac{k}{n}v}$ by, resp. $W_{\frac{k}{n},v}$, $B_v - B_{(1-v)\frac{k}{n}}$ and $\sqrt{v}B_{\frac{k}{n}}$.

We now observe that, for $i = 0, \dots, n-1$, $j = 0, 1, \dots, n-1$ and $q > 1$:

$$\frac{\partial^2 L}{\partial x_i \partial y_j} = \frac{-q(q-1)\nu^2}{n^2} e^{\nu x_i - \frac{\nu^2 is}{2n}} e^{\nu y_j - \frac{\nu^2 jt}{2n}} \left|\frac{1}{n}\sum_{k=0}^{n-1} e^{\nu x_k - \frac{\nu^2 ks}{2n}} - e^{\nu y_k - \frac{\nu^2 kt}{2n}}\right|^{q-2} \leq 0. \tag{8.76}$$

We then conclude for $q > 1$ by applying Slepian's Lemma, thanks to Lemma 8.2. The case $q = 1$ is obtained by passing to the limit. □

Exercise 8.1 (Comparison of integrals of log-normal r.v.'s). Let $(B_s, s \geq 0)$ be a Brownian motion started at 0 and $\nu \in \mathbb{R}$. Let us define:

$$\left(Y_t := \int_0^t \left(e^{\nu B_s - \frac{\nu^2}{2}s} - 1\right) ds = t\int_0^1 \left(e^{\nu B_{ut} - \frac{\nu^2}{2}ut} - 1\right) du, t \geq 0\right)$$
$$\left(Y_t^{(S)} := t\int_0^1 \left(e^{\nu W_{u,t} - \frac{\nu^2}{2}ut} - 1\right) du, t \geq 0\right)$$
$$\left(Y_t^{(R)} := t\int_0^1 \left(e^{\nu (B_t - B_{(1-u)t}) - \frac{\nu^2}{2}ut} - 1\right) du, t \geq 0\right)$$
$$\left(Y_t^{(Sc)} := t\int_0^1 \left(e^{\nu\sqrt{t}B_u - \frac{\nu^2}{2}ut} - 1\right) du, t \geq 0\right).$$

1) Prove that $(Y_t, t \geq 0)$ is a peacock and that $(Y_t^{(R)}, t \geq 0)$ is a martingale (use the results of Chapters 1 and 3).

2) Prove that:

$$Y_t \overset{(1.d)}{=} Y_t^{(S)} \overset{(1.d)}{=} Y_t^{(R)} \overset{(1.d)}{=} Y_t^{(Sc)}$$

3) Show that:

$$(Y_t, t \geq 0) \underset{u.o}{\geq} (Y_t^{(S)}, t \geq 0) \underset{u.o}{\geq} (Y_t^{(R)}, t \geq 0)$$

and

$$(Y_t^{(Sc)}, t \geq 0) \underset{u.o}{\geq} (Y_t^{(S)}, t \geq 0) \underset{u.o}{\geq} (Y_t^{(R)}, t \geq 0).$$

(*Hint*: use Theorem 8.1 and Lemma 8.2.)

Exercise 8.2 (Comparison of integrals of some Brownian martingales). Let $h : \mathbb{R} \longrightarrow \mathbb{R}$ be an increasing (or decreasing) function such that there exist $C > 0$ and $a > 0$ such that:

$$\forall x \in \mathbb{R}, \quad |h(x)| \leq Ce^{a|x|}.$$

Let $\varphi(t,x)$ defined for $t \in [0,1]$ by:

$$\varphi(t,x) = P_{1-t}h(x)$$

where $(P_u, u \geq 0)$ denotes the heat semi-group. Recall that φ is a space-time harmonic function for Brownian motion on the time interval $[0,1]$.

1) Prove that, for $t \in [0,1]$, $\varphi(t,\centerdot)$ is an increasing (or decreasing) function, which is of class $\mathscr{C}^1$ if $t \in [0,1[$.

2) We define:

$$\begin{aligned}
&\left(X_t := \int_0^1 \varphi(ut, B_{ut})du, 0 \leq t \leq 1\right) \\
&\left(X_t^{(S)} := \int_0^1 \varphi(ut, W_{u,t})du, 0 \leq t \leq 1\right) \\
&\left(X_t^{(R)} := \int_0^1 \varphi(ut, B_t - B_{(1-u)t})du, 0 \leq t \leq 1\right) \\
&\left(X_t^{(Sc)} := \int_0^1 \varphi(ut, \sqrt{t}B_u)du, 0 \leq t \leq 1\right).
\end{aligned} \tag{8.77}$$

Prove that:

a) $(X_t^{(S)}, t \geq 0)$ is a martingale with respect to the filtration $(\mathscr{G}_t := \sigma(W_{u,s}; u \geq 0, s \leq t), 0 \leq t \leq 1)$,

b) $X_t \overset{(1.d)}{=} X_t^{(S)} \overset{(1.d)}{=} X_t^{(R)} \overset{(1.d)}{=} X_t^{(Sc)}$,

c) $(X_t, 0 \leq t \leq 1) \underset{u.o}{\geq} (X_t^{(S)}, 0 \leq t \leq 1) \underset{u.o}{\geq} (X_t^{(R)}, 0 \leq t \leq 1)$ and
$(X_t^{(Sc)}, 0 \leq t \leq 1) \underset{u.o}{\geq} (X_t^{(S)}, 0 \leq t \leq 1) \underset{u.o}{\geq} (X_t^{(R)}, 0 \leq t \leq 1)$.

3) Prove that Theorem 8.5 is true for the processes defined by (8.77).

8.5 The Stochastic and Convex Orders

The purpose of this section is to study a link between the stochastic and convex orders.

Definition 8.6. Let μ and ν be two probability measures on $\mathbb{R}_+$. We shall say that μ is stochastically greater than ν, and we write:

$$\mu \overset{(st)}{\geq} \nu$$

if for every $t \geq 0$,

$$F_\mu(t) := \mu([0,t]) \leq F_\nu(t) := \nu([0,t]).$$

In Chapter 1, Theorem 1.4, we proved that if $(M_t, t \geq 0)$ is a martingale in H^1_{loc} (thus a peacock), and $\alpha : \mathbb{R}_+ \longrightarrow \mathbb{R}_+$ a continuous and increasing function such that $\alpha(0) = 0$, then the process

$$\left(\frac{1}{\alpha(t)} \int_0^t M_u d\alpha(u)\,,\, t \geq 0\right)$$

is a peacock. In other words, for every $0 \leq s \leq t$:

$$\int_0^{+\infty} M_u \left(\frac{1}{\alpha(t)} 1_{[0,t]}(u)\right) d\alpha(u) \overset{(c)}{\geq} \int_0^{+\infty} M_u \left(\frac{1}{\alpha(s)} 1_{[0,s]}(u)\right) d\alpha(u).$$

Now, it is clear that:

$$\left(\frac{1}{\alpha(t)} 1_{[0,t]}(u)\right) d\alpha(u) \overset{(st)}{\geq} \left(\frac{1}{\alpha(s)} 1_{[0,s]}(u)\right) d\alpha(u),$$

and this leads to the following question: which processes $(X_t, \geq 0)$ satisfy, for every couple of probabilities (μ, ν) such that $\mu \overset{(st)}{\geq} \nu$, the property:

$$A^{(\mu)} := \int_0^{+\infty} X_u \mu(du) \overset{(c)}{\geq} A^{(\nu)} := \int_0^{+\infty} X_u \nu(du)\ ? \tag{8.78}$$

Note that such a process $(X_t, t \geq 0)$ must be a peacock. Indeed, taking for $0 \leq s \leq t$, $\mu = \delta_t$ and $\nu = \delta_s$, we deduce from (8.78) that $X_t \overset{(c)}{\geq} X_s$, i.e. $(X_t, t \geq 0)$ is a peacock. Here is a partial answer to this question:

Theorem 8.6. *Let $(X_t, t \geq 0)$ be an integrable right-continuous process satisfying both following conditions:*

i) For every bounded and increasing function $\phi : \mathbb{R} \longrightarrow \mathbb{R}_+$ and every $0 \leq s \leq t$, $\mathbb{E}[\phi(X_t)|\mathscr{F}_s]$ is an increasing function of X_s.

ii) For every $n \in \mathbb{N}^*$, every $0 \leq t_1 < \cdots < t_{n+1}$ and every $\phi : \mathbb{R}^n \to \mathbb{R}$ in $\mathscr{E}_n$ (see Definition 1.11), we have:

$$\mathbb{E}\left[\phi(X_{t_1}, \ldots, X_{t_n})(X_{t_{n+1}} - X_{t_n})\right] \geq 0.$$

Let μ and ν two probability measures on $\mathbb{R}_+$ such that $\mu \overset{(st)}{\geq} \nu$. Moreover, we assume that either:

$$\mu \text{ and } \nu \text{ have compact supports, and for every compact } K \subset \mathbb{R}_+, \\ \sup_{t \in K} |X_t| \text{ is integrable,}$$

or:

$$\sup_{t \geq 0} |X_t| \text{ is integrable.}$$

Then:

$$A^{(\mu)} := \int_0^{+\infty} X_u \, \mu(du) \overset{(c)}{\geq} A^{(\nu)} := \int_0^{+\infty} X_u \, \nu(du).$$

Remark 8.3. *a*) Observe that condition *ii*) implies that the process $(X_t, t \geq 0)$ is a peacock. Indeed, if ψ is a convex function of $\mathscr{C}^1$ class, then, for $0 \leq s \leq t$:

$$\mathbb{E}[\psi(X_t)] - \mathbb{E}[\psi(X_s)] \geq \mathbb{E}[\psi'(X_s)(X_t - X_s)] \geq 0.$$

In particular, $\mathbb{E}[X_t]$ does not depend on t.

b) Note also that condition *i*) implies that $(X_t, t \geq 0)$ is Markovian.

Before proving Theorem 8.6, we shall give some examples of processes which satisfy both conditions *i*) and *ii*).

Example 8.1. Let X be a r.v. such that for every $t \geq 0$, $\mathbb{E}[e^{tX}] < \infty$. We define $(\xi_t^X = \exp(tX - h_X(t)), t \geq 0)$ where $h_X(t) = \log \mathbb{E}[e^{tX}]$. Then, $(\xi_t^X, t \geq 0)$ satisfies the conditions of Theorem 8.6. Indeed, condition *i*) is obvious, and condition *ii*) follows from:

$$\mathbb{E}\left[\phi(\xi_{t_1}^X, \dots, \xi_{t_n}^X)(\xi_{t_{n+1}}^X - \xi_{t_n}^X)\right] \\ \geq \phi(e^{t_1 \beta_n - h_X(t_1)}, \dots, e^{t_n \beta_n - h_X(t_n)}) \mathbb{E}\left[\xi_{t_{n+1}}^X - \xi_{t_n}^X\right] = 0,$$

where $\beta_n = \dfrac{h_X(t_{n+1}) - h_X(t_n)}{t_{n+1} - t_n}$. In particular, we recover that, if $\alpha : \mathbb{R}_+ \longrightarrow \mathbb{R}_+$ is a continuous and strictly increasing function such that $\alpha(0) = 0$, then the process

$$\left(\frac{1}{\alpha(t)} \int_0^t e^{uX - h_X(u)} d\alpha(u) \, , \, t \geq 0\right)$$

is a peacock, see Theorem 1.7.

Example 8.2 (Martingales). Clearly, martingales satisfy condition *ii*). Here are some examples of martingales satisfying also condition *i*):

a) Let $(X_t, t \geq 0)$ be an integrable process with independent and centered increments. Then

$$\mathbb{E}[\phi(X_t)|\mathscr{F}_s] = \mathbb{E}[\phi(X_s + X_t - X_s)|\mathscr{F}_s] = \mathbb{E}[\phi(x+Z)],$$

where $X_s = x$ and $Z \overset{(law)}{=} X_t - X_s$, is an increasing function of x.

b) Let $(L_t, t \geq 0)$ be an integrable right-continuous process with independent increments, and such that, for every $\lambda, t \geq 0$, $\mathbb{E}[e^{\lambda L_t}] < \infty$. Then, the process

$$\left(X_t := e^{\lambda L_t - h_{L_t}(\lambda)}, t \geq 0\right) \quad \text{where } h_{L_t}(\lambda) = \log \mathbb{E}\left[e^{\lambda L_t}\right]$$

is a martingale which, as in item *a*), satisfies condition *i*).

c) Let $(X_t, t \geq 0)$ be a diffusion process which satisfies an equation of type

$$X_t^{(x)} = x + \int_0^t \sigma(X_s^{(x)})\, dB_s.$$

Then condition *i*) follows from the stochastic comparison theorem (see Subsection 1.4.3.1).

Example 8.3 ("Well-reversible" diffusions). We adopt here the terminology of Subsection 1.4.3. Let $(Z_t, t \geq 0)$ be a "well-reversible" diffusion satisfying (1.115) and such that b is an increasing function. Then $(X_t := Z_t - \mathbb{E}[Z_t], t > 0)$ satisfies both conditions *i*) and *ii*). Indeed, condition *i*) is clearly satisfied from Lemma 1.5. As for condition *ii*), setting $h(t) = \mathbb{E}[Z_t]$, we have, with $0 < t_1 < \ldots < t_{n+1}$, by time reversal at t_{n+1}:

$$\begin{aligned}
&\mathbb{E}\left[\phi\left(X_{t_1}, \ldots, X_{t_n}\right)\left(X_{t_{n+1}} - X_{t_n}\right)\right] \\
&\quad = \overline{\mathbb{E}}\left[\phi\left(\overline{X}^{(t_{n+1})}_{t_{n+1}-t_1}, \ldots, \overline{X}^{(t_{n+1})}_{t_{n+1}-t_n}\right)\left(\overline{X}^{(t_{n+1})}_0 - \overline{X}^{(t_{n+1})}_{t_{n+1}-t_n}\right)\right] \\
&\quad = \overline{\mathbb{E}}\left[\mathbb{E}\left[\phi\left(\overline{X}^{(t_{n+1})}_{t_{n+1}-t_1}, \ldots, \overline{X}^{(t_{n+1})}_{t_{n+1}-t_n}\right)|\overline{\mathscr{F}}_{t_{n+1}-t_n}\right]\left(\overline{X}^{(t_{n+1})}_0 - \overline{X}^{(t_{n+1})}_{t_{n+1}-t_n}\right)\right] \\
&\quad = \overline{\mathbb{E}}\left[\widetilde{\phi}\left(\overline{X}^{(t_{n+1})}_{t_{n+1}-t_n}\right)\left(\overline{X}^{(t_{n+1})}_0 - \overline{X}^{(t_{n+1})}_{t_{n+1}-t_n}\right)\right] \\
&\quad\quad \text{(where } \widetilde{\phi} \text{ is an increasing function),} \\
&\quad = \mathbb{E}\left[\widetilde{\phi}\left(X_{t_n}\right)\left(X_{t_{n+1}} - X_{t_n}\right)\right].
\end{aligned}$$

Now from (1.115):

$$\begin{aligned}
&\mathbb{E}\left[\widetilde{\phi}\left(Z_{t_n} - h(t_n)\right)\left(\int_{t_n}^{t_{n+1}} \sigma(s, Z_s) dB_s + \int_{t_n}^{t_{n+1}} b(s, Z_s) ds - h(t_{n+1}) + h(t_n)\right)\right] \\
&\quad = \int_{t_n}^{t_{n+1}} \mathbb{E}\left[\widetilde{\phi}\left(Z_{t_n} - h(t_n)\right)\left(b(s, Z_s) - h'(s)\right)\right] ds \\
&\quad = \int_{t_n}^{t_{n+1}} \mathbb{E}\left[\widetilde{\phi}\left(Z_{t_n} - h(t_n)\right)\left(\widetilde{b}(s, Z_{t_n}) - h'(s)\right)\right] ds
\end{aligned}$$

where $x \longmapsto \widetilde{b}(s,x) := \mathbb{E}[b(s,Z_s)|Z_{t_n} = x]$ is an increasing function such that $\mathbb{E}\left[\widetilde{b}(s,Z_{t_n})\right] = \mathbb{E}[b(s,Z_s)] = h'(s)$. Denoting by $\widetilde{b}_s^{-1}$ its right-continuous inverse, we finally obtain:

$$\mathbb{E}\left[\widetilde{\phi}\,(X_{t_n})\,(X_{t_{n+1}} - X_{t_n})\right] \geq \int_{t_n}^{t_{n+1}} \widetilde{\phi}\left(\widetilde{b}_s^{-1}(h'(s)) - h(t_n)\right)\mathbb{E}\left[\widetilde{b}(s,Z_{t_n}) - h'(s)\right] ds = 0.$$

Example 8.4. Let $(B_t, t \geq 0)$ be a Brownian motion started from 0 and φ be a strictly increasing odd function of $\mathscr{C}^2$ class such that φ' is convex and $\mathbb{E}[|\varphi(B_t)|] < \infty$ for every $t \geq 0$. From Theorem 1.3, the process $(\varphi(B_t), t \geq 0)$ is a peacock. As a consequence of Example 8.3 and of Theorem 8.6 applied with $\mu(du) = \dfrac{1}{\alpha(t)} 1_{[0,t]}(u) d\alpha(u)$ and $\nu(du) = \dfrac{1}{\alpha(s)} 1_{[0,s]}(u) d\alpha(u)$, where $\alpha : \mathbb{R}_+ \longrightarrow \mathbb{R}_+$ is a continuous and strictly increasing function such that $\alpha(0) = 0$, we deduce that the process

$$\left(\frac{1}{\alpha(t)} \int_0^t \varphi(B_u) d\alpha(u)\ ,\ t \geq 0\right)$$

is also a peacock. Indeed, from Itô's formula, $(\varphi(B_u), u \geq 0)$ satisfies (1.115) with $b = \frac{1}{2}\varphi'' \circ \varphi^{-1}$ increasing.

Proof. 1) Since

$$A^{(\mu)} := \int_0^{\infty} X_s\, d\mu(s) = \int_0^{\infty} X_s\, dF_\mu(s) = \int_0^1 X_{F_\mu^{-1}(u)}\, du,$$

it suffices, by approximation of dF_μ with a linear combination of Dirac measures (as in the proof of Theorem 1.12), to show that for every $n \in \mathbb{N}^*$, for every $a_1, a_2, \dots, a_n$ and for every $t_1 \geq s_1, \dots, t_n \geq s_n$,

$$\sum_{i=1}^n a_i X_{t_i} \overset{(c)}{\geq} \sum_{i=1}^n a_i X_{s_i}. \tag{8.79}$$

2) Let $\psi : \mathbb{R} \to \mathbb{R}$ in **C**. By convexity, we have:

$$\begin{aligned}\psi\left(\sum_{i=1}^n a_i X_{t_i}\right) &= \psi\left(\sum_{i=1}^n a_i X_{s_i} + \sum_{i=1}^n a_i (X_{t_i} - X_{s_i})\right)\\ &\geq \psi\left(\sum_{i=1}^n a_i X_{s_i}\right) + \psi'\left(\sum_{i=1}^n a_i X_{s_i}\right) \sum_{j=1}^n a_j (X_{t_j} - X_{s_j}).\end{aligned}$$

Then, taking the expectation leads to:

$$\begin{aligned}\mathbb{E}\left[\psi\left(\sum_{i=1}^n a_i X_{t_i}\right)\right] \geq \mathbb{E}&\left[\psi\left(\sum_{i=1}^n a_i X_{s_i}\right)\right]\\ &+ \mathbb{E}\left[\psi'\left(\sum_{i=1}^n a_i X_{s_i}\right) \sum_{j=1}^n a_j (X_{t_j} - X_{s_j})\right].\end{aligned} \tag{8.80}$$

We set $\phi(x_1,\dots,x_n) := \psi'\left(\sum_{i=1}^{n} a_i x_i\right)$. Thus, $\phi \in \mathscr{E}_n$. Let j be fixed and assume that:

$$0 \le s_1 < \dots < s_j < \dots < s_{j+r} < t_j < s_{j+r+1} < \dots < s_n.$$

We write:

$$\begin{aligned}
&\phi(X_{s_1},\dots,X_{s_n})(X_{t_j} - X_{s_j})\\
&\quad= \phi(X_{s_1},\dots,X_{s_n})(X_{t_j} - X_{s_{j+r}} + X_{s_{j+r}} - \dots + X_{s_{j+1}} - X_{s_j})\\
&\quad= \phi(X_{s_1},\dots,X_{s_n})(X_{t_j} - X_{s_{j+r}}) + \sum_{k=0}^{r-1} \phi(X_{s_1},\dots,X_{s_n})(X_{s_{j+k+1}} - X_{s_{j+k}})
\end{aligned}$$

and we study the expectation of each term separately. From condition i), we obtain by iteration:

$$\begin{aligned}
&\mathbb{E}\left[\phi(X_{s_1},\dots,X_{s_n})(X_{t_j} - X_{s_{j+r}})\right]\\
&\qquad= \mathbb{E}\left[\widetilde{\phi}(X_{s_1},\dots,X_{s_{j+r}},X_{t_j})(X_{t_j} - X_{s_{j+r}})\right]\\
&\qquad\ge \mathbb{E}\left[\widetilde{\phi}(X_{s_1},\dots,X_{s_{j+r}},X_{s_{j+r}})(X_{t_j} - X_{s_{j+r}})\right]\\
&\qquad\ge 0 \quad \text{from condition } ii).
\end{aligned}$$

The other terms can be dealt with in the same way. □

Exercise 8.3 (Convex comparison of integrals of log-normal r.v.'s). Let $(B_u, u \ge 0)$ be a Brownian motion started from 0, and let σ and τ be two probability measures on $\mathbb{R}_+$ such that $\sigma \overset{\text{(st)}}{\ge} \tau$. Our purpose is to prove, without using Theorem 8.6, that:

$$A^{(\sigma)} := \int_0^\infty e^{B_s - \frac{s}{2}} \sigma(ds) \overset{\text{(c)}}{\ge} A^{(\tau)} := \int_0^\infty e^{B_s - \frac{s}{2}} \tau(ds). \tag{8.81}$$

1) Let F_σ (resp. F_τ) the distribution function of σ (resp. τ):

$$F_\sigma(a) := \sigma([0,a]), \quad F_\tau(a) := \tau([0,a]) \qquad \text{for every } a \ge 0.$$

Let us denote by F_σ^{-1} (resp. F_τ^{-1}) the right-continuous inverse of F_σ (resp. F_τ). Show that, for every $a \ge 0$:

$$F_\sigma^{-1}(a) \ge F_\tau^{-1}(a). \tag{8.82}$$

2) Show that

$$A^{(\sigma)} = \int_0^1 e^{B_{F_\sigma^{-1}(\lambda)} - \frac{1}{2}F_\sigma^{-1}(\lambda)} d\lambda \quad \text{and} \quad A^{(\tau)} = \int_0^1 e^{B_{F_\tau^{-1}(\lambda)} - \frac{1}{2}F_\tau^{-1}(\lambda)} d\lambda.$$

3) Prove, by using Exercise 1.45, that $A^{(\sigma)} \overset{(c)}{\ge} A^{(\tau)}$.

4) In this question, we consider more generally, in place of $(B_u, u \geq 0)$, a Gaussian centered process $(G_u, u \geq 0)$ whose covariance is denoted by γ:

$$\gamma(u,v) = \mathbb{E}[G_u G_v].$$

We assume that γ is continuous and, for every $v \geq 0$, the function $\gamma(\centerdot, v) : u \geq 0 \longmapsto \gamma(u,v)$ is increasing. We set:

$$G^{(\sigma)} := \int \exp\left(G_s - \frac{1}{2}\gamma(s,s)\right)\sigma(ds)$$

$$\text{and} \quad G^{(\tau)} := \int \exp\left(G_s - \frac{1}{2}\gamma(s,s)\right)\tau(ds).$$

Prove that $G^{(\sigma)} \overset{(c)}{\geq} G^{(\tau)}$.

5) *Example*: let $(G_u, u \geq 0)$ be the fractional Brownian motion of Hurst index $H \in]0,1[$. Its covariance is: $\gamma_H(u,v) = \frac{1}{2}\left(u^{2H} + v^{2H} - |u-v|^{2H}\right)$. Prove that $\gamma_H(\centerdot, v)$ is increasing for every $v \geq 0$ if and only if $H \geq \frac{1}{2}$.

Notes and Comments

The notions and results about the upper orthant and lower orthant orders are taken from M. Shaked and J. Shanthikumar [157] (and [158]). Section 8.5 is taken from A. Bogso *et al.* [33]. All other results in this chapter are new.

Open Problems

1) Looking for martingales associated to $\left(\frac{e^{tX}}{\mathbb{E}[e^{tX}]}, t \geq 0\right)$

Let X be a r.v. such that, for every $t \geq 0$, $\mathbb{E}[e^{tX}] < \infty$. Find a martingale associated to the peacock $\left(Z_t := \frac{e^{tX}}{\mathbb{E}[e^{tX}]}, t \geq 0\right)$. Here are some particular cases we know how to deal with:

a) X is a symmetric Bernoulli r.v. (see Exercise 1.40), and more generally, by multiplication and passing to the limit, when $X = \sum_{i=0}^{+\infty} a_i \varepsilon_i$, where $(\varepsilon_i)_{i \geq 0}$ is a sequence of i.i.d. symmetric Bernoulli r.v.'s, and $(a_i)_{i \geq 0}$ a sequence of reals such that $\sum_{i=0}^{+\infty} |a_i| < \infty$.

b) X admits a density f such that $x \longmapsto \int_x^{+\infty} f(u)du$ is log-concave, applying Azéma-Yor SE (see Exercise 7.12).

c) X admits a strictly positive log-concave density, thanks to the SDE method (see Theorem 6.10).

d) X is a self-decomposable r.v., then, an associated martingale is given by

$$M_t = \frac{e^{U_t}}{\mathbb{E}[e^{tX}]}$$

where $(U_t, t \geq 0)$ is the 1-Sato process attached to X (see Theorem 5.5). A particular case of this situation is given by $X = -\tau^{(\alpha)}$, where $\tau^{(\alpha)}$ is a positive stable r.v. of index $\alpha \in]0,1]$, i.e. such that $\mathbb{E}\left[e^{-\lambda \tau^{(\alpha)}}\right] = e^{-\lambda^\alpha}$ $(\lambda \geq 0)$. Here, $(U_t = -\gamma_{t^\alpha}^{(\alpha)}, t \geq 0)$ where $(\gamma_t^{(\alpha)}, t \geq 0)$ is a stable subordinator of index α.

2) Looking for processes obtained from integration over one parameter of two-parameter processes

a) Describe a class of processes $(Y_{t,\lambda};\, t \geq 0, \lambda \geq 0)$ such that, for every positive and finite (resp. signed and finite) measure σ:

$$\left(Z_t^{(\sigma)} := \int \frac{e^{Y_{t,\lambda}}}{\mathbb{E}[e^{Y_{t,\lambda}}]} \sigma(d\lambda), t \geq 0 \right) \quad \text{is a peacock.}$$

See Section 2.1 for a sum-up of most of the examples presented in this monograph.

b) Let us suppose that $(Z_t^{(\sigma)}, t \geq 0)$ is a peacock for every finite and signed measure σ. Does there exist a sheet attached to $(Y_{t,\lambda};\, t \geq 0, \lambda \geq 0)$, i.e.: a process $(\widetilde{Y}_{t,\lambda};\, t, \lambda \geq 0)$ such that:

- for every $t \geq 0$: $(\widetilde{Y}_{t,\lambda}, \lambda \geq 0) \overset{(law)}{=} (Y_{t,\lambda}, \lambda \geq 0)$,
- for every $s \leq t$, $(\widetilde{Y}_{t,\lambda} - \widetilde{Y}_{s,\lambda}, \lambda \geq 0)$ is independent of $\mathscr{G}_s := \sigma(\widetilde{Y}_{u,\mu};\, u \leq s, \mu \geq 0)$?

3) Semimartingales which are peacocks

Let $(X_t = M_t + V_t, t \geq 0)$ be a semimartingale given via its Doob-Meyer decomposition. Under which conditions on X (or M and V) is $(X_t, t \geq 0)$ a peacock? (See Exercise 1.29 for a partial answer, see also Bergenthum and Rüschendorf [23]).

4) Looking for a "martingale sheet" to construct martingales associated to $\mathbf{F}_1$-type peacocks

a) Construct a general "martingale sheet" $(\widetilde{M}_{u,t}; 0 \leq u \leq 1, t \geq 0)$ which allows to associate a martingale to any $\mathbf{F}_1$-type peacock, i.e. to:

$$\left(\frac{1}{t} \int_0^t M_s\, ds = \int_0^1 M_{ut}\, du, t \geq 0 \right), \quad \text{with } M \in H^1_{\text{loc}}.$$

In other words, does there exist a process $(\widetilde{M}_{u,t};\, 0 \leq u \leq 1, t \geq 0)$ such that, for every $0 \leq u \leq 1$, $(\widetilde{M}_{u,t},\, t \geq 0)$ is a $(\mathscr{M}_t, t \geq 0)$-martingale, with $\mathscr{M}_t = \sigma(\widetilde{M}_{u,s};\, 0 \leq u \leq 1, s \leq t)$ and such that, for every $t \geq 0$,

$$(M_{ut}, u \leq 1) \overset{(law)}{=} (\widetilde{M}_{u,t}\,,\, u \leq 1)\ ?$$

b) Let $f \in L^2_{\text{loc}}(\mathbb{R}_+, dx)$, $(B_s, s \geq 0)$ a Brownian motion started from 0 and define the martingale:

$$M_s^{(f)} := \exp\left(\int_0^s f(u) dB_u - \frac{1}{2} \int_0^s f^2(u) du \right), \quad s \geq 0.$$

We set $G_s := \int_0^s f(u) dB_u \sim \mathscr{N}(0, \int_0^s f^2(u) du)$. Let $(\Gamma_{\lambda,t};\, \lambda, t \geq 0)$ be the Gaussian sheet attached to the family of Gaussian processes $(G_{\lambda t}, \lambda \geq 0)_{t \geq 0}$ (such a sheet exists from Theorem 2.16) and define $\widetilde{M}_{u,t} := \exp\left(\Gamma_{u,t} - \frac{1}{2}\int_0^{ut} f^2(s)ds\right)$. Then it is not difficult to see that $(\widetilde{M}_{u,t};\ u,t \geq 0)$ is a "martingale sheet" which satisfies the conditions of Point a). Now, what can be said when f is a predictable process such that, for every $t \geq 0$, $\mathbb{E}\left[\int_0^t f^2(s,\omega)ds\right] < \infty$?

5) Looking for a "Markov sheet" to construct martingales associated to functionals of a conditionally monotonous Markov process

Let $(Y_\lambda, \lambda \geq 0)$ be a conditionally monotonous Markov process, which satisfies some suitable integrability conditions. Then, from Theorem 1.12, the process

$$Z_t^{(\sigma)}(Y) := \int_0^{+\infty} \frac{e^{tY_\lambda}}{\mathbb{E}[e^{tY_\lambda}]}\sigma(d\lambda), \quad t \geq 0$$

is, for every positive and finite measure σ, a peacock. Does there exist:

- A Markov sheet $(Y_{\lambda,t};\ \lambda,t \geq 0)$ attached to $(tY_\lambda, \lambda \geq 0)_{t \geq 0}$, i.e. a process such that:
  - $(Y_{\bullet,t};\ t \geq 0)$ is Markovian,
  - for every $t \geq 0$, $(tY_\lambda, \lambda \geq 0) \overset{(law)}{=} (Y_{\lambda,t};\ \lambda \geq 0)$.
- For every positive and finite measure σ, a space-time harmonic function $H^{(\sigma)}$ for the Markov process $(Y_{\bullet,t}, t \geq 0)$ such that $Z_t^{(\sigma)}(Y) \overset{(1.d)}{=} H^{(\sigma)}(t, Y_{\bullet,t})$? If such a construction is possible, then $(H^{(\sigma)}(t, Y_{\bullet,t}), t \geq 0)$ is a martingale associated to $(Z_t^{(\sigma)}(Y), t \geq 0)$.

6) Looking for a Doob-Meyer decomposition type for peacocks

Let $(M_t, t \geq 0)$ a martingale and $\psi : \mathbb{R} \longrightarrow \mathbb{R}$ a convex function. Under some reasonable integrability properties, the process $(\psi(M_t), t \geq 0)$ is a submartingale which admits a Doob-Meyer decomposition: $(\psi(M_t) = N_t^\psi + V_t^\psi, t \geq 0)$, where $(N_t^\psi, t \geq 0)$ (resp. $(V_t^\psi, t \geq 0)$) is a martingale (resp. an increasing process).

Now, let $(X_t, t \geq 0)$ be a peacock. By analogy with the case where X is a martingale, under which conditions does there exist a peacock $(\widetilde{N}_t^\psi, t \geq 0)$ and an increasing process $(\widetilde{V}_t^\psi, t \geq 0)$ such that:

$$\psi(X_t) = \widetilde{N}_t^\psi + \widetilde{V}_t^\psi \ ?$$

Observe that we have directly: $\psi(X_t) \overset{(1.d)}{=} \psi(M_t) = N_t^\psi + V_t^\psi$, where $(M_t, t \geq 0)$ is a martingale associated to $(X_t, t \geq 0)$, and $(N_t^\psi + V_t^\psi, t \geq 0)$ its Doob-Meyer decomposition (if it exists!). See Exercise 3.3 for an example.

7) Looking for other versions of Kellerer's theorem

a) Let $(X_{t,\lambda}; t, \lambda \geq 0)$ be a two-parameter peacock. Does there exist an associated two-parameter martingale $(M_{t,\lambda};\ t, \lambda \geq 0)$? (See Exercise 2.3 for the definition and some examples of such peacocks and associated martingales.)

b) A $\mathbb{R}^n$-valued process $(X_t, t \geq 0)$ is called a peacock if:

- it is integrable,
- for every convex function $\psi : \mathbb{R}^n \longrightarrow \mathbb{R}$, $t \longmapsto \mathbb{E}[\psi(X_t)] \leq +\infty$ is an increasing function.

Clearly, from Jensen's inequality, a $\mathbb{R}^n$-valued martingale is a peacock. Conversely, is a $\mathbb{R}^n$-valued peacock a $\mathbb{R}^n$-valued 1-martingale? Some examples of this situation are presented in Theorems 5.4 and 5.16, with $n = 2$ and $\mathbb{R}^2 \simeq \mathbb{C}$ (see also Strassen [163]).

List of Symbols

Equalities and Inequalities

$X \sim \mu$	The r.v. X has law μ, p. 266.
$X \overset{(law)}{=} Y$	The r.v.'s X and Y have the same law, p. xxix.
$X_t \overset{(1.d)}{=} Y_t$	The two processes $(X_t, t \geq 0)$ and $(Y_t, t \geq 0)$ have the same one-dimensional marginal distributions, p. xxix.
$(X_t, t \geq 0) \overset{(law)}{=} (Y_t, t \geq 0)$	The two processes $(X_t, t \geq 0)$ and $(Y_t, t \geq 0)$ have the same law, p. xxix.
$(X_1, \dots, X_k) \overset{(law)}{=} (Y_1, \dots, Y_k)$	The two random vectors have the same law.
$X \overset{(c)}{\geq} Y$	The r.v. X is greater than Y in the convex order, p. 1.
$\mu \overset{(c)}{\geq} \nu$	The probability law μ is greater than the probability ν in the convex order.
$X \overset{(st)}{\geq} Y$	The r.v. X is greater than Y for the stochastic order, p. 4.
$X \underset{u.o}{\geq} Y$	The random vector $X = (X_1, \dots, X_n)$ is greater than $Y = (Y_1, \dots, Y_n)$ for the upper orthant order, p. 338.
$X \underset{l.o}{\geq} Y$	The random vector $X = (X_1, \dots, X_n)$ is greater than $Y = (Y_1, \dots, Y_n)$ for the lower orthant order, p. 344.

Sets and Spaces

$\sigma(X_u, u \in A)$	σ-algebra generated by the set of r.v.'s $\{X_u, u \in A\}$, p. 7.
$\boldsymbol{C}$	Set of convex functions of $\mathscr{C}^2$ class $\psi : \mathbb{R} \longrightarrow \mathbb{R}$ such that ψ'' has compact support, p. 1.

H^1_{loc}	Space of martingales $(M_t, t \geq 0)$ such that, for every $t \geq 0$, $\mathbb{E}\left[\sup_{s\leq t}\lvert M_s\rvert\right] < \infty$, p. 26.
$\mathbf{F}_1$	Set of peacocks of the form $(X_t := \frac{1}{\alpha(t)}\int_0^t M_s d\alpha(s), t \geq 0)$ where $M \in H^1_{\text{loc}}$, and $\alpha : \mathbb{R}_+ \longrightarrow \mathbb{R}_+$ is an increasing and continuous function such that $\alpha(0) = 0$, p. 27.
$\mathbf{F}_2$	Set of peacocks of the form $(X_t := \int_0^t M_s d\alpha(s), t \geq 0)$ where $M \in H^1_{\text{loc}}$, $M_o = 0$, and $\alpha : \mathbb{R}_+ \longrightarrow \mathbb{R}_+$ is an increasing and continuous function such that $\alpha(0) = 0$, p. 27.
$\mathbf{F}'_2$	Set of peacocks of the form $(\alpha(t)X, t \geq 0)$ where X is a centered r.v. and $\alpha : \mathbb{R}_+ \longrightarrow \mathbb{R}_+$ is an increasing function, p. 33.
$\mathscr{E}_n$	Set of bounded Borel functions $\phi : \mathbb{R}^n \longrightarrow \mathbb{R}$ which are increasing with respect to each of their arguments, p. 64.
$\mathbf{I}$	Set of increasing functions $\varphi : \mathbb{R} \longrightarrow \mathbb{R}$, p. 18.
$\mathbf{J}$	Set of increasing functions $\varphi : \mathbb{R} \longrightarrow \mathbb{R}$ which are convex on $\mathbb{R}_+$ and concave on $\mathbb{R}_-$, p. 18.
$\mathscr{P}_2^{(\text{loc})}$	Set of predictable processes $(h(u,(B_a, a \leq u)), u \geq 0)$ such that, for every $t \geq 0$, $\mathbb{E}\left[\int_0^t h^2(u,(B_a, a \leq u))du\right] < \infty$, p. 159.
$\mathbb{S}_n$	Space of symmetric $n \times n$ matrices, p. 47.
$\mathbb{S}_n^+$	Cone of symmetric positive $n \times n$ matrices, p. 47.

Processes

$\mathscr{G}_K^{(M)}$	Last passage time at level K of the martingale $(M_t, t \geq 0)$, p. 27.
$(\mathscr{E}_t^{(\nu)}, t \geq 0)$	The ν-exponential Brownian martingale, p. 28.
$(B_t^H, t \geq 0)$	Fractional Brownian motion with Hurst index H, p. 52.
$(R_t, t \geq 0)$	Bessel process, p. 188.
$(R_{K,\alpha}(t), t \geq 0)$	Perturbed Bessel process, p. 202.
$(Q_t^{(\delta)}, t \geq 0)$	Squared Bessel process of dimension δ, p. 82.
$(W_{s,t}; s \geq 0, t \geq 0)$	Standard Brownian sheet, p. 92.
$(W_{s,t}^{(L)}; s \geq 0, t \geq 0)$	Lévy sheet extending the Lévy process L, p. 116.
$(G_{\lambda,t}; \lambda \in \Lambda)_{t\geq 0}$	A family of Gaussian processes, p. 56.

$(\Gamma_{\lambda,t};\lambda \in \Lambda, t \geq 0)$	Gaussian sheet attached to the family of Gaussian processes $(G_{\lambda,t};\lambda \in \Lambda)_{t\geq 0}$, p. 132.
$(c_{j,k}(t);\ j,k = 1\dots n, t \geq 0)$	Covariance matrix of a Gaussian random vector $(G_t^{(1)},\dots,G_t^{(n)})$, p. 47.
$(C_s^X, s \geq 0)$	Background driving Lévy process of a self-decomposable r.v. X, p. 183.
$(S_t^{H,X}, t \geq 0)$	H-Sato process attached to a self-decomposable r.v. X, p. 183.
$(S_t^X, t \geq 0)$	1-Sato process attached to a self-decomposable r.v. X, p. 183.
$(\widehat{S}_{\lambda,t}^X;\ \lambda \geq 0, t \geq 0)$	Sato sheet attached to a self-decomposable r.v. X, p. 216.
$\left(\widehat{S}_{\lambda,t}^{(L)};\ \lambda \geq 0, t \geq 0\right)$	Sato sheet extending a self-decomposable Lévy process, p. 217.
$D(a,b;[s,t])$	Number of downcrossings on the spatial interval $[a,b]$ between the times s and t, p. 261.
$\left(\frac{1}{t}\int_0^t e^{\nu B_s - \frac{\nu^2 s}{2}} ds, t \geq 0\right)$	Guiding example, p. 28.
$\left(X_t^{(S)} := \int_0^1 e^{\nu W_{u,t} - \frac{\nu^2 ut}{2}} du, t \geq 0\right)$	Guiding example where we replace $(B_{ut}, u \geq 0)$ by $(W_{u,t}, u \geq 0)$, p. 346.
$\left(X_t^{(R)} := \int_0^1 e^{\nu(B_t - B_{(1-u)t}) - \frac{\nu^2 ut}{2}} du, t \geq 0\right)$	Guiding example where we replace $(B_{ut}, u \geq 0)$ by $(B_t - B_{(1-u)t}, u \geq 0)$, p. 346.
$\left(X_t^{(Sc)} := \int_0^1 e^{\nu\sqrt{t}B_u - \frac{\nu^2 ut}{2}} du, t \geq 0\right)$	Guiding example where we replace $(B_{ut}, u \geq 0)$ by $(\sqrt{t}B_u, u \geq 0)$, p. 346.

Measures

$\mathbb{P}_{\mid \mathscr{F}_t}$	Restriction of the probability $\mathbb{P}$ to the σ-algebra $\mathscr{F}_t$, p. 25.
μ, σ	Measures on $\mathbb{R}_+$.
$\overline{\mu}(K) := \mu([K,+\infty[)$	Tail function attached to μ, p. 291.
$\overline{\overline{\mu}}(K) := \int_K^{+\infty} \overline{\mu}(a)da$	Double tail function attached to μ, p. 291.
$\psi_\mu(x) := \frac{1}{\overline{\mu}(x)} \int_{[x,+\infty[} y\mu(dy)$	Hardy-Littlewood function attached to μ, p. 287.
$D_\mu(a) := \frac{a}{\psi_\mu(a)}$	Madan-Yor function attached to μ, p. 290.

Abbreviations

BDLP	Background Driving Lévy Process, p. 183.
CSBP	Continuous Space Branching Process, p. 170.
SE	Skorokhod Embedding, p. 265.
STHBS	Space Time Harmonic Function for the Brownian Sheet, p. 96.

List of Exercises

Chapter 1

Chapter 2

Chapter 3

Chapter 4

Chapter 5

Chapter 6

Chapter 7

Chapter 8

References

1. Adler, R.J., Monrad, D., Scissors, R.H., Wilson, R.: Representations, decompositions and sample function continuity of random fields with independent increments. Stochastic Process. Appl. **15**(1), 3–30 (1983)
2. Albin, J.M.P.: A continuous non-Brownian motion martingale with Brownian motion marginal distributions. Statist. Probab. Lett. **78**(6), 682–686 (2008). DOI 10.1016/j.spl.2007.09.031. URL http://dx.doi.org/10.1016/j.spl.2007.09.031
3. Alili, L., Dufresne, D., Yor, M.: Sur l'identité de Bougerol pour les fonctionnelles exponentielles du mouvement brownien avec drift. In: Exponential functionals and principal values related to Brownian motion, Bibl. Rev. Mat. Iberoamericana, pp. 3–14. Rev. Mat. Iberoamericana, Madrid (1997)
4. Alòs, E., Mazet, O., Nualart, D.: Stochastic calculus with respect to Gaussian processes. Ann. Probab. **29**(2), 766–801 (2001). DOI 10.1214/aop/1008956692. URL http://dx.doi.org/10.1214/aop/1008956692
5. An, M.: Log-concave probability distributions: Theory and statistical testing. SSRN pp. i–29 (May 1997)
6. Azéma, J., Gundy, R.F., Yor, M.: Sur l'intégrabilité uniforme des martingales continues. In: Séminaire de Probabilités, XIV (Paris, 1978/1979), *Lecture Notes in Math.*, vol. 784, pp. 53–61. Springer, Berlin (1980)
7. Azéma, J., Yor, M.: Une solution simple au problème de Skorokhod. In: Séminaire de Probabilités, XIII (Univ. Strasbourg, Strasbourg, 1977/78), *Lecture Notes in Math.*, vol. 721, pp. 90–115. Springer, Berlin (1979)
8. Azéma, J., Yor, M.: Le problème de Skorokhod: compléments à "Une solution simple au problème de Skorokhod". In: Séminaire de Probabilités, XIII (Univ. Strasbourg, Strasbourg, 1977/78), *Lecture Notes in Math.*, vol. 721, pp. 625–633. Springer, Berlin (1979)
9. Azéma, J., Yor, M.: Etude d'une martingale remarquable. In: Séminaire de Probabilités, XXIII, *Lecture Notes in Math.*, vol. 1372, pp. 88–130. Springer, Berlin (1989)
10. Azencott, R.: Densité des diffusions en temps petit: développements asymptotiques. I. In: Seminar on probability, XVIII, *Lecture Notes in Math.*, vol. 1059, pp. 402–498. Springer, Berlin (1984)
11. Baker, D., Donati-Martin, C., Yor, M.: A sequence of Albin type continuous martingales, with Brownian marginals and scaling. In: Séminaire de Probabilités XLIII, *Lecture Notes in Math.*, vol. 2006, pp. 441–448. Springer, Berlin (2011)
12. Baker, D., Yor, M.: A brownian sheet martingale with the same marginals as the arithmetic average of geometric brownian motion. Elect. J. Prob. **14**(52), 1532–1540 (2009)
13. Barenblatt, G.I.: On some unsteady motions of a liquid and gas in a porous medium. Akad. Nauk SSSR. Prikl. Mat. Meh. **16**, 67–78 (1952)

14. Barlow, M., Pitman, J., Yor, M.: Une extension multidimensionnelle de la loi de l'arc sinus. In: Séminaire de Probabilités, XXIII, *Lecture Notes in Math.*, vol. 1372, pp. 294–314. Springer, Berlin (1989)
15. Barrieu, P., Schoutens, W.: Iterates of the infinitesimal generator and space-time harmonic polynomials of a Markov process. J. Comput. Appl. Math. **186**(1), 300–323 (2006)
16. Bass, R.: Skorokhod embedding via stochastic integrals. In: Séminaire de probabilités, XVII, *Lecture Notes in Math.*, vol. 986, pp. 221–224. Springer, Berlin (1983)
17. Bellaïche, C.: Géodésiques et diffusions en temps petit, *Astérisque*, vol. 84. Société Mathématique de France, Paris (1981). Probability Seminar, University of Paris VII, Paris
18. Ben Arous, G.: Développement asymptotique du noyau de la chaleur hypoelliptique hors du cut-locus. Ann. Sci. École Norm. Sup. (4) **21**(3), 307–331 (1988)
19. Benachour, S., Chassaing, P., Roynette, B., Vallois, P.: Processus associés à l'équation des milieux poreux. Ann. Scuola Norm. Sup. Pisa Cl. Sci. (4) **23**(4), 793–832 (1997) (1996)
20. Bender, C.: An Itô formula for generalized functionals of a fractional Brownian motion with arbitrary Hurst parameter. Stochastic Process. Appl. **104**(1), 81–106 (2003). DOI 10.1016/S0304-4149(02)00212-0. URL http://dx.doi.org/10.1016/S0304-4149(02)00212-0
21. Bénilan, P.: A strong regularity L^p for solution of the porous media equation. In: Contributions to nonlinear partial differential equations (Madrid, 1981), *Res. Notes in Math.*, vol. 89, pp. 39–58. Pitman, Boston, MA (1983)
22. Bentata, A., Cont, R.: Mimicking the marginal distributions of a semimartingale. arXiv:0910.3992 (Oct 2009)
23. Bergenthum, J., Rüschendorf, L.: Comparison of semimartingales and Lévy processes. Ann. Probab. **35**(1), 228–254 (2007). DOI 10.1214/009117906000000386. URL http://dx.doi.org/10.1214/009117906000000386
24. Bertoin, J.: Lévy processes, *Cambridge Tracts in Mathematics*, vol. 121. Cambridge University Press, Cambridge (1996)
25. Bertoin, J., Fujita, T., Roynette, B., Yor, M.: On a particular class of self-decomposable random variables: the durations of Bessel excursions straddling independent exponential times. Probab. Math. Statist. **26**(2), 315–366 (2006)
26. Bertoin, J., Le Jan, Y.: Representation of measures by balayage from a regular recurrent point. Ann. Probab. **20**(1), 538–548 (1992)
27. Biagini, F., Hu, Y., Oksendal, B., Zhang, T.: Stochastic calculus for fractional Brownian motion and applications. Probability and its Applications (New York). Springer, London Ltd., London (2008). DOI 10.1007/978-1-84628-797-8. URL http://dx.doi.org/10.1007/978-1-84628-797-8
28. Biagini, F., Oksendal, B., Sulem, A., Wallner, N.: An introduction to white-noise theory and Malliavin calculus for fractional Brownian motion. Proc. R. Soc. Lond. Ser. A Math. Phys. Eng. Sci. **460**(2041), 347–372 (2004). DOI 10.1098/rspa.2003.1246. URL http://dx.doi.org/10.1098/rspa.2003.1246. Stochastic analysis with applications to mathematical finance
29. Biane, P.: Comparaison entre temps d'atteinte et temps de séjour de certaines diffusions réelles. In: Séminaire de Probabilités, XIX, 1983/84, *Lecture Notes in Math.*, vol. 1123, pp. 291–296. Springer, Berlin (1985)
30. Bingham, N.H., Goldie, C.M., Teugels, J.L.: Regular variation, *Encyclopedia of Mathematics and its Applications*, vol. 27. Cambridge University Press, Cambridge (1989)
31. Björk, T.: Arbitrage theory in continuous time. Oxford Finance. Oxford University Press (2009)
32. Blumenthal, R.M., Getoor, R.K.: Markov processes and potential theory. Pure and Applied Mathematics, Vol. 29. Academic Press, New York (1968)
33. Bogso, A., Profeta, C., Roynette, B.: Some examples of peacocks in a markovian set-up. In: Séminaire de Probabilités XLIV, *Lecture Notes in Math.* Springer (To appear, 2012)
34. Bondesson, L.: Generalized gamma convolutions and related classes of distributions and densities, *Lecture Notes in Statistics*, vol. 76. Springer, New York (1992)
35. Borodin, A.N., Salminen, P.: Handbook of Brownian motion—facts and formulae, second edn. Probability and its Applications. Birkhäuser, Basel (2002)
36. Bougerol, P.: Exemples de théorèmes locaux sur les groupes résolubles. Ann. Inst. H. Poincaré Sect. B (N.S.) **19**(4), 369–391 (1983)

37. Boylan, E.: Local times for a class of Markoff processes. Illinois J. Math. **8**, 19–39 (1964)
38. Breiman, L.: Probability. Addison-Wesley Publishing Company, Reading, Mass. (1968)
39. Cairoli, R., Walsh, J.B.: Stochastic integrals in the plane. Acta Math. **134**, 111–183 (1975)
40. Calais, J.Y., Génin, M.: Sur les martingales locales continues indexées par $]0,\infty[$. In: Séminaire de probabilités, XVII, *Lecture Notes in Math.*, vol. 986, pp. 162–178. Springer, Berlin (1983)
41. Carasso, L., El Karoui, N., Obłój, J.: On Azéma-Yor processes, their optimal properties and the Bachelier Drawdown equation (To appear in Annals of Probability, 2011)
42. Carmona, P., Coutin, L., Montseny, G.: Stochastic integration with respect to fractional Brownian motion. Ann. Inst. H. Poincaré Probab. Statist. **39**(1), 27–68 (2003). DOI 10.1016/S0246-0203(02)01111-1. URL http://dx.doi.org/10.1016/S0246-0203(02)01111-1
43. Carr, P., Ewald, C.O., Xiao, Y.: On the qualitative effect of volatility and duration on prices of Asian options. Finance Research Letters **5**(3), 162–171 (September 2008)
44. Chaumont, L., Doney, R.A.: Pathwise uniqueness for perturbed versions of Brownian motion and reflected Brownian motion. Probab. Theory Related Fields **113**(4), 519–534 (1999). DOI 10.1007/s004400050216. URL http://dx.doi.org/10.1007/s004400050216
45. Chaumont, L., Vostrikova, L.: Reflection principle and Ocone martingales. Stochastic Process. Appl. **119**(10), 3816–3833 (2009). DOI 10.1016/j.spa.2009.07.009. URL http://dx.doi.org/10.1016/j.spa.2009.07.009
46. Chaumont, L., Yor, M.: Exercises in probability, *Cambridge Series in Statistical and Probabilistic Mathematics*, vol. 13. Cambridge University Press, Cambridge (2003). A guided tour from measure theory to random processes, via conditioning
47. Ciesielski, Z., Taylor, S.J.: First passage times and sojourn times for Brownian motion in space and the exact Hausdorff measure of the sample path. Trans. Amer. Math. Soc. **103**, 434–450 (1962)
48. Coddington, E.A., Levinson, N.: Theory of ordinary differential equations. McGraw-Hill, New York (1955)
49. Daduna, H., Szekli, R.: A queueing theoretical proof of increasing property of Pólya frequency functions. Statist. Probab. Lett. **26**(3), 233–242 (1996). DOI 10.1016/0167-7152(95)00015-1. URL http://dx.doi.org/10.1016/0167-7152(95)00015-1
50. Dalang, R., Walsh, J.: The sharp Markov property of Lévy sheets. Ann. Probab. **20**(2), 591–626 (1992)
51. Deheuvels, P., Martynov, G.V.: A Karhunen-Loeve decomposition of a Gaussian process generated by independent pairs of exponential random variables. J. Funct. Anal. **255**(9), 2363–2394 (2008). DOI 10.1016/j.jfa.2008.07.021. URL http://dx.doi.org/10.1016/j.jfa.2008.07.021
52. Delbaen, F., Schachermayer, W.: Arbitrage possibilities in Bessel processes and their relations to local martingales. Probab. Theory Related Fields **102**(3), 357–366 (1995). DOI 10.1007/BF01192466. URL http://dx.doi.org/10.1007/BF01192466
53. Doeblin, W.: Sur l'équation de Kolmogorov. C. R. Acad. Sci. Paris Sér. I Math. **331**(Special Issue), 1059–1128 (December 2000). DOI 10.1016/S0764-4442(01)01847-X. URL http://dx.doi.org/10.1016/S0764-4442(01)01847-X
54. Doney, R.A., Warren, J., Yor, M.: Perturbed Bessel processes. In: Séminaire de Probabilités, XXXII, *Lecture Notes in Math.*, vol. 1686, pp. 237–249. Springer, Berlin (1998)
55. Doney, R.A., Zhang, T.: Perturbed Skorohod equations and perturbed reflected diffusion processes. Ann. Inst. H. Poincaré Probab. Statist. **41**(1), 107–121 (2005). DOI 10.1016/j.anihpb.2004.03.005. URL http://dx.doi.org/10.1016/j.anihpb.2004.03.005
56. Doob, J.L.: Generalized sweeping-out and probability. J. Functional Analysis **2**, 207–225 (1968)
57. Dubins, L.E., Émery, M., Yor, M.: On the Lévy transformation of Brownian motions and continuous martingales. In: Séminaire de Probabilités, XXVII, *Lecture Notes in Math.*, vol. 1557, pp. 122–132. Springer, Berlin (1993)
58. Dufresne, D.: The distribution of a perpetuity, with applications to risk theory and pension funding. Scand. Actuar. J. (1-2), 39–79 (1990)
59. Dunford, N., Schwartz, J.T.: Linear operators. Part I. Wiley Classics Library. John Wiley & Sons Inc., New York (1988). General theory. With the assistance of William G. Bade and Robert G. Bartle, Reprint of the 1958 original, A Wiley-Interscience Publication

60. Dupire, B.: Pricing with a smile. Risk Magazine **7**, 17–20 (1994)
61. Duquesne, T., Le Gall, J.F.: Random trees, Lévy processes and spatial branching processes. Astérisque (281), vi+147 (2002)
62. Efron, B.: Increasing properties of Pólya frequency functions. Ann. Math. Statist. **36**, 272–279 (1965)
63. Eisenbaum, N., Kaspi, H., Marcus, M.B., Rosen, J., Shi, Z.: A Ray-Knight theorem for symmetric Markov processes. Ann. Probab. **28**(4), 1781–1796 (2000)
64. Elworthy, K.D., Li, X.M., Yor, M.: On the tails of the supremum and the quadratic variation of strictly local martingales. In: Séminaire de Probabilités, XXXI, *Lecture Notes in Math.*, vol. 1655, pp. 113–125. Springer, Berlin (1997). DOI 10.1007/BFb0119298. URL http://dx.doi.org/10.1007/BFb0119298
65. Elworthy, K.D., Li, X.M., Yor, M.: The importance of strictly local martingales; applications to radial Ornstein-Uhlenbeck processes. Probab. Theory Related Fields **115**(3), 325–355 (1999). DOI 10.1007/s004400050240. URL http://dx.doi.org/10.1007/s004400050240
66. Föllmer, H.: Martin boundaries on Wiener space. In: Diffusion processes and related problems in analysis, Vol. I (Evanston, IL, 1989), *Progr. Probab.*, vol. 22, pp. 3–16. Birkhäuser Boston, Boston, MA (1990)
67. Föllmer, H., Wu, C.T., Yor, M.: On weak Brownian motions of arbitrary order. Ann. Inst. H. Poincaré Probab. Statist. **36**(4), 447–487 (2000). DOI 10.1016/S0246-0203(00)00133-3. URL http://dx.doi.org/10.1016/S0246-0203(00)00133-3
68. Getoor, R.K.: The Brownian escape process. Ann. Probab. **7**(5), 864–867 (1979)
69. Getoor, R.K., Sharpe, M.J.: Excursions of Brownian motion and Bessel processes. Z. Wahrsch. Verw. Gebiete **47**(1), 83–106 (1979). DOI 10.1007/BF00533253. URL http://dx.doi.org/10.1007/BF00533253
70. Girsanov, I.: An example of non-uniqueness of the solution of the stochastic equation of K. Itô. Theor. Probability Appl. **7**, 325–331 (1962)
71. Gradshteyn, I.S., Ryzhik, I.M.: Table of integrals, series, and products, sixth edn. Academic Press Inc., San Diego, CA (2000). Translated from the Russian, edited and with a preface by A. Jeffrey and D. Zwillinger
72. Gyöngy, I.: Mimicking the one-dimensional marginal distributions of processes having an Itô differential. Probab. Theory Relat. Fields **71**(4), 501–516 (1986). DOI 10.1007/BF00699039. URL http://dx.doi.org/10.1007/BF00699039
73. Hall, W.J.: On the Skorokhod embedding theorem. Tech. rep., Stanford University, Dept. of Stat. (1968)
74. Hall, W.J.: Embedding submartingales in Wiener processes with drift, with applications to sequential analysis. J. Appl. Probability **6**, 612–632 (1969)
75. Harrison, J.M.: Ruin problems with compounding assets. Stochastic Processes Appl. **5**(1), 67–79 (1977)
76. Haussmann, U.G., Pardoux, É.: Time reversal of diffusions. Ann. Probab. **14**(4), 1188–1205 (1986)
77. Hewitt, E., Savage, L.J.: Symmetric measures on Cartesian products. Trans. Amer. Math. Soc. **80**, 470–501 (1955)
78. Hirsch, F., Profeta, C., Roynette, B., Yor, M.: Constructing self-similar martingales via two Skorokhod embeddings. In: Séminaire de Probabilités, XLIII, *Lecture Notes in Math.*, vol. 2006, pp. 451–503. Springer, Berlin (2011)
79. Hirsch, F., Roynette, B., Yor, M.: Applying Itô's motto: "look at the infinite dimensional picture" by constructing sheets to obtain processes increasing in the convex order. Periodica Mathematica Hungarica **61(1-2)**, 195–211 (2010)
80. Hirsch, F., Roynette, B., Yor, M.: Unifying constructions of martingales associated with processes increasing in the convex order , via Lévy and Sato sheets. Expositiones Mathematicae **28**(4), 299–324 (2010)
81. Hirsch, F., Roynette, B., Yor, M.: From an Itô type formula for Gaussian processes to integrals of log-normal processes increasing in the convex order. Prépublication Equipe d'Analyse et Probabilités, Université d'Evry **288** (to appear in J. Mat. Soc. Japan, 2011.)

82. Hirsch, F., Yor, M.: Looking for martingales associated to a self-decomposable law. Electronic Journal of Probability **15 (29)**, 932–961 (2010)
83. Hirsch, F., Yor, M.: A construction of processes with one-dimensional martingale marginals, based upon path-space Ornstein-Uhlenbeck processes and the Brownian sheet. J. Math. Kyoto Univ. pp. 389–417 (49-2, 2009)
84. Hirsch, F., Yor, M.: A construction of processes with one-dimensional martingale marginals, associated with a Lévy process, via its Lévy sheet. J. Math. Kyoto Univ. pp. 785–815 (49-4, 2009)
85. Hobson, D.: The Skorokhod Embedding Problem and Model-Independent Bounds for Option Prices, Paris-Princeton Lectures on Mathematical Finance 2010, *Lecture Notes in Math.*, vol. 2003, pp. 267–318. Springer (2011)
86. Ikeda, N., Watanabe, S.: Stochastic differential equations and diffusion processes, *North-Holland Mathematical Library*, vol. 24, second edn. North-Holland Publishing Co., Amsterdam (1989)
87. Ismail, M.E.H., Kelker, D.H.: Special functions, Stieltjes transforms and infinite divisibility. SIAM J. Math. Anal. **10**(5), 884–901 (1979). DOI 10.1137/0510083. URL http://dx.doi.org/10.1137/0510083
88. Itô, K.: Selected papers. Springer, New York (1987). Edited and with an introduction by S. R. S. Varadhan and D. W. Stroock
89. Itô, K., Nisio, M.: On stationary solutions of a stochastic differential equation. J. Math. Kyoto Univ. **4**, 1–75 (1964)
90. Jacod, J., Protter, P.: Time reversal on Lévy processes. Ann. Probab. **16**(2), 620–641 (1988)
91. Jacod, J., Shiryaev, A.N.: Limit theorems for stochastic processes, *Grundlehren der Mathematischen Wissenschaften [Fundamental Principles of Mathematical Sciences]*, vol. 288, second edn. Springer, Berlin (2003)
92. James, L.F., Roynette, B., Yor, M.: Generalized gamma convolutions, Dirichlet means, Thorin measures, with explicit examples. Probab. Surv. **5**, 346–415 (2008). DOI 10.1214/07-PS118. URL http://dx.doi.org/10.1214/07-PS118
93. Janson, S.: Gaussian Hilbert spaces, *Cambridge Tracts in Mathematics*, vol. 129. Cambridge University Press, Cambridge (1997)
94. Jeanblanc, M., Pitman, J., Yor, M.: Self-similar processes with independent increments associated with Lévy and Bessel processes. Stochastic Process. Appl. **100**, 223–231 (2002)
95. Jeulin, T.: Semi-martingales et grossissement d'une filtration, *Lecture Notes in Math.*, vol. 833. Springer, Berlin (1980)
96. Jeulin, T.: Sur la convergence absolue de certaines intégrales. In: Seminar on Probability, XVI, *Lecture Notes in Math.*, vol. 920, pp. 248–256. Springer, Berlin (1982)
97. Jeulin, T., Yor, M.: Inégalité de Hardy, semimartingales, et faux-amis. In: Séminaire de Probabilités, XIII (Univ. Strasbourg, Strasbourg, 1977/78), *Lecture Notes in Math.*, vol. 721, pp. 332–359. Springer, Berlin (1979)
98. Jeulin, T., Yor, M.: Sur les distributions de certaines fonctionnelles du mouvement brownien. In: Séminaire de Probabilités, XV (Univ. Strasbourg, Strasbourg, 1979/1980), *Lecture Notes in Math.*, vol. 850, pp. 210–226. Springer, Berlin (1981)
99. Jeulin, T., Yor, M.: Grossissements de filtrations: exemples et applications, *Lecture Notes in Math.*, vol. 1118. Springer, Berlin (1985). Papers from the seminar on stochastic calculus held at the Université de Paris VI, Paris, 1982/1983
100. Kac, M., Siegert, A.J.F.: An explicit representation of a stationary Gaussian process. Ann. Math. Statistics **18**, 438–442 (1947)
101. Kac, M., Siegert, A.J.F.: On the theory of noise in radio receivers with square law detectors. J. Appl. Phys. **18**, 383–397 (1947)
102. Kahane, J.P.: Une inégalité du type de Slepian et Gordon sur les processus gaussiens. Israel J. Math. **55**(1), 109–110 (1986). DOI 10.1007/BF02772698. URL http://dx.doi.org/10.1007/BF02772698
103. Kalashnikov, A.S.: Some problems of the qualitative theory of second-order nonlinear degenerate parabolic equations. Uspekhi Mat. Nauk **42**(2(254)), 135–176, 287 (1987)
104. Kamae, T., Krengel, U.: Stochastic partial ordering. Ann. Probab. **6**(6), 1044–1049 (1978)

105. Karatzas, I., Shreve, S.E.: Brownian motion and stochastic calculus, *Graduate Texts in Mathematics*, vol. 113, second edn. Springer, New York (1991)
106. Kellerer, H.G.: Markov-Komposition und eine Anwendung auf Martingale. Math. Ann. **198**, 99–122 (1972)
107. Kent, J.: Some probabilistic properties of Bessel functions. Ann. Probab. **6**(5), 760–770 (1978)
108. Kesavan, S.: Symmetrization & Applications, *Series in Analysis*, vol. 3. World Scientific Publishing Co. Pte. Ltd., Hackensack, NJ (2006)
109. Knight, F.B.: Random walks and a sojourn density process of Brownian motion. Trans. Amer. Math. Soc. **109**, 56–86 (1963)
110. Knight, F.B.: On the sojourn times of killed Brownian motion. In: Séminaire de Probabilités, XII (Univ. Strasbourg, Strasbourg, 1976/1977), *Lecture Notes in Math.*, vol. 649, pp. 428–445. Springer, Berlin (1978)
111. Kolmogoroff, A.: Über die analytischen Methoden in der Wahrscheinlichkeitsrechnung. Math. Ann. **104**(1), 415–458 (1931). DOI 10.1007/BF01457949. http://dx.doi.org/10.1007/BF01457949
112. Krylov, N.V.: On the relation between differential operators of second order and the solutions of stochastic differential equations. Steklov Seminar 1984 pp. 214–229 (1985)
113. Le Gall, J.F.: Some aspects of planar Brownian motion. In: Cours de l'Ecole d'Eté de St.-Flour (1990), *Lecture Notes in Math.*, vol. 1527, pp. 112–234. Springer, Berlin (1992)
114. Le Gall, J.F.: Spatial branching processes, random snakes and partial differential equations. Lectures in Mathematics ETH Zürich. Birkhäuser, Basel (1999)
115. Le Gall, J.F., Yor, M.: Excursions browniennes et carrés de processus de Bessel. C. R. Acad. Sci. Paris Sér. I Math. **303**(3), 73–76 (1986)
116. Le Gall, J.F., Yor, M.: Enlacements du mouvement brownien autour des courbes de l'espace. Trans. Amer. Math. Soc. **317**(2), 687–722 (1990). DOI 10.2307/2001484. URL http://dx.doi.org/10.2307/2001484
117. Lebedev, N.N.: Special functions and their applications. Revised English edition. Translated and edited by Richard A. Silverman. Prentice-Hall Inc., Englewood Cliffs, N.J. (1965)
118. Ledoux, M., Talagrand, M.: Probability in Banach spaces, *Ergebnisse der Mathematik und ihrer Grenzgebiete (3) [Results in Mathematics and Related Areas (3)]*, vol. 23. Springer, Berlin (1991). Isoperimetry and processes
119. Lelièvre, T., Rousset, M., Stoltz, G.: Long-time convergence of an adaptive biasing force method. Nonlinearity **21**(6), 1155–1181 (2008). DOI 10.1088/0951-7715/21/6/001. URL http://dx.doi.org/10.1088/0951-7715/21/6/001
120. Lowther, G.: Fitting martingales to given marginals. http://arxiv.org/abs/0808.2319v1 (2008)
121. Lukacs, E.: Characteristic functions. London: Charles Griffin and Company. X (1970)
122. Madan, D., Roynette, B., Yor, M.: Option Prices as Probabilities. Finance Research Letters **5**, 79–87 (2008)
123. Madan, D., Yor, M.: Making Markov martingales meet marginals: with explicit constructions. Bernoulli **8**(4), 509–536 (2002)
124. Mansuy, R.: On a one-parameter generalization of the Brownian bridge and associated quadratic functionals. J. Theoret. Probab. **17**(4), 1021–1029 (2004). DOI 10.1007/s10959-004-0588-8. URL http://dx.doi.org/10.1007/s10959-004-0588-8
125. Mansuy, R., Yor, M.: Random times and enlargements of filtrations in a Brownian setting, *Lecture Notes in Mathematics*, vol. 1873. Springer, Berlin (2006)
126. Mansuy, R., Yor, M.: Aspects of Brownian motion. Springer, Berlin (2008)
127. Meyer, P.A.: Probabilités et potentiel. Publications de l'Institut de Mathématique de l'Université de Strasbourg, No. XIV. Actualités Scientifiques et Industrielles, No. 1318. Hermann, Paris (1966)
128. Meyer, P.A.: Processus de Poisson ponctuels, d'après K. Ito. In: Séminaire de Probabilités, V (Univ. Strasbourg, année universitaire 1969–1970), pp. 177–190. *Lecture Notes in Math.*, vol. 191. Springer, Berlin (1971)
129. Meyer, P.A.: Sur une transformation du mouvement brownien dûe à Jeulin et Yor. In: Séminaire de Probabilités, XXVIII, *Lecture Notes in Math.*, vol. 1583, pp. 98–101. Springer, Berlin (1994). DOI 10.1007/BFb0073836. URL http://dx.doi.org/10.1007/BFb0073836

130. Meziane, K., Yen, J.Y., Yor, M.: Some examples of Skorokhod embeddings obtained from the Azéma-Yor algorithm (In preparation, 2010)
131. Millet, A., Nualart, D., Sanz, M.: Integration by parts and time reversal for diffusion processes. Ann. Probab. **17**(1), 208–238 (1989)
132. Mishura, Y.S.: Stochastic calculus for fractional Brownian motion and related processes, *Lecture Notes in Mathematics*, vol. 1929. Springer, Berlin (2008). DOI 10.1007/978-3-540-75873-0. URL http://dx.doi.org/10.1007/978-3-540-75873-0
133. Neveu, J.: Processus aléatoires gaussiens. Séminaire de Mathématiques Supérieures, No. 34 (Été, 1968). Les Presses de l'Université de Montréal, Montreal, Que. (1968)
134. Nilsen, T., Paulsen, J.: On the distribution of a randomly discounted compound Poisson process. Stochastic Process. Appl. **61**(2), 305–310 (1996). DOI 10.1016/0304-4149(95)00076-3. URL http://dx.doi.org/10.1016/0304-4149(95)00076-3
135. Nualart, D.: The Malliavin calculus and related topics, second edn. Probability and its Applications (New York). Springer, Berlin (2006)
136. Obłój, J.: The Skorokhod embedding problem and its offspring. Probab. Surv. **1**, 321–390 (electronic) (2004)
137. Pal, S., Protter, P.: Analysis of continuous strict local martingales via *h*-transforms. Stochastic Process. Appl. **120**, 1424–1443 (2010)
138. Petit, F., Yor, M.: Itô's formula and the marginals of certain submartingales. In: A.N. Shiryaev, A. Sulem (eds.) Journés sur les mathématiques financières, Proceeding of INRIA, pp. 164–167 (March 1999)
139. Pierre, M.: Le problème de Skorokhod: une remarque sur la démonstration d'Azéma-Yor. In: Séminaire de Probabilités, XIV (Paris, 1978/1979), *Lecture Notes in Math.*, vol. 784, pp. 392–396. Springer, Berlin (1980)
140. Pierre, M.: Uniqueness for a Fokker-Planck equation. Private communication (article in preparation, 2011)
141. Pipiras, V., Taqqu, M.S.: Integration questions related to fractional Brownian motion. Probab. Theory Related Fields **118**(2), 251–291 (2000). DOI 10.1007/s440-000-8016-7. URL http://dx.doi.org/10.1007/s440-000-8016-7
142. Pitman, J.: One-dimensional Brownian motion and the three-dimensional Bessel process. Advances in Appl. Probability **7**(3), 511–526 (1975)
143. Pitman, J.: The distribution of local times of a Brownian bridge. In: Séminaire de Probabilités, XXXIII, *Lecture Notes in Math.*, vol. 1709, pp. 388–394. Springer, Berlin (1999). DOI 10.1007/BFb0096528. URL http://dx.doi.org/10.1007/BFb0096528
144. Pitman, J., Yor, M.: Bessel processes and infinitely divisible laws. In: Stochastic integrals (Proc. Sympos., Univ. Durham, Durham, 1980), *Lecture Notes in Math.*, vol. 851, pp. 285–370. Springer, Berlin (1981)
145. Pitman, J., Yor, M.: A decomposition of Bessel bridges. Z. Wahrsch. Verw. Gebiete **59**(4), 425–457 (1982)
146. Prékopa, A.: On logarithmic concave measures and functions. Acta Sci. Math. (Szeged) **34**, 335–343 (1973)
147. Profeta, C.: Pénalisation, pseudo-inverses et peacocks dans un cadre markovien. Ph.D. thesis, Université Henri Poincaré, Nancy (2010)
148. Profeta, C., Roynette, B., Yor, M.: Option Prices as Probabilities. Springer-Finance (2010)
149. Ray, D.: Sojourn times of diffusion processes. Illinois J. Math. **7**, 615–630 (1963)
150. Revuz, D., Yor, M.: Continuous martingales and Brownian motion, *Grundlehren der Mathematischen Wissenschaften [Fundamental Principles of Mathematical Sciences]*, vol. 293, third edn. Springer, Berlin (1999)
151. Rogers, L.C.G.: Williams' characterisation of the Brownian excursion law: proof and applications. In: Séminaire de Probabilités, XV (Univ. Strasbourg, Strasbourg, 1979/1980), *Lecture Notes in Math.*, vol. 850, pp. 227–250. Springer, Berlin (1981)
152. Roynette, B., Vallois, P., Yor, M.: Limiting laws associated with Brownian motion perturbed by normalized exponential weights. I. Studia Sci. Math. Hungar. **43**(2), 171–246 (2006). DOI 10.1556/SScMath.43.2006.2.3. URL http://dx.doi.org/10.1556/SScMath.43.2006.2.3

153. Roynette, B., Yor, M.: Couples de Wald indéfiniment divisibles. Exemples liés à la fonction gamma d'Euler et à la fonction zêta de Riemann. Ann. Inst. Fourier (Grenoble) **55**(4), 1219–1283 (2005)
154. Russo, F., Vallois, P.: The generalized covariation process and Itô formula. Stochastic Process. Appl. **59**(1), 81–104 (1995). DOI 10.1016/0304-4149(95)93237-A. URL http://dx.doi.org/10.1016/0304-4149(95)93237-A
155. Sato, K.: Lévy processes and infinitely divisible distributions, *Cambridge Studies in Advanced Mathematics*, vol. 68. Cambridge University Press, Cambridge (1999). Translated from the 1990 Japanese original, Revised by the author
156. Schoutens, W.: Stochastic Processes And Orthogonal Polynomials, *Lecture Notes in Statistics*, vol. 146. Springer (2000)
157. Shaked, M., Shanthikumar, J.: Stochastic orders and their applications. Probability and Mathematical Statistics. Academic Press, Boston (1994)
158. Shaked, M., Shanthikumar, J.: Stochastic orders. Springer Series in Statistics. Springer, New York (2007)
159. Shanthikumar, J.G.: On stochastic comparison of random vectors. J. Appl. Probab. **24**(1), 123–136 (1987)
160. Shiga, T., Watanabe, S.: Bessel diffusions as a one-parameter family of diffusion processes. Z. Wahrscheinlichkeitstheorie und Verw. Gebiete **27**, 37–46 (1973)
161. Skorokhod, A.V.: Studies in the theory of random processes. Translated from the Russian by Scripta Technica, Inc. Addison-Wesley Publishing Co., Inc., Reading, Mass. (1965)
162. Stein, C.: Approximate computation of expectations. Institute of Mathematical Statistics Lecture Notes—Monograph Series, 7. Institute of Mathematical Statistics, Hayward, CA (1986)
163. Strassen, V.: The existence of probability measures with given marginals. Ann. Math. Statist. **36**, 423–439 (1965)
164. Stroock, D.W., Varadhan, S.R.S.: Multidimensional diffusion processes, *Grundlehren der Mathematischen Wissenschaften [Fundamental Principles of Mathematical Sciences]*, vol. 233. Springer-Verlag, Berlin (1979)
165. Talagrand, M.: Spin glasses: a challenge for mathematicians, *Ergebnisse der Mathematik und ihrer Grenzgebiete. 3. Folge. A Series of Modern Surveys in Mathematics [Results in Mathematics and Related Areas. 3rd Series. A Series of Modern Surveys in Mathematics]*, vol. 46. Springer, Berlin (2003). Cavity and mean field models
166. Talagrand, M.: Mean field models for spin glasses. Volume I, *Ergebnisse der Mathematik und ihrer Grenzgebiete. 3. Folge. A Series of Modern Surveys in Mathematics [Results in Mathematics and Related Areas. 3rd Series. A Series of Modern Surveys in Mathematics]*, vol. 54. Springer, Berlin (2011). Basic examples
167. Vallois, P.: Le problème de Skorokhod sur **R**: une approche avec le temps local. In: Séminaire de probabilités, XVII, *Lecture Notes in Math.*, vol. 986, pp. 227–239. Springer, Berlin (1983)
168. Vares, M.E., Yor, M.: A tribute to Kiyosi Itô : Introducing the volume. In: Stochastic Process. Appl., vol. 120, pp. 585–589 (2010). DOI 10.1016/j.spa.2010.01.008. URL http://dx.doi.org/10.1016/j.spa.2010.01.008. Special issue edited by Marc Yor and Maria E. Vares
169. Watanabe, S.: On time inversion of one-dimensional diffusion processes. Z. Wahrscheinlichkeitstheorie und Verw. Gebiete **31**, 115–124 (1974/75)
170. Westwater, J.: On Edwards' model for long polymer chains. Comm. Math. Phys. **72**(2), 131–174 (1980). URL http://projecteuclid.org/getRecord?id=euclid.cmp/1103907655
171. Westwater, J.: On Edwards' model for polymer chains. II. The self-consistent potential. Comm. Math. Phys. **79**(1), 53–73 (1981). URL http://projecteuclid.org/getRecord?id=euclid.cmp/1103908888
172. Westwater, J.: On Edwards' model for polymer chains. III. Borel summability. Comm. Math. Phys. **84**(4), 459–470 (1982). URL http://projecteuclid.org/getRecord?id=euclid.cmp/1103921283
173. Widder, D.V.: The heat equation. Academic Press [Harcourt Brace Jovanovich Publishers], New York (1975). Pure and Applied Mathematics, Vol. 67

174. Williams, D.: Decomposing the Brownian path. Bull. Amer. Math. Soc. **76**, 871–873 (1970)
175. Williams, D.: Path decomposition and continuity of local time for one-dimensional diffusions. I. Proc. London Math. Soc. (3) **28**, 738–768 (1974)
176. Williams, D.: Probability with martingales. Cambridge University Press, Cambridge (1991)
177. Yor, M.: Sur le balayage des semi-martingales continues. In: Séminaire de Probabilités, XIII (Univ. Strasbourg, Strasbourg, 1977/78), *Lecture Notes in Math.*, vol. 721, pp. 453–471. Springer, Berlin (1979)
178. Yor, M.: Compléments aux formules de Tanaka-Rosen. In: Séminaire de probabilités, XIX, 1983/84, *Lecture Notes in Math.*, vol. 1123, pp. 332–349. Springer, Berlin (1985)
179. Yor, M.: Renormalisation et convergence en loi pour les temps locaux d'intersection du mouvement brownien dans $\mathbf{R}^3$. In: Séminaire de probabilités, XIX, 1983/84, *Lecture Notes in Math.*, vol. 1123, pp. 350–365. Springer, Berlin (1985)
180. Yor, M.: Some aspects of Brownian motion. Part I, Some special functionals. Lectures in Mathematics ETH Zürich. Birkhäuser, Basel (1992)

Index

B&SS – Bocconi & Springer Series

THE ONLINE VERSION OF THE BOOKS PUBLISHED IN THE SERIES IS AVAILABLE ON SpringerLink

1. G. Peccati, M.S. Taqqu
   Wiener Chaos: Moments, Cumulants and Diagrams – A survey with computer implementation
   2011, XIV+274 pp, ISBN 978-88-470-1678-1

2. A. Pascucci
   PDE and Martingale Methods in Option Pricing
   2011, XVIII+720 pp, ISBN 978-88-470-1780-1

3. F. Hirsch, C. Profeta, B. Roynette, M. Yor
   Peacocks and Associated Martingales, with Explicit Constructions
   2011, XXXII+384 pp, ISBN 978-88-470-1907-2

For further information, please visit the following link:
http://www.springer.com/series/8762

MIX
Papier aus verantwortungsvollen Quellen
Paper from responsible sources
FSC® C105338

If you have any concerns about our products,
you can contact us on
ProductSafety@springernature.com

In case Publisher is established outside the EU,
the EU authorized representative is:
Springer Nature Customer Service Center GmbH
Europaplatz 3, 69115 Heidelberg, Germany

Printed by Libri Plureos GmbH
in Hamburg, Germany